Aspects of the life record

Major events	Dominant forms[b]		Systems
	Homo		Quaternary[c]
			Tertiary[c]
Grasses become abundant	Mammals		
		Flowering plants	
Horses first appear			
Extinction of dinosaurs			Cretaceous
Birds first appear			Jurassic
Dinosaurs first appear	Reptiles	Conifer and cycad plants	Triassic
			Permian
	Amphibia		
Coal-forming swamps			Pennsylvanian
			Mississippian
		Spore-bearing land plants	Devonian
	Fish		Silurian
Vertebrates first appear (fish)			Ordovician
	Marine invertebrates		
		Marine plants	
First abundant fossil record (marine invertebrates)			Cambrian
	Primitive marine plants and invertebrates		Precambrian
	One-celled organisms		

[c] Some geologists prefer to use the term Cenozoic for the Quaternary and the Tertiary.
[d] In most European and some American literature Pennsylvanian and Mississippian are combined in a period called the Carboniferous.
[e] Subdivisions not firmly established.

PHYSICAL GEOLOGY

Sixth Edition

PRENTICE-HALL, INC.

ENGLEWOOD CLIFFS, N.J. 07632

L. Don Leet
Late, Harvard University

Sheldon Judson
Princeton University

Marvin E. Kauffman
Franklin and Marshall College

PHYSICAL GEOLOGY

Library of Congress Cataloging in Publication Data

LEET, L. DON (LEWIS DON).
    Physical geology.

    Includes bibliographies and index.
    1. Physical geology.   I. JUDSON, SHELDON.
II. KAUFFMAN, MARVIN EARL.   III. Title.
QE28.2.L43 1982   551   81-15760
ISBN 0-13-669762-3   AACR2

PHYSICAL GEOLOGY, SIXTH EDITION

L. Don Leet, Sheldon Judson, and Marvin E. Kauffman

Printed in the United States of America

10 9 8 7 6 5 4 3 2 1

Designer, Walter Behnke
Production, Nicholas Romanelli
Manufacturing buyer, John Hall

PRENTICE-HALL INTERNATIONAL, INC., *London*
PRENTICE-HALL OF AUSTRALIA PTY. LIMITED, *Sydney*
PRENTICE-HALL OF CANADA, LTD., *Toronto*
PRENTICE-HALL OF JAPAN, INC., *Tokyo*
PRENTICE-HALL OF SOUTHEAST ASIA PTE. LTD., *Singapore*
WHITEHALL BOOKS LIMITED, Wellington, *New Zealand*

Cover photo by David Muench:
A view of Lake Powell at the
Utah-Arizona border.

SUMMARY CONTENTS

COMPLETE CONTENTS vii

PREFACE xv

1 TIME AND A CHANGING EARTH 1

2 MINERALS 15

3 ORIGIN AND OCCURRENCE
OF IGNEOUS ROCKS 37

4 WEATHERING AND SOILS 67

5 ORIGIN AND OCCURRENCE
OF SEDIMENTS
AND SEDIMENTARY ROCKS 91

6 METAMORPHISM
AND METAMORPHIC ROCKS 112

7 GEOLOGIC TIME 130

8 DEFORMATION 144

9 EARTHQUAKES AND THE EARTH'S
INTERIOR 177

10 CONTINENTS, OCEANS, PLATES,
AND DRIFT 215

11 MASS MOVEMENT OF SLOPE
MATERIAL 241

12 RUNNING WATER 255

13 UNDERGROUND WATER 288

14 GLACIATION 304

15 OCEANS 331

16 WIND AND OTHER DESERT
PROCESSES 352

17 ENERGY 368

18 USEFUL MATERIALS 385

APPENDIX A **The planetary system 401**

APPENDIX B **Minerals 416**

APPENDIX C **Periodic chart 432**

APPENDIX D **Earth data 433**

APPENDIX E **Topographic and geologic maps 437**

Glossary 441

Index 461

COMPLETE CONTENTS

SUMMARY CONTENTS v

PREFACE xv

1 TIME AND A CHANGING EARTH 1

1.1 Time 2

Absolute and relative time 2.   Radioactivity 3.
Divisions of geologic time 3.   Uniformitarianism 3.

1.2 Earth materials and the rock cycle 4

The three rock families 5.   The rock cycle 5.

1.3 Plate tectonics, sea-floor spreading, and continental drift 8.

Lithosphere, asthenosphere, crust, and mantle 8.
Vertical movements and isostasy 8.   Horizontal
movements 9.   Plates of the earth 9.   The rock cycle
again 11.

1.4 Some practical considerations 11

OUTLINE 13.   SUPPLEMENTARY READINGS 13.   QUESTIONS 14.

2 MINERALS 15

2.1 Particles and elements 16

Electric charge 16.   Atoms 16.   Atomic structure 17.
Atomic size 17.   Atomic mass 17.   Ions 17.
Elements 17.   Isotopes 18.   Organization of matter
summarized 19.

2.2 Distribution of elements in nature 19

Big-bang theory 19.   Steady-state theory 19.
Current theories 20.   Element abundance 20.

2.3 A first look at minerals 20

Mineral composition 20.   Characteristics of minerals 20.
Crystal form 20.   Cleavage 22.   Striations 22.
Hardness 23.   Specific gravity 23.   Color 23.   Streak 24.
Other physical properties 24.

2.4 Mineral structure 24

Compounds 25.   Ionic bond 25.   Covalent bond 26.
Metallic bond 27.

2.5 The most common minerals 27

Silicates 28.   Ferromagnesians 28.
Nonferromagnesians 30.   Oxide minerals 32.   Sulfide
minerals 32.   Carbonate and sulfate minerals 32.
Halides 32.   Mineraloids 33.

2.6 Organization and association of minerals 33

Associations of minerals 34.

2.7 Reshuffling the elements and minerals 34

OUTLINE 34.   SUPPLEMENTARY READINGS 36.   QUESTIONS 36.

3 ORIGIN AND OCCURRENCE OF IGNEOUS ROCKS 37

3.1 Volcanoes 37

Evolution of volcanoes and types of volcanism 39.
Volcanic eruptions 40.   Composition of lavas 40.
Temperature of lavas 41.   Volcanic gases 41.
Pyroclastic debris 41.   Fiery clouds 42.   Worldwide
effects 42.   Formation of calderas 42.   Classification of
volcanoes 44.   History of some volcanoes 45.
Vesuvius 45.   Krakatoa 46.   Parícutin 46.   Mount
St. Helens 47.   Hawaii 48.   Disappearing islands 49.
Distribution of active volcanoes 50.

3.2 Basalt plateaus 50

3.3 Igneous activity and earthquakes 51

Depth of magma source 52.

3.4 The earth's heat 52

Cause of igneous activity 52.

3.5 Masses of igneous rocks 53

Plutons 53.   Tabular plutons 53.   *Sills 53.*   *Dikes 53.*
Lopoliths 54.   Massive plutons 54.   *Laccoliths 54.*   *Batholiths 55.*

3.6 Formation of igneous rocks 56

Crystallization of magma 56.   Bowen's reaction
principle 57.   Rate of crystallization 57.

3.7 Texture of igneous rocks 58

Granular texture 58.   Aphanitic texture 58.   Glassy
texture 58.   Porphyritic texture 59.

3.8 Types of igneous rocks 60

Light-colored igneous rocks 60.   Dark-colored igneous
rocks 61.   Intermediate types 61.   Composition 61.
Texture 61.   Pegmatite 62.   Igneous rocks on the
moon 63.
OUTLINE 63.   SUPPLEMENTARY READINGS 65.   QUESTIONS 65.

4 WEATHERING AND SOILS 67

4.1 Energy sources 68

Potential energy 69.   Kinetic energy 69.   Energy and
weathering 69.

4.2 Types of weathering 70

Mechanical weathering 70.   Expansion and contraction 70.
Frost action 70.   Exfoliation 71.   Other types of
mechanical weathering 72.   Chemical weathering 73.
Particle size 73.   Other factors 74.   Chemical weathering
of igneous rocks 74. Weathering of quartz 75.
Weathering of feldspars 75.   Weathering of
ferromagnesians 77.   Summary of weathering products 78.

4.3 Rates of weathering 78

Rate of mineral weathering 78.   Depth and rapidity of
weathering 78.

4.4 Rates of erosion 80

Differential erosion 81.

4.5 Soils 82

Soil classification 82.   *C* Horizon 85.   *B* Horizon 85.   *A*
Horizon 86.   Some soil types 86.   Pedalfers 86.
Pedocals 87.   Tropical soil (laterites) 88.   Paleosols 88.
OUTLINE 89.   SUPPLEMENTARY READINGS 90.   QUESTIONS 90.

5 ORIGIN AND OCCURRENCE OF SEDIMENTS AND SEDIMENTARY ROCKS 91

5.1 Amounts of sediments and sedimentary rocks 92

5.2 Formation of sedimentary rocks 93

Origin of material 93.   Sedimentation 94.   Methods of
transportation 94.   Processes of sedimentation 94.
Mineral composition of sedimentary rocks 95.   Clay 95.
Quartz 95.   Calcite 95.   Other materials in sedimentary
rocks 95.   Texture 96.   Clastic texture 96.   Nonclastic
texture 97.   Lithification 97.   Cementation 98.
Compaction and desiccation 98.   Crystallization 98.
Diagenesis 98.

5.3 Types of sedimentary rocks 98

Classification 98.   Detrital sedimentary rocks 99.
Conglomerate 99.   Sandstone 99.   Mudstone, pelite, and

shale 99.   Chemical sedimentary rocks 100.
Limestone 100.   Dolomite 100.   Evaporites 100.   Organic
rocks 101.   Coal 102.   Relative abundance of
sedimentary rocks 102.

5.4 Features of sedimentary rocks   102

Bedding 102.   Ripple marks, mud cracks, and sole
marks 103.   Nodules, concretions, and geodes 104.
Fossils 106.   Color of sedimentary rocks 106.
Sedimentary facies 107.

5.5 Plate tectonics and sedimentation   108

Geosynclines 109.   Plate boundaries and
sedimentation 109.
OUTLINE 109. SUPPLEMENTARY READINGS 110. QUESTIONS 110.

6 METAMORPHISM AND METAMORPHIC ROCKS   112

6.1 Metamorphism   112

Agents of metamorphism 113.   Heat 113.   Pressure 113.
Chemically active fluids 113.

6.2 Types of metamorphism   114

Contact metamorphism 114.   Contact metamorphic
minerals 114.   Regional metamorphism 115.   Regional
metamorphic minerals 115.   Regional metamorphic
zones 115.   Regional metamorphic facies 116.   Low-
grade regional metamorphism in progress 117.
Geothermal field 119.   Outside the geothermal field 120.

6.3 Metamorphic rocks   120

Texture of metamorphic rocks 120.   Types of
metamorphic rocks 121.   Slate 121.   Phyllite 121.
Schist 121.   Amphibolite 122.   Gneiss 122.   Marble 122.
Quartzite 123.

6.4 Metamorphism and plate boundaries   123

Convergent-boundary metamorphism 123.   Divergent-
boundary metamorphism 124.   Transform- (parallel-)
boundary metamorphism 125.

6.5 Origin of granite   125

6.6 Geothermometers   126

Temperature of coal formation 126.   Conodont color
changes 128.
OUTLINE 128. SUPPLEMENTARY READINGS 129. QUESTIONS 129.

7 GEOLOGIC TIME   130

7.1 Relative time   130

The law of superposition 130.   Correlation of
sedimentary rocks 131.   Correlation by physical
features 131.   Correlation by fossils 132.

7.2 Absolute time   133

Radioactivity 133.   Fission-track dating 136.
Radioactivity and sedimentary rocks 136.   Sedimentation
and absolute time 136.

7.3 The geologic column   137

7.4 The geologic-time scale   138

Absolute dates in the geologic-time scale 139.
Magnetostratigraphy 139.
OUTLINE 141. SUPPLEMENTARY READINGS 143. QUESTIONS 143.

8 DEFORMATION   144

8.1 Deformation of the earth's crust   144

Abrupt movements 145.   Vertical displacements 146.
Horizontal displacements 147.   Slow movements 148.

8.2 Stress, strain, and strength of rocks   149

Stress and strain 149.   Elastic deformation 150.
Retardation of recovery 150.   Plastic deformation 150.   Time
factor 151.   Types of stress 151.   Strength of rocks 151.
Isostasy and equilibrium 152.   Gravity 153.
Acceleration of gravity 154.   Gravity anomalies—
evidence for inequilibrium 154.   Why do we still have
mountains? 155.

8.3 Structural features   156

Folds 157.   Mechanism of folding 158.   Monoclines 158.
Anticlines and synclines 159.   Structural domes 161.
Joints 162.   Columnar jointing 163.   Sheeting 164.   Joints
and dikes formed during movements of magma 164.
Faults 165.   Normal faults 167.   Thrust faults 167.
Strike-slip (transcurrent) faults 167.   Transform faults 169.
Unconformities 169.   Angular unconformity 170.
Disconformity 171.   Nonconformity 171.   Unconformities in
the Grand Canyon 171.   Broad surface features 173.
Broad warps 174.
OUTLINE 174. SUPPLEMENTARY READINGS 175. QUESTIONS 176.

9 EARTHQUAKES AND THE EARTH'S INTERIOR   177

9.1 Seismology   177

9.2 Effects of earthquakes   177

Fire 179.   Damage to structures 179.   Seismic sea
waves 179.   Seismic-sea-wave warning system 181.

Landslides 182.   Cracks in the ground 182.   Land movements 184.   Liquefaction 184.   Sound 184.

9.3 Interpreting earthquakes   184

Worldwide standard seismograph network 184. Earthquake focus 185.   Earthquake intensity 186. Earthquake magnitude and energy 187.

9.4 Distribution of earthquakes   188

Seismicity of Alaska and Aleutian Islands 189.   The Prince William Sound earthquake of 1964 191. *Aftershocks 194.*   Seismic and volcanic activity 195.

9.5 Cause of earthquakes   196

Elastic-rebound hypothesis 197.   Earthquakes not necessarily associated with elastic rebound 197.

9.6 Earthquakes resulting from plate movements   198

Clusters of severe earthquakes 199.   Guatemala 199. China 199.   Turkey 200.

9.7 Early instrumental observations   200

9.8 Modern seismology   201

Modern instruments 202.

9.9 Earth waves   202

Body waves 203.   Surface waves 203.

9.10 Records of earthquake waves   203

Time-distance graphs 204.   Travel times 204.   Less than 11,000 km 204.   Beyond 11,000 km 205.   Locating earthquakes 206.

9.11 Earthquake prediction and control   207

Earthquake precursors 207.   Dilatancy model 207. Control of earthquakes 208.

9.12 Structure of the earth   208

Crust 209.   Crust of the continents 209.   Crust under oceans 210.   Crust under Hawaii 210.   Mantle 210. Lithosphere and asthenosphere 211.   Core 211. OUTLINE 212.   SUPPLEMENTARY READINGS 214. QUESTIONS 214.

10   CONTINENTS, OCEANS, PLATES, AND DRIFT   215

10.1 Origin and permanence of continents and ocean basins   215

Constant freeboard of continents 218.   Age of ocean basins 218.

10.2 Magnetism and paleomagnetism   218

Earth's magnetism 218.   Secular variation of the magnetic field 219.   The geomagnetic pole 219. Megnetosphere 220.   Cause of the earth's magnetism 221. Paleomagnetism 221.   Virtual geomagnetic poles 222. Geomagnetic intensity, variations, and reversals 222. Results of paleomagnetic studies 223.

10.3 Continenta! drift   224

Evidence for continental drift 224.   Wegener's theory 224.   Shape of continents 225.   Evidence from paleomagnetism 225.   Sea-floor spreading 225.   Evidence from ancient climates 227.   *Late Paleozoic glaciation 227. Ancient evaporite deposits and coral reefs 228.*   Plants, reptiles, and continental drift 229.   Other evidence 230. Reliability of evidence 230.   Summary of the problem 230.

10.4 Plate tectonics   231

The world system of plates 231.   Makeup of plates 232.   Classification of plate boundaries 232. Movement of plates 234.   Convection cells 235.   Plume theory 235.   Recycling the sea floor 235.   Obduction 235. Dating the movements 235.   Summary speculations 236. OUTLINE 239.   SUPPLEMENTARY READINGS 240. QUESTIONS 240.

11   MASS MOVEMENT OF SLOPE MATERIAL   241

11.1 Factors of mass movement   241

Behavior of material 243.

11.2 Rapid movements   243

Landslides 243.   Slump 243.   Rock slides 243.   Debris slides 246.   Warnings of landslides 247.   Mudflows 248. Earthflows 248.   Talus 248.

11.3 Slow movements   249

Creep 249.   Solifluction 250.   Rock glaciers 252.

11.4 Direct action of water on hill slopes   253

Impact of raindrops 253.   Rain wash 253.   Solution 253. OUTLINE 254.   SUPPLEMENTARY READINGS 254. QUESTIONS 254.

12   RUNNING WATER   255

12.1 World distribution of water   255

12.2 Hydrologic cycle   255

12.3 Precipitation and stream flow   257

Laminar and turbulent flow 258.   Velocity, gradient, and discharge 258.   The hydrograph 259.

12.4 Economy of a stream 260

Adjustments of discharge, velocity, and channel 260. Floods 261. Urbanization and suburbanization 264. Changes in vegetative cover 265. Base level of a stream 265. Adjustment to a changing base level 265.

12.5 Work of running water 266

Transportation 266. Solution 266. Suspension 266. Bed load 267. Changes in water quality 267. Solid load 268. Dissolved load 269. Thermal pollution 269. Erosion 270. Direct lifting 270. Abrasion, impact, and solution 271. Deposition 272.

12.6 Features of valleys 273

Cross-valley profiles 273. Drainage basins, networks, and patterns 273. Water gaps 274. Enlargement of valleys 276. Features of narrow valleys 278. Features of broad valleys 278. Meanders 279. Braided streams 280. Natural levees 280. Flood plains and their deposits 281. Deltas and alluvial fans 282. Stream terraces 283.

12.7 Of rivers and continents 285

OUTLINE 286. SUPPLEMENTARY READINGS 286. QUESTIONS 287.

13 UNDERGROUND WATER 288

13.1 Basic distribution 288

Zones of saturation and aeration 288. The water table 289.

13.2 Movement of underground water 290

Porosity 291. Permeability 291.

13.3 Groundwater in nature 292

Simple springs and wells 292. Artesian water 294. Thermal springs 294. Geysers 295.

13.4 Recharge of groundwater 296

13.5 Caves and related features 297

13.6 Some groundwater problems caused by human use 301

Pollution 301. Saltwater invasion 301. Land subsidence 301.
OUTLINE 302. SUPPLEMENTARY READINGS 303. QUESTIONS 303.

14 GLACIATION 304

14.1 Formation of glacier ice 304

Classification of glaciers 305. Distribution of modern glaciers 307. Nourishment and wastage of glaciers 307. Glacier movement 307. Temperatures of glaciers 310.

14.2 Results of glaciation 310

Movement of material 310. Erosional effects 311. Cirques 311. Horns, arêtes, and cols 312. Glaciated valleys 312. Asymmetric rock knobs and hills 315. Types of glacial deposits 315. Deposits of unstratified drift 315. *Moraines 315. Drumlins 317. Erratics and boulder trains 318.* Deposits of stratified drift 318. *Outwash sand and gravel 318. Kettles 319. Eskers and crevassse fillings 320. Kames and kame terraces 320. Varves 321.* Comparison of valley- and continental-glaciation features 321. Indirect effects of glaciation 321.

14.3 Development of the glacial theory 322

The beginnings 322. Agassiz 322. Proof of the glacial theory 322. Theory of multiple glaciation 323.

14.4 Extent of Pleistocene glaciation 323

14.5 Pre-Pleistocene glaciations 325

14.6 Causes of glaciation 325

Long-term changes in climate 326. Causes of long-term climatic fluctuation 326. *Increasing continentality 326. Continental drift 327.* Other causes 327. Climatic changes of shorter duration 327. Causes of short-term climatic fluctuation 327. Summary 328

14.7 Implications for the future 328

OUTLINE 329. SUPPLEMENTARY READINGS 329. QUESTIONS 330.

15 OCEANS 331

15.1 Ocean water 331

Origin of ocean water 331. The nature of seawater 332.

15.2 Currents 333

Tidal currents 333. Density currents 334. Major surface currents 335.

15.3 The ocean basins 337

15.4 Topography of the ocean floor 337

The continental margins 337. Floors of the ocean basins 339. Oceanic ridges 339. Midoceanic ridges 339. Aseismic ridges 339. Seamounts and deep-sea trenches 339. Seamounts 339. Deep-sea trenches 340.

15.5 Sediments of the oceans 340

Deposits on the continental margins 340. Deposits on the deep-sea floors 341.

15.6 Shorelines 343

The processes 343. Wind-formed waves 343. Wave refraction and coastal currents 344. Shoreline features 345. Features caused by erosion 346. Features caused by deposition 346.

15.7 The changing sea level 348

OUTLINE 350. SUPPLEMENTARY READINGS 351. QUESTIONS 351.

16 WIND AND OTHER DESERT PROCESSES 352

16.1 Distribution of deserts 352

16.2 Work of the wind 354

Movement of material 354. Dust storms and sandstorms 354. *Movement of sand grains 355. Movement of dust particles 356.* Erosion 357. Abrasion 357. Deflation 358. Deposition 358. Loess 358. Sand deposits 360. *The wind shadow 360. Wind shadow of a dune 361. Shoreline dunes 362. Barchans 362. Parabolic dunes 362. Longitudinal dunes 362. Stratification in dunes 363.*

16.3 Other desert processes 365

Weathering and soils 365. Water 365. OUTLINE 367. SUPPLEMENTARY READINGS 367. QUESTIONS 367.

17 ENERGY 368

17.1 Natural resources 369

17.2 Coal 370

17.3 Oil and natural gas 371

Origin 371. Accumulation 372. The search for petroleum 375. Oil shale and tar sands 376. Oil shale 376. Tar sands 378. Some conclusions 378.

17.4 Water power 379

17.5 Nuclear power 379

Fission 379. Fusion 380.

17.6 Geothermal power 380

17.7 Solar energy 382

17.8 Other sources of energy 382

OUTLINE 382. SUPPLEMENTARY READINGS 383. QUESTIONS 383.

18 USEFUL MATERIALS 385

18.1 Mineral deposits 385

Mining 385. Geologic processes involved 386. Igneous activity 386. Hydrothermal activity 386. Weathering 386. Placer deposits 386. Rock formation 386.

18.2 Metals 386

Geochemically abundant metals 386. Iron 386. Aluminum 388. Manganese 389. Titanium 389. Magnesium 389. Geochemically scarce metals 389. Copper 390. Lead and zinc 390. Nickel 390. Molybdenum 391. Silver 391. Gold 391. Platinum 392. Chromium 392. Tin 392. Tungsten 393.

18.3 Agricultural and chemical minerals 393

Fertilizers 393. Nitrogen 393. Potassium 393. Phosphorous 394. Sulfur 394. Minerals for the chemical industry 394. Diamonds 395.

18.4 Building materials 395

Building stone 396. Crushed rock, sand, and gravel 396. Cement 396. Clay 396. Plaster 397. Glass 397. Asbestos 397.

18.5 Earth materials and society 397

OUTLINE 398. SUPPLEMENTARY READINGS 398. QUESTIONS 399.

APPENDIX A THE PLANETARY SYSTEM 401

A.1 The sun 401

Planets and satellites 402. Earth's moon 404. Moon rocks 406. Lunar history 406.

A.2 Mercury 407

A.3 Venus 408

A.4 Mars 408

The Martian landscape 408.

A.5 Asteroids 411

A.6 Jupiter 411

A.7 Saturn 413

A.8 Beyond Saturn 414

APPENDIX B MINERALS 415

B.1 Mineral characteristics 415

Crystal symmetry and systems 415.   Hardness 415.
Specific gravity 416.   Magnetism 416.   Pyroelectricity 416.
Piezoelectricity 417.   Luster 418.   Fluorescence and
phosphorescence 418.   Fusibility 418.   Solubility 418.
Fracture 418.   Tenacity 419.

B.2 Important minerals 419

Silicates 419.   Plagioclase feldspars 419.   Pyroxenes
and amphiboles 419.

APPENDIX C PERIODIC CHART 432

APPENDIX D EARTH DATA 433

APPENDIX E TOPOGRAPHIC AND GEOLOGIC MAPS 437

E.1 Topographic maps 437

Definitions 437.   Conventional symbols 438.   Locating
points 438.   Contour sketching 438.   Topographic
profiles 439.

E.2 Geologic maps 439

Definitions 439.   Construction of a geologic map 440.
Geologic cross sections 440.

GLOSSARY 441

INDEX 461

PREFACE

Our understanding of the earth—its chemical and physical characteristics, its history, and how it has changed during 4.5 billion years—is derived chiefly from studies during the nineteenth and twentieth centuries. Modern geology began to develop in Europe and the British Isles late in the eighteenth century. Since then geologists have explored the earth, seeking answers to questions about this planet. In recent years we have found that some of our answers have come from space exploration—from our moon, our sister planets and their moons, our sun, and even the distant stars.

Since the first edition of *Physical Geology* appeared nearly three decades ago, discoveries in the earth sciences have been spectacular. Yet, however impressive they have been, we clearly realize that these contemporary discoveries depend heavily on data, principles, and techniques developed over the preceding decades. Therefore, our goal in this edition is to continue to maintain a concise and balanced synthesis of both new and traditional knowledge.

We have benefited from the comments of users of the fifth edition of *Physical Geology* and have introduced a number of changes into this edition, resulting in a more effective and attractive book.

Chapter 1 sets the stage for our introduction to physical geology: the earth's makeup, how continents grow and contract, and how ocean basins form, expand, and eventually disappear. The fundamental concept of time is introduced in the first chapter and expanded on in Chapter 7, which has been moved forward to a more useful location in the sequence of topics. Chapter 1 also introduces the rock cycle, a model which can serve as an outline for much of this book. And in this initial chapter we first see the grand scheme of global tectonics, a subject treated more fully in Chapter 10.

Traditional approaches to matter and minerals in Chapter 2 give the student a firm base for the discussion of earth materials in Chapters 3 through 6. Chapter 6, on metamorphism and metamorphic rocks, has been extensively revised and concludes with an up-to-date review of geothermometers, by which we can interpret much about past conditions of temperatures.

The newly generated interest in volcanic activity occasioned by the eruptions of Mt. St. Helens is treated in Chapter 3. And our continuing concern regarding earthquakes finds expression in Chapter 9, which not only relates these earth movements to the global tectonic framework but also to our search for valid methods of earthquake prediction.

This edition maintains the traditional treatment of many topics, including weathering, now presented as a separate chapter (Chapter 4) from sedimentary rocks (Chapter 5); mass movement (Chapter 11); running water (Chapter 12); underground water (Chapter 13); glaciation (Chapter 14); and deserts and wind (Chapter 15). In this edition, we return to a more extensive treatment of the oceans and again devote an entire chapter (Chapter 15) to that subject.

The continuing search for sources of energy is discussed in a revised Chapter 17, which reviews coal, oil, gas, oil shale, tar sands, geothermal, and nuclear potentials in energy production. Other natural resources, particularly a wide variety of minerals, become increasingly important as demands increase and supplies dwindle. We have, therefore, completely revised Chapter 18 on useful minerals.

We have found that users of previous editions have generally not had time to include various aspects of space exploration in their programs. Nevertheless, the information gained from space missions has proved valuable as supplementary material. Therefore, we have moved our discussion of planetary bodies other than the earth to Appendix A, which includes information gathered by various vehicles, including the *Voyager* missions to Saturn.

In addition to the new appendix on our solar system, we have retained the appendices on minerals, the periodic chart, tables on earth data, and material on topographic and geologic maps.

We have revised the several study aids of past editions. These include the outlines and supplementary readings at the end of each chapter and a glossary that now includes over 1,100 entries. For the first time, we have provided a series of review questions for each chapter. Where simple numerical questions are used we have included the answers. A Student Handbook/Study Guide and an Instructor's Manual are also available for this text, prepared by Richard Gerfin and Robert Koch.

Much of the material on man's effect on the environment is now included in appropriate chapters rather than isolated as a separate chapter.

Many persons have played important roles during the preparation of this edition. Particularly helpful reviews of a preliminary manuscript were received from David E. Dunn, University of New Orleans; Bryce M. Hand, Syracuse University; Wallace C. Fallow, Furman University; and William Parrott, Stockton State College. Staff members of Prentice-Hall, Inc., have provided much appreciated professional support, advice, and patience, and we extend special thanks to Logan Campbell and Nicholas Romanelli. Sheldon Judson appreciates the editorial assistance of Joan Wyckoff. And we acknowledge the patient understanding and helpful support of friends and members of our families.

The death of L. Don Leet prior to the preparation of the fifth edition deprives us of the renewed contributions that he would certainly have made were he still with us.

SHELDON JUDSON
MARVIN E. KAUFFMAN

PHYSICAL GEOLOGY

TIME
AND
A CHANGING
EARTH

Geology is the science of the earth, an organized body of knowledge about the globe on which we live—about the mountains, plains, and ocean deeps, about the history of life from slime-born amoeba to humanity, and about the succession of physical events that accompanied this orderly development of life (Figure 1.1).

Geology helps us unlock the mysteries of our environment. Geologists explore the earth from the ocean floors to the mountain peaks to discover the origins of our continents and the encircling seas. They try to explain a land surface so varied that in less than an hour a traveler can fly over the Grand Canyon of the Colorado River, over the dammed waters of Lake Meade, over Death Valley, the lowest point in the contiguous United States, and over Mount Whitney, the highest point. They probe the action of glaciers that crawled over the land and then melted away over half a billion years ago and of some that even today cling to high valleys and cover most of Greenland and Antarctica, the remnants of a recent but presently receding Ice Age.

Geologists search for the record of life from the earliest one-celled organisms of ancient seas to the complex plants and animals of the present. This story, from simple algae to seed-bearing trees and from primitive protozoans to highly organized mammals, is told against the ever-changing physical environment of the earth.

For the earth has not always been as we see it today, and it is changing (but slowly) before our eyes. The highest mountains are built of materials that once lay beneath the oceans. Fossil remains of animals that swarmed the seas millions of years ago are now dug from lofty crags. Every continent is partially covered with sediments that were once laid down on the ocean floor, evidence of an intermittent rising and settling of the earth's surface.

In this chapter we take a preliminary look at some of the important concepts in the study of our changing earth. In subsequent chapters we shall discuss at greater length the subjects touched on here. Therefore you must read this chapter with the understanding that

1.1 The earth seen from the *Apollo 8* spacecraft during the first circumlunar voyage. The atmosphere, which partially obscures the earth here, makes possible not only life but oceans, running water, glaciers, winds, and the familiar processes of weathering. [NASA.]

the assertions we make will be more fully explained at appropriate later points. This chapter, however, is intended to provide a framework within which to organize your thinking about much of that more detailed material: It serves as a kind of map by which to chart our exploration of physical geology.

1.1 TIME

In 1654, a biblical scholar, James Ussher, archbishop of Armagh, declared that the earth had been created at 9:00 A.M. on October 26, 4004 B.C.—an event and a time presumably synonymous in his mind with the birth of the universe. Since his announcement, over 300 years, or nearly 6 percent of all the time assigned by the archbishop to earth history, have passed, and our notion of earth time has changed dramatically. We now know that the earth is about 4.5 billion years old—nearly a million times the age attributed to it in the seventeenth century. We are less certain about the universe, of which our earth is a part. But most current evidence suggests that the universe is more than three times the age of the earth.

In the following paragraphs we briefly discuss some of the ways in which geologic time is divided and measured, deferring until Chapter 7 a more detailed consideration of the subject.

ABSOLUTE AND RELATIVE TIME

An initial and casual reaction to the notion of time is that we can mark it off without much difficulty, even though we recognize, as Thomas Mann wrote in *The Magic Mountain*, that

time has no division to mark its passage, there is never a thunder-storm or blare of trumpets to announce the beginning of a new month or year. Even when a new century begins it is only we mortals who ring bells and fire off guns.[1]

Nevertheless, our modes—seconds, years, millennia, and the rest—are all we have to work with; and consequently we can consider geologic time from two points of view: as relative or as absolute. **Relative time**—that is, whether one event in earth history came before or after another event—disregards years. On the other hand, whether a geologic event took place a few thousand years ago, a billion years ago, or at some date even farther back in earth history is reported in **absolute time.**

Relative and absolute time in earth history have their counterparts in human history. In tracing the history of the earth, we may wish to know whether some event, such as a volcanic eruption, occurred before or after another event, such as a rise in sea level, and how these two events are related in time to a third event, perhaps a mountain-building episode. In human history, too, we try to determine the relative position of events in time. In studying United States history, we find it important to know that the Revolution-

[1]Thomas Mann, *The Magic Mountain*, trans. H. T. Lowe-Porter, p. 225, Modern Library, Inc., New York, 1955.

ary War preceded the Civil War and that the Canadian-American boundary was fixed some time between these two events.

Sometimes events in both earth history and human history can be established only in relative terms. Yet our record becomes increasingly precise as we fit more and more events into an actual chronological calendar: If we did not know the date of the United States–Canadian boundary treaty—if we knew only that it was signed between the two wars—we could place it between 1783 and 1861. (Recorded history, of course, provides us with the actual date, 1846.)

Naturally, we should like to be able to date geologic events with similar precision. But so far this has been impossible, and the accuracy in determining the dates of human history—that is, written human history—will likely never be achieved in geologic dating. Still, we can determine approximate dates for many geologic events, which are probably of the correct order of magnitude. We can say that dinosaurs became extinct about 63 million years ago and that about 11,000 years ago the last continental glacier began to recede from New England and the area bordering the Great Lakes.

Radioactivity Radioactive elements (those whose nuclei spontaneously emit particles to produce new elements, as discussed in Section 7.2) have provided the most effective means of measuring absolute time. The rate at which a given radioactive element decays is (so far as we have been able to determine) unaffected by changes in physical conditions or by time. So if we know the amount of original radioactive material (the **parent**) that remains, the rate of radioactive decay, and the amount of new elements (the **daughters**) that has formed, then we can calculate the time elapsed since radioactive decay began. Of course, the calculation is not quite so simple, as we discuss in Section 7.2. Nevertheless, from the time that the first radioactive age determinations were made (in 1907) to now, we have learned enough about the techniques and pitfalls of radioactive dating to be confident in the thousands of dates now available, particularly those made during the last two decades. A variety of elements has proved useful, with a time range from that of carbon 14, which can be used to date events that occurred a few hundred to a few tens of thousands of years ago, to that of an element such as uranium 238, which has the potential of dating events several times greater than the age of the earth.

DIVISIONS OF GEOLOGIC TIME

The application of radioactive elements to the measurement of geologic time became useful only after geologists had already constructed a calendar of geologic events. This calendar, still in use, was based on the ages of rock units relative to each other. Such relative ages were determined by conclusions drawn from a number of phenomena, including the superposition of younger rocks on older rocks, the cutting of older rocks by more recently formed rocks, and the progressive evolutionary stages of plant and animal life, as represented by remains in some rocks. The determination methods are discussed in detail in Section 7.1. It suffices to say here, however, that the arrangement of rock units and the earth events they record in the geologic calendar, as determined before the twentieth century, have been confirmed by the absolute dates of later radioactive dating.

The rock units in their proper chronological order make up the **geologic column,** which is reproduced on the front endpaper of this book.

UNIFORMITARIANISM

Modern geology was born in 1785, when James Hutton (1726–1797), a Scottish medical man, gentleman farmer, and geologist (Figure 1.2), formulated the principle now known to geologists as the **doctrine of uniformitarianism.** This principle simply means that the physical processes operating in the present to modify the earth's surface have also operated in the geologic past, that there is a uniformity of processes past and present.

1.2 Many concepts of present-day geology stem directly from observations by James Hutton.

Here is an example. We know from observations that modern glaciers deposit a distinctive type of debris made up of rock fragments that range in size from submicroscopic particles to boulders weighing several tons. This debris is jumbled, and many of the large fragments are scratched and broken. We know of no agent other than glacier ice that produces such a deposit. Now suppose that in the New England hills or across the plains of Ohio or in the deep valleys of the Rocky Mountains we find deposits that in every way resemble glacial debris but find no glacier in the area. We can still assume that the debris was deposited by now-vanished glaciers. On the basis of evidence like this, geologists have worked out the concept of the great Ice Age (Chapter 14).

Such an example can be multiplied many times. Today most earth features and rocks exposed at the earth's surface are explained as the result of past processes similar to those of the present; we shall find many conclusions of physical geology based on the conviction that modern processes have also operated in the past.

Armed with Hutton's concept of uniformitarianism, nineteenth-century geologists were able to explain earth features on a logical basis. But the very logic of the explanation gave rise to a new concept for students of the earth. Past processes presumably operated at the same slow pace as those of today. Consequently, as we have implied in the preceding sections, very long periods of time must have been available for those processes to accomplish their tasks: It was apparent that a great deal of time was needed for a river to cut its valley or for hundreds or thousands of feet of mud and sand to be deposited on an ocean bottom, hardened into solid rock, and raised far above the level of the sea.

The concept of almost unlimited time in earth history is thus a necesssary outgrowth of the application of the principle that **the present is the key to the past.** For example, geologists know that mountains as high as the modern Rockies once towered over what are now the low uplands of northern Wisconsin, Michigan, and Minnesota. But only the roots of these mountains are left, the great peaks having long since disappeared. Geologists explain that the ancient mountains were destroyed by rain and running water, creeping glaciers and wind, and landslides and slowly moving rubble and that these processes acted essentially as they do in our present-day world.

Now think of what this explanation means. We know from firsthand observation that streams, glaciers,

and winds have some effect on the surface of the earth. But can such feeble forces level whole mountain ranges? Instinct and common sense tell us that they cannot. This is where the factor of time comes into the picture. True, the small, almost immeasurable amount of erosion that takes place in a human lifetime has little effect; yet when the erosion during one lifetime is multiplied by thousands and millions of lifetimes, mountains can be worn away. Time makes possible what seems impossible.

The distinguished American geologist Adolph Knopf has written:

If I were asked as a geologist what is the single greatest contribution of the science of geology to modern civilized thought, the answer would be the realization of the immense length of time. So vast is the span of time recorded in the history of the earth that it is generally distinguished from the more modest kinds of time by being called "geologic time."[2]

1.2 EARTH MATERIALS AND THE ROCK CYCLE

Geology is based on the study of rocks, and we seek to know their composition, their distribution, how they are formed and destroyed, and why they are lifted up into continental masses and depressed into ocean basins.

Rock is the most common of all the materials on earth. It is familiar to everyone. We may recognize it as the gravel in a driveway, the boulders in a stream, the cliffs along a ridge. And it is common knowledge that firm bedrock is exposed at the earth's surface or lies beneath a thin cover of soil or loose debris (Figure 1.3).

The solid rock and the soil and debris above it are made up of minerals (Figure 1.4). Each type (and over two thousand different minerals have been described) has its own chemical composition and physical characteristics to identify it (see Chapter 2). But a bare handful, a dozen more or less, make up the bulk of the minerals that we see in the upper layers of the earth. If we were to examine the rock material of these layers from place to place, we should begin to see differences among different samples, including, for instance, the types of mineral in each rock; and on the basis of simple observation we should begin to place rocks into different classes.

[2] Adolph Knopf, *Time and Its Mysteries*, ser. 3, p. 33, New York University Press, New York, 1949.

1.4 Rock, debris and soil particles are composed of minerals. Each type of mineral exhibits its own chemical composition and physical characteristics. Here is a cluster of crystals of the mineral pyrite, FeS_2. The larger crystals measure approximately 1 cm across. [Willard Starks.]

rocks exposed at the surface that actually cooled some distance beneath it. We see such rocks today only because erosion has stripped away the rocks that covered them during their formation.

Most sedimentary (Latin *sedimentum*, "settling") rocks are made up of particles derived from the breakdown of preexisting rocks. Usually these particles are transported by gravity, water, wind, or ice to new locations, where they are deposited in new arrangements. For example, waves beating against a rocky shore may provide the sand grains and pebbles for a nearby beach. If these beach deposits were to be hardened, we should have sedimentary rock. One of the most characteristic features of sedimentary rocks is the layering of the deposits that go to make them up (Figure 1.6).

Metamorphic rocks compose the third large family of rocks. Metamorphic (from the Greek words *meta*, "change," and *morphē*, "form") refers to the fact that the original rock has been changed from its primary form to a new form. Earth pressures, heat, and chemically active fluids beneath the surface may all be involved in changing an originally sedimentary or igneous rock into a metamorphic rock (Figure 1.7).

THE THREE ROCK FAMILIES

Observations, then, have led geologists to divide the earth's rocks into three main groups based on mode of origin: **igneous**, **sedimentary**, and **metamorphic**. Later on, we shall discuss each type in detail, but here is a short explanation of all three.

Igneous rocks, the ancestors of all other rocks, take their name from the Latin *ignis*, "fire." These "fire-formed" rocks were once a hot, molten mass known as **magma**, which subsequently cooled into firm, hard rock. Thus the lava flowing across the earth's surface from an erupting volcano soon cools and hardens into an igneous rock (Figure 1.5). There are other igneous

THE ROCK CYCLE

We have suggested that there are definite relationships among sedimentary, igneous, and metamorphic rocks. With time and changing conditions, any one of

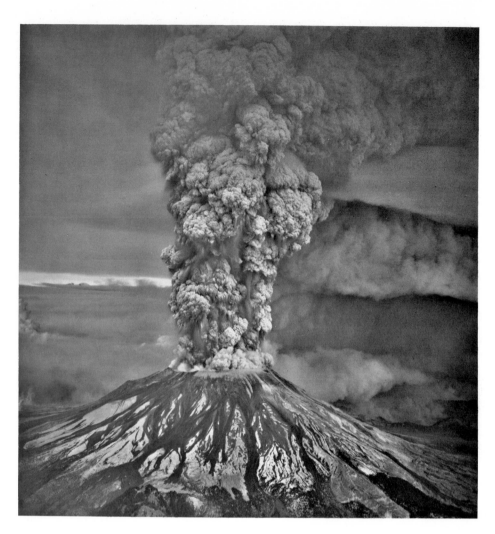

1.5 Mount St. Helens, Washington, about noon on May 18, 1980, seen from the southwest. The cloud of ash and steam here reaches 4 to 5 km into the atmosphere. This volcano is one of a chain that extends from Lassen Peak, northern California, through Oregon and Washington to Mt. Garibaldi in British Columbia and represents features built by igneous material periodically extruded through the earth's crust. [U.S. Geological Survey.]

the rock types may be changed into some other form. These relationships form a cycle, as shown in Figure 1.8, which is simply a way of tracing out the various paths that earth materials follow. The outer circle represents the complete cycle; the arrows within the circle represent shortcuts in the system that can be, and often are, taken. Notice that the igneous rocks are shown as having been formed from a magma and as providing one link in a continuous chain. From these parent rocks, through a variety of processes, all other rocks can be derived.

First, weathering attacks the solid rock, which either has been formed by the cooling of a lava flow at the surface or is an igneous rock that was formed deep beneath the earth's surface and then was exposed by erosion. The products of weathering are the materials that will eventually go into the creation of new rocks—

sedimentary, metamorphic, and even igneous. Landslides, running water, wind, and glacier ice all help to move the materials from one place to another. In the ideal cycle, this material seeks the ocean floors, where layers of soft mud, sand, and gravel are consolidated into sedimentary rocks. If the cycle continues without interruption, these new rocks may in turn be deeply buried and subjected to pressures caused by overlying rocks, to heat, and to forces developed by earth movements. The sedimentary rocks may then change in response to these new conditions and become metamorphic rocks. If these metamorphic rocks undergo continued and increased pressure and heat, they may eventually lose their identity and melt into a magma. When this magma cools, we have an igneous rock again; we have come full circle.

But notice too that the complete rock cycle may be **interrupted.** An igneous rock, for example, may never be exposed at the surface and hence may never be converted to sediments by weathering. Instead, it may be subjected to pressure and heat and converted directly into a metamorphic rock without passing

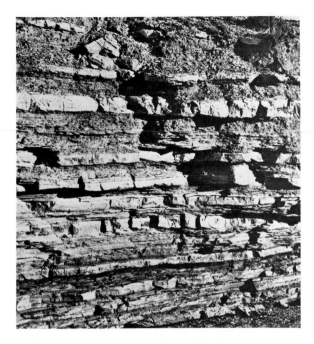

1.6 Layering is a characteristic of sedimentary rocks. Here, in Pueblo County, Colorado, beds of limestone alternate with layers of shale, which are slightly recessed because they weather more rapidly than the limestone. [G. K. Gilbert.]

1.7 This aerial photograph shows the twisted pattern of deformed and metamorphosed rocks. The beds, which show as bands 100 to 300 m wide, were once horizontal sedimentary rocks. They were deeply buried beneath the surface, tilted and folded by earth forces, and then exposed to view by subsequent erosion. [Royal Canadian Air Force.]

through the intermediate sedimentary stage. Other interruptions may take place if sediments or sedimentary rock or metamorphic rock are attacked by weathering before they continue to the next stage in the larger, complete cycle.

This concept of the rock cycle was probably first stated in the late eighteenth century by James Hutton, whom we have already mentioned.

We are thus led to see a circulation in the matter of this globe, and a system of beautiful economy in the works of nature. This earth, like the body of an animal, is wasted at the same time that it is repaired. It has a state of growth and augmentation; it has another state, which is that of diminution and decay. This world is thus destroyed in one part, but it is renewed in another; and the operations by which this world is thus constantly renewed are as evident to the scientific eye, as are those in which it is necessarily destroyed.[3]

We can consider the rock cycle to be a kind of outline of physical geology, as a comparison of Figure 1.8 with the Contents of this book will show.

[3]James Hutton, *Theory of the Earth*, vol. 2, p. 562, Edinburgh, 1795. Hutton's theory of the earth was first presented as a series of lectures before the Royal Society of Edinburgh in 1785. These lectures were published in book form in 1795. Seven years later, Hutton's concepts were given new impetus through a more readable treatment, called *Illustrations of the Huttonian Theory*, by John Playfair.

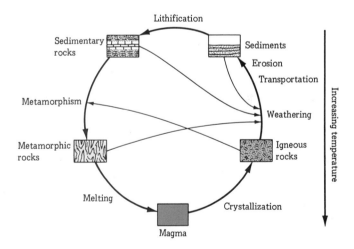

1.8 The rock cycle, shown diagrammatically. If uninterrupted, the cycle will continue completely around the outer margin of the diagram, from magma through igneous rocks, sediments, sedimentary rocks, and metamorphic rocks and back again to magma. The cycle may be interrupted, however, at various points along its course to follow the path of one of the arrows through the interior of the diagram.

1.3 PLATE TECTONICS, SEA-FLOOR SPREADING, AND CONTINENTAL DRIFT

To generations of geologists it has been clear that the earth is a dynamic, changing body. As we have pointed out, new rocks, sedimentary rocks, are made from the weathered debris of older rocks, and they can be crumpled, metamorphosed, and lifted into high mountain chains; or old rocks can be melted and the resulting magma cooled to form igneous rocks. Indeed, these changes can be traced in nature and pictured as we have in the rock cycle in Figure 1.8. But only recently have we been able to fit the rocks and their alterations into a worldwide, integrated system and to explain in a general way the origin of continents and ocean basins and of mountain ranges and continental plains and the location of volcanoes and earthquake belts. Two of the processes involved are referred to as **plate tectonics** and **sea-floor spreading.** Although we shall later extensively discuss these processes—particularly in Chapter 10—we take a preliminary look at them here: They include the movement of several large plates that, fitted together, form the rigid rind of the earth. This movement causes the growth as well as the closure of ocean basins and the creation of earthquakes, volcanoes, and mountain building along the plate boundaries. The movement also accounts for the shifting of continents (**continental drift**) over the last several hundred million years. The processes focus on the outer 200 km (kilometers) of the earth, a subject to which we shall also return, and a brief sketch of what we know about this zone is presented below.

LITHOSPHERE, ASTHENOSPHERE, CRUST, AND MANTLE

The outer 50 to 100 km of the earth is a rigid shell of rock, called the **lithosphere** (from the Greek *lithos,* "rock," and "sphere"). Yet as observations from deep mines tell us, the temperature of the earth increases by around 15°C (Celsius) with each kilometer of depth. So at a depth of about 70 km the temperature averages 1000°C, at which rock will slowly flow if pressure is applied. This "soft" zone from 70 to 100 km is called the **asthenosphere** (Greek *asthenēs,* "weak"). As we shall see, its existence helps explain some of the earth's major movements, both vertical and horizontal.

The lithosphere and asthenosphere are distinguished by temperature, but we can also divide the outer part of the earth into shells on the basis of composition. We speak, therefore, of the skin of the earth as the **crust** and of the bulk of the earth beneath the crust as the **mantle.** A wide variety of rock types constitutes the earth's crust, and we can make direct observations of most of the types. A discontinuous cover of sedimentary rocks, a few meters to a few kilometers thick, overlies igneous and metamorphic rocks. Beneath the continents the crust is thicker (averaging about 35 km) than that beneath the oceans (about 5 km). A dark-colored, relatively heavy igneous rock called **basalt** dominates the crust beneath the oceans, and we quite naturally call it **oceanic crust.** Its density, about 3 t/m^3 (tonnes per cubic meter), contrasts with the lighter-weight **continental crust,** with a density of about 2.6 t/m^3. A large portion of the continental crust is composed of the igneous rock called **granite,** which not only is less dense than basalt but also is light gray to pink in color. Beneath the rocks of the crust at a depth of 5 to 50 km lies the mantle, made of rocks with a density of about 3.3 t/m^3; the asthenosphere lies within the upper mantle (Figure 1.9).

VERTICAL MOVEMENTS AND ISOSTASY

Precise surveying shows that when a sufficiently large lake forms behind a dam, it will depress the earth's crust slightly. We also know that when glaciers expanded during the Ice Age, their weight depressed large areas of the crust. Conversely, when the glaciers waned and disappeared, the land rose, recovering its earlier elevation; and even today some areas of Scandinavia and Canada are still rising in response to the melting of the last glaciers. The ancient geologic record provides other examples. Studies of thick sequences of sedimentary rocks show numerous situations in which thousands of meters of sediments accumulated in shallow marine environments. There is no way to explain a continuous pile of shallow-water sediments unless the basin in which they had accumulated was shallow from the beginning and kept sinking slowly as additional sediments were added.

In all these examples of **vertical movement,** portions of the earth's crust behave as if they were floating in a soft, slowly flowing zone. And that is what we believe happens. The crust and the uppermost mantle that lies above the asthenosphere have some strength. But they cannot resist the pull of gravity, and they respond to the addition or removal of a load. So if material such as water, ice, or sediments are added at the surface, that overloaded area (and the column of rigid rock beneath) sinks slightly into the asthenosphere;

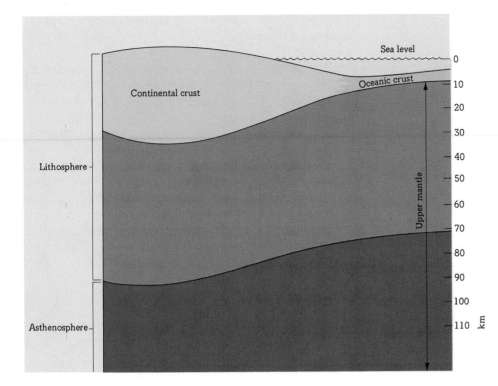

1.9 The relationships among continental crust, oceanic crust, upper mantle, lithosphere, and asthenosphere. (See also Figure 9.38.)

conversely, if a load is removed, an area of the crust floats upward. This floating balance of the crust is called **isostasy** (meaning "equal standing"; Figure 1.10). It explains why the thick, lightweight continental crust stands high in relation to the ocean basins underlain by thinner, heavier oceanic crust.

1.10 Isostasy is a condition of balance, of floating equilibrium, illustrated here by blocks of wood of varying height floating in water. Each block sinks to a different depth in the water, but the amount above the surface is proportionally the same for each—in this example about one-third. For each block the amount above the water is balanced by the part below the water. Large blocks of the earth's lithosphere behave similarly, floating in the asthenosphere.

HORIZONTAL MOVEMENTS

The geologic record also reveals that large portions of the earth's crust have moved horizontally. And if we accept the idea of continental drift, then we can say that the entire surface of the earth is subject to lateral movements, measured in thousands of kilometers. This notion brings us back to our preliminary look at plate tectonics and sea-floor spreading.

Plates of the earth If we plot earthquakes on a world map, we find them concentrated in narrow, well-defined belts (as in Figure 1.11). These belts are also marked by extensive volcanic activity. Between the belts are large areas, both on the continents and in the ocean basins, where earthquakes and volcanoes, although not entirely absent, are infrequent. We now know that the zones of intense activity mark the boundaries between rigid **plates** of the earth's crust—plates forming a jigsaw puzzle whose pieces constantly jostle one another. It is this movement that is associated with volcanoes and earthquakes.

We are now able to demonstrate (as will be discussed in Section 10.3) that plates move at rates be-

1.3 PLATE TECTONICS, SEA-FLOOR SPREADING, AND CONTINENTAL DRIFT

Eurasian

Arabian

African

Philippine

Australindian

Antarctic

Gorda

American

Eurasian

Pacific

African

Caribbean

Cocos

Nazca

American

Antarctic

1.11 Earthquakes are concentrated in narrow bands around the earth. The boundaries of the earth's major plates follow these bands. (See also Figure 10.19.)

tween 1 and 9 cm/year (centimeters per year). The relative motion along plate margins can be any of three types: two plates may slide parallel to each other but in opposite directions; two plates may diverge from each other; or two plates may converge. In fact, each of these movements is found along the margin of any given plate.

What goes on along the boundary of diverging or converging plates?

If two plates are pulling apart, it must mean that something is being added to their separating margins. For example, the broad rise that runs down the center of the Atlantic Ocean—the Midatlantic Ridge—marks the axis of separation between the Old and New Worlds, with the American plate moving generally westward as the Eurasian and African plates drift eastward. The crest of the Midatlantic Ridge is marked by a precipitously walled **rift valley** along its entire length. Along the valley, new volcanic material (basalt) wells upward from the mantle. The newly formed oceanic crust is then slowly transported laterally away from the ridge (sea-floor spreading), allowing room for still younger material to be added along the axial valley.

If the plates grow along a diverging boundary, they must be destroyed elsewhere to allow space for the new material. This happens along the converging boundaries of plates. Look, for example, at the western boundary of the American plate, where that plate, which includes both the southwestern Atlantic Ocean basin and the South American continent, collides along most of its western margin with a large Pacific Ocean plate, the Nazca plate. The convergence of the two plates is marked by the Andes Mountains, intense earthquakes, and volcanic activity; in addition a deep, narrow trench characterizes the ocean floor just west of the South American shore (see Figure 1.12).

The leading edge of the American plate is riding over the leading edge of the Nazca plate. The horizontal movements of the crust are piling up an excess of rock material where the two plates collide, and isostasy operates to ensure that this thickened column of the earth's crust not only sinks lower into the asthenosphere but stands higher above sea level.

Now look at the trench just off South America. The ocean floor here is depressed much below the average deep ocean floor. The leading edge of the Nazca plate is being shoved downward and eastward beneath the American plate, pulling part of the oceanic crust and

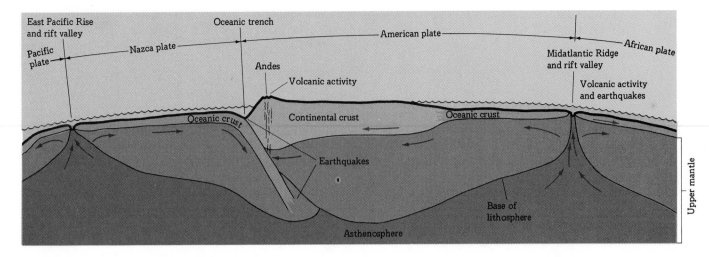

1.12 The relationships among the African, American, and Nazca plates. See text for discussion.

floor with it. Along the dipping zone of contact between the two plates, earthquakes occur as the plates slip by each other. The bulk of the Nazca plate continues downward under the leading edge of the American plate until temperature and pressure are so great that, along its front edge, it begins to lose its rigidity and is reincorporated into the mantle in the asthenosphere. During this process, igneous activity goes on. The thermal energy needed to generate molten material comes from the friction generated between the plates, from the heat of radioactivity in the thickened rock pile, and from the heat generated by the weight of the thickened crust.

As a hypothesis first seriously introduced in 1912 by Alfred Wegener, a German meteorologist, continental drift received little acceptance until the mid-1960s, when the reality of plate tectonics and sea-floor spreading began to emerge. Wegener had pictured the continents as moving about in individual units; but after the concept of continental drift had become scientifically respectable, it was demonstrated that the moving units included not only continents but portions of ocean basins. It is a plate that moves, and any earth feature—whether continent or ocean—that it includes moves with it.

THE ROCK CYCLE AGAIN

In later sections we shall return to the subject of plate tectonics and sea-floor spreading, expanding the bare outline presented above and laying out the data that have led us to this model of crustal movement. For the moment, however, accept the reality of plate tectonics,

and apply it to the concept of the rock cycle that we introduced in Section 1.2 and Figure 1.8.

In describing the rock cycle, we pictured it without reference to world geography and without providing a general mechanism for the cycle. The plate-tectonic model furnishes some of this detail. We can visualize sediments moved by water, wind, and ice from the continents to the oceans, only to be brought back to the continents by the slowly moving ocean floors. Again, plate movements, particularly along plate boundaries, create temperatures and pressures that lead to metamorphism and igneous activity; and especially along converging boundaries, continents grow in thickness and elevation, allowing erosion to eat deeper and deeper into the thickened wedge of continental rocks and creating more sediments for transfer back into the oceans (see Figure 1.13).

1.4 SOME PRACTICAL CONSIDERATIONS

In this opening chapter we have touched on some major concepts in geology and, in doing so, have sought to provide a general guide to the details in the chapters to follow. You may want to come back to this chapter from time to time for the perspective we hope it provides.

But before beginning more detailed discussion, there is an additional important point to emphasize. It

Sediments and sedimentary rock

Igneous rock

Metamorphic rock

Hot
Cold
Mantle

Volcano

Erosion

Trench

Seawater

Oceanic crust

Rise

Continental
crust

Base of
continental
lithosphere

Base of oceanic
lithosphere

1.13 A cross section through a growing ocean plate, which plunges beneath the edge of a continent, shows that the igneous, sedimentary, and metamorphic components of the rock cycle are present in both continental and oceanic areas. At locations where the rock cycle has converted one rock type to another, the newly formed rock is shown in color. [Modified from Sheldon Judson, Kenneth S. Deffeyes, and Robert B. Hargraves, *Physical Geology*, p. 28, Prentice-Hall, Inc., Englewood Cliffs, N. J. 1976.]

is this: Geology has had and will continue to have practical applications for a wide range of human activities. The processes that have gone on within the earth and at its surface have created the natural resources of fuels and metallic and nonmetallic deposits. The very search for these resources has provided us with the data and techniques that allow us to understand much about the operation of the earth. And this new knowledge, directly and indirectly, will help us to find additional resources and to better understand already discovered resources.

But the practical aspects of the subject are by no means limited to the search for natural resources, important as this may be. As the 1960s drew to a close, people began to realize that human activity could have startling, unexpected, and often disastrous effects on the earth's environment. The next decade saw the beginning of the search for answers to the problems that our society has generated when it interferes with natural systems. As we entered the 1980s, recognition that our natural resources were in many cases limited and finite became generally widespread. And we began to recognize as well that the earthly environment produces events from floods to earthquakes, events that we cannot forestall and yet can be disastrous in human terms.

With burgeoning population, the human demands on resources and environment have produced a myriad of social, political, and economic stresses. Amelioration of these stresses will demand the best from many disciplines as different as astrophysics and zoology, philosophy and economics, and sociology and chemistry. Here we point out, however, that geology brings some unique information to many of these problems. The search for new resources is an obvious area for geologic expertise. In the area of environmental protection the geologist brings certain skills cultivated over years of study of natural systems. For example: What is the probability of a flood of a given magnitude? What dangers are involved in the underground disposal of wastes? What are the factors that lead to erosion of the shoreline? Can earthquakes or volcanic eruptions be predicted? Can they be controlled?

Thus from time to time we draw attention to certain geological factors that bear directly on human activity. And we devote Chapters 17 and 18 to the subjects of energy and useful materials. There will also be many other places where the reader will make the connection between geological information and the world of human activity.

Geology is the science of the earth and deals with the materials that compose it, the history of life, and the physical events of earth history.

Geologic time encompasses the age of the earth, about 4.5 billion years.

> **Absolute time and relative time** are used to divide geologic time. Relative time—whether an event in earth history came before or after another event—disregards years. Absolute time measures the age of a geologic event in years.

> > **Radioactivity** provides most of the absolute dates in geology and depends on the breakdown, at a constant rate, of unstable elements to stable ones.

> **Divisions of geologic time** are listed in a **geologic-time scale,** and the rock units representing these time divisions are shown in a **geologic column,** as reproduced on the front endpaper of this book.

> **Uniformitarianism** is the principle that holds that the earth processes now going on also operated in the past.

Earth materials are composed of minerals, which in turn make up rocks.

> **The three rock families** are **igneous, sedimentary,** and **metamorphic.**

> **The rock cycle** traces the changes as rocks are transformed from one type into another.

Plate tectonics, sea-floor spreading, and **continental drift** describe the motion of large pieces of the outer 50 to 100 km of the earth.

> **Lithosphere, asthenosphere, crust,** and **mantle** are terms applied to certain portions of the outer 200 km of the earth.

> **Vertical movements** of the crust often reflect the floating balance reached by portions of the earth's lithosphere as it sinks or rises in the asthenosphere, a process called **isostasy.**

> **Horizontal movements** are often the result of plate tectonics and sea-floor spreading.

> > **Plates of the earth** are defined by narrow bands of earthquakes and volcanic activity. Plates move relative to each other in converging, diverging, and parallel directions. Plates grow at diverging boundaries and are destroyed at converging boundaries.

> **The rock cycle** can be related to the events accompanying plate tectonics and sea-floor spreading.

Practical applications of geology include the search for natural resources and provide basic data on the natural systems of our environment.

Eicher, Donald L. *Geologic Time,* Prentice-Hall, Inc., Englewood Cliffs, N. J., 1968. *A short book that describes the various ways by which we measure geologic time.*

Marvin, Ursula B. *Continental Drift,* Smithsonian Institution Press, Washington, 1973. *An easy-to-read but authoritative account of the history of the idea of continental drift.*

McPhee, John *Basin and Range,* Farrar, Straus & Giroux, New York, 1981. *This short book, which originally appeared in serial form in* The New Yorker *magazine, is an imaginative, well-written account of the geology along transcontinental Interstate Highway 80. Geographic emphasis is on the Basin and Range, the desert land that extends from Salt Lake City to the Sierra Nevada Mountains, from southern Oregon to Mexico. The book discusses geology, its modern practitioners, and the history of the science. Read it as an introduction to, or a summary of, much of what is covered in this volume.*

Sullivan, Walter *Continents in Motion,* McGraw-Hill Book Company, New York, 1974. *The science editor of* The New York Times *tells the story of ideas about a changing earth, which led to the present concepts of plate tectonics, sea-floor spreading, and continental drift.*

1 Geologic events may be dated in relative or absolute terms. Give one or more examples of each type.

2 Who was James Hutton? Why is he considered important in the development of the science of geology?

3 What is uniformitarianism? Give one or more examples.

4 List the three major rock families, and in a brief sentence describe each.

5 Describe, with the help of a diagram, how the rock families are interrelated by the rock cycle.

6 The outer layer of the earth is called the **crust.** What are the two types of crust and what are the differences between them?

7 Distinguish between crust and mantle.

8 Distinguish between lithosphere and asthenosphere. How do they relate to crust and mantle?

9 What is sea-floor spreading? Plate tectonics? Draw a cross section of the upper 150 km of the earth to show the relation and motions of two or more plates.

10 If the basin of the South Atlantic Ocean is about 200 million years old and was formed by sea-floor spreading, what has been the average annual separation of South America from Africa? (Hint: You need to know the present distance between the two continents.)

(We got 25 mm per year. Using 5,000 km as the present distance between the two continents, we set up the problem as follows):

$$\frac{5 \times 10^9 \text{ mm (distance)}}{2 \times 10^8 \text{ yr (time)}} = 25 \text{ mm/year (distance/year)}$$

11 If you answered Question 10, then you should be able to predict how far apart Africa and South America will be in 60 million years. What is your prediction? What assumption or assumptions have you made in arriving at this prediction?

12 How does the concept of the rock cycle as shown in Figure 1.8 fit into the model for plate tectonics? A neat, well-labeled diagram would be helpful in answering this question.

2

MINERALS

Today we find at the earth's surface those proverbial kinds of matter sometimes called "animal, vegetable, and mineral." As we now know, the earth has not always had all of these kinds of matter. Animals and plants are fairly recent newcomers to the earth as compared to minerals, whereas minerals (and rocks) have made up the solid surface of the earth for most of its history (see Figures 2.1 and 2.2).

Rocks are made of minerals; minerals are made of elements; and elements are made of other smaller particles. Minerals and rocks are basic to our study of geology, but a brief look at the origin of the elements and particles that form minerals will help our understanding of the meaning of matter.

As suggested in the opening paragraph, the word **mineral** has different meanings in everyday usage; but in our discussion it will be used to refer to a **naturally occurring element or compound of elements that has been formed by various processes.** Minerals are everywhere about us. Almost any small plot of ground will offer numerous samples. They can occur in several forms, such as rocky ledges, the soil of plowed fields, or the sands of a river bank. Even some of the most common types are valuable enough to make mining

them commercially worthwhile; and the rarer minerals of gold and silver have provided the basis of wealth and power since the dawn of civilization.

The physical universe is composed of what we call **matter;** yet one of the most elusive problems in science is to define "matter" precisely. It has long been customary to refer to the states of matter as solid, liquid, or gas. We say that matter has physical properties, such as color or hardness, as well as chemical properties, which govern its ability to change or to react with other bits of matter. But all this simply tells us *about* matter, not what matter is.

Many centuries ago, Greek philosophers speculated on this problem, arguing over whether the smallest pieces into which anything could be divided were just miniatures of the original—microscopic drops of water, extremely small grains of sand, infinitesimal pieces of salt—or whether at some point down the line certain particles might be found that were joined to-

2.1 Calcite crystals from Wisconsin; specimen is 25 cm long. [M. E. Kauffman.]

gether differently to form water, sand, salt, and all the other substances that make up what we think of as the material world.

Today, we know that the second explanation is in a general way the true one. These particles are called **atoms.** So if we are to deal with matter, we must first learn about the atoms and their combinations. This information will then give us a basic understanding of what minerals and rocks actually are, why they differ, and how they can be changed.

2.1 PARTICLES AND ELEMENTS

ELECTRIC CHARGE

All matter behaves as if it were essentially electrical in nature. Some of the earliest ideas about what we now call **electricity** sprang from a very simple experiment—the rubbing together of a piece of amber and a piece of fur. After the rubbing, both the fur and the amber were found to be capable of picking up light pieces of other materials, such as feathers or wool. The interesting thing was that materials **attracted** by the amber were **repelled** by the fur. So scientists decided that there must be two kinds of electricity: one we now call **positive,** and the other **negative.** This fact was first dis-

covered by the Greeks about 600 B.C. Then, late in the sixteenth century, William Gilbert, personal physician to Elizabeth I, proposed that the power responsible for this phenomenon be called electricity, from the Greek word *ēlektron,* meaning "amber." And in technical terms we say that **like electric charges repel each other and unlike charges attract each other.**

ATOMS

Just what happens when amber is rubbed with fur? Particles pass from the fur to the amber, and the amber becomes charged—we say "negatively." So we reason that the particles that bring about this condition must also be negatively charged. These negatively charged particles are called **electrons,** and they are fundamental constituents of atoms, of which all matter is composed.

Atoms cannot be assembled from electrons alone, however; for all electrons are negative and would not stick together by themselves. So scientists reasoned that there had to be some positively charged particles. At last they were found and were called **protons.** In addition to having a positive charge, a proton differs from an electron in another respect: It acts like a much heavier unit of matter. Because any quantity of matter is arbitrarily described by a number called its **mass,** the proton is said to have greater mass than the electron. In fact, **we define the fundamental particles of atoms in terms of mass and electric charge** (see Table 2.1).

Nuclear physicists found a third particle, with a

2.2 Quartz (large crystals) and dolomite (smaller off-white crystals) from Pennsylvania. Largest crystal is 4.5 cm long. [M. E. Kauffman.]

mass about equal to that of the proton but with no electric charge. This electrically neutral particle they called the **neutron** (Table 2.1).

TABLE 2.1

Fundamental particles

	Electric charge	Mass, u^a
Electron	−1	0.00055
Proton	+1	1.00760
Neutron	0	1.00890

^{a} Atomic mass unit.

Protons, neutrons, and electrons are the fundamental particles that combine to form atoms.

Nobody has yet come forward with a completely adequate explanation of exactly what these fundamental particles are, but we do know that **energy is intimately involved** in their makeup and in their combinations.

Atomic structure   All the information that we have about the structure of atoms has been established by indirect observation. Using doubly-charged positive particles called **alpha particles** (two protons and two neutrons bound together) to bombard atoms, physicists have made some interesting discoveries. For example, if these particles are shot at a target, such as a piece of metal, made up of billions of atoms, no more than 1 alpha particle in 10,000 hits anything inside the target. It has been concluded, therefore, that the target must be largely open space.

Repeated tests have shown that an atom contains a nucleus of protons and neutrons surrounded by a cloud of electrons. They have also revealed that a normal atom has as many electrons as it has protons. The number of protons plus the number of neutrons in an atom constitutes its **mass number** (the neutrons contribute to the mass of the atom, but they do not affect its electrical charge). The electrons, being negatively charged, spin very rapidly around the nucleus (like planets about the sun); otherwise they would be pulled in by the attraction of the positive protons. In fact; so small is the atom that the electrons complete many billions of round trips per second, at a speed of hundreds of kilometers per second. Again like the planets revolving around the sun, electrons revolve around the nucleus at different distances.

Atomic size   The electrons form a protective shield around the nucleus and define the size of the atom (see Figure 2.2). In describing atomic dimensions, we use a special unit of length, the **angstrom** (Å or sometimes A), which is a hundred-millionth of a centimeter, that is, 0.00000001 cm, usually written 1×10^{-8} cm. Atomic nuclei have diameters that range from a ten-thousandth to a hundred-thousandth of an angstrom—that is, from 1×10^{-4} to 1×10^{-5} Å. Atoms of the most common elements have diameters of about 2 Å, which is roughly 20,000 to 200,000 times the diameter of the nucleus. Again we see that matter consists mostly of open space. If the sun were truly the nucleus of our atomic model, the diameter of the atom would be greater than the diameter of the entire solar system.

Atomic mass   Although the nucleus occupies only about a trillionth of the volume of an atom, it contains 99.95 percent of the atom's mass. In fact, if it were possible to pack a cubic centimeter with nothing but protons, this small cube would weigh more than 100 million t at the earth's surface.

Ions   An **ion** is an electrically unbalanced form of an atom or group of atoms. An atom is electrically neutral. But if it loses an electron from its outermost shell, the portion that remains behind has an extra unmatched positive charge. This unit is known as a **cation**, a positively charged ion. If the outermost shell gains an electron, the ion has an extra negative charge and is known as an **anion**, a negatively charged ion. As we shall see later, more than one electron may be lost or gained, leading to the formation of ions with two or more units of electrical charge.

ELEMENTS

An atom is the smallest unit of an element. **Each element is a special combination of protons, neutrons and electrons.** The distinguishing feature of each element is the number of protons in its nucleus. An element has an **atomic number** corresponding to this number, and it is given a name and a symbol.

Element 1 is a combination of one proton and one electron (see Figure 2.3). Long before its atomic structure was known, this element was named **hydrogen,** or "water former" (from the Greek roots *hydōr*, "water," and *-genēs*, "born") because water is formed when hydrogen burns in air. The symbol of hydrogen is H. Because it has a nucleus with only one proton, hydrogen assumes place 1 in the table of elements.

Element 2 consists of two protons (plus two neutrons in the most common form) and two electrons (see Figure 2.4). It was named **helium,** with the symbol He,

Hydrogen

Diagram of a hydrogen atom, which consists of one proton and one electron. This is the simplest atom.

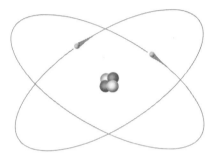

Helium

2.4 Diagrammatic representation of an atom of helium. The nucleus consists of two protons and two neutrons, and accordingly has a mass number of 4. There are two electrons (negative charges) to balance the positive charges on the two protons.

Since there are two protons in the nucleus, this atom is number 2 in the table of elements. The nucleus of helium (two protons + two neutrons) without any accompanying electrons is an alpha particle.

from the Greek *hēlios,* "the sun," because it was identified in the solar spectrum before being isolated on the earth. Because of the two protons in its nucleus, helium takes place 2 in the table of elements.

Each addition of a proton, with a matching electron to maintain electrical balance, produces another element. Neutrons seem to be included more or less indiscriminately, although there are about as many neutrons as protons in the common form of many of the first 20 or so elements.

Of the total of 103 elements, 77 are classed as metals, 17 as nonmetals, and 9 as metalloids. These are keyed in the Periodic Chart (Appendix C). A metal typically shows a peculiar luster, called **metallic luster,** is a good conductor of electricity or heat, is opaque, and

may be fused, drawn into wire, or hammered into sheets. A nonmetal lacks some or all of these properties. A metalloid has some metallic and some nonmetallic properties.

The elements found in nature begin with the lightest, hydrogen (place 1), and end with the heaviest, the metal uranium (place 92). Technetium (43), promethium (61), astatine (85), and francium (87) are not readily found in nature; they have been produced only for an instant during radioactive decay.

Isotopes Elements have forms that, although essentially identical in chemical and physical properties, have different masses. Such forms are called **isotopes**—from the Greek *isos,* "equal" or "the same," and *topos,* "place"—because each form has the same number of protons in the nucleus and occupies the same position in the table of elements. Isotopes show differences in mass as a result of differences in the number of neutrons in their nuclei. For example, hydrogen with one proton and no neutrons in its nucleus has a mass number of 1. When a neutron is present, however, the atom is an isotope of hydrogen with a mass number of 2, called **deuterium** (Figure 2.5). All elements have the capacity to form different isotopes.

2.5 Schematic of deuterium, an isotope of hydrogen formed by the addition of a neutron to the nucleus. It has a mass number of 2.

Information about atomic structure is commonly expressed symbolically by a subscript number prefixed to the element's symbol to indicate the atomic number and by a superscript for the mass number. thus hydrogen would be represented by $_1^1H$, and its isotope deuterium by $_1^2H$.

Many of the elements as they occur in nature are mixtures of isotopes. In general, regardless of the source of the element, the proportion of isotopes composing it is constant unless special steps have been taken to separate them. Thus chlorine, whether occurring naturally or prepared in the laboratory, has an atomic weight of 35.457, which represents the average weight of a mixture of about three parts of $_{17}^{35}Cl$ to one part of $_{17}^{37}Cl$.

Matter is the substance of the physical universe. It is composed of atoms. Atoms are combinations of the fundamental particles protons, neutrons, and electrons (except for the common isotope of hydrogen, which is a combination of one proton and one electron). An atom is the smallest unit of an element. In nature, 88 elements have been found; 4 others have been identified as short-lived transients; and 11 have been artificially prepared.

Atoms are rarely found alone in nature. They are found in combination with each other and with other atoms. Atoms of different elements combine by exchanging or sharing electrons to form compounds. The smallest unit of a substance is a molecule. The 88 naturally occurring elements have been found combined in nature, forming numerous compounds. They have also been combined in the laboratory to form the millions of compounds we use today.

Water is the only substance to occur naturally in the three states of solid, liquid, and gas.

2.2 DISTRIBUTION OF ELEMENTS IN NATURE

We have said that all the elements consist of just two essential nuclear particles, or building blocks: protons and neutrons. (Remember that the third essential atomic particle, the electron, is *not* part of the *nucleus.*) The nucleus of the simplest element—hydrogen—is a single proton. One of the earliest notions concerning elements was that they consist of varying combinations of hydrogen atoms. Though considerably more complicated, most modern theories on the origin of elements are based on this essential idea.

BIG-BANG THEORY

One of the more popular theories about how elements (and the universe) began was first proposed by George Gamow, in 1952, then at George Washington University and from 1956 at the University of Colorado. He and his fellow workers developed the **big-bang theory.** Astronomical observations that the entire universe is expanding led to the conclusion that about 10 to 20 billion years ago the cosmos began in a big bang, or explosion, from a very dense "core" made up mainly of neu-

trons (electrons would be pressed into protons under the tremendous pressure). As this large ball expanded, some of the neutrons decayed to protons and electrons. Each proton captured a neutron and this mass captured another neutron, thus building up the masses. By an extremely rapid series of events, all the elements were formed in this first stage of the universe's history; indeed, it is suggested that the whole process of element formation occurred within a few minutes after the initial explosion. The material thus created continued to expand and eventually clumped into galaxies, stars, and planets.

Although there are a number of problems with Gamow's concept, many contemporary theories are based on this same model modified in various technical ways.

STEADY-STATE THEORY

A second group of theories on the origin of the elements relies on a more complex series of events. Fred Hoyle of Cambridge University, among others, concluded, in 1951, that the elements were built in the hot interior of stars rather than in a primordial explosion. Nuclear reactions and transformations are known to be occurring constantly in the stars. The long-lasting energy of the sun and other stars is explained by two chains of nuclear reactions, which together produce tremendous amounts of energy and build new nuclei: **proton-proton fusion** and the **carbon cycle.**

Basically the steady-state theories propose that an initial potential universe consisted of cold but turbulent hydrogen atoms in a dilute gaseous state. Part of the gas condensed into stars because of gravitational attraction. Under this gravitational force the stars contracted and their interiors grew very hot and dense. At temperatures in the millions of degrees the protons moved so fast as to fuse and form deuterons (hydrogen 2), which combined with other protons to form helium 3. Pairs of helium 3 nuclei fused to produce helium 4, releasing two surplus protons for each fusion. The overall result of these proton-proton chains was the formation of single atoms of helium from sets of four atoms of hydrogen.

Carbon 12 nuclei form from such combinations in the hydrogen-helium core of a star. They may combine with more helium nuclei to complete the carbon cycle with the formation of oxygen 16, neon 20, and perhaps magnesium 24 and others.

Most astronomers today find that the evidence favors a specific single time and place for the beginning of the elements and the universe. Current thinking follows the view that the lighter elements were formed at the moment of the "big bang," during which time hydrogen may have been converted in part to helium, giving a ratio of about 80 percent H and 20 percent He a few minutes after this initial event. Elements from atomic number 3 (lithium) through atomic number 26 (iron) are believed to form inside massive stars as they evolve. The heaviest elements, from atomic number 27 (cobalt) on, are thought to form in supernovae. In a real sense, then, we have a combination of ideas associated with both the big-bang and the steady-state theories that dominate our thought today concerning the origin of elements and the universe.

ELEMENT ABUNDANCE

The relative abundance of elements found on the earth contrasts markedly with the abundance of elements detected in the universe.

About 90 percent by weight of the earth consists of four elements: iron (35 to 40 percent), oxygen (25 to 30 percent), silicon (13 to 15 percent), and magnesium (about 10 percent). Four other elements are present amounting to about 1 percent: nickel, calcium, aluminum, and sulfur. Another seven elements amount to from 0.1 to 1 percent: sodium, potassium, chromium, cobalt, phosphorus, manganese, and titanium. The earth is made up almost entirely of these fifteen elements, with all the remaining elements contributing to less than 0.1 percent of its weight.

Now let us look at the abundances in the universe. Hydrogen and helium, respectively, make up around 76 and 23 percent of the weight of the universe. Thus *all* the remaining elements amount to only a little more than 1 percent of the weight of the universe. The sun, like the universe as a whole, is mostly hydrogen and helium, although it has slightly lower relative amounts of the remaining common elements.

Apparently, in the formation of the earth and similar planets of the solar system, the lighter gases—hydrogen and helium especially—were lost very quickly. This process left the earth with a residue of unusual relative abundances of elements compared with the abundances of elements in the universe as a whole.

2.3 A FIRST LOOK AT MINERALS

At the beginning of this chapter we defined a mineral as a natural element or compound formed by various processes. Now we shall see specifically what this definition means.

MINERAL COMPOSITION

More than 2,000 minerals are known. Some are composed of only one element, such as diamond, [composed entirely of carbon (C)], copper (Cu), silver (Ag), gold (Au), mercury (Hg), and sulfur (S). (See the Periodic Table, Appendix C, for a list of properties and symbols of all the elements.) Others are made up of relatively simple compounds of elements. For example, the mineral halite is composed of two elements, sodium (Na) and chlorine (Cl), in equal amounts. The chemical symbol for halite, NaCl, indicates that every sodium ion present is matched by one chlorine ion. The mineral pyrite, sometimes known as "fool's gold," is also composed of two elements, iron (Fe) and sulfur, but in this mineral there are two atoms of sulfur for each atom of iron, a relationship expressed by the chemical symbol FeS_2. Other minerals have a more complex composition.

A mineral's composition can vary slightly: An occasional replacement by other elements whose atoms are of similar size need not create a new mineral. We can therefore say that **every mineral is composed of elements in definite or slightly varying proportions.**

CHARACTERISTICS OF MINERALS

All the properties of minerals are determined by the composition and internal atomic arrangement of their elements. Although we can identify minerals on the basis of their chemical properties, physical properties are the ones most often used. Such physical properties include crystal form, hardness, specific gravity, cleavage, color, streak, and striations.

Crystal form When a mineral grows without interference or obstacle, it will be bounded by plane surfaces symmetrically arranged and will acquire a characteristic **crystal form,** which is the external expression of its internal crystalline structure. The faces of crystals are defined by surface layers of atoms. These faces lie at angles to one another, angles that have definite characteristic values, with the same value for all specimens of the same mineral. The size of the faces may vary from specimen to specimen, but the angles between the faces remain constant.

Every crystal consists of atoms arranged in a three-dimensional pattern that repeats itself regularly. A combination of atoms whose repetition can produce a given mineral is called its **unit cell**. In a crystal of copper all the atoms are alike and are arranged in a cube. This arrangement is limited by the size of the copper atoms. By repetition the entire crystal is built up (see Figure 2.6). The principal surface layers correspond to the faces of a cube; the smaller surface layer obtained by cutting off a corner of a cube is called an **octahedral face**. Native copper found in deposits of copper ore is often found in the form of crystals with cubic and octahedral faces. Remember that the atoms in the figure are greatly enlarged: If the crystal had sides only 0.1 mm (millimeter) long, there would be about 400,000 atoms in a row along each edge.

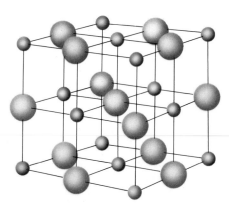

2.7 Arrangement of sodium ions with positive electrical charge, Na^+, and chlorine ions with negative electrical charge, Cl^-, to form the ionic compound $NaCl$, common salt. Na^+ has a radius of 0.97 Å, and Cl^- a radius of 1.81 Å.

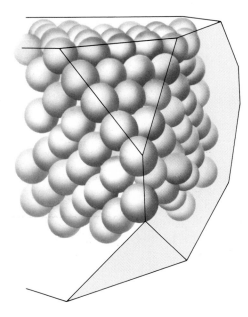

2.6 An atomic view of a copper crystal, showing small octahedral faces and large cubic faces. [After Linus Pauling, *General Chemistry*, p. 23, W. H. Freeman and Company, San Francisco, copyright © 1953.]

ular grains because its growth has been constricted. Even in these irregular grains, however, the atoms are arranged according to their typical crystalline structure. And where conditions have permitted the mineral to develop freely, it forms crystals that are always six-sided prisms. Whether an individual crystal of quartz is only 1 mm long or 25 cm, the faces of the prism always meet at the same angle (see Figure 2.8). This is an example of the constancy of interfacial angles. Although faces of a crystal may be of different sizes

2.8 Quartz crystals. Regardless of the shape or size of crystals, the angles between true crystal faces remain the same. Transverse striations on prism faces are most clearly seen on the lower face; but striations are present on the other prism faces too. The large crystal is about 10 cm long. [M. E. Kauffman.]

In crystals of halite there are ions of two different kinds arranged in the regular pattern shown in Figure 2.7. The smaller ones are those of sodium, and the larger ones are those of chlorine. The atomic arrangement is limited by the size of the ions. Halite is a cube made up of six chlorine ions surrounding one sodium ion, and each chlorine ion is surrounded by six sodium ions, all being equidistant. Repetition of the unit cell in three dimensions forms the cubic halite crystal.

The mineral quartz occurs in many rocks as irreg-

2.3 A FIRST LOOK AT MINERALS

21

and shapes because of malformation, their similarity is frequently demonstrated by natural striations, etchings, or growths. On some crystals the similarity of faces of a form can be seen only after etching with acid.

There are six different systems of crystals, classified according to the angles between their faces and described in Appendix B.

Some elements or compounds may develop several different crystal forms, producing different minerals. This assumption of two or more crystal forms by the same substance is called **polymorphism.** Each of these crystal forms results from the conditions under which it was made. For example, carbon's two common forms are graphite and diamond: The crystal form of diamond is an eight-sided solid, an octahedron; the crystal form of graphite is a flat crystal with six sides. Although both minerals are composed of carbon, the difference in their crystal forms comes from the arrangement of their carbon atoms—one pattern in diamond, another in graphite. The crystal form of pyrite is a cube (Figure 2.9), and that of marcasite is a flattened tubular shape. Both are composed of FeS_2. Here again the reason for the difference in crystal form lies in the internal arrangement of the atoms.

Every mineral has a characteristic crystal form, which is the external shape produced by its crystalline structure.

Scanning electron microscopes The crystalline form of even submicroscopic crystals can now be observed by use of a scanning electron microscope, which focuses beams of electrons across a specimen. These electrons are electrically or magnetically moved sequentially back and forth until the reflected and emitted electrons produce an image of the sample. A great depth of field permits examination of three-dimensional opaque objects, with enormous magnifications (as much as 100,000 times). (See Figure 2.10.)

Cleavage Cleavage is the tendency of a mineral to break in certain preferred directions along smooth plane surfaces. Cleavage planes are governed by the internal arrangement of the atoms; they represent the directions in which the atomic bonds are relatively weak. Because cleavage is the breaking of a crystal between atomic planes, it is a directional property, and any plane throughout a crystal parallel to a cleavage surface is a potential cleavage plane. Moreover, the cleavage is always parallel to crystal faces or pos-

2.9 Cubic crystals of pyrite. Striations are clear on the large specimen. Note that those in adjacent faces are perpendicular to each other. The small specimen, about 5 cm wide, consists of three intergrown cubes. [Walter R. Fleischer, Harvard Mineralogical Collection.]

sible crystal faces because faces and cleavage both reflect the same crystalline structure.

Cleavage is a **direction** of weakness, and a mineral sample tends to break along planes parallel to this direction. This weakness may be due to a weaker type of atomic bond, as we have indicated, to greater atomic spacing, or frequently to a combination of the two. Graphite has a platy cleavage because of relatively weak bonds between the carbon layers. Diamond has but one bond joining all its carbon atoms; its cleavage takes place along planes having the largest atomic spacing.

Cleavage may be perfect, as in the micas, more or less obscure, as in beryl and apatite, or entirely lacking, as in quartz.[1]

Striations A few common minerals have parallel, threadlike lines or narrow bands, called **striations,** running across their crystal faces or cleavage surfaces. These lines can be clearly seen on crystal faces of quartz and pyrite (see Figure 2.9).

Once again, this propety is a reflection of the inter-

[1]See Appendix B and later parts of this chapter for certain specific mineral names and characteristics.

2.10 Scanning electron micrograph of books of chamosite, an iron alumino-silicate, and rosettes of hematite, Fe_2O_3., in oolitic ironstone from Aswan, Egypt. [Deba Bhattacharyya.]

nal arrangement of the atoms and the conditions of growth of the crystals.

Hardness Hardness is another physical property governed by the internal atomic arrangement of the mineral elements. (The stronger the binding forces between the atoms, the harder the mineral.) The degree of hardness is determined by the relative ease or difficulty with which one mineral is scratched by another or by a fingernail or knife. It might be called the mineral's "scratchability." Hardness (H) ranges from 1 through 10:

To illustrate, if you pick up a piece of granite and

Range	Scratchability
$H < 2.5$	Will leave mark on paper; can be scratched by fingernail
$2.5 < H < 3$	Cannot be scratched by fingernail; can be scratched by penny
$3 < H < 5.5$	Cannot be scratched by penny; can be scratched by knife
$5.5 < H < 7$	Cannot be scratched by knife; can be scratched by quartz
$7 < H$	Cannot be scratched by quartz

try to scratch one of its light-colored grains with a steel knife blade, this mineral simply refuses to be scratched. But if you drag one of these grains across a piece of glass, the glass is easily scratched. Clearly, then, these particular mineral grains in granite are harder than either steel or glass. But if you have a piece of topaz handy, you can reveal their vulnerability; for although they are harder than steel or glass, they are not as hard as topaz.

Minerals differ widely in hardness (see Appendix B): Some are so soft that they can be scratched with a fingernail; some are so hard that a steel knife is required to sratch them; diamond, the hardest mineral known, cannot be scratched by any other substance.

Specific gravity Every mineral has an average mass per unit volume. This characteristic mass is usually described in comparison with the mass of the same volume of water. The number that represents this ratio is known as the **specific gravity** of the mineral. **Density** is given in grams per cubic centimeter (g/cm^3) or tonnes per cubic meter (t/m^3) and is equal to specific gravity in numerical value.

The specific gravity of a mineral increases roughly with the mass of its constituent elements and with the closeness with which these elements are packed together in their crystalline structure. Most rock-forming minerals have a specific gravity of around 2.7, although the average specific gravity of metallic minerals is 5. Pure gold has the highest specific gravity, 19.3.

It is not difficult to acquire a sense of relative weight by which to compare specific gravities. We can learn to distinguish between two bags of equal size, one filled with feathers and one filled with lead, and experience in hefting stones has given most of us a sense of the "normal" weight of rocks.

Color Although **color** is not a reliable property in identifying most minerals, it is strikingly characteristic for a few. These include the intense azure blue of azurite, the bright green of malachite, and the pale yellow of sulfur. Magnetite is iron black, and galena is lead gray. The color of other minerals, such as quartz, can be quite variable because of slight impurities.

Minerals containing iron are usually "dark-colored." In geologic usage, **dark** includes dark gray, dark green, and black. Minerals that contain aluminum as a predominant element are usually "light-colored," a term that includes purples, deep red, and some browns.

Streak The **streak** of a mineral is the color it displays in finely powdered form. The streak may be different from the color of the hand specimen. Although the color of a mineral may vary between wide limits, the streak is usually constant.

One of the simplest ways of determining the streak of a mineral is to rub a specimen across a piece of unglazed porcelain known as a **streak plate.** The color of the powder left behind on the streak plate helps to identify some minerals. Because the streak plate has a hardness of 7, it cannot be used to identify minerals with greater hardness.

Hematite, Fe_2O_3, may be reddish brown to black; its streak, however, is light to dark blood-red, which becomes black on heating. Limonite, $FeO(OH) \cdot nH_2O$, sometimes known as "brown hematite" or "bog-iron ore," has a color that is dark brown to black but a streak that is yellowish brown. Cassiterite, SnO_2 (tin-stone), is usually brown or black, but it has a white streak.

Other physical properties Minerals have other physical properties that may be helpful in identification: magnetism, electrical properties, luster, fluorescence, fusibility, solubility, fracture, and tenacity (see Appendix B).

2.4 MINERAL STRUCTURE

In the section on crystal form we have seen that atoms of elements in a mineral assume an orderly pattern, its **crystalline structure.** (For example, in halite the ions of sodium are in alternation with the ions of chlorine.)

Each mineral has a unique crystalline structure that will distinguish it from another mineral, even if the two are composed of the same element or elements. Let us again consider the minerals diamond and graphite. Each is composed of only one element, carbon (Figure 2.11). In diamond, each atom of carbon is bonded (see next section) to four neighboring carbon atoms. This complete joining of all its atoms results in very stable bonds and is the reason why diamond is so hard. In graphite, each atom of carbon is bonded in a plane to three neighboring atoms. This bonding forms sheets, or layers, of carbon piled one on another, but the sheets can easily be separated. Thus graphite is a soft substance. Pyrite and marcasite are another pair

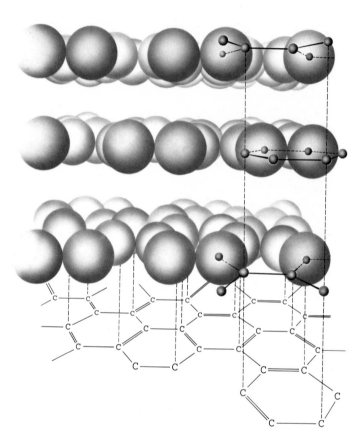

(a)

(b)

2.11 Different arrangements of atoms of carbon produce the crystalline structure of (a) diamond and (b) graphite. [After Linus Pauling, *General Chemistry*, W. H. Freeman and Company, San Francisco, copyright © 1953.]

of minerals with identical composition, FeS_2, but with different crystalline structures. In pyrite, ions of iron are equally spaced in all directions. In marcasite, they are not equally spaced. The difference in spacing accounts for their being two different minerals (Figure 2.12).

(a)   (b)

2.12 Representation of X-ray pattern of atoms. (a) Pyrite, showing orderly arrangement characteristic of crystalline structure. The mineral is composed of iron and sulfur, FeS_2. Large spots are atoms of iron; small ones, atoms of sulfur. Each atom of iron is bonded to two atoms of sulfur, and spacing of iron atoms is the same in both directions of the plane of the photograph. Magnification, approximately 2.2 million ×. (b) Marcasite, FeS_2. Note difference between horizontal and vertical spacing of iron atoms. Compare with pyrite. Magnification, about 2.8 million ×. [Martin J. Buerger.]

Other minerals may have more complicated crystalline structures. They may contain more elements, and these are joined together in more complex patterns. The color, shape, and size of any given mineral may vary from one sample to another, but the internal atomic arrangement of its component elements is **identical** in all specimens of a particular mineral.

So we find it necessary to include in our definition of a mineral not only that it is a naturally occurring element or compound with a diagnostic chemical composition but also that it has a unique orderly internal atomic arrangement of its elements.

COMPOUNDS

Compounds are combinations of atoms of different elements. **Organic** compounds are those in which the carbon atom (or carbon and oxygen) plays a large part—that is, those mostly formed by life processes. Others are **inorganic** compounds. The method by which the atoms are held together (or "bonded") varies, as described below.

IONIC BOND

The number of electrons in the outer shell of an atom determines the manner and ease with which the atom can join with other atoms to form compounds. Compounds may be formed by different atoms losing or gaining electrons. For example, an atom of sodium has only one electron in its outermost shell but eight in the next shell, whereas chlorine has seven in its outermost shell. When a sodium atom and a chlorine atom join, the sodium's outermost electron slips into the vacancy in the chlorine's outermost shell. Chlorine gains the electron and becomes a negative ion, Cl^-, whereas sodium, by losing it, becomes a positive ion, Na^+ (see Figure 2.13). Opposite charges attract; therefore a bond results, forming an **ionic compound,** the mineral halite, or common table salt, a most abundant substance. ($NaCl$ means that one atom of sodium is joined with one atom of chlorine.) The unit is a **molecule** of salt. A molecule is the smallest unit of a compound.

When electrons are lost or added to form an ion, electrical forces between the nucleus and the electrons are thrown out of balance, and this affects the radius of the ion. For example, if sodium loses the electron from its shell 3, it has only two shells of electrons left. This is the same number of electrons as Ne (neon) has, but the radius of Na^+ is 0.97 Å, whereas that of an atom of Ne with the same number of electrons and shells is 1.60 Å.

TABLE 2.2

Radii of atoms and ions^a

		Atom		Ion	
At. no.	Element	Symbol	Radius, Å	Symbol	Radius, Å
8	Oxygen	O	0.60	O^{2-}	1.32
10	Neon	Ne	1.60		
11	Sodium	Na	1.86	Na^+	0.97
12	Magnesium	Mg	1.60	Mg^{2+}	0.66
13	Aluminum	Al	1.43	Al^{3+}	0.51
14	Silicon	Si	1.17	Si^{4+}	0.42
15	Phosphorus	P	1.08	P^{5+}	0.35
16	Sulfur	S	1.04	S^{6+b}	0.30
17	Chlorine	Cl	1.07	Cl^{7+b}	0.27

^{a} Radii from Cornelius S. Hurlbut, Jr., *Dana's Manual of Mineralogy*, 18th ed., John Wiley & Sons, Inc., New York, 1971.
^{b} S^{6+} and Cl^{7+} are not the usual ionic forms. The elements would normally gain electrons to become S^{2-} and Cl^-.

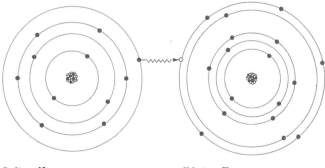

Sodium, Na   Chlorine, Cl

2.13 In the formation of the compound halite, an atom of sodium, which has only one electron in its outermost shell, joins an atom of chlorine, which needs one electron to fill its outermost shell. The sodium's lone outermost electron slips into the vacant place to fill the chlorine's outermost shell.

The excess of unbalanced positive charge in Na^+ pulls in its remaining electrons and shrinks the apparent radius of the ion. In fact, if two full shells of electrons always meant the same radius, the positive ions of the geologically important elements 11 through 17 would have about the same radius after all shell 3 electrons were lost. However, as more and more outer electrons are removed, the increasing relative force of the central positive charge pulls in the remaining electrons until the radii decrease progressively from 0.97 Å for Na^+ to 0.27 Å for Cl^{7+} (see Table 2.2).

COVALENT BOND

Atoms can combine in various ways to form compounds, as, for example, in the formation of a water molecule. The oxygen atom has six electrons in its outermost shell and needs two more to achieve the stable number of eight. If two hydrogen atoms, each with its single electron, approach an oxygen atom, the hydrogen electrons are shared near the vacant slots in the outermost shell of the oxygen atom (see Figure 2.14). So the hydrogen and oxygen nuclei are sharing their electrons in a way called a **covalent bond**. Again, the result is a compound that is totally different from the elements themselves. This compound is water, whose symbol, H_2O, represents the elements that make it up and the proportions in which they are present—a com-

Hydrogen, H

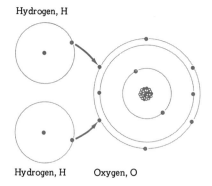

Hydrogen, H   Oxygen, O

2.14 Two hydrogen atoms and one oxygen join to form water, H_2O, by a covalent bond. In this bond the hydrogen electrons do double duty in a sense, filling the two empty places in the outer shell of oxygen yet remaining at normal distance from their nuclei. The result is the formation of a molecule of water, the smallest unit that displays the properties of that compound.

bination so perfect that water is one of nature's most stable compounds.

Because the oxygen atom has essentially gained two electrons, it becomes negatively charged. And each hydrogen atom, acting as though it had lost its electron, takes on a positive charge. As a result, a molecule of water acts like a small rod, one end positively charged and the other negatively (see Figure 2.15). These ends are referred to as a **positive pole** and a **negative pole** because of the molecule's similarity to a bar magnet. The molecule, then, is a **dipole** ("two-pole"); and water is known as a **dipolar compound.** This polarity gives water special properties that make it an extremely important agent in geological processes. The mechanism by which water dissolves salt (see Figure 2.16) is an illustration of the ease with which water dissolves various substances and participates in weathering and other geological activities.

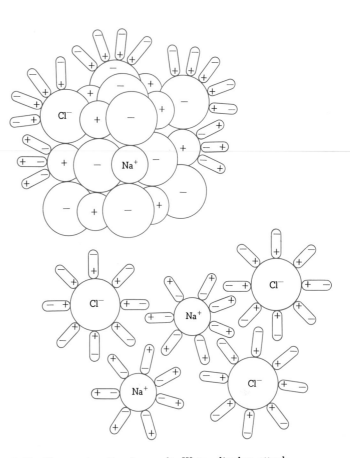

2.16 How water dissolves salt. Water dipoles attach themselves to the ions that compose the salt and overcome the ionic attractions that hold the salt together as a solid. Each sodium and chlorine ion is then convoyed by a number of water dipoles into the body of the liquid.

2.15 The dipolar character of water: The oxygen has, in effect, gained two electrons, hence a double negative charge, whereas the hydrogen atoms have each lost the effective service of an electron and represent positive charges. Accordingly, the water molecule acts like a small rod positively charged on one end and negatively charged on the other. Some possible combinations of water molecules are suggested here.

METALLIC BOND

The atoms of metallic elements have a special kind of bonding, which is responsible for metal being a good conductor of heat and electricity. In a piece of metal, the atoms of a single element are closely packed together; their outermost electrons are not shared or exchanged but are free to move around and connect with any atoms in the solid. These wandering electrons, shared by any of the atoms in the metal, create the **metallic bond.**

The relative freedom of movement of the electrons in this relationship accounts for the high level of electrical conductivity. When there is no current, electrons jump randomly from atom to atom. When an electrical potential is applied to a good conductor, such as a copper wire, electrons flow through the wire without stopping to attach themselves to any particular atoms. They thus convey electricity (or heat) throughout the whole wire.

2.5 THE MOST COMMON MINERALS

Although there are more than 2,000 minerals known, only a limited number compose most of the rocks of the earth's crust. These can be grouped into closely related families of minerals. As we pointed out in Section 2.3, minerals are homogeneous crystalline materials but not necessarily pure substances. Most **rock-forming minerals,** or minerals found most abundantly in the rocks of the crust, have variable compositions caused by ionic substitution of some elements for other elements. These substitutions are distributed at random throughout the crystalline structure. Such replacement is called **solid solution.** The resultant mineral is homogeneous and has the same physical properties throughout, but its composition is slightly variable.

TABLE 2.3

Silicate classification

Arrangement of SiO$_4$ tetrahedra	Si/O ratio	Rock-forming minerals
Isolated	1:4	Olivine
Rings	1:3	Beryl (beryllium aluminosilicate)
Chains (single)	1:3	Augite (pyroxene family)
Chains (double)	4:11	Hornblende (amphibole family)
Sheets	2:5	Biotite (black mica)
		Muscovite (white mica)
Frameworks	3:8	Orthoclase (potassium feldspar)
	3:8	Plagioclase (calcium-sodium feldspar)
	1:2	Quartz

SILICATES

More than 90 percent of the rock-forming minerals are silicates, compounds containing silicon, oxygen, and one or more metals. Each silicate mineral has as its fundamental unit the **silicon-oxygen tetrahedron** (see Figure 2.17). This is a combination of one "small" silicon ion, with a radius of 0.42 Å, surrounded as closely as geometrically possible by four "large" oxygen ions, each with a radius of 1.40 Å, forming a tetrahedron.

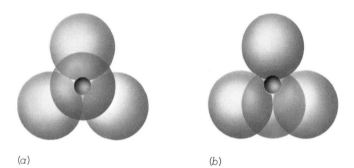

(a)   (b)

2.17 The silicon-oxygen tetrahedron, $(SiO_4)^{4-}$: (a) from above, (b) from the side. This is the most important complex ion in geology because it is the central building unit of nearly 90 percent of the minerals of the earth's crust.

The oxygen ions contribute an electric charge of 8− to the tetrahedron, and the silicon ion contributes 4+. Therefore the silicon-oxygen tetrahedron is a complex ion with a net charge of 4−. Its symbol is $(SiO_4)^{4-}$.

The most common of the rock-forming silicates are olivine, augite, hornblende, biotite, muscovite, feldspars, and quartz. They are listed and classified in Table 2.3.

Ferromagnesians In the first four of these rock-forming silicates—olivine, augite, hornblende, and biotite—the silicon-oxygen tetrahedra are joined by ions of iron and magnesium. These silicate minerals are known as **ferromagnesians,** from the joining of the Latin *ferrum,* "iron," and "magnesium." All four ferromagnesians

are very dark or black and have a higher specific gravity than do the other rock-forming minerals.

Olivine is composed of isolated silicon-oxygen tetrahedra held together by positive ions of magnesium and iron. Because there are no planes of weakness and its elements are so firmly held together by the ionic bonds, olivine exhibits no cleavage but rather fractures when struck a blow. It is a relatively hard mineral, 6.5 to 7.

Olivine is an example of a mineral that undergoes compositional changes. Its formula is $(Mg, Fe)_2SiO_4$. The proportions of magnesium and iron vary; so their symbols are in parentheses. The proportion of silicon and oxygen remains constant. Iron and magnesium substitute for one another quite freely in olivine's crystalline structure as they each have two electrons in the outer shell and their ionic radii are almost identical, 0.97 and 0.99 Å. The general tendency for two or more ions to substitute for one another is known as **isomorphism.** When the ions substitute in a continuous series (like magnesium and iron in olivines) the process is called a **solid-solution** mineral series. The end members of the olivine series are forsterite and fayalite: When the positive ions are all magnesium, the mineral is forsterite, Mg_2SiO_4; when they are all iron, the mineral is fayalite, Fe_2SiO_4.

Olivine is geologically important as it makes up several percent of the crustal rocks at the surface and is believed to predominate in the heavier and deeper-seated rocks. Its specific gravity is 3.27 to 4.37, increasing with iron content (Figure 2.18). Its color ranges from olive to grayish green, sometimes brown. This mineral usually occurs in grains or granular masses.

Augite has a crystalline structure based on single chains of tetrahedra,[2] as shown in Figure 2.19, joined by ions of iron, magnesium, calcium, sodium, and alu-

[2] In the remainder of this chapter, "tetrahedra" is equivalent to "silicon-oxygen tetrahedra."

Specific gravity of olivine.

2.19 Single chain of tetrahedra viewed (a) from above and (b) from an end. Each silicon ion (small black sphere) has two of the four oxygen ions of its tetrahedron bonded exclusively to itself, and it shares the other two with neighboring tetrahedra fore and aft. The resulting individual chains are in turn bonded to one another by positive metallic ions. Because these bonds are weaker than the silicon-oxygen bonds that form each chain, cleavage develops parallel to the chains.

(a)

(b)

minum. It is dark green to black, with a colorless streak. Its hardness is 5 to 6, and its specific gravity ranges from 3.2 to 3.4. It has rather poor cleavage along two planes almost at **right** angles to each other. This cleavage angle is important in distinguishing augite from hornblende. (A good way to recall it is to remember that "augite" rhymes with "right.") The cleavage results from the contrast between the stronger covalent bonding *within* the chains of silica

tetrahedra, and the weaker ionic bonding *between* adjacent chains and the intervening cations.

Augite is the commonest variety of a group of minerals designated by the family name **pyroxenes** (see Appendix B) and is characterized by a fundamental crystalline structure built on single chains of tetrahedra. Pyroxenes crystallize and form a series whose members are closely analogous chemically to members of the amphibole group (see below).

Hornblende has a crystalline structure based on double chains of tetrahedra, as shown in Figure 2.20, joined by the iron and magnesium ions common to all ferromagnesians and by ions of calcium, sodium, and aluminum. Hornblende's color is dark green to black, like that of augite; its streak, colorless; its hardness, 5 to 6; and its specific gravity, 3.2. Two directions of good

2.20 Double chain of tetrahedra viewed (a) from above and (b) from an end. The doubling of the chain of Figure 2.19 is accomplished by the sharing of oxygen atoms by adjacent chains.

(a)

(b)

2.5 THE MOST COMMON MINERALS

29

2.21 Cleavage of hornblende. The top "roof" of the hornblende specimen and the slanted surfaces are cleavage planes. Throughout each specimen, easiest breaking is parallel to these surfaces. On the front face of the specimen are some "steps" outlined by cleavage planes. Such steps are the most common manifestation of cleavage, which seldom produces pieces as large as this one 15 cm wide. [M. E. Kauffman.]

cleavage meet at angles of approximately 56° (degrees) and 124°, which helps distinguish hornblende from augite (see Figure 2.21).

Hornblende is an important and widely distributed rock-forming mineral. It is the commonest mineral of a group called the **amphiboles,** which closely parallel the pyroxenes in composition but contain hydroxyl, OH. Amphiboles are characterized by double chains of tetrahedra. The minerals classified as hornblendes are in reality a complex solid-solution series from anthophyllite to hornblende (see Appendix B).

Biotite, black mica, is a potassium-magnesium-iron-aluminum silicate, $K(Mg,Fe)_3[AlSi_3O_{10}(OH)_2]$. (It was named for a distinguished French physicist, J. B. Biot.) Like other micas, it is constructed of tetrahedra in sheets, as shown in Figure 2.22. Each silicon ion shares three oxygen ions with adjacent silicon ions to form a pattern like wire netting. The fourth, unshared oxygen ion of each tetrahedron stands above the plane of all the others. The basic structural unit of mica consists of two of these sheets of tetrahedra with their flat surfaces facing outward and their inner surfaces held together by positive ions. In biotite the ions are magnesium and iron. These basic double sheets of mica, in turn, are loosely joined together by positive ions of potassium.

Layers of biotite or of any of the other micas can be peeled off easily (see Figure 2.23) because there is a distinct contrast between the stronger covalent bonding *within* a sheet, and the weaker ionic bonding *between* adjacent sheets and with the intervening cations. In thick blocks, biotite is usually dark green or brown to black. Its hardness is 2.5 to 3, and its specific gravity is 2.8 to 3.2.

Nonferromagnesians The other common rock-forming silicate minerals, **nonferromagnesians** do *not* contain

2.22 Tetrahedral sheets. Each tetrahedron is surrounded by three others, and each silicon ion has one of the four oxygen ions to itself, sharing the other three with its neighbors.

2.23 Mica cleavage. The large block (or "book") is bounded on the sides by crystal faces. Cleavage fragments lying in front of the block are of different thicknesses, as indicated by their degrees of transparency. The reference grid is made of 5-cm squares. [Walter R. Fleischer.]

and crystal habit. They all show good cleavages in two directions, which make an angle of 90° or close to 90° with each other. Their hardness is about 6, and their specific gravity ranges from 2.55 to 2.76.

In feldspars all the oxygen ions in the tetrahedra are shared by adjoining silicon ions in a three-dimensional network. However, in the centers of one-quarter to one-half of the tetrahedra, aluminum ions with a radius of 0.51 Å and an electric charge of 3+ have replaced silicon ions with their radius of 0.42 Å and electric charge of 4+. The negative electric charge resulting from this substitution is corrected by the entry of K^+, Na^+, or Ca^{2+} into the crystalline structures.

Two feldspars are orthoclase with potassium and plagioclase with sodium or calcium (see Table 2.4). **Orthoclase** is named from the Greek *orthos*, "straight," and *klasis*, "a breaking," because the two dominant cleavages intersect at a right angle when a piece of orthoclase is broken. Aluminum replaces silicon in every fourth tetrahedron, and positive ions of potassium correct the electrical unbalance. The introduction

TABLE 2.4	Diagnostic ion	Name	Symbol	Formula[a]
The feldspars	K^+	Orthoclase (potassium feldspar)	Or	$K(AlSi_3O_8)$
	Na^+	Albite (sodium plagioclase)	Ab	$Na(AlSi_3O_8)$
	Ca^{2+}	Anorthite (calcium plagioclase)	An	$Ca(Al_2Si_2O_8)$

[a] In these formulas the symbols inside the parentheses indicate the tetrahedra. The symbols outside the parentheses indicate the diagnostic ions that are worked in among the tetrahedra.

magnesium or iron (such minerals as muscovite, feldspars, and quartz). They are all marked by their light colors and relatively low specific gravities, ranging from 2.6 to 2.9.

Muscovite is white mica, so named because it was once used as a substitute for glass in old Russia (Muscovy). It has the same basic crystalline structure as biotite, but in muscovite each pair of tetrahedra sheets is tightly cemented together by ions of aluminum. As in biotite, however, the double sheets are held together loosely by potassium ions, along which cleavage readily takes place. It has a formula of, essentially, $KAl_2[AlSi_3O_{10}(OH)_2]$. In thick blocks the color of muscovite is light yellow, brown, green, or red. Its hardness ranges from 2 to 2.5, with specific gravity 2.8 to 3.1.

Feldspars are the most abundant rock-forming silicates. Their name comes from the German *feld*, "field," and "spath," a term used by miners for various nonmetallic minerals. The name reflects the abundance of these minerals: "field minerals," or minerals found in any field. Feldspars make up nearly 54 percent of the minerals in the earth's crust.

The feldspars are silicates of aluminum with potassium, sodium, or calcium. Crystals of feldspars of different systems resemble each other closely in angles

of aluminum in the tetrahedra cannot be regarded as solid solution: Aluminum is not a vicarious constituent of orthoclase whose percentage varies from sample to sample or which may be wholly replaced by silicon; it is an essential constituent and is not replaceable by silicon without breakdown of the crystalline structure. Orthoclase is white, gray, or pinkish; its streak is white; and its specific gravity is 2.57.

Plagioclase ("oblique-breaking") feldspars have cleavage planes that intersect at about 86°. One of the cleavage planes is marked by striations. Plagioclase feldspars may be colorless, white, blue-gray, or black, and some samples show a striking play of colors called **opalescence**.

The plagioclase feldspars albite and anorthite are end members of a solid-solution series. The relative amounts of calcium and sodium vary in proportion to the amount of aluminum in order to maintain electrical neutrality; the more calcium, the greater the amount of aluminum. Here the variation in amount of aluminum may be properly regarded as ionic substitution. The number of ions of calcium that replace sodium is equaled by the number of aluminum ions replacing silicon in the silicon-oxygen framework. Sodium can substitute quite easily for calcium in a mineral's crystalline structure because their ionic radii are almost identical, 0.97 and 0.99 Å, even though there is a difference in electrons present in the outer shell—one in sodium and two in calcium.

In albite, aluminum replaces silicon in every fourth tetrahedron, and positive ions of sodium correct the electrical unbalance. The specific gravity of albite is 2.62. In anorthite, aluminum replaces silicon in every second tetrahedron, and positive ions of calcium correct the electrical unbalance. The specific gravity of anorthite is 2.76.

Although the plagioclase feldspars have different compositions, they have the same crystal form and for that reason are called **isomorphous**. At the same time they can be thought of as solutions of albite in anorthite in varying proportions, or the reverse. For that reason, they have been called examples of **solid solution**. This term, however, does not imply that in the intermediate minerals there are molecules of either albite or anorthite as there would be of, say, sodium and chlorine in a solution of salt in water. Intermediate minerals simply have compositions that are conveniently described as mixtures of the pure end members of the series.

Quartz is the most common rock-forming silicate mineral that is composed "exclusively" of tetrahedra. Every oxygen ion is shared by adjacent silicon ions, which means that there are two ions of oxygen for every ion of silicon. This relationship is presented by the formula SiO_2. Because of its crystalline structure, quartz is a relatively hard mineral, 7. Because there are no planes of weakness in its crystalline structure, it does not exhibit cleavage but fractures along curved surfaces when struck a blow. Being composed of fairly light elements, its specific gravity is 2.65.

Of all the rock-forming minerals, quartz is most nearly a pure chemical compound. However, spectrographic analyses show that even in its most perfect crystals there are traces of other elements. Quartz usually appears smoky to clear in color, but less common varieties include purple or violet amethyst, massive rose-red or pink rose quartz, smoky yellow to brown smoky quartz, and milky quartz. These color differences are caused by the other elements present as minor impurities.

OXIDE MINERALS

Oxide minerals are formed by the direct union of an element with oxygen. These have relatively simple formulas compared to the complicated silicates. The oxide minerals are usually harder than any other class except the silicates, and they are heavier than others except the sulfides. Within the oxide class are the chief ores of aluminum, iron, tin, chromium, and manganese. So we have the common oxide minerals ice, H_2O, corundum, Al_2O_3, hematite, Fe_2O_3, magnetite, Fe_3O_4, and cassiterite, SnO_2.

SULFIDE MINERALS

Sulfide minerals are formed by the direct union of an element with sulfur. The elements that occur most commonly in combination with sulfur are iron, copper, lead, zinc, silver, and mercury. Some of these sulfide minerals occur as valuable ores, such as pyrite, FeS_2, chalcocite, Cu_2S, galena, PbS, and sphalerite, ZnS.

CARBONATE AND SULFATE MINERALS

In discussing silicates we found them to be built around the complex ion $(SiO_4)^{4-}$, that is, the silicon-oxygen tetrahedron. Two other complex ions are also of great importance in geology. One of these consists of a single carbon ion with three oxygen ions packed around it, the complex ion $(CO_3)^{2-}$. Compounds in which this ion appears are called **carbonates**. For example, the combination of a calcium ion with a carbon-oxygen ion produces calcium carbonate, $CaCO_3$ (Figure 2.24), known in its mineral form as **calcite**. This mineral is the principal component of the common sedimentary rock limestone. The other complex ion is $(SO_4)^{2-}$, a combination of one sulfur ion and four oxygen ions. This complex ion combines with other ions to form **sulfates**, for example $CaSO_4$, the mineral **anhydrite**, and its hydrous form, $CaSO_4 \cdot nH_2O$, the mineral **gypsum**.

HALIDES

A variety of minerals results from the combination of certain elements (positive ions) with the halogen elements (chlorine, iodine, bromine, and fluorine): the very common rock salt, halite, $NaCl$, its poisonous partner, sylvite, KCl, and many other chlorides, iodides, bromides, and fluorides. Halides occur as precipitates from evaporating ponds, salt flats, and natural brines.

TABLE 2.5

Summary of the organization of some common minerals[a]

Elemental minerals	Compound minerals					
	Silicates [elements + $(SiO_4)^{4-}$]	Oxides (elements + O)	Sulfides (elements + S)	Carbonates [elements + $(CO_3)^{2-}$]	Sulfates [elements + $(SO_4)^{2-}$]	Halides (elements + Cl^-, I^-, Br^-, or F^-)
Copper	Ferromagnesian	Cassiterite	Chalcocite	Calcite	Anhydrite	Halite
Diamond	Augite	Corundum	Galena	Dolomite	Gypsum	Sylvite
Gold	Biotite	Hematite	Pyrite	Magnesite		
Graphite	Hornblende	Ice	Sphalerite			
Platinum	Olivine	Magnetite				
Silver	Nonferromagnesian					
Sulfur	Feldspars					
	Orthoclase					
	Plagioclase					
	Albite					
	Anorthite					
	Muscovite					
	Quartz					

[a]Minerals may be either elements or compounds, but not all elements or compounds are minerals.

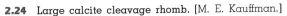

2.24 Large calcite cleavage rhomb. [M. E. Kauffman.]

MINERALOIDS

Some substances do not yield definite chemical formulas upon analysis and show no sign of crystallinity. These are said to be **amorphous** and have been called **mineraloids**. A mineral may exist in a crystalline phase with a definite composition and crystalline structure, or under certain conditions of formation practi-cally the same substance may occur as a mineraloid. Mineraloids are formed under conditions of low pressure and temperature and are commonly substances originating during the process of weathering of the materials of the earth's crust. They characteristically occur in mammillary, botryoidal, stalactitic, and similar-shaped masses. Their ability to absorb other substances accounts for their wide variations in chemical composition. Bauxite, limonite, and opal are examples of mineraloids.

2.6 ORGANIZATION AND ASSOCIATION OF MINERALS

We know that minerals are special combinations of elements or compounds (see summarizing Table 2.5), and now we can complete our definition of a mineral: (1) It is a naturally occurring element or compound in the solid state;[3] (2) it has a diagnostic composition; (3) it

[3]Some mineralogists do not restrict the definition to the solid state but include such substances as water and mercury.

has a unique crystalline structure; and (4) it exhibits certain physical properties as a result of its composition and crystalline structure.

ASSOCIATIONS OF MINERALS

Certain groups of minerals are most commonly found associated in specific geological settings. For example, some minerals formed at high temperatures commonly occur in igneous rocks that crystallize deep within the earth. Other mineral assemblages reflect their origin under the high temperatures *and* increased pressures associated with metamorphic events. Finally, there are some mineral groupings that usually develop under conditions approaching the atmospheric temperatures and pressures found at the earth's surface. Such mineral associations are useful indicators of past conditions at the time of either their origin or their recombination to form new minerals, where temperature, pressure, or chemical environment has changed.

2.7 RESHUFFLING THE ELEMENTS AND MINERALS

As we noted in Chapter 1, rocks continuously undergo changes during their cyclic passage through time, and the minerals making up rocks are thus changed. Likewise, the elements constituting these minerals are rearranged in combinations with new elements in varying proportions.

One location where much of this reorganization occurs is along plate boundaries within the earth's lithosphere (see Section 1.3). For example, at the trailing edge of a plate, new material is welling up from the deeper portion of the earth's mantle, solidifying and being added as new plate material.

On the other hand, at the leading edge of a lithospheric plate other possibilities exist. The plate may be moving down into the mantle along a subduction zone. Here the increased temperature and pressure cause the minerals in the rocks to undergo changes into other minerals and even to be melted if carried deep enough.

Another possibility exists at the leading edge of a lithospheric plate. The plate may carry a continent upon its back, and this continental mass may encounter another similar mass carried by a plate approaching from the opposite direction. The resulting continent-to-continent collision produces both uplift of mountains and downwarp of "roots" under those mountains (see Chapter 8). Elements, minerals, and rocks found in the mountains will undergo changes during uplift, including weathering and erosion. Similarly, those materials found in the roots of the mountains will undergo changes resulting from increased temperature and pressure: melting or remelting of the constituents.

In this way earth materials are constantly undergoing a reshuffling in terms of their combinations, relative proportions, and associations. New minerals are formed from old, new concentrations of elements and minerals develop where they did not exist previously, and old minerals are broken down into their constituent parts (elements or combinations of elements) and recombined in differing arrangements of elements, minerals, and rocks.

OUTLINE

Minerals are naturally occurring elements or compounds.

Matter, the substance of the physical universe, is described in terms of atoms with mass and electric charge.

Electric charge is of two kinds, arbitrarily called **positive** and **negative.**

Atoms are composed of protons, neutrons, and electrons.

Atomic structure is expressed by a nucleus of protons and neutrons surrounded by a cloud of electrons.

For the most common elements, **atomic size,** determined by the protective shield of electrons, is about 2 Å in diameter around a nucleus of 10^{-5} to 10^{-4} Å.

Atomic mass is 99.95 percent in the nucleus.

Ions are electrically unbalanced forms of atoms or groups of atoms.

Elements are particular combinations of protons, neutrons, and electrons.

> **Isotopes** are forms of elements with the same number of protons in the nucleus but different mass because of their different number of neutrons.

Gamow's **big-bang theory** assumes that all elements developed very rapidly in an explosive event about 10 to 20 billion years ago.

Hoyle's **steady-state theory** concludes that elements are constantly being formed in the hot interior of stars from the hydrogen and helium fuels.

Mineral identification is accomplished by examination of chemical and physical properties, including crystal form, cleavage, striations, hardness, specific gravity, color, and streak.

> **Crystal form** is the external shape produced by a mineral's crystalline structure.
>
> **Cleavage** is the tendency of a mineral to break in certain preferred directions along smooth plane surfaces.
>
> **Striations** are parallel, threadlike lines or narrow bands running across crystal faces or cleavage surfaces.
>
> **Hardness** is governed by the internal atomic arrangement of the elements of a mineral.
>
> **Specific gravity** is a number that compares the weight of a given volume of a mineral with the weight of the same volume of water.
>
> **Color** is not a reliable property for identifying most minerals, but minerals containing iron are usually dark-colored and those containing aluminum are light-colored.
>
> **Streak** is the color of a mineral in finely powdered form.
>
> **Other physical properties** include magnetism, electrical properties, luster, fluorescence, fusibility, solubility, fracture, and tenacity.

Mineral structure is the internal orderly arrangement of atoms, which is unique for each mineral.

> **Compounds** are combinations of atoms of different elements.
>
> > A **molecule** is the smallest unit of a compound.
>
> **Ionic compounds** occur when electrons are lost or added to atoms.
>
> **Covalent bonding** results from sharing electrons rather than from gaining or losing them.
>
> **Metallic bonds** are responsible for metals being such good conductors of heat and electricity.

Rock-forming minerals are 90 percent silicates, based on a complex ion called the silicon-oxygen tetrahedron.

> **Ferromagnesians** olivine, augite, hornblende, and biotite contain iron and magnesium.
>
> **Nonferromagnesians** include muscovite, the feldspars, and quartz.
>
> **Quartz** is the only rock-forming silicate mineral composed exclusively of tetrahedra.
>
> **Oxide minerals** are formed by the direct union of an element with oxygen.
>
> **Sulfide minerals** are formed by the direct union of an element with sulfur.
>
> **Carbonate minerals** are built around the complex ion $(CO_3)^{2-}$.
>
> **Sulfate minerals** are built around the complex ion $(SO_4)^{2-}$.
>
> **Halides** form from combinations of positive ions with chlorine, iodine, bromine, and fluorine.
>
> **Mineraloids** do not have definite compositions or crystalline structure.

Organization of minerals lies in naturally occurring combinations of elements or compounds in the solid state, each with diagnostic composition and unique crystalline structure as well as certain physical properties.

> **Associations of minerals** commonly occur in specific geologic settings and reflect the conditions of temperature, pressure, or chemical environment at the time of their origin or when recombined into new minerals.

New combinations of reshuffled elements and compounds produce new minerals, especially along edges of lithospheric plates.

SUPPLEMENTARY READINGS

Fowler, William A. "The Origin of the Elements," *Sci. Am.,* vol. 195, no. 5, 1956. *Discusses Gamow and Hoyle theories.*

Frye, Keith *Modern Mineralogy,* Prentice-Hall, Inc., Englewood Cliffs, N. J., 1974. *Somewhat advanced treatment from chemical fundamentals to a complete survey of mineralogy.*

Hurlbut, Cornelius S., Jr. *Minerals and Man,* Random House, Inc., New York, 1969. *For the layman a lively, balanced presentation of expert information on the nature, origin, and properties of 150 of the world's principal minerals, together with stories of how they have been used throughout the ages; with 217 illustrations, including 160 in color.*

Hurlbut, Cornelius S., Jr. *Dana's Manual of Mineralogy,* 18th ed., John Wiley & Sons, Inc., New York, 1971. *Latest edition of the textbook that has dominated this field for most of this century.*

Loeffler, Bruce M., and Roger G. Burns "Shedding Light on the Color of Gems and Minerals," *Am. Sci.,* vol. 64, no. 6, 1976. *Gives good explanation for variations in color of minerals due to selective absorption of different wavelengths of light.*

Nassau, Kurt "The Causes of Color," *Sci. Am.,* vol. 243, no. 4, 1980. *Discusses the electronic responses to different wavelengths of light and the ultimate colors produced by differential absorption of parts of the energy spectrum.*

QUESTIONS

1 How do we know the probable arrangements of protons, neutrons, and electrons in atoms? Draw a diagram to illustrate these arrangements.

2 Why do we assume that matter consists of particles that behave essentially as electrical particles?

3 Use another common analogy to describe the relative size of protons and neutrons when compared to the elements orbiting around this nucleus.

4 How do isotopes of one particular element behave similarly and how do they behave differently from one another?

5 What is the principal property of minerals that controls most of their physical properties? Explain your answer.

6 How are atoms combined with one another?

7 What is the basic building block of the silicates, the most common of all the rock-forming minerals?

8 Explain how solid solution works. Give examples from a common mineral family.

9 When Al^{+3} substitutes for some Si^{+4} ions in silicate minerals, how is electrical neutrality maintained? Give a specific example.

10 Upon examining the periodic table in Appendix C, explain why certain elements with similar physical properties occur at systematic intervals within that table.

11 How do associations of minerals reflect the conditions of temperature at the time of their origin?

12 Why are edges of lithospheric plates likely locations for the occurrence of newly combined elements and compounds?

3

ORIGIN
AND OCCURRENCE
OF IGNEOUS
ROCKS

Most of us have seen pictures of volcanic eruptions, where molten rock derived from deep within the earth comes up to the surface and is poured or blown out onto the ground, where it solidifies again into rock. When extruded through extensive fractures in the earth's surface, fissure eruptions build large plateaus. In other cases, molten material escapes through vents. Ejected solid and liquid matter builds up landforms around these vents that are called **volcanoes.**

As noted in Chapter 1, much of the earth's igneous activity is concentrated along the margins of lithospheric plates. Molten material rises along the mid-oceanic-ridge zones of ocean basins and the rifted margins of continents. Other igneous masses are associated with the convergent edges of moving plates along subduction zones (see Figure 1.12). Later we shall look at the character of these igneous intrusions and extrusions as they relate to plate boundaries. In this chapter we shall also examine the movement of plates over zones of hot rising material and consider the resulting surface features.

3.1 VOLCANOES

A volcano may grow in size until it becomes a mountain (Figure 3.1). Normally cone-shaped, it has a pit at the summit, which may be either a crater or a caldera. A **crater** is a steep-walled depression out of which volcanic materials are ejected. Its floor is seldom over 2 km in diameter; its depth may be as much as 200 m. A crater may be at the top of a volcano or on its flank. The much larger <u>caldera</u> is a basin-shaped depression, more or less circular, with a diameter many times greater than that of the included volcanic vent or vents

37

(a)   (b)

3.1 (a) Mount St. Helens and Spirit Lake as most southwest Washington residents remember them prior to the explosive eruption on May 18, 1980. (b) After the eruption. [World Wide Photos.]

(see Figure 3.2). Most calderas, in fact, are more than 1,500 m in diameter; some are several kilometers across and several hundred meters deep.

Between eruptions a volcano's vent may become choked with rock congealed from a past eruption. Sometimes small jets of gas come out through cracks in this rock plug. A volcano is built by, and remains active because of, materials coming from a large deep-seated reservoir of molten rock. While in the ground, this molten rock is called **magma**. When extruded on the surface, it is called **lava**.

The largest volcano on earth is Mauna Loa, part of the island of Hawaii. It is 600 km around the base, and its summit towers nearly 10 km above the surrounding ocean bottom. This and the rest of the island represent

3.2 Aerial photograph of Halema'uma'u Crater on the floor of Kilauea Caldera, Hawaii. Note parking lot, highway, and cars near top for sense of scale. The crater is about 1 km across and 200 m deep; it is in mild eruption with fresh lava mantling the crater floor. [U.S. National Park Service.]

accumulations from eruptions that have gone on for more than 1 million years. The largest volcano yet found anywhere is Olympus Mons ("Mount Olympus") on Mars. Photographed originally from the *Mariner 9* spacecraft, this volcano may be over 23 km high, with a caldera 65 km across.

EVOLUTION OF VOLCANOES AND TYPES OF VOLCANISM

Many volcanoes start their eruptive cycles with <u>basaltic</u>[1] lava flows and then <u>andesitic</u>[2] lava flows, and then for thousands of years they produce only fragmental materials that build up their cones. In their final stages, a cone-disrupting explosion occurs with ejection of only fragmental materials. The cycle ends when lava congeals in the volcanic vent. First eruptions consist of <u>basaltic low-viscosity lava</u>, which flows readily. With succeeding eruptions, the lava becomes more silicic and more viscous. Eventually, the lava becomes so viscous that even gases escape with difficulty, great pressures building up in the magma with eruptions of fragmental pyroclastic debris resulting.

There exists a distinction between volcanoes on continents and those in oceanic areas. Although highly fluid basalt occurs in both regions, associated volcanic rocks are markedly different.

Highly viscous magmas are commonly associated with continents and islands near continental borders. These magmas also tend to have large amounts of gas and therefore tend to produce much more explosive volcanic eruptions than do the less viscous oceanic magmas.

Oceanic magmas characteristically contain more silicon in relation to the alkalies (sodium and potassium) and are called **tholeiitic basalts**. The initial bases of most of the Hawaiian volcanoes consist of this material. Most young eruptions in oceanic basins consist of basalt, whereas most young eruptions on continents show increases in alkali content and are andesitic (hence the "andesite line," separating typical oceanic from typical continental geology; see Figure 3.3).

Nearly all samples of lava dredged from the deep ocean floor are tholeiitic basalt. Most of the samples taken from the volcanic islands along the Midatlantic Ridge are **alkalic basalt**.

Volcanic eruptions on stable portions of continental masses tend to be like those of the ocean basins, although more explosive. The volcanism in folded mountain belts, on the other hand, tends to begin with basalts, sodium-rich basalts called **spilites**, and

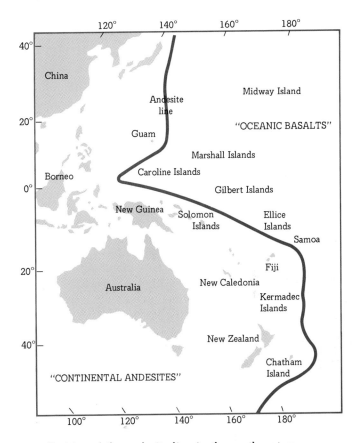

3.3 Position of the andesite line in the southwestern Pacific Ocean. This line marks the border of the Pacific basin in a geologic sense. On the Pacific side of the line, young eruptive rocks are basaltic; on the other side they are principally andesitic. Islands east of the line are isolated or grouped volcanic peaks; west of the line they have the characteristic structure of folded continental mountain ranges. [From various sources.]

[1]Igneous rocks are commonly divided into **felsic** rocks, dominated by light minerals, and **mafic** rocks, dominated by dark minerals. These terms come from the fact that quartz and feldspars, both rich in silica, are dominant in felsic rocks (*fel* for feldspar) whereas the ferromagnesian minerals (pyroxenes, amphiboles, and olivines) are dominant in the mafic rocks (*ma* from magnesium and *f* for ferrous). The most common mafic igneous rock is basalt (its coarse-grained equivalent is called **gabbro**) composed of calcium-rich plagioclase plus pyroxene. The most common felsic rock is granite (its fine-grained equivalent is called **rhyolite**), made up of quartz, orthoclase feldspar, some sodium-rich plagioclase, plus minor amounts of ferromagnesian minerals and muscovite. (We also refer to the granitic rocks being more **silicic**, that is, containing more silica, and the basaltic rocks being less silicic, or containing less silica.) **Ultramafic rocks** are even lower in silica than the basalts and gabbros and contain principally pyroxene and olivine. Other rock terms are defined later in this chapter and appear in the classification chart, Figure 3.28.
[2]Andesite is a volcanic rock composed essentially of the plagioclase mineral andesine, with varying amounts of pyroxene, hornblende, or biotite.

serpentinized peridotites (ultrabasic rocks). Volcanic activity may be associated with periods of mountain building and often also occurs after the folding ceases.

VOLCANIC ERUPTIONS

In its reservoir far below the earth's surface, magma is composed of elements in solution. Some of these elements are vaporized as the magma approaches the surface. These volatile components play an extremely important role in igneous activity and are the primary agents in producing a **volcanic eruption.** As the magma nears the surface, volatiles tend to separate from the other components and migrate through them to the top of the moving mass. The gases accumulate if the volcanic vent is blocked. Pressure builds up until it can no longer be confined. Then it pushes out. With temperatures of 1000°C or higher, the gases expand several thousandfold as they escape, shattering rock that blocks the vent and throwing the rock fragments and magma into the air. After the explosion, the magma still in the ground is left poorer in volatile components but may be fluid enough to pour out.

Volcanic eruptions are of different types, from relatively quiet outpourings of lava to violent explosions accompanied by showers of volcanic debris. The type of eruption depends to a great extent on the magma's composition and its volatile content. Lavas rich in ferromagnesians but low in silica (SiO_2) tend to flow relatively quietly, whereas those high in silica are viscous and explosive.

COMPOSITION OF LAVAS

Lavas are of three principal types, classified primarily by their proportions of silica: acid, intermediate, and basic. The acid lavas have 70 percent or more of silica; intermediate lavas, 50 to 70 percent; and basic lavas, less than 50 percent.

All lavas have some similarities in components, but no two volcanoes erupt lavas of exactly the same composition. In fact, the composition may vary from one eruption to another in the same volcano. Generally, however, over the course of its history, a volcano starts with less silicic lava, grades into intermediate, and ends its activity with dominantly silicic lava. Volcanoes erupting primary basaltic lavas are confined to ocean basins. In the Pacific Ocean four such volcanoes are on the Hawaiian Islands, and three are on the Galápagos. In the Atlantic Ocean, they range from Iceland along the Midatlantic Ridge to Tristan da Cunha. Intermediate-type lavas are erupted by volcanoes

ringing the Pacific and constituting the island arcs. In the Mediterranean Sea, a number of Italian volcanoes erupt lavas believed to have been contaminated by the ingestion of large quantities of limestone; hence their compositions differ from most other volcanoes.

The composition of volcanic eruptions will therefore vary, depending on the source of the magma and the nature of the rocks through which it passes on its upward journey to the earth's surface. Magma generated entirely within material of uniform composition will be composed of that single rock type (such as basic types along the midoceanic ridges). Magmas resulting from melting of the lithospheric crust that has been carried down along subduction zones will produce intermediate-type rocks (such as along the perimeter of the Pacific Ocean).

The surfaces of lava flows commonly have one of two contrasting shapes, for which the Hawaiian names of **pahoehoe** and **aa** are used. In the pahoehoe type, sometimes known as "corded" or "ropy," the surface is smooth and billowy and frequently molded into forms that resemble huge coils of rope (Figure 3.4). In the aa type, the surface of the lava is covered with a random mass of angular, jagged blocks. The surface that develops depends on the viscosity of the lava, with viscosity increasing as lavas become more acidic. Pahoehoe is the shape produced on the surface of a basic lava, and aa is the shape produced when the

3.4 Tongues of pahoehoe lava overflow a levee around a perched lava lake. Plates of crusted lava from the lake surface have been pushed on top of the levee, which consists mostly of pahoehoe tongues. The fountain feeding the lake is at upper left. [R. T. Holcomb, U.S. Geological Survey.]

flow becomes sluggish because of an increase in silica content.

Temperature of lavas Measurements of lava temperature have been made at some volcanic vents. These show that basic lavas have an average temperature of around 1100°C. The hottest lavas known are the Hawaiian lavas measured at Kilauea. Measurements were made in the lava lake Halema'uma'u (Figure 3.2) by the use of Seger cones (ceramic material constructed to melt at various temperatures). The cones were placed in pipes, and then the pipes were lowered into the lake of lava. At a depth of 13 m the lava had a temperature of 1175°C. This decreased to 860°C at a depth of 1 m and increased again to 1000°C at the lava surface. Temperatures as high as 1350°C were recorded above the lava surface, where volcanic gases were reacting with atmospheric oxygen.

In 1910 Frank A. Perret recorded temperatures between 900 and 1000°C from the lava at Mount Etna in Sicily and in 1916 to 1918 observed lavas in Mount Vesuvius from 1015 to 1040°C. Minimum temperatures observed on lavas are around 750°C.

VOLCANIC GASES

As one might expect, taking an accurate sampling of volcanic gases is not an easy job. It is also difficult to decide whether the gases have come exclusively from the magma or partly from the surrounding rocks. We can, however, make a few generalizations from measurements taken at Kilauea. Close to 70 percent of the volume of gases collected directly from a molten lake of lava at Kilauea was steam. Next in abundance were carbon dioxide, nitrogen, and sulfur gases, with smaller amounts of carbon monoxide, hydrogen, and chlorine.

Even when gases other than steam make up only a small percentage of the total volume, their absolute quantities may be large. For example, in 1919 during the cooling of material erupted in 1912 from Mount Katmai in Alaska, the total amount of hydrochloric acid released was estimated at 1,250,000 t, and the total amount of hydrofluoric acid was approximately 200,000 t.

When any igneous rock is heated, it yields some quantity of gases. Water vapor predominates, and measurements indicate that it constitutes about 1 percent of fresh—that is, unweathered—igneous rocks. Estimates of the average water content of actual magma range from about 1 to 8 percent, with the weight of opinion centering around 2 percent. A silicate melt will not hold more than about 11 percent of volatiles under any circumstances.

PYROCLASTIC DEBRIS

Fragments blown out by explosive eruptions and subsequently deposited on the ground are called **pyroclastic debris**. The finest of these constitute **dust**, which is made up of pieces of the order of 10^{-4} cm in diameter. When volcanic dust is blown into the upper atmosphere, it can remain there for months, traveling great distances. The following fragments settle around or near the volcanic crater:

Ash Fragments consisting of sharply angular glass particles. Smaller than cinders.

Cinders Small, slaglike, solidified pieces of magma 0.5 to 2.5 cm across. Cinder **cones** may develop from the accumulation of these fragments (see Figure 3.5).

Lapilli Pieces about the size of walnuts.

Blocks Coarse, angular pieces of the cone or masses broken away from rock that blocks the vent.

Bombs Rounded masses that congeal from magma as it travels through the air.

3.5 A cinder cone (approximately 0.5 km in diameter) with basaltic lava spreading from its base. Some older cones and flows can also be seen. San Francisco Mountains area, Coconino County, Arizona. [From Harvard University, Kirk Bryan Library of Geomorphology.]

3.6 Two examples of pumice; fine-textured and coarse-textured.

Pumice Pieces of magma up to several centimeters across that trap bubbles of steam or other gases as they are thrown out (see Figure 3.6). After these solidify, they are honeycombed with gas-bubble holes, which give them sufficient buoyancy to float on water.

Fiery clouds During the Katmai eruption, a great avalanche of incandescent ash mixed with steam and other gases was extruded. Heavier than air, this highly heated mixture rolled down the mountain slope. Masses of such material are called **fiery clouds** (sometimes referred to by the French equivalent, *nuées ardentes*). The volume of material extruded was so great that it covered a valley of 140 km² to a depth of 30 m. For the next 10 years, the steam and gases kept erupting from this extruded material through a great number of holes, called **fumaroles,** and the area was given the name Valley of Ten Thousand Smokes.

Fiery clouds have characterized eruptions at Mount Pelée on Martinique Island in the West Indies to such an extent that they have come to be known as Peléan types of eruption. Other characteristics are magma expelled as pumice and ash, and in the final stage of eruption, a mass of viscous lava is accumulated in a domelike form when the gas content is so reduced that it no longer shatters the magma on reaching the surface.

At a few minutes before 8:00 A.M. on May 8, 1902, a gigantic explosion occurred through one side of Pelée. A fiery cloud at temperatures around 800°C swept down the mountainside and engulfed the city of Saint Pierre, wiping out its 25,000 inhabitants and many ref-

ugees from other parts of the island who had gathered there during the preceding days, when the eruption was building up with minor explosions and earthquakes (Figures 3.7 and 3.8); estimates of the death toll ran as high as 40,000. By the middle of October, a dome of lava too stiff to flow had formed in the crater, and from it a spine was extruded like a great blunted needle, with a diameter of 100 to 200 m and at its maximum a height of 300 m above the crater floor.

Worldwide effects In 1783, Asama in Japan and Laki in Iceland had explosive volcanic eruptions. Large quantities of dust were blown into the upper atmosphere. The sun's effectiveness in heating the earth's surface was so reduced that the winter of 1783 and 1784 was one of the severest on record. Benjamin Franklin was the first person to connect the unusual weather with the volcanic eruptions, publishing his ideas in May 1784.

During 1814 and 1815, the earth's temperature was reduced following volcanic eruptions of Mayon and Mount Tambora on Sumbawa Island, east of Java. The eruption of Tambora threw so much dust into the air that for 3 days there was absolute darkness over a distance of 500 km. With this dust and the dust erupted from Mayon in the atmosphere, the amount of the sun's heat reaching the earth's surface was significantly reduced. The year 1815 became known as the "year without a summer," marked throughout the world by long twilights and spectacular sunsets caused by the dust in the stratosphere.

FORMATION OF CALDERAS

Calderas may be formed by explosion, by collapse, or by a combination of both (see Figure 3.9). It is often difficult to determine just which mechanism is responsible.

The caldera on Kilauea was probably formed by the collapse of the summit. As great quantities of magma escaped from the reservoir beneath the volcano, support for the summit was withdrawn, and large blocks of it fell in, forming the caldera.

Crater Lake, in southern Oregon, lies in a basin that is an almost perfect example of the caldera shape. This is circular, with a diameter of a little more than 8.5 km and a maximum depth of 1,200 m, and it is surrounded by a cliff that rises 750 to 1,200 m. Crater Lake itself is about 600 m deep. The caldera was formed when the top of a symmetrical volcanic cone, Mount Mazama, vanished during an eruption. Geologists have studied the deposits on the slopes and have tried to piece together the history of the basin.

First, a composite cone was slowly built to a height of around 3,600 m. Then glaciers formed, moving down from the crest and grooving the slopes as they traveled. An explosive eruption occurred 6,600 years ago, and the caldera was formed. Later activity built up a small cone inside the caldera, which now protrudes above the surface of Crater Lake as Wizard Island.

3.8 Partially melted goblet, 15 cm tall, as affected by the fiery cloud event on Martinique in May 1902. [Willard Starks.]

3.9 Sequence of events proposed by one hypothesis for the formation of a caldera. (a) An eruption begins with fiery clouds and dust clouds distributing materials on slopes and surrounding country. (b) Eruption continues; part of cone is blown away; and lava flows join in draining the magma reservoir. (c) Most of cone collapses into the reservoir; later activity forms cinder cone in caldera. [After H. Williams, "Calderas and Their Origin," *Bull. Univ. Calif. Dept. Geol. Sci.*, vol. 25, pp. 239–346, 1941.]

Not all observers agree on the origin of the Crater Lake caldera itself, however. The question is whether all or nearly all the missing material from the cone was actually blown out during an eruption or whether the

caldera was created when the summit collapsed. The answer to the mystery should be provided by an analysis of the unconsolidated material found in the vicinity. Does this material consist of pyroclastics formed during an eruption, or does it consist of the broken remnants of Mazama's blown-off summit? The problem is that the summit itself originally included pyroclastics and congealed lavas from earlier eruptions, and it is difficult to distinguish between the two. One investigator has concluded that, of the 70 km³ of Mazama that disappeared, only 8 km³ are represented in the materials now lying on the immediate slopes and that the rest of it was dropped into the volcano when the roof of an underlying chamber collapsed.[3] This chamber may have been partially emptied by the ejection of large volumes of material during an eruption. H. Williams found evidence that ash spread over a radius of nearly 50 km. Others have reported Mount Mazama ash as far east and northeast as Alberta and eastern Montana, and some of the magma may have worked its way beneath the surface into adjoining areas. This explanation of the Crater Lake caldera has been challenged on the grounds that it is based on an invalid distinction between "old" and "new" pyroclastics in the debris that covers the area.

CLASSIFICATION OF VOLCANOES

Volcanoes are classified according to the materials that have accumulated around their vents. Thus we have shield volcanoes, composite volcanoes, and cinder cones.

When the extruded material consists almost exclusively of lava poured out in quiet eruptions from a central vent or from closely related fissures, a dome builds up that is much broader than it is high, with slopes seldom steeper than 10° at the summit and 2° at the base. Such a dome is called a **shield volcano.** The five volcanoes of the island of Hawaii are shields.

Sometimes a cone is built up of a combination of pyroclastic material and lava flows around the vent. This form is called a **composite volcano** and is characterized by slopes of close to 30° at the summit, tapering off to 5° near the base. Mayon, on Luzon, is one of the finest examples of a composite cone (Figures 3.10 and 3.11).

A single volcano may develop as a shield volcano during part of its history and as a composite volcano later. Mount Etna is an example of such a volcano (see Figure 3.12).

Finally, small cones consisting mostly of pyroclastic debris, particularly cinders, are called **cinder cones.** They achieve slopes of 30° to 40° and seldom

[3]H. Williams, "Calderas and Their Origin," *Bull. Univ. Calif. Dept. Geol. Sci.*, vol. 25, pp. 239–346, 1941.

3.10 Fiery clouds of eruption of Mount Mayon, Philippine Islands, April 27, 1968. [Smithsonian Institution.]

3.11 Profile of one of the world's most nearly perfect composite cones: Mayon, on Luzon. [Harvard University, Gardner Collection.]

exceed 500 m in height. Many cinder cones have flows of basalt issuing from their base (Figure 3.5). Parícutin, in Mexico, is an example of a cinder cone that has developed in modern times, as we shall describe.

HISTORY OF SOME VOLCANOES

A volcano is considered **active** if there is some historical record of its having erupted. If a volcano has not erupted in historical time and yet shows only minor erosional alteration, it is considered **dormant,** or merely "sleeping," and capable of renewed activity. If a volcano not only has not erupted within historic time but also shows wearing away by erosion and no signs of activity (such as escaping steam or local earthquakes), it is considered **extinct.**

Vesuvius Vesuvius, on the shore of the Bay of Naples, has supplied us with a classic example of the reawakening of a dormant volcano (Figure 3.13). At the time of Christ, Vesuvius was a vine-clad mountain, Mount Somma, a vacation spot in southwest Italy favored by wealthy Romans. For centuries it had given no sign of its true nature. Then, in A.D. 63, a series of strong earthquakes shook the area; and around noon on August 24, A.D. 79, Somma started to erupt. The catastrophe of that August day lay silent for nearly 17 centuries. When the remains of Herculaneum and Pompeii—cities that had been buried by the eruption—

3.12 Mount Etna viewed from the sea near Catania, Sicily. The flat slopes to the left are those of a shield volcano. When Etna changed its eruptive habit late in its history, the explosive ejection of fragmental material built a 300-m pyroclastic cone on the summit of the broad shield of lava flows. These pyroclastics form the irregular and steepened slopes nearest the smoking vent. [Vittorio Sella.]

were uncovered, a story to grip the world's imagination transpired: Roman sentries had been buried at their posts; family groups, in the supposed safety of subterranean vaults, had been cast in molds of volcanic mud cemented to a rocklike hardness, along with their jewels, candelabra, and the food they had hoped would sustain them through the emergency.

Somma is believed to have erupted first about 10,000 years ago as a submarine volcano in the Bay of Naples. It then emerged as an island and finally filled in so much of the bay around it that it became a part of the mainland. It is the youngest volcano found in that vicinity.

There must have been an exceedingly long interval of quiet before the eruption of the year 79 because no earlier historical records of volcanic activity exist. But in 79, part of the old Somma cone was destroyed, and the cone now called Vesuvius started. During this eruption, Pompeii was buried by pyroclastic debris; people were asphyxiated by gases from the ash and suffocated from the dust. Herculaneum was overwhelmed by mud flows of water-soaked ash deposited to a considerable depth.

Eruptions of pyroclastic debris occurred at intervals after 79. The longest period of quiescence lasted 494 years and was followed by an eruption in 1631, which poured out the first lava in historic time. This rejuvenation is attributed to chemical changes accompanying the magmatic ingestion of great quantities of limestone in a reservoir with a roof about 5 km below sea level (see Figure 3.14).

Krakatoa One of the world's greatest explosive eruptions took place in 1883 at Krakatoa, in Sunda Strait

between Java and Sumatra. Krakatoa had once been a single island, a volcanic mountain built up from the sea bottom. At a remote period in the past it split apart during an eruption. By 1883, after a long period of rebuilding, three cones had risen above sea level and had merged. These cones, named Rakata, Danan, and Perboewatan, and various unnamed shoals made up the outline of Krakatoa.

On August 26, 1883, a series of explosions began. The next day, at 10:20 A.M. a gigantic explosion blew the two cones Danan and Perboewatan to bits. A part of the island that had formerly stood 800 m high was left covered by 300 m of water. The noise of the eruption was heard on Rodrigues Island, 5,000 km across the Indian Ocean, and a wave of pressure in the air was recorded by barographs around the world. A great flood of water created by the activity drowned 36,500 persons in the low coastal villages of western Java and southern Sumatra. Columns of ash and pumice soared kilometers into the air, and fine dust rose to such heights that it was distributed around the globe and took more than 2 years to fall. During that time sunsets were abnormally colored all over the world. A reddish-brown circle, known as "bishop's ring," which was seen around the sun under favorable conditions, gave evidence not only of the continued presence of dust in the upper air but of the approximate size of the pieces—just under 0.002 mm. Since 1883, Krakatoa has revealed from time to time that it is in the process of actively rebuilding.

Parícutin About 320 km west of Mexico City (19.50°N, 102.05°W), Parícutin sprang into being on February 20, 1943. Nine years later, it had become quite inactive, but during its life it was studied more closely than any other newborn vent in history, until the eruption of Mount St. Helens in 1980.

Many stories of Parícutin's first hours have been told. According to the version now generally regarded spewn over a tremendously large area, killing trees, livestock, and at least 70 humans.

Mount St. Helens is located along a chain of volcanoes extending from northern California through Oregon and Washington into southern British Columbia. This chain marks the boundary between two colliding plates, the American plate and the very small Gorda or Juan de Fuca plate, lying just offshore (See Figure 1.11). Volcanism has been occurring along the Cascade Mountains for over 250 million years, but most of the volcanic rocks are less than 50 million years old. Mount St. Helens is relatively young, with a history going back about 40,000 years. During the past 4,500 years, Mount St. Helens has been the most explosive volcano in the conterminous United States. Its eruption should not have been such a surprise!

Hawaii The Hawaiian Islands are peaks of volcanoes projecting above the ocean and strung out along a line extending roughly 2,400 km in a northwesterly direction. The Marquesas, Society, Tuamotu, Tubuai, Samoan, and other volcanic groups of the South Pacific form lines roughly parallel to the Hawaiian Islands.

At the northwestern end of the Hawaiian chain are the low Kure and Midway islands. At the southeastern end is Hawaii, the largest of the group, 140 km long and 122 km wide, and the tallest deep-sea island in the world. It is composed of five volcanoes—Kohala, Hualalai, Mauna Kea, Mauna Loa, and Kilauea (see Figure 3.15). Each has developed independently, and each has its own geologic history. The dimensions of the volcanoes are listed in Table 3.1 and represent por-

TABLE 3.1

Volcanoes of the island of Hawaii[a]

Name	Area, km²	Portion of island, %	Summit elevation, m
Mauna Loa	5,290	50.5	4,100
Mauna Kea	2,390	22.8	4,135
Kilauea	1,430	13.7	1,225
Hualalai	755	7.2	2,475
Kohala	610	5.8	1,650

[a]From H. T. Stearns and G. A. Macdonald, "Geology and Groundwater Resources of the Island of Hawaii," *Hawaii Div. of Hydrog. Bull. 9*, p. 24, 1946.

tions of the volcanoes above sea level at the present time; they do not take into account buried slopes.

Kohala has been extinct for many years, but Mauna Kea shows evidence of having been active in the recent geological past, although not within recorded history.

Investigation of the Hawaiian Island chain suggests that it is the result of the relative northwesterly motion of the Pacific plate over a zone of melting in the earth's mantle (a **plume**), which causes crustal extension and produces a "hot spot" in the earth's crust (according to J. Tuzo Wilson,[4] and W. Jason Morgan[5]). Lavas of basalt composition pour out at these **hot spots**, forming islands, which in turn become inactive as motion of the plate continues (Figure 3.16). Radiometric ages of the Hawaiian Island basalts support this hypothesis, with the ages becoming younger toward the hot spot (the southeastern end of the chain) and older away from it. The Emperor seamount chain is a continuation of the Hawaiian Islands, and indicates even older ages and a shift in the direction of plate migration over this hot spot (see Figure 3.17 and Section 15.2, Aseismic Ridge).

Measurements made on tilt meters show that both Mauna Loa and Kilauea swell up during the period when magma is rising from below. The uplift reaches a maximum just before an eruption, but after the eruption the mountains shrink back again. Such measurements on Kilauea[6] show that slow swelling over a period of months, about 16 on the average, produces a maximum uplift of 40 to 135 cm. Flank eruptions of the volcano allow sudden collapse of the summit over a period of days, amounting to 20 to 50 cm of subsidence per eruption. The swelling extends over an area about 10 km in diameter centered on, or just south of, Kilauea

3.15 Schematic drawing of the five volcanoes that have been built up from the sea floor to merge and form the island of Hawaii (viewed here from the southeast).

[4]"Continental Drift," *Sci. Am.*, vol. 208, no. 4, 1963.
[5]"Hot-spots, Plate Motions, and Polar Wandering," *Geol. Sci. Am. Abstr*, vol. 4, no. 3, p. 202, 1972.
[6]R. W. Decker, D. P. Hill, and T. L. Wright, "Deformation Measurements on Kilauea Volcano, Hawaii," *Bull. Volcanol.*, vol. 29, pp. 721–732, 1966.

3.13 Vesuvius in eruption, 190[ ] snow-clad slopes mark the mode[ ] cano. To the left of this and in th[ ] ground is the jagged remnant of [ ] Somma. [A. & C. Caggiano.]

Somma caldera
Vesuvius
Pompeii
Amalfi
Sea level
Clay, marl, sandstone, limestone, dolomite
Marl
Tertiary
Cretaceous
Triassic

0

5

0   5   10 km

3.14 Schematic reconstruction [ ] posed relationships in the S[ ] Vesuvius volcano. [After J. [ ] Umbgrove, *The Pulse of the Ear*[ ] ed., p. 74, Martinus Nijhoff, The [ ] 1947.]

as the most reliable, the volcano began about noon as a thin wisp of smoke rising from a cornfield that was being plowed by Dionisio Pulido. By 4 P.M., explosions were occurring every few seconds, dense clouds of ash were rising, and a cone had begun to build up. Within 5 days the cone was 100 m high, and after 1 year it had risen to 425 m. Two days after the eruption began, the first lava flowed from a fissure in the field about 300 m north of the center of the cone. At the end of 7 weeks, this flow had advanced about 1.5 km. Some 15 weeks after the first explosion, lava had also begun to flow from the flanks of the cone itself.

After 9 years of activity, Parícutin abruptly stopped its eruptions and became just another of the many small "dead" cones in the neighborhood. The histories of these other cones, parasites of Toncítaro or of neighboring major volcanoes, undoubtedly parallel the story of Parícutin.

Mount St. Helens A sleeping giant awoke on M[ ] 1980, with an enormous blast that blew off the [ ] Mount St. Helens in the state of Washington (se[ ] ure 3.1). Warnings of potential severe volcanic [ ] had occurred in the form of earthquakes and ver[ ] steam for nearly two months prior to the eruptio[ ] last previous eruption of this formerly dormant v[ ] began in 1831 and was followed by 15 to 20 y[ ] lava and mud flows, which built the surround[ ] pography.

Formerly standing at a height of 2,950 m [ ] sea level, the top of the majestic cone was beve[ ] at 2,560 m as rock, ash, and clouds of gases[ ]

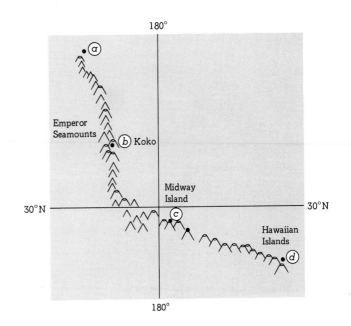

3.16 Schematic section and plan view of a plume similar to that postulated for producing the volcanoes of the Hawaiian Islands. As the lithospheric plate moves slowly over a cylindrical upwelling of magmatic material, a new volcano develops from time to time, older volcanoes becoming inactive as the plate moves away from this hot spot.

3.17 Bend in seamount line (Emperor and Hawaiian chains) marks change in movement of Pacific plate over hot spot in mantle. Volcanoes of progressively younger age developed as the plate moved first in a northwesterly direction and later in a more westerly direction. An age of 70 million years has been obtained from rocks near the northwestern extremity of the Emperor chain (a), 46 million years for Koko seamount (b), 20 million years for Midway Island (c), and less than 1 million years for the volcano Kilauea on Hawaii (d).

caldera. From the area and the amount of swelling involved, it was computed that in the 1960 eruption each centimeter of uplift or subsidence represented deformation of 1 million m³ of earth materials. Total volumes of 20 to 30 million m³ have been computed for larger eruptions. Records of tilt can also be used to forecast an impending eruption.

When Kilauea is active, magma rises up within the mountain and floods out as lava into a pit in the floor of the caldera. Occasionally the lava flows out over the rim of the pit onto the floor and gradually raises its level. Usually, however, the lava is confined to the pit, forming what is termed a **lava lake.** This lava lake may last for years and then disappear completely for equally long periods. The level of the lake falls when lava flows from the flanks of the volcano, both above and below sea level. From time to time the

system is drained, and the caldera floor collapses. Then the magma rises again, lava floods into the caldera, and the process is repeated.

Disappearing islands Submarine volcanoes, like Krakatoa, that build themselves up above sea level, blow off their heads, and then rebuild, produce the so-called disappearing islands of the Pacific Ocean. In 1913, for example, Falcon Island (20.4°S, 175.6°W), in the South Pacific, suddenly disappeared after an explosive eruption. On October 4, 1927, accompanied by a series of violent explosions, it just as suddenly reappeared. The island of Bogosloff (about 56°N, 168°W), in the Aleutians, was first reported in 1826 and has been playing hide-and-seek with map makers ever since.

On December 11, 1967, a volcanic eruption started at Metis Shoal of the Tonga Archipelago. After 20 days, a new island about 700 m long and 100 m wide had poked up 20 m above the sea. From the start of the activity the island stayed above the surface for a total of 58 days. By February 1, it was a few "jagged rocks

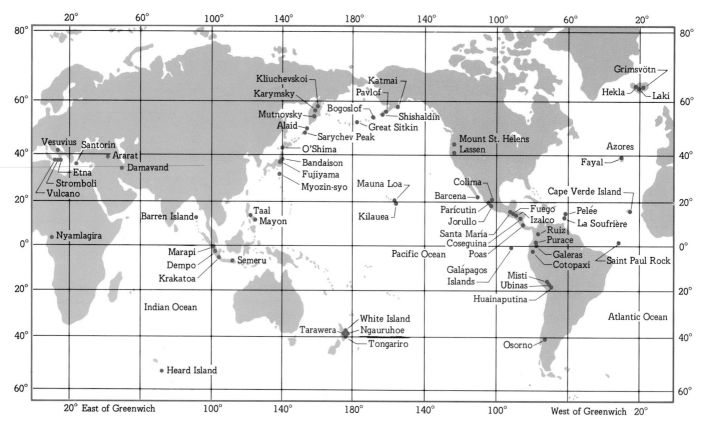

3.18 Location of some active volcanoes [Various sources.]

and water washing across." On February 19, there were "very high breakers on subsurface rocks" at the site; and by April 1, the shoal was completely under water, and there were no observed breakers.[7]

DISTRIBUTION OF ACTIVE VOLCANOES

We find evidence of volcanic eruptions in rocks of all ages. Apparently, igneous activity has been going on throughout geologic time and has occurred in the highest mountain ranges, on the bottom of the ocean, and on open plains. There are more than 500 active volcanoes in the world today. Some are located in Figure 3.18.

[7]Event Report, "The Submarine Volcanic Eruption and Formation of a Temporary Island at Metis Shoal, Tonga Islands," Center for Short-Lived Phenomena, Smithsonian Institution, Washington, D.C., June 15, 1968.

3.2 BASALT PLATEAUS

On June 11, 1783, after a series of violent earthquakes on Iceland near Mount Skapta, an immense outpouring of lava began along a 16-km line, the Laki Fissure. Lava poured into the Skapta River, drying up the water and overflowing the stream's channel, which was 150 to 200 m deep and 60 m wide in places. Soon the Skapta's tributaries were dammed, and many villages in adjoining areas were flooded. The lava flow was followed by another 1 week later and a third on August 3. So great was the volume of the lava flows that they filled a former lake and an abyss at the foot of a waterfall. They spread out in great tongues 20 to 25 km wide and 30 m deep. As the lava flows diminished and the Laki Fissure began to choke up, 22 small cones formed along its length, relieving the waning pressures and serving as outlets for the final extrusion of debris.

This is one of the few authenticated instances within historical time of the mechanism known as **fissure eruption,** or **lava flood.** There is strong evidence, however, that floods of this sort occurred on a gigantic scale in the geological past. The rocks produced by lava floods are known as **flood basalts,** or **plateau basalts,** because of the tendency to form great plateaus. The low viscosity required for lava to flood freely over such great areas is characteristic only of lavas that have basaltic composition.

Iceland itself is actually a remnant of extensive lava floods that have been going on for over 50 million years and that have blanketed 0.5 million km². The congealed lava is believed to be at least 2,700 m thick in this area. The Antrim Plateau of northeastern Ireland, the Inner Hebrides, the Faeroes, and southern Greenland are also remnants of this great North Atlantic, or Britoarctic, Plateau.

Of equal magnitude is the Columbia Plateau in Washington, Oregon, Idaho, and northeastern California (see Figure 3.19). In some sections, more than

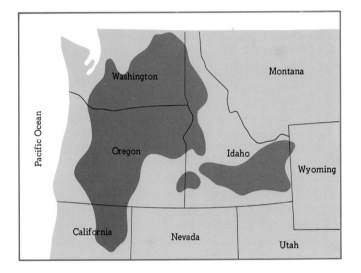

3.19 Map of Columbia Plateau, showing areas built by fissure eruptions.

1,500 m of rock have been built up by a series of fissure eruptions. Individual eruptions deposited layers ranging from 3 to 30 m thick, with an occasional greater thickness. In the canyon of the Snake River, Idaho, granite hills from 600 to 750 m high are covered by 300 to 450 m of basalt from these flows. The Columbia Plateau has been built up during the past 30 million years. The principal activity took place a million years ago in northeastern California and Idaho, but some flows in the Craters of the Moon National Monument in southern Idaho, probably the most recent of United States fissure eruptions, are believed to have occurred within the last 250 to 1,000 years.

Other extensive areas built up by fissure eruptions include north-central Siberia, the Deccan Plateau of India, Ethiopia, around Victoria Falls on the Zambezi River in Africa, and parts of Australia.

3.3 IGNEOUS ACTIVITY AND EARTHQUAKES

Most volcanic eruptions are associated with earthquakes. In recent years, records of this seismic activity have given warnings of impending eruptions. The relationship between local earthquakes and eruptions is important enough to be described in detail in the following case.[8]

Raoul Island, a volcanic island, is 8 km across and is the largest of the Kermadec group, located approximately 1,000 km northeast of New Zealand. It lies in a very active zone of earthquake activity extending from Tonga Island to New Zealand.

Unusual seismic activity began on November 10, 1964, when the first of a swarm of local earthquakes was recorded. In the 10 weeks before this there had been only 1 earthquake. Then, within 4 h (hours), over 80 earthquakes per hour were being recorded. On November 11, earthquakes became less frequent, but the ground began to shake continuously, a phenomenon called **volcanic tremor.** Within 1 day the volcanic tremors were large enough to mask records of small individual earthquakes. By November 13, the tremors dropped off to about one-half the maximum intensity, and earthquakes were again being recorded at about 30 to 40 per hour. At 21 h 57 min (minutes) Greenwich time, the largest earthquake of the series occurred. Raoul's seismograph was temporarily out of order, but this earthquake was large enough to be recorded on other seismographs, some as far away as North America. The United States Coast and Geodetic Survey determined it to be a moderate-sized earthquake and located its center at 20 km west of the island at a depth of 77 km. Volcanic tremor and frequency of earthquakes decreased then, until on November 20, only 10 to 15 per hour were being recorded.

Suddenly, on November 20, just before 18 h Greenwich time, the amplitude of volcanic tremor increased, and there was a "sound of a big landslide" accompanied by a cloud of steam, followed by "another roaring noise," which heralded the appearance

[8]R. D. Adams and R. R. Dibble, "Seismological Studies of the Raoul Island Eruption, 1964," *New Zealand J. Geol. Geophys.,* vol. 10, no. 6, 1967.

of a great column of black mud that shot up to 1,000 m in the center of the cloud with rocks flying out of the column and falling back into the crater. During the eruption a new crater 100 m in diameter was blown out within the main crater. Within 1 h the level of volcanic tremor settled down, with about 10 earthquakes per hour.

The Raoul eruption followed the not unusual pattern of occurring when the associated earthquake activity was on the wane after an extremely sudden rise. Seismographs are being watched carefully in the hope that they will provide forewarning of any future outbursts of volcanic activity.

DEPTH OF MAGMA SOURCE

Magma supplying eruptive materials of active volcanoes comes from deep within the earth. The depths of earthquakes and earth tremors associated with eruptions are giving us some clues to the depths of the magma.

Swarms of earthquakes and "harmonic tremor" with foci 60 km below the summit of Kilauea preceded the 1959 eruptions, which suggests that 60 km may be the depth at which magma enters the "plumbing system" of Kilauea.

3.4 THE EARTH'S HEAT

The principal requirement for igneous activity is heat, and evidence has accumulated that there is a great deal of heat in the earth's interior. That the earth's temperature increases with depth was first noted as early as the seventeenth century, when German miners were reporting a steady rise in temperature with increasing depth in their mine shafts. Further proof has now come from the temperature measurements made in other mines and from deep drill holes in the earth.

The increase in temperature with depth is called the **geothermal gradient.** Reported rates of increase have varied from place to place even in areas far removed from igneous activity. The geothermal gradient can range from less than 10°C/km to as much as

50°C/km. Excluding seasonal temperature changes that affect only the top few meters of the earth's surface, a geothermal gradient of 30°C/km seems to be about average. These figures get us "into hot water," so to speak, very quickly. Within 50 km of depth, a geothermal gradient of 30°C/km would result in a temperature of 1500°C, believed to be high enough to melt any known rock. We believe the geothermal gradient diminishes with depth after several tens of kilometers.

Heat travels extremely slowly through soil and rock. Measurements and calculations have shown that the earth's **heat flow,** which is the product of the temperature gradient and the thermal conductivity of earth materials, is reasonably constant. Its average value is 1.2 μcal (microcalories)/cm²-s (second), that is, 1.2×10^{-6} cal/cm²-s. Present indications are that there is little difference in the heat flow from a number of different land and sea areas. The average heat flow represents about 40 cal/cm²-year, which does not seem like much. But in its steady, quiet way it totals more energy in a year than all that released by seismic and igneous activity.

CAUSE OF IGNEOUS ACTIVITY

Igneous activity requires magma, and magma requires that rocks be melted. Where does this melted rock come from? Earthquake records indicate that most magma comes from the mantle. The heat to melt solid rock is apparently produced in two ways: (1) radioactive decay generates heat energy as a product of the transformation of one element into another; (2) movement of rock masses past one another produces enormous amounts of frictional energy manifested as heat. These two mechanisms can account for the total heat necessary to generate magmas.

The ultimate causes of igneous activity are related to the same internal forces that elevate mountains, cause earthquakes, and result in metamorphism. Volcanic outpourings of lava and gases originate near the mantle's upper boundary. Hot magma squeezes upward, driven by internal forces that seem to cause constant motion within the earth (see Chapter 10 for more discussion of convection cells and convective overturn within the mantle). This magma may remain for relatively long periods in shallow chambers within the crust, or it may erupt at once through fissures in the earth's surface. The complex interaction of many factors determines when and how an eruption is triggered. Geologists are just now learning more about these phenomena through geophysical and geochemical studies combined with detailed field examination of critical regions.

3.5 MASSES OF IGNEOUS ROCKS

Igneous rocks are formed from the solidification of magma. From time to time, molten rock within the earth's crust works its way out to the surface, as we have discussed. There it may pour or be blown forth from vents or fissures and solidify into rock, or it may remain trapped within the crust, where it slowly cools and solidifies. Rocks of igneous origin compose 95 percent of the outermost 10 km of the globe. During our discussion of igneous activity, we dealt primarily with the extrusion of lava and pyroclastic debris and with some of the landforms that result: basalt plateaus and volcanoes. These are surface masses of igneous rock.

When magma within the crust loses its mobility and ceases activity, it solidifies in place, forming igneous rock masses of varying shapes and sizes. Today, such rocks can be seen at the surface on continents where previously overlying rocks have been worn away by erosion. For example, the first internal part of a volcano to be exposed by erosion is the plug that formed when magma solidified in the vent. Revealed next are the channels through which the magma moved to the surface. Finally, in some regions of ancient activity, the crust has been so elevated (see Chapter 8) and eroded that the reservoir that once stored the magma can now be seen as a solid rock mass at the surface. Solidified offshoots of magma from the reservoir, having intruded themselves into other rocks within the crust, are included in these masses, which were not necessarily connected with the eruption of a volcano.

PLUTONS

All igneous rock masses that were formed when magma solidified within the earth's crust are called **plutons.** When rocks have a definite layering, we may speak of the magma that invades them as **concordant** if its boundaries are parallel to the layering, or **discordant** if its boundaries cut across the layering.

Plutons are classified according to their size, shape, and relationship to surrounding rocks. They include sills, dikes, lopoliths, laccoliths, and batholiths (Figure 3.20).

Tabular plutons A pluton with a thickness that is small relative to its other dimensions is called a **tabular** pluton.

Sills A tabular concordant pluton is called a **sill.** It may be horizontal, inclined, or vertical, depending on the attitude of the rock structure with which it is concordant.

Sills range in size from sheets less than 1 cm in thickness to tabular masses hundreds of meters thick. Because a sill is an intrusive form—that is, it has found its way into already existing rocks—it is always younger than the rocks that surround it. A sill must not be confused with an ordinary lava flow that has been buried by other rocks later on. There are fairly reliable ways of distinguishing between the two types: A buried lava flow usually has a rolling or wavy top pocked by the scars of vanished gas bubbles and showing evidence of erosion, whereas a sill has a more even and unweathered surface; also, a sill may contain fragments of rock that were broken off when the magma forced its way into the surrounding structures.

The Palisades along the west bank of the Hudson River near New York are the remnants of a sill that was several hundred meters thick. Here the magma was originally intruded into flat-lying sedimentary rocks. These are now inclined at a low angle toward the west.

Dikes A tabular discordant pluton is called a **dike** (see Figure 3.21). Dikes originated when magma forced its way through the fractures of adjacent rocks.

The width of individual dikes ranges from a few centimeters to many meters. The Medford dike near Boston, Massachusetts, is 150 m wide in places. Just how far we can trace the course of a dike across the countryside depends in part on how much of it has been exposed by erosion. In Iceland, dikes 15 km long are common, and many can be traced for 50 km; at least one is known to be 100 km long.

As magma forces its way upward, it sometimes pushes out a cylindrical section of the crust. Today, as a result, we find exposed at the surface some roughly circular or elliptical masses of rock that outline the cylindrical sections of the crust. These solidified bodies of magma are called **ring dikes.** Large ring dikes may be many kilometers around and hundreds or thousands of meters deep. Ring dikes have been mapped with widths of 500 to 1,200 m and diameters ranging from 2 to 25 km.

Some dikes occur in concentric sets. These originated in fractures that outline an inverted cone, with the apex pointing down into the former magma source. These dikes are called **cone sheets.** In Scotland, the

3.20 Plutons and landforms associated with igneous activity.

dip of certain cone sheets suggests an apex approximately 5 km below the present surface of the earth. Dikes are also found in approximately parallel groups called **dike swarms.**

Lopoliths Lopoliths are tabular concordant plutons shaped like a spoon, with both the roof and the floor sagging downward.

A well-known example is the Duluth lopolith, which crops out on both sides of Lake Superior's western end and appears to continue beneath the lake. It has been computed to be 250 km across and 15 km deep, with a volume of 200,000 km³.

Most lopoliths are composed of rock that has been differentiated into alternating layers of dark and light minerals, presenting the appearance of thinly bedded sedimentary rock.

Massive plutons Any pluton that is not tabular in shape is classified as a **massive** pluton.

Laccoliths A massive concordant pluton that was created when magma pushed up the overlying rock structures into a dome is called a **laccolith.**

A classic development of laccoliths is found in the Henry, La Sal, and Abajo Mountains of southeastern Utah, where their features are exposed on the Colorado Plateau, a famous geological showplace.

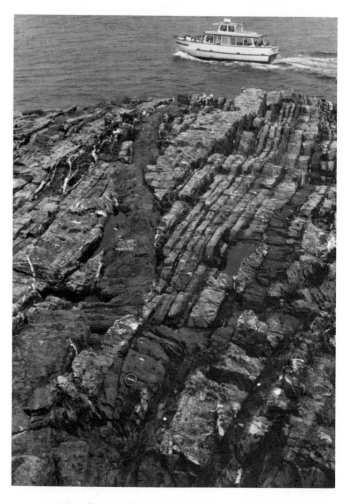

3.21 Basalt dike (approximately 1-m thick black body) cutting through nearly vertical bedded rocks along the coast of Maine. [Logan Campbell.]

3 ORIGIN AND OCCURRENCE
OF IGNEOUS ROCKS

54

Batholiths A large discordant pluton that increases in size as it extends downward is called a **batholith.** The term "large" in this connection is generally taken to mean a surface exposure of more than 100 km². A pluton that has a smaller surface exposure but exhibits the other features of a batholith is called a **stock.**

Batholiths are exposed thousands of meters above sea level, where they have been lifted by forces operating in the earth's crust. Thousands of meters of rock that covered the batholiths have been stripped away by the erosion of millions of years. We can observe these roots of mountain ranges in the White Mountains of New Hampshire and in the Sierra Nevada.

Although batholiths provide us with some valuable data, they also raise a host of unsolved problems. All these problems bear directly on our understanding of igneous processes and the complex events that accompany the folding, rupture, and eventual elevation of sediments to form mountains. (We shall discuss these mountian-forming processes in Chapters 8 and 10.)

We can summarize what we know about batholiths as follows:

1 Batholiths are located in mountain ranges (or regions that once were parts of mountain ranges). Although in some mountain ranges no batholiths are exposed at the present time, we never find batholiths that are not associated with mountain ranges. In any given mountain range, the number and size of the batholiths seem to be directly related to the intensity of the folding and crumpling that has taken place. This does not mean, however, that the batholiths caused the folding and crumpling. Actually, there is convincing evidence to the contrary, as we shall see in some of the following features.

2 Batholiths usually run parallel to the axes of mountain ranges.

3 Batholiths have been intruded across the folds, indicating that they were formed after the folding of the mountains—although the folding may have continued after the batholiths were formed.

4 Batholiths have irregular dome-shaped roofs. This characteristic shape is related to **stoping,** one of the mechanisms by which magma moves upward into the crust. As the magma moves upward, blocks of rock are broken off from the structures into which it is intruding. At low levels, when the magma is still very hot, the stoped blocks may be melted and assimilated by the magma reservoir. Higher in the crust, as the magma approaches stability and its heat diminishes, the stoped blocks are frozen in the intrusion as **xenoliths,** that is, "strange rocks" (see Figure 3.22).

5 Batholiths are composed primarily of granite or granodiorite (a plutonic rock consisting of quartz, calcic plagioclase, orthoclase, and some mafic constituents such as pyroxene, hornblende, and/or biotite).

6 Batholiths give the impression of having replaced the rocks into which they have intruded instead of having pushed them aside or upward. But if that is what really took place, what happened to the great volumes of rock that the batho-

3.22 Xenolith in Mount Airy granite, North Carolina, which provides evidence in support of magmatic origin of this granite because the inclusion could not have retained its sharp edges and separate identity during granitization. [T. M. Gathright II.]

liths appear to have replaced? Here we come up against the problem of the origins of batholiths—in fact, against the whole mystery of igneous activity. Some observers have been led even to question whether granitic batholiths were formed from true magmas at all. The suggestion has been made that the batholiths may have been formed through a process called **granitization,** in which solutions from magmas move into solid rocks, exchange ions with them, and convert them into rocks that have the characteristics of granite but have never actually existed as magma. We shall return to this highly controversial proposal in Chapter 6.

7 Batholiths contain a great volume of rock. The Sierra Nevada batholith of California is 650 by 60 to 100 km, and a partially exposed batholith in southern California and Baja California is probably 1,600 by 100 km.

8 Gravity measurements have been interpreted[9] as indicating that the downward extent, or "thickness," of many batholiths is some 10 to 15 km.

Figures 3.23 and 3.24 summarize the conventional representation of the concordant and discordant plutons.

[9]M. H. P. Bott and Scott B. Smithson, "Gravity Investigations of Subsurface Shape and Mass Distribution of Granite Batholiths," *Bull. Geol. Soc. Am.,* vol. 78, pp. 859–878, 1967.

3.5 MASSES OF IGNEOUS ROCKS

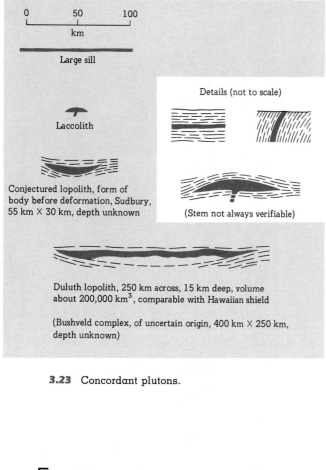

3.23 Concordant plutons.

3.6 FORMATION OF IGNEOUS ROCKS

Igneous rocks at the surface today have been formed from magma. As we pointed out earlier in this chapter, molten rock in the ground is magma, and when it is extruded on the surface, it is lava. And when solidified pieces of magma are blown out, they are pyroclastic debris.

Pyroclastic debris eventually becomes hardened into rock through the percolation of ground water. In one sense, rocks formed in this way could be classified as sedimentary; but because they consist of solidified pieces of magma, we shall include them in our discussion of igneous rocks. Volcanic ash that has hardened into rock is called **tuff.** If many relatively large angular blocks of congealed lava are embedded in a mass of ash and then hardened to rock, the rock is called **volcanic breccia.** If such included pieces are mainly rounded fragments, the rock is called **volcanic conglomerate.**

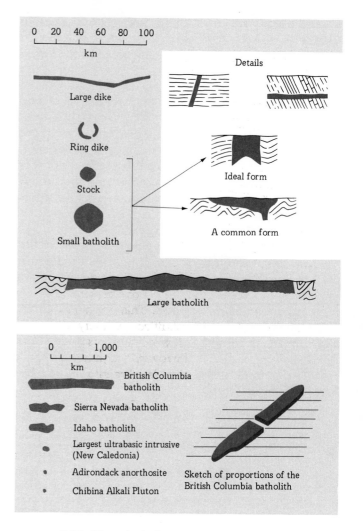

3.24 Discordant plutons.

Magmas result from the partial melting of portions of the mantle or by heating of parts of the crust that have been carried below the earth's surface during subduction along plate boundaries. These magmas rise toward the surfaces because they have greater buoyancy than the surrounding cooler, or more viscous, or more solid rocks.

Magma, extruded as lava at the surface, cools and solidifies to form igneous rocks. The offshoots of magma that work their way into surrounding rock below the surface cool more slowly and solidify. Even the magma reservoir eventually cools and solidifies, but it takes much longer because it is a larger mass. **All igneous rocks have been formed from the solidification of magma.**

CRYSTALLIZATION OF MAGMA

Magma solidifies through the process of **crystallization.** At first, magma is a **melt,** a liquid solution of elements at high temperature. After a decrease occurs in the heat that keeps the magma liquid, the melt starts to solidify. Bit by bit, mineral grains begin to grow. As this growth goes on, gases are released. Now we no longer have a complete liquid but rather a liquid mixed with solid and gaseous materials.[10] As the temperature continues to fall, the mixture solidifies until igneous rock is formed.

Igneous rocks may consist of interlocking grains of a single mineral or a mixture of several. As pointed out in Section 2.5, the most common rock-forming minerals are the silicates olivine, augite, hornblende, biotite, muscovite, orthoclase, plagioclase, and quartz.

BOWEN'S REACTION PRINCIPLE

Magma is a solution of elements, but it does not crystallize in the way ordinary solutions do. Most solutions of a given composition always crystallize into a solid of the same composition. This happens regardless of conditions during solidification. If crystallization of a magma were similar, it would always yield rock of the same composition. However, magma of a given composition may be able to crystallize into a number of different kinds of rock.

In 1922, N. L. Bowen[11] of the Geophysical Laboratory of the Carnegie Institution of Washington proposed that the differences in end products depend on the rate at which the magma cools and on whether early-formed minerals remain in or settle out of the remaining liquid during its crystallization. He suggested that, as a magma cools, the first-formed minerals undergo continuous modification with the liquid remaining after they crystallized. He called this process **reaction**. Reaction is the key to magmatic crystallization.

Based on a number of laboratory experiments with silicate melts, Bowen was able to arrange the rock-forming minerals of igneous rocks into **reaction series.** He found that an important characteristic of a silicate melt is that, as one mineral develops at a certain temperature, it will be converted upon further cooling into a different mineral by reaction with the liquid around it. The reaction series are of two different types, continuous and discontinuous. In the **continuous** reaction series, some early-formed minerals are converted into new minerals by continuously changing their composition but not their crystalline structure. In the **discontinuous** reaction series, some early-formed minerals react with the melt to change to new minerals with different compositions and a different crystalline structure.

Taking a magma with olivine basaltic composition, Bowen arranged the rock-forming minerals into a discontinuous reaction series of the ferromagnesian minerals and a continuous reaction series of the feldspars. As this magma cools, olivine and the calcium feldspar, anorthite, are the first minerals to form. If no minerals settle out, the melt may solidify to basalt or gabbro. If some of the early-formed minerals settle out, however, in a process called **fractionation**, the reaction process will continue further and the remaining minerals react with the remaining melt. Augite becomes hornblende, and the calcium feldspar first becomes a calcium-sodium feldspar and then the sodium feldspar, albite. The greater the degree of fractionation, the more extensive the reaction process. With a high degree of fractionation the whole reaction series is gone through and later-formed minerals will be rich in silica (see Figure 3.25). Thus, according to Bowen's reaction principle, the fractional crystallization of an olivine basaltic magma can lead to the formation of successively more siliceous rocks until finally a rock of granitic composition is reached.

Bowen's reaction principle explains how an olivine basaltic magma may solidify as one rock type or may produce several rock types. It might crystallize to a gabbro that is composed largely of ferromagnesian minerals, or it might produce a granite that is composed predominantly of sialic minerals. The rock formed depends on the extent to which early-formed minerals were removed from further reaction with the melt and on the rate of cooling.

Rate of crystallization The rate at which a magma crystallizes influences the extent to which fractionation and reaction take place. When magma cools rapidly, there is no time for minerals to settle or to react with the remaining liquid. This occurs when a partially crystallized magma is extruded onto the surface or injected into thin dikes or sills. But when a large body of magma cools slowly, deep within the crust, a high degree of fractionation or chemical reaction may take place. So the rate of crystallization varies with depth. For example, a magma consisting largely of nonferromagnesians at 1100°C, exposed to the air on top and ranging in thickness from 1 to 10,000 m, would solidify as shown in Table 3.2.

There are some objections to Bowen's hypothesis. According to his reaction principle, only about 5 or 10 percent of an original basic magma could solidify into granite or granodiorite (a composition between that of granite and of diorite). This extremely small percent-

[10]To reflect this changing picture, we define a magma as any naturally occurring silicate melt, whether or not it contains suspended crystals or dissolved gases.

[11]N. L. Bowen, "The Reaction Principle in Petrogenesis," *J. Geol.*, vol. 30, pp. 177–198, 1922.

3.25 Bowen's reaction series.

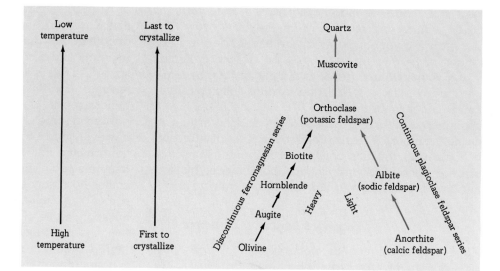

TABLE 3.2

Magma crystallization and depth[a]

Thickness, m	Time required, yr
1	0.033
10	3
100	300
1,000	30,000
10,000	3,000,000

[a]R. A. Daly, *Igneous Rocks and the Depths of the Earth*, p. 63, McGraw-Hill Book Company, New York, 1933.

age cannot account for the large masses of these rocks found on the continents. Where are the earlier-formed more basic rocks? Although there is overwhelming experimental evidence that the reaction series is correct, we do know that not all granites need be derived from mafic magmas. We shall examine this problem further in Chapter 6, under the topic of granitization.

3.7 TEXTURE OF IGNEOUS ROCKS

Texture, a term derived from the Latin *texere,* "to weave," is a physical characteristic of all rocks. The term refers to the general appearance of rocks. In referring to the texture of igneous rocks, we mean specifically the size, shape, and arrangement of their interlocking mineral grains.

GRANULAR TEXTURE

If magma has cooled at a relatively slow rate, it will have had time to develop grains that the unaided eye can see in hand specimens. Rocks composed of such large mineral grains are called **granular** or **phaneritic** (Greek *phaneros,* "visible") (see Figure 3.26).

The rate of cooling, however, though important, is not the only factor that affects the texture of an igneous rock. For example, if a magma is of low viscosity—that is, if it is thin and watery and flows readily—large, coarse grains may form even though the cooling is relatively rapid; for in a magma of this sort the ions can move easily and quickly into their rock-forming mineral combinations.

APHANITIC TEXTURE

The rate at which a magma cools depends on the size and shape of the magma body, as well as on its depth below the surface. For example, a small body of magma with a large surface area—that is, a body that is much longer and broader than it is thick—surrounded by cool, solid rock loses its heat more rapidly than would the same volume of magma in a spherical reservoir. And because rapid cooling usually prevents large grains from forming, the igneous rocks that result have **aphanitic** textures. Individual minerals are present but are too small to be identified without a microscope.

GLASSY TEXTURE

If magma is suddenly ejected from a volcano or a fissure at the earth's suface, it may cool so rapidly that there is no time for minerals to form at all. The result is

3.26 Enlarged photograph of a piece of granular igneous rock, taken through a slice that has been ground to translucent thinness (known as a *thin section*). The photograph shows the rock to be composed of interlocking crystals of different minerals. (Field of view approximately 0.5 cm in diameter.)

3.27 Enlarged photograph of a thin section of porphyritic igneous rock. (Largest crystal approximately 0.125 cm by 0.25 cm.)

a **glass,** which by a rigid application of our definition is not really a rock but is generally treated as one. Glass is a special type of solid in which the ions are not arranged in an orderly manner. Instead, they are disorganized, like the ions in a liquid. And yet they are frozen in place by the quick change of temperature.

PORPHYRITIC TEXTURE

Occasionally, a magma cools at variable rates—slowly at first, then more rapidly. It may start to cool under conditions that permit large mineral grains to form in the early stages, and then it may move into a new environment where more rapid cooling freezes the large grains in a **groundmass** of finer-grained texture (see Figure 3.27). The large minerals are called **phenocrysts.** The resulting texture is said to be **porphyritic. Porphyry,** from the Greek word for "purple," was originally applied to rocks containing phenocrysts in a dark-red or purple groundmass.

In rare cases, magma may be suddenly expelled at the surface after large mineral grains have already formed. Then the final cooling is so rapid that the phenocrysts become embedded in a glassy groundmass.

3.8 TYPES OF IGNEOUS ROCK

Several systems have been proposed for the classification of igneous rocks. All are artificial in one detail or another, and all rely on certain characteristics that cannot be determined in the field or from hand specimens. For our present purposes we shall emphasize texture and composition. Such a classification is entirely adequate for an introductory study of physical geology and even for many advanced phases of geology.

This classification appears in tabular form in Figure 3.28, together with a graph that shows the proportions of silicates in each type of igneous rock. The graph demonstrates the *continuous progression from rock types in which light-colored minerals predominate to rock types in which dark-colored minerals predominate.* The names of rocks are arbitrarily assigned on the basis of average mineral composition and texture. Sometimes intermediate types are indicated by such names as granodiorite. There are many more igneous rocks than are shown in this figure.

LIGHT-COLORED IGNEOUS ROCKS

The igneous rocks on the left side of Figure 3.28 are light both in color and in specific gravity. They are sometimes referred to as **sialic** rocks.

It has been estimated that granites and granodiorites together comprise 95 percent of the igneous rocks of the continents that have solidified from magma. The origin and history of some granites are still under debate, but we use the term here to indicate composition and texture, not origin.

Granite is a granular rock (see Figure 3.29). Its mineral composition is as follows:

2 parts orthoclase feldspar + 1 part quartz
+ 1 part plagioclase feldspars + small amount
of ferromagnesians = **GRANITE**

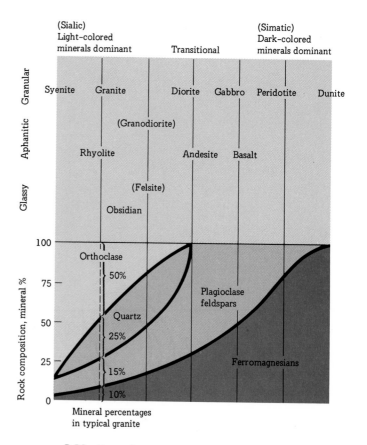

3.28 General composition of igneous rocks is indicated by a line from the name to the composition chart. Granite and rhyolite consist of about 50 percent orthoclase, 25 percent quartz, and 25 percent plagioclase feldspars and ferromagnesian minerals. Granite is the most important granular rock, and basalt the most important aphanitic rock. [Composition chart modified from L. Pirsson and A. Knopf, *Rocks and Rock Minerals*, p. 144, John Wiley & Sons, Inc., New York, 1926.]

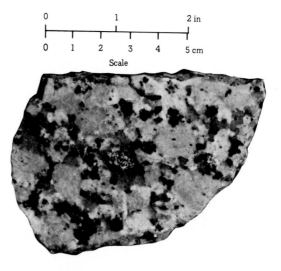

3.29 Polished slab of granite showing very coarse texture; gray quartz, white plagioclase feldspar, pink (off-white) potassium feldspar (orthoclase), and black ferromagnesian minerals. [M. E. Kauffman.]

Rocks with the same mineral composition as granite but with an aphanitic rather than granular texture are called **rhyolite**.

The glassy equivalent of granite is **obsidian** (see Figure 3.30). Although this rock is listed near the left side of the composition chart, it is usually pitch black in appearance. Actually, though, pieces of obsidian thin enough to be translucent turn out to be smoky white against a light background.

3.30 Obsidian, the glassy equivalent of granite. (Specimen approximately 10 cm across.) [Robert Navias.]

DARK-COLORED IGNEOUS ROCKS

Of the total volume of rock formed from magma that has poured out onto the earth's surface, it is estimated that 98 percent are basalts and andesites.

A popular synonym for basalt is **trap rock**, from a Swedish word meaning "step." This name refers to the tendency of certain basalts to form columns that look like stairways in some outcrops (see Figure 3.31). These are a product of the cooling process and become evident after weathering.

Basalt has aphanitic texture. Its mineral composition is as follows:

1 part plagioclase feldspars + 1 part ferromagnesians = **BASALT**

The granular equivalent of basalt is **gabbro**. **Peridotite** is a granular igneous rock that is composed largely of ferromagnesian minerals.

INTERMEDIATE TYPES

Composition Igneous-rock compositions blend continuously from one to another as we go from the light to the dark side of the classification chart. **Andesite** is the name given to aphanitic igneous rocks that are intermediate in composition between granite and basalt. These rocks were first identified in the Andes mountains of South America—hence the name. Andesites are mostly found in areas around the Pacific Ocean. The granular equivalent of andesite is **diorite**.

Texture Going from the top to the bottom of the chart in Figure 3.28, we find that the rock textures grade continuously from granular to aphanitic, whereas the composition remains the same. For example, if we read down along the first vertical rule, we find that granite, rhyolite, and obsidian become progressively

3.31 Columnar jointing, a pattern sometimes found in basalt, outlines this series of columns in a volcanic flow along Yellowstone River near Tower Falls, Yellowstone National Park. [U.S. National Park Service.]

3.32 Orthoclase phenocrysts in a granite porphyry. (Specimen approximately 10 cm across.) [Robert Navias.]

mately intergrown grains of potassium feldspar and quartz, which form essentially a single unit. The quartz is darker than the feldspar, and the overall pattern suggests the wedge-shaped figures of the writing of ancient Assyria, Babylonia, and Persia. As a result, this has become known as **graphic structure** (from the Greek *graphein*, "to write") (see Figure 3.33).

3.33 Graphic granite. (Specimen approximately 7 cm by 11 cm.) [Walter R. Fleischer.]

finer-grained, although all three have essentially the same composition. The same is true of gabbro and basalt.

In addition to these textures, any of the rocks may have porphyritic texture. Essentially, this means that a given rock has grains of two distinctly different sizes: conspicuously large phenocrysts embedded in a finer-grained groundmass. When the phenocrysts constitute less than 25 percent of the total, the adjective **porphyritic** is used to modify the rock name, as in porphyritic granite or porphyritic andesite. When the phenocrysts constitute more than 25 percent, the rock is called a **porphyry** (see Figure 3.32). The composition of a porphyry and the texture of its groundmass are indicated by using rock names as modifiers, as granite porphyry or andesite porphyry.

PEGMATITE

The solutions that develop late in the cooling of a magma are called **hydrothermal**. These crystallize into exceptionally granular igneous rock, called **pegmatite** (from the Greek *pēgmat-*, "fastened together"), which embodies the chief minerals to form from the hydrothermal solutions: potassium feldspar and quartz. Pegmatites are distinguished solely on the basis of extraordinary grain size. A few pegmatites consist of inti-

Pegmatite is found in dikes at the margins of batholiths and stocks. The dikes range in length from a few centimeters to a few hundred meters and contain crystals of very large size. In fact, some of the largest crystals known have been found in pegmatite. Crystals of spodumene (a lithium mineral) that measure 12 m in length have been found in the Black Hills of South Dakota; crystals of beryl (a silicate of beryllium and aluminum) that measure 1 by 5 m have been discovered in Albany, Maine. Great masses of potassium feldspar, weighing over 2,000 t, yet showing the characteristics of a single crystal, have been mined from pegmatite in the Karelo-Finnish Soviet Socialist Republic.

Nearly 90 percent of all pegmatite is **simple pegmatite**, composed of quartz, orthoclase, and unimportant percentages of micas. It is more generally

3 ORIGIN AND OCCURRENCE
OF IGNEOUS ROCKS

called **granite pegmatite** because the composition is that of granite and the texture that of pegmatite. The remaining 10 percent includes **complex pegmatites.** The major components of complex pegmatites are the same sialic minerals that we find in simple pegmatites, but in addition they contain a variety of rare minerals: lepidolite, tourmaline (best known as a semiprecious gem), topaz (also a gem), tantalite, and uraninite (pitchblende).

Simple pegmatites are common in some regions and complex pegmatites in others. In southwestern New Zealand, for example, pegmatites are uniformly simple, but throughout the Appalachian regions of North America, complex pegmatites are more abundant. Figure 3.34 shows the areas in New England where pegmatites are found.

IGNEOUS ROCKS ON THE MOON

Rocks brought back from the moon by the *Apollo* missions are very much like terrestrial rocks except that they have no water in their compositions and almost no free oxygen. Most of the rocks from the moon are igneous in origin[12]—basalt lavas from the maria; gabbro, norite (a gabbro consisting of plagioclase and orthorhombic pyroxene), and anorthosite (composed essentially of plagioclase with more calcium than sodium) from the highlands.

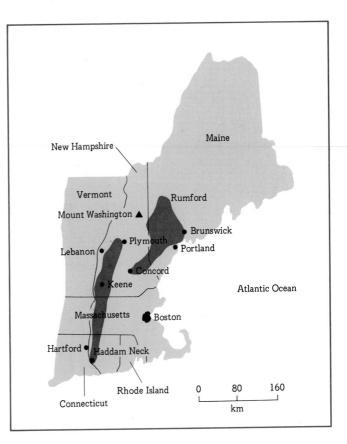

3.34 Map of regions where pegmatites are extensively developed, east of the Connecticut Valley in New England. Pegmatite areas are in dark color.

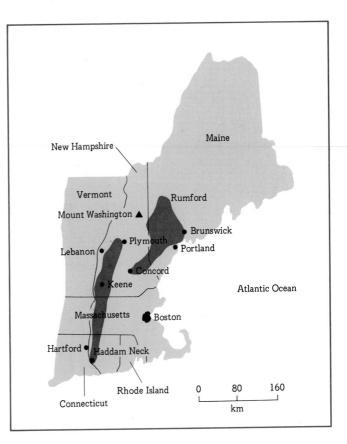

Igneous activity consists of movements of molten rock inside and outside the earth and the variety of effects associated with these movements.

Volcanoes are surface piles of material that have accumulated around vents during successive eruptions.

    Evolution of volcanoes starts with basaltic lava flows, then andesitic lava flows, and then ejection of fragmental materials; the cycle ends when acidic lava congeals in the volcanic vent. Not all volcanoes evolve in this same pattern.

    A **volcanic eruption** is started and maintained by gases, of which steam is the most important.

    Composition and temperatures of lavas have been measured from 750 to 1175°C.

    Volcanic gases are two-thirds steam but include carbon monoxide and dioxide, nitrogen, sulfur, hydrogen, fluorine, and chlorine.

    Pyroclastic debris consists of fragments blown out by explosive eruptions and includes ash, cinders, lapilli, blocks, bombs, and pumice.

        Fiery clouds of ash mixed with gases are heavier than air and flow down a volcano's side.

        Worldwide effects include volcanic dust in the stratosphere and reduction of solar energy reaching the earth's surface.

OUTLINE

[12] A few rocks are made of angular particles formed by fragmentation of surface material by meteorite impact and subsequent compaction into a coherent rock (breccia) by shock compression during cratering by later meteorite impacts.

Formation of calderas may be by explosion, by collapse, or by a combination of both.

Classification of volcanoes according to the materials that have accumulated around their vents designates them as **shield, composite,** and **cinder cones.**

History of some volcanoes shows they may be dormant for long periods and again become active.

Vesuvius was dormant for centuries and then in A.D. 79 started to erupt again.

Krakatoa erupted in 1883 with one of the world's greatest explosions.

Disappearing islands are submarine volcanoes that blow their heads off.

Parícutin sprang into being on February 20, 1943, and was active for 9 years.

Mount St. Helens abruptly came alive May 18, 1980, and became the most widely studied volcano in history.

Hawaiian Islands are peaks of volcanoes projecting above sea level and strung out along a line running 2,400 km to the northwest.

Distribution of active volcanoes locates many in the circum-Pacific, the Alpine-Himalayan, and the mid-oceanic belts and others in the Pacific Ocean, the Atlantic Ocean, the Indian Ocean, Africa, and the Antarctic, for a total of more than 500.

Basalt plateaus are built by fissure eruptions.

Igneous activity and earthquakes seem to be closely related.

Depth of magma source suggested by associated earthquakes may be 60 km at Kilauea and 77 km at Raoul Island.

The earth's heat is indicated by the thermal gradient to be more than adequate for igneous activity. **Speculations on the origin of the earth's heat** have included original heat and radioactivity.

Cause of igneous activity seems to be the same internal force that elevates mountains, causes earthquakes, and causes metamorphism, but it is not surely known.

Igneous rocks are formed from the solidification of molten matter.

Masses of igneous rocks are called **plutons,** which are classified according to size, shape, and relationships to surrounding rocks.

Sills are concordant tabular plutons.

Dikes are discordant tabular plutons.

Lopoliths are tabular concordant plutons shaped like a spoon.

Laccoliths are massive concordant plutons with domed tops.

Batholiths are massive discordant plutons 10 to 40 km thick.

Igneous rocks at the surface today were formed from magma.

Magma solidifies through the process of crystallization.

Bowen's reaction series are incorporated in a hypothesis accounting for all igneous rocks coming from an olivine basaltic magma.

Limitations of Bowen's hypothesis include failure to account for large undifferentiated masses of granite.

The rate of crystallization is an important control over the rocks that form.

Texture of igneous rocks is determined by the size, shape, and arrangement of their interlocking mineral grains.

Granular texture includes large mineral grains from slow-cooling or low-viscosity magma.

Aphanitic texture from rapid cooling consists of individual minerals so small that they cannot be identified without the aid of a microscope.

Glassy texture results from ions disorganized as in a liquid but frozen in place by quick cooling.

Porphyritic texture is a mixture of large mineral grains in an aphanitic or glassy groundmass.

Types of igneous rock are arbitrarily defined in terms of texture and composition.

Light-colored igneous rocks, sometimes called **sialic,** are dominated by granites and granodiorites.

Dark-colored igneous rocks (basalts and andesites) constitute 98 percent of rock formed from magma that has poured out onto the earth's surface.

Intermediate types of composition are given arbitrary names, such as **andesite** and **diorite,** because igneous rock compositions blend continuously from one to another from the light to the dark side of the classification chart.

Intermediate types of texture are also given arbitrary names such as **granite, rhyolite,** and **obsidian,** because rocks of a given composition grade continuously from granular to aphanitic to glassy texture.

Pegmatite is an exceptionally granular rock formed by hydrothermal solutions late in the cooling of a magma.

Igneous rocks on the moon are similar in composition to basalt, gabbro, norite, and anorthosite. Moon rocks have no water and almost no free oxygen.

SUPPLEMENTARY READINGS

Carmichael, I. S. E., F. J. Turner, and J. Verhoogen *Igneous Petrology,* McGraw-Hill Book Company, New York, 1974. *An advanced-level treatment of all aspects of igneous rocks.*

Dietrich, R. V., and B. J. Skinner *Rocks and Rock Minerals,* John Wiley & Sons, Inc., New York, 1979. *Textbook and reference for studying igneous, metamorphic, and sedimentary rocks in hand-specimens and thin sections.*

Green, J., and N. M. Short (eds.) *Volcanic Landforms and Surface Features: A Photographic Atlas and Glossary,* Springer-Verlag New York Inc., 1971. *A beautifully illustrated systematic treatment of volcanoes and volcanism.*

Kruger, C. *Volcanoes,* G. P. Putnam's Sons, New York, 1971. *Finely illustrated survey of the history and science of volcanoes, from early mythology to recent concepts.*

Macdonald, G. A. *Volcanoes,* Prentice-Hall, Inc., Englewood Cliffs, N. J., 1972. *A relatively moderate-level introduction to volcanic activity, volcanic types, and characteristics of eruption.*

Pitcher, W. S., and A. R. Berger *The Geology of Donegal: A Study of Granite Emplacement and Unroofing,* John Wiley & Sons, Inc., New York, 1972. *A regionally specialized discussion of granitic plutons from emplacement through uplift, erosion, and resulting sedimentary residues.*

QUESTIONS

1 What are some of the kinds of activity commonly found to occur prior to the eruption of a volcano?

2 What mechanisms are responsible for the kinds of activity described in Question 1? What is actually happening within the earth beneath the site of the eventual volcano?

3 Why do the volcanic eruptions differ from one location on a continent to another location on that same continent? Why do continental and oceanic eruptions differ?

4 What is the difference between active, dormant, and dead volcanoes?

5 What are the principal reasons for human deaths associated with volcanic eruptions, such as the eruption of Mount St. Helens?

6   Why do most volcanoes occur in certain specific belts around the earth? Why are a few found at other locations?

7   What is the basic cause of igneous activity?

8   Is the temperature of the earth's crust essentially the same around the globe or does it differ? Why or why not? How do we know? Define geothermal gradient.

9   Explain in step-by-step fashion what occurs during the fractional crystallization of an olivine basaltic magma according to Bowen's hypothesis.

10   How does the texture of an igneous rock permit us to interpret much about its history of development?

11   What have we learned about the similarity or difference between igneous rocks on the moon when compared with those found on the earth?

4

WEATHERING
AND SOILS

Our initial view of the earth's surface provides us with clues to the great variety of forces acting to shape its configuration. In this chapter we will examine the effects of such forces as they relentlessly work on the materials at and near the surface of the earth.

We now know that such surface activity is not confined to the earth. Gravity is a universally acting force. Wind plays an exceptionally active role on Mars, where huge wind-swept sand dune fields have been photographed. Ice likewise has shaped the polar regions of this Martian surface, where ice-formed ridges of material are much in evidence and suggest advances and retreats of polar ice on a seasonal basis and perhaps on longer term cycles, similar to our Great Ice Age.

Pioneer spacecraft missions to Venus, *Voyager* missions to Jupiter and Saturn, and similar probes to all parts of our Solar System indicate that some processes are similar on other planets, but the products differ with differing surface conditions of temperature, differing atmospheric concentrations, and the presence or absence of water in its various forms.

The blurred inscription on a gravestone, the crumbling foundation of an ancient building, the broken rock exposed along a roadside—all tell us that rocks are subject to constant destruction. Marked changes of temperature, moisture soaking into the ground, the ceaseless activity of living things—all work to destroy rock material. This process of destruction we call **weathering,** and we define it as the changes that take place in minerals and rocks at or near the surface of the earth in response to the atmosphere, to water, and to plant and animal life. Later on we shall extend this definition slightly, but for the time being, it will serve our purpose.

Weathering leaves its mark everywhere about us (Figure 4.1). The process is so common, in fact, that we tend to overlook the way in which it functions and the significance of its results. It plays a vital role in the rock cycle; for by attacking the exposed material of the earth's crust—both solid rock and unconsolidated deposits—and converting it into different forms, it produces the raw materials for new rocks (see Chapter 1).

4.1 Terraces and templelike forms of the Grand Canyon have developed on alternating resistant and nonresistant sedimentary rocks. Differential weathering of rocks of varying resistance helps explain the topography in this region. The Colorado River flows in the steep-walled inner gorge cut in igneous and metamorphic rocks. [M. E. Kauffman.]

Solid rock (or **bedrock**) underlies all parts of the land surface. In some localities, it is exposed at the surface. In other areas, especially in tropical regions, it might be covered by as much as a hundred meters of soil and loose rock fragments (collectively called the **regolith**). Soil is the surface accumulation of sand, clay, and decayed plant material (called **humus**). It supports the growth of vegetation.

Weathering includes those processes that break down the bedrock. This is accomplished in two different ways: (1) disintegration by the physical or mechanical breaking down of bedrock to make smaller and smaller particles; (2) decomposition by chemical change in the minerals making up the bedrock.

Erosion, as we will discuss in subsequent chapters, includes those processes by which weathered material is moved from one place to another or is removed entirely from the land surface. Agents of erosion include gravity, water, ice, and the wind. Weathering and erosion work together to level the earth's surface, removing materials from higher elevations and depositing them at lower elevations.

The products of weathering are the raw materials of sediments and sedimentary rocks. Streams, glaciers, wind, and ocean currents move the weathered materials to new localities and deposit them as sand, gravel, or mud. Sometimes, however, the products of weathering remain right where they are formed and are incorporated into the rock record. Certain ores, for example, such as those of aluminum, are actually old weathering zones (see Chapter 18).

Before we examine the processes of weathering, however, we pause a moment to look at the energy sources responsible for their action.

4.1 ENERGY SOURCES

Energy is a word that sounds familiar enough. We speak of energy-producing foods, of the energy needed to climb stairs, of energy radiated to us from the sun. Matter and energy are intimately related, but what is "energy"?

Energy is the capacity for producing motion. No form of matter is entirely devoid of motion. The motion may be in things we see, such as an automobile speeding down the highway, or in things we cannot see, such as atoms and molecules. All motion is produced by energy in one form or another. In fact, we can think of the entire universe as a great bundle of energy. Heat and light from the sun represent energy in one form; the revolution of the earth around the sun represents energy in another form; and the chemical transformation of food into heat and body activity represents energy in yet another form. So energy manifests itself in different forms, which have descriptive names: potential, kinetic, heat, chemical, electrical, atomic, and radiant.

In geological processes, through the motion of running water, energy sculptures the land; through the deformation of rocks, it builds mountains; and through the rupture of earth materials, it produces earthquakes. In fact, energy is involved in every geological process.

Potential energy is stored energy waiting to be used. Uranium, petroleum, coal, and natural gas are eagerly sought after because the potential energy they contain can be effectively released and put to work. Water in clouds or in lakes and reservoirs at high altitudes has potential energy that is released when the water falls or runs downhill. A boulder poised on a hilltop has potential energy. Less obvious, but just as real, is the potential energy stored in the nucleus of every atom and in the molecules of every compound.

KINETIC ENERGY

The potential energy of a 10-t boulder resting on a hillside is transformed into the energy of movement when the boulder is dislodged and rolls downhill (see Figure 4.2). This energy of movement is called **kinetic. Every moving object possesses kinetic energy.**

Potential
energy

Kinetic
energy

4.2 The potential energy of a 10-t boulder resting on a hillside is transformed into kinetic energy, the energy of movement, when the boulder is dislodged and rolls downhill.

The amount of kinetic energy possessed by an object depends on the mass of the object and on the speed with which it moves:

$$E_K = \frac{Mv^2}{2}$$

A 10-t boulder rolling down a hill has more energy at the bottom of the slope than does a pebble that has rolled down the same slope at the same speed. On the other hand, this same 10-t boulder moving a few meters per hour would have less energy than if it were hurtling along at 30 m/s.

Running water owes much of its effectiveness as a geological agent to kinetic energy; so do waves beating on the coast. Massive glaciers creeping down a mountain slope do geological work by means of kinetic energy.

The energy that drives the weathering process comes both from within the earth and from outside. From time to time, motions originating inside the earth elevate some portions of its surface above other portions. These are the same motions that express themselves in earthquakes (Chapter 9) and mountain building (Chapter 8). Whatever their cause, they arrange the earth's surface materials so that gravity can be effective in the breakdown of rock material. Thus rock raised in mountain building has potential energy, which may be transformed to kinetic energy if gravity is strong enough to pull it downward to a lower level. The shattering of this rock as it falls is truly a form of weathering.

We already have seen that there is a measurable amount of heat that flows from the earth's interior to its surface and is there dissipated. Far more heat is received at the earth's surface from the sun. It is the distribution of this heat that causes the differential heating of the atmosphere and of the oceans. This differential heating brings about circulation of atmosphere and ocean, causes our weather and climate, and determines the pattern of organic activity. All these in turn help to modify the earth's surface materials—in short, to determine the process of weathering.

How much solar energy is available at the earth's surface? The sun radiates about 100,000 cal/cm²-min from its surface. A plane at the earth's outer surface perpendicular to the sun's rays would receive approximately 2 cal/cm²-min, which is 2×10^{-5} the amount of heat generated by an equivalent area on the sun's surface. This figure is the **solar constant.**

If our hypothetical plane perpendicular to the sun's rays at the edge of the outer atmosphere has the same diameter as the earth, geometry tells us that its area must be one-quarter that of the earth. Then we can express the solar constant for the earth's surface as 0.5 cal/cm²-min. This is an average for the entire globe. But of this energy, approximately one-third is unavailable to heat either the atmosphere or the earth. It is reflected back into space from clouds or from the earth, or it is scattered into space from particles within the atmosphere. Therefore, approximately 0.3 cal/cm²-min is what is available to heat the atmosphere and the earth.

It is obvious that this energy is not evenly distributed over the earth's surface. Less is available toward either pole than is available toward the equator. We shall have occasion to return to this distribution later in this chapter as we consider the process of chemical weathering.

4.2 TYPES OF WEATHERING

There are two general types of weathering: **mechanical** and **chemical.** It is difficult to separate these two types in nature because they often go hand in hand, although in some environments one or the other predominates. Still, for our purposes here it is more convenient to discuss them separately.

MECHANICAL WEATHERING

Mechanical weathering, which is also referred to as **disintegration,** is the process by which rock is broken down into smaller and smaller fragments as the result of energy developed by physical forces. For example, when water freezes in a fractured rock, sufficient energy may develop from the pressure caused by expansion of the frozen water to split off pieces of the rock. Or a boulder moved by gravity down a rocky slope may be shattered into smaller fragments.

Expansion and contraction Changes in temperature, if they are rapid enough and great enough, may bring about the mechanical weathering of rock. In areas where bare rock is exposed at the surface and is unprotected by a cloak of soil, forest or brush fires can generate heat adequate to break up the rock. The rapid and violent heating of the exterior zone of the rock causes it to expand; and if the expansion is sufficiently great, flakes and larger fragments of the rock are split off. Lightning often starts such fires and, in rare instances, may even shatter exposed rock by means of a direct strike.

The debate continues concerning whether variations in temperature from day to night or from summer to winter are enough to cause mechanical weathering. Theoretically, such variations cause disintegration. For instance, we know that the different minerals forming a granite expand and contract at different rates as they react to rising and falling temperatures. We expect that even minor expansion and contraction of adjacent minerals would, over long periods of time,

weaken the bonds between mineral grains and that it would thus be possible for disintegration to occur along these boundaries. In deserts we may find fragments of a single stone lying close beside one another. Obviously, the stone has split. But how? Many think the cause lies in expansion and contraction caused by heating and cooling.

But laboratory evidence to support these speculations is inconclusive. In one laboratory experiment, coarse-grained granite was subjected to temperatures ranging from 14.5 to 135.5°C every 15 min. This alternate heating and cooling eventually simulated 244 years of daily heating and cooling; yet the granite showed no signs of disintegration. Perhaps experiments extended over longer periods of time would produce observable effects. In any event, we are still uncertain of the mechanical effect of daily or seasonal temperature changes; if these fluctuations bring about the disintegration of rock, they must do so very slowly.

Frost action Frost is much more effective than heat in producing mechanical weathering. When water trickles down into the cracks, crevices, and pores of a rock mass and then freezes, its volume increases about 9 percent. This expansion of water as it passes from the liquid to the solid state sets up pressures that are directed outward from the inside of the rock and **frost wedging** results. These pressures can dislodge fragments from the rock's surface.

The dislodged fragments of mechanically weathered rock are angular, and their size depends largely on the nature of the bedrock from which they have been displaced. Usually the fragments are only a few centimeters in maximum dimension, but in some places—along the cliffs bordering Devil's Lake, Wisconsin, for instance—they reach sizes of up to 3 m.

A second type of mechanical weathering produced by freezing water is **frost heaving.** This action usually occurs in fine-grained, unconsolidated deposits rather than in solid rock. Much of the water that falls as rain or snow soaks into the ground, where it freezes during the winter months. If conditions are right, more and more ice accumulates in the zone of freezing as water is added from the atmosphere above and drawn upward from the unfrozen ground below, much as a blotter soaks up moisture. In time, lens-shaped masses of ice are built up, and the soil above them is heaved upward. Frost heaving of this sort is common on poorly constructed roads, and lawns and gardens are often soft and spongy in the springtime as a result of the soil's heaving up during the winter.

Certain conditions must exist before either type of frost action can take place: There must be an adequate supply of moisture; the moisture must be able to enter the rock or soil; and temperatures must move back and forth across the freezing line. As we might expect, frost

4.3 Frost action in Acadia National Park, Maine, has aided in widening the cracks and causing rounding of bedrock edges and corners. [U.S. National Park Service.]

action is more pronounced in high mountains and in moist regions where temperatures fluctuate across the freezing line, either daily or seasonally (see Figure 4.3).

Exfoliation Exfoliation is a mechanical weathering process in which curved plates of rock are stripped from a larger rock mass by the action of physical forces. This process produces two features that are fairly common in the landscape: large, domelike hills, called **exfoliation domes,** and rounded boulders, usually referred to as **spheroidally weathered boulders.** It seems likely that the forces that produce these two forms originate in different ways.

Let us look first at the manner in which exfoliation domes develop. Fractures, or parting planes, called **joints,** occur in many massive rocks. These joints are broadly curved and run more or less parallel to the rock surface. The distance between joints is only a few centimeters near the surface, but it increases to several meters as we move deeper in the rock (see Chapter 8). Under certain conditions, one after another of the curved slabs between the joints is spalled, or sloughed, off the rock mass. Finally, a broadly curved hill of bedrock develops, as shown in Figure 4.4.

Just how these slabs of rock come into being in the first place is still a matter of dispute. Most observers

believe that, as erosion strips away the surface cover, the downward pressure on the underlying rock is reduced. Then, as the rock mass begins to expand upward, lines of fracture develop, marking off the slabs that later fall away. Precise measurements made on granite blocks in New England quarries provide some support for this theory. Selected blocks were accurately measured and then removed from the quarry face, away from the confining pressures of the enclosing rock mass. When the free-standing blocks were measured again, it was found that they had increased in size by a small but measurable amount. Massive rock does expand, then, as confining pressures are reduced, and this slight degree of expansion may be enough to start the exfoliation process.

Well-known examples of exfoliation domes are Stone Mountain, Georgia, the domes of Yosemite Park, California, and Sugar Loaf in the harbor of Rio de Janeiro, Brazil.

Now let us look at smaller examples of exfoliation: spheroidally weathered boulders. These boulders have been rounded by the spalling off of a series of concentric shells of rock (Figures 4.5 to 4.7). But here the shells develop from pressures set up within the rock by chemical weathering rather than from the lessening

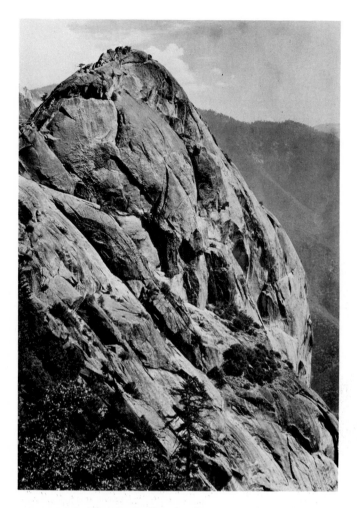

of pressure from outside by erosion. We shall see later that, when certain minerals are chemically weathered, the resulting products occupy a greater volume than does the original material, and it is this increase in volume that creates the pressures responsible for spheroidal weathering. Because most chemical weathering takes place in the portions of the rock most exposed to air and moisture, it is there that we find the most expansion and hence the greatest number of shells. Spheroidally weathered boulders are sometimes produced by the crumbling off of concentric shells. If the cohesive strength of the rock is low, individual grains are partially weathered and dissociated, and the rock simply crumbles away. The underlying process is the same in both cases, however.

Certain types of rocks are more vulnerable to spheroidal weathering than are others. Igneous rocks such as granite, diorite, and gabbro are particularly susceptible because they contain large amounts of the mineral feldspar, which, when chemically weathered, produces new minerals of greater volume.

Other types of mechanical weathering Plants also play a role in mechanical weathering. The roots of trees and shrubs growing in rock crevices sometimes exert sufficient pressure to dislodge previously loosened fragments of rock, much as tree roots heave and crack sidewalk pavements (Figure 4.8).

More important, though, is the mechanical mixing of the soil by ants, worms, rodents, and other small animals. Constant activity of this sort makes the soil particles more susceptible to chemical weathering (see below) and may even assist in the mechanical breakdown of the particles.

4.4 Moro Rock, Sequoia National Park, California, is an example of an exfoliation dome. The massive granite in this dome has developed a series of partings, or joints, more or less parallel to the surface. Rock slabs spall off the dome, giving it its rounded aspect. The jointing probably originated as the granite expanded after the erosion of the overlying material. [G. A. Grant, U.S. National Park Service.]

4.5 Near Clark, Wyoming, only isolated patches of partially weathered granite remain as most of the bedrock wears away, producing rounded masses, which tend to undergo further spheroidal weathering. [M. E. Kauffman.]

4.6 Spheroidally weathered granite boulders almost completely isolated from the bedrock. [Sheldon Judson.]

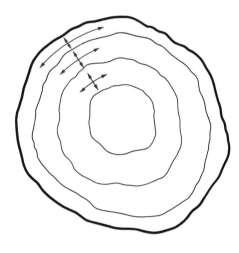

4.7 This cross section through a spheroidally weathered boulder suggests the stresses set up within the rock. The stress is thought to develop as a result of the change in volume as feldspar is converted to clay. (See also Figures 4.5 and 4.6.)

CHEMICAL WEATHERING

Chemical weathering, sometimes called **decomposition,** is a more complex process than mechanical weathering. As we have seen, mechanical weathering merely breaks rock material down into smaller and smaller particles, without changing its composition. Chemical weathering, however, actually transforms the original material into something different. The chemical weathering of the mineral feldspar, for example, produces the clay minerals, which have a different composition and different physical characteris-

4.8 A white birch tree growing in a crevice pries a large block from a low rock cliff in Hermosa Park, Colorado. [U.S. Geological Survey.]

tics from those of the original feldspar. Sometimes the products of chemical weathering have no mineral form at all, as in the salty solution that results when halite is dissolved.

Particle size The size of the individual particles of rock is an extremely important factor in chemical weathering because substances can react chemically only where they come into contact with one another. The greater the surface area of a particle, the more vulnerable it is to chemical attack. If we were to take a pebble, for example, and grind it up into a fine powder, the total surface area exposed would be greatly increased. As a result, the materials that make up the pebble would undergo more rapid chemical weathering.

Figure 4.9 shows how the surface area of a 1-cm (or any other unit) cube increases as we cut it up into smaller and smaller cubes. The initial cube has a surface area of 6 cm² and a volume of 1 cm³. If we divide the cube into smaller cubes, each 0.5 cm on a side, the total surface area increases to 12 cm², although, of course, the total volume remains the same. Further

4.9 Relationships of volume, particle size, and surface area. In this illustration a cube 1 cm (or any other unit) on a side is divided into smaller and smaller units. The volume remains unchanged, but as the particle size decreases, the surface area increases. Because chemical weathering is confined to surfaces, the more finely a given volume of material is divided, the greater is the surface area exposed to chemical activity and the more rapid is the process of chemical weathering.

subdivision into 0.25-cm cubes increases the surface to 24 cm². And if we divide the original cube into units 0.125 cm on a side, the surface area increases to 96 cm². As we have seen, this same process is performed by mechanical weathering. It reduces the size of the individual particles of rock, increases the surface area exposed, and thus promotes more rapid chemical weathering.

Other factors The rate of chemical weathering is affected by other factors as well—the composition of the original mineral, for example. As we shall see later, a mineral such as quartz, SiO_2, responds much more slowly to chemical weathering than does a mineral such as olivine, $(Fe, Mg)_2SiO_4$.

Climate also plays a key role in chemical weathering. Moisture, particularly when it is accompanied by warmth, speeds up the rate of chemical weathering; conversely, dryness slows it down. Finally, plants and animals contribute directly or indirectly to chemical weathering because their life processes produce oxygen, carbon dioxide, and certain acids that enter into chemical reactions with earth materials.

The interrelation of some of these factors is shown in Figure 4.10. This is a generalized section from the pole to the equator and shows the fluctuation of precipitation, temperature, and amount of vegetation. At the same time, the figure shows the relative depth of weathering as these three factors vary. Thus weathering is most pronounced in the equatorial zone, where the factors of precipitation, temperature, and vegetation reach a maximum. Weathering is least in the

desert and semidesert areas of the subtropics and in the far north. A secondary zone of maximum weathering exists in the zone of temperate climates. Here both the precipitation and the vegetation can reach secondary maxima.

CHEMICAL WEATHERING OF IGNEOUS ROCKS

In Chapter 3 we found that the most common minerals in igneous rocks are silicates and that the most important silicates are quartz, the feldspars, and certain ferromagnesian minerals. Let us see how chemical weathering acts on each of these three types.

Weathering of quartz Chemical weathering affects quartz very slowly, and for this reason we speak of quartz as a relatively stable mineral. When a rock such as granite, which contains a high percentage of quartz, decomposes, a great deal of unaltered quartz is left behind. The quartz grains (commonly called **sand grains**) found in the weathered debris of granite are the same as those that appeared in the unweathered granite.

When these quartz grains are first set free from the mother rock, they are sharp and angular; but because even quartz slowly responds to chemical weathering, the grains become more or less rounded as time passes. After many years of weathering, they look as

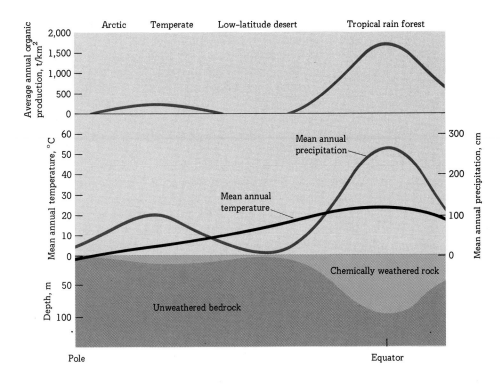

Arctic   Temperate   Low–latitude desert   Tropical rain forest

4.10 Variation in temperature, precipitation, and organic matter from the poles to the tropics is related to the depth of chemical weathering.

Mean annual precipitation

Mean annual temperature

Chemically weathered rock

Unweathered bedrock

Pole

Equator

though they had been abraded and worn by the action along a stream bed or a beach. And yet the change may have come about solely through chemical action. Indeed, the presence of silica in natural waters of lakes and rivers reminds us that the silicate minerals are soluble and that some of this may come from the chemical weathering of quartz.

Weathering of feldspars   In the Bowen reaction series (Section 3.6), we saw that when a magma cools to form an igneous rock such as granite, feldspars crystallize before quartz. When granite is exposed to weathering at the earth's surface, the feldspars are also the first minerals to be broken down. Mineralogists and soil scientists still do not understand the precise process by which feldspars weather, and some of the end products of this action—the clay minerals—offer many puzzles. But the general direction and results of the process seem fairly clear.

Aluminum silicate, derived from the chemical breakdown of the original feldspar, combines with water to form hydrous aluminum silicate, which is the basis for another group of silicate minerals, the clays. Let us examine the decomposition of orthoclase, a good example of the chemical weathering of the feldspar group of silicates. In this instance, a source of hydrogen ions is necessary to the weathering process and two substances play an important role in producing them: carbon dioxide and water. The atmosphere contains small amounts of carbon dioxide, and the soil

contains much greater amounts. Because carbon dioxide is extremely soluble in water, it unites with rainwater and water in the soil to form the weak acid H_2CO_3, carbonic acid. This ionizes to form hydrogen and bicarbonate ions as follows:

Water *plus* carbon *yields* carbonic *yields* hydrogen *plus* bicarbonate
            dioxide   acid   ion   ion

$$H_2O \ + \ CO_2 \ \rightarrow \ H_2CO_3 \rightarrow \ H^+ \ + \ (HCO_3)^-$$

And when orthoclase comes into contact with hydrogen ions, the following reaction takes place:

Orthoclase   *plus* hydrogen *plus* water *yields*
                ions

$$2K(AlSi_3O_8) + \ 2H^+ \ \ \ + \ H_2O \rightarrow$$

           clay   *plus* potassium *plus* silica
                                      ions

$$Al_2Si_2O_5(OH)_4 + \ \ 2K^+ \ \ \ \ + \ \ 4SiO_2$$

In this reaction, the hydrogen ions from the water force the potassium out of the orthoclase, disrupting its crystal structure. The hydrogen ion combines with the aluminum silicate radical of the orthoclase to form the new clay minerals. (The process by which water combines chemically with other molecules is called **hydra-**

4.2   TYPES OF WEATHERING

75

tion.) The disruption of the orthoclase crystal yields a second product, potassium ions. These may join with the bicarbonate ions formed by the ionization of the carbonic acid to form potassium bicarbonate. The third product, silica, is formed by the silicon and oxygen that are left after the potassium has combined with the aluminum silicate to form the clay mineral.

The action of living plants may also bring about the chemical breakdown of orthoclase. A plant root in the soil is negatively charged and is surrounded by a swarm of hydrogen ions, H^+. If there happens to be a fragment of orthoclase lying nearby, these positive ions may change places with the potassium of the orthoclase and disrupt its crystal structure (see Figure 4.11). Once again, a clay mineral is formed, as in the equation above.

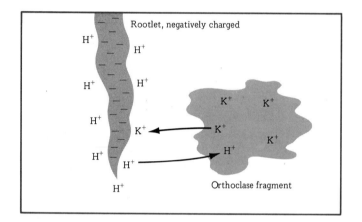

4.11 The conversion of orthoclase to a clay mineral by plant roots. In this diagram, a swarm of hydrogen ions (positive) is shown surrounding a negatively charged plant rootlet. The suggestion has been made that a hydrogen ion from the rootlet may replace a potassium ion in a nearby orthoclase fragment and there bond with the oxygen within the original mineral, to begin the conversion of the orthoclase to clay. The potassium ion thus ejected replaces the hydrogen ion along the negatively charged rootlet and is eventually utilized in plant growth. [After W. D. Keller and A. F. Frederickson, "Role of Plants and Colloidal Acids in the Mechanism of Weathering," *Am. J. Sci.*, vol. 250, p. 603, 1952.]

Now let us look more closely at each of the three products of the decomposition of orthoclase: first, the clay minerals. At the start, these minerals are very finely divided. In fact, they are sometimes of **colloidal** size, a size variously estimated as between 0.2 and 1 μm (micrometer). Immediately after it is formed, the

aluminum silicate may possibly be amorphous; that is, its atoms are not arranged in any orderly pattern. It seems more probable, however, that even at this stage, the atoms are arranged according to the definite pattern of a true crystal. In any event, as time passes, the small individual particles join together to form larger crystals, which, when analyzed by such means as X-rays, exhibit the crystalline pattern of true minerals.

There are many different clay minerals and each has its own chemical behavior, physical structure, and evolution. Most of the clay minerals fall into three major groups: **kaolinite, montmorillonite,** and **illite.** Kaolinite is derived from the Chinese *Kao-ling*, "High Hill," the name of the mountain from which the first kaolinite was shipped to Europe for ceramic uses; the mineral montmorillonite was first described from samples collected near Montmorillon, a town in west-central France; and the name "illite" was selected by geologists of the Illinois Geological Survey in honor of their state. Like the micas shown in Figure 2.22, the clay minerals are built up of silicon-oxygen tetrahedra linked together in sheets. These sheets combine in different ways with sheets composed of aluminum atoms and hydroxyl molecules. (For this reason we refer to the clay minerals as **hydrous aluminum silicates.**) In addition, montmorillonite may contain magnesium and some sodium and calcium, and illite contains potassium, occasionally with some magnesium and iron. Some clay minerals alter to other clay minerals. For example, under deep burial the important clay mineral illite is produced from other clays.

We still do not understand exactly what factors determine which clay minerals will form when a feldspar is weathered. Climate is important, for we know that kaolinite tends to form as a result of the intense chemical weathering in warm, humid climates and that illite and montmorillonite seem to develop more commonly in cooler, drier climates. The history of the rock also seems to be influential. For example, when a soil forms from a sedimentary rock in which a clay has been incorporated, we often find that the soil contains the same type of clay as does the parent rock. The analysis of a number of sedimentary rocks and of the soils developed on these rocks has shown that, when illite is present in the original rock, it is usually the dominant clay in the soil, regardless of climate. Clearly, then, **both environment and inheritance** seem to influence the type of clay that will develop from the chemical weathering of a feldspar.

Let us look back for a minute to the equation for the decomposition of orthoclase. Notice that the second product is potassium ions. We might expect that these would be carried off by water percolating through the ground and that all the potassium would eventually

TABLE 4.1

Chemical-weathering products of common rock-forming silicate minerals

Mineral	Composition	Important decomposition products	
		Minerals	Others
Quartz	SiO_2	Quartz grains	Some silica in solution
Feldspars:			
Orthoclase	$K(AlSi_3O_8)$	Clay	Potassium carbonate (soluble)
		Silica	Some silica in solution
Albite (sodium plagioclase)	$Na(AlSi_3O_8)$	Clay	Some silica in solution
Anorthite (calcium plagioclase)	$Ca(Al_2Si_2O_8)$	Silica	Sodium and calcium carbonates (soluble)
		Calcite	
Ferromagnesians:			
Biotite	Fe, Mg, Ca silicates of Al	Clay	Calcium and magnesium carbonates (soluble)
Augite		Hematite	
Hornblende		Limonite	Some silica in solution
		Silica	
		Calcite	
Olivine	$(Fe, Mg)_2SiO_4$	Hematite	Iron and magnesium carbonates (soluble)
		Limonite	
		Silica	Some silica in solution

find its way to the rivers and finally to the sea. Yet analyses show that not nearly so much potassium is present in river and ocean water as we should expect. What happens to the rest of it? Some of it is used by growing plants before it can be carried away in solution, and some of it is absorbed by clay minerals or even taken into their crystal structure.

The third product resulting from the decomposition of orthoclase is silica, which appears either in solution (for even silica is slightly soluble in water) or as very finely divided quartz in the size range of the colloids. In the colloidal state, silica may exhibit some of the properties of silica in solution.

So far we have been talking about the weathering of only orthoclase feldspar. But the products of the chemical weathering of plagioclase feldspars are very much the same. Instead of potassium carbonate, however, either sodium or calcium carbonate is produced, depending on whether the feldspar is the sodium albite or the calcium anorthite (see Table 4.1). As we found in Section 2.5 (see Silicates), the plagioclase feldspars almost invariably contain both sodium and calcium. The carbonates of sodium and calcium are soluble in water and may eventually reach the sea. We should note here, however, that calcium carbonate also forms the mineral calcite (see Carbonate and Sulfate Minerals, Section 2.5). Calcite, in turn, forms the greater part of limestone (a sedimentary rock) and marble (a metamorphic rock). Both limestone and marble are discussed in subsequent chapters.

Weathering of ferromagnesians Now let us turn to the chemical weathering of the third group of common minerals in igneous rocks: the ferromagnesian silicates. The chemical weathering of these minerals produces the same products as the weathering of the feldspars: clay, soluble salts, and finely divided silica. But the presence of iron and magnesium in the ferromagnesian minerals makes possible certain other products as well.

The iron may be incorporated into one of the clay minerals or into an iron carbonate mineral. Usually, however, it unites with oxygen to form hematite, Fe_2O_3, one of the most common of the iron oxides. Hematite commonly has a deep red color, and in powdered form it is always red; this characteristic gives it its name, from the Greek *haimatitēs*, "bloodlike." Sometimes the iron unites with oxygen and a hydroxyl ion to form goethite, $FeO(OH)$, generally brownish in color. (Goethite was named after the German poet Goethe, because of his lively scientific interests.) Chemical weathering of the ferromagnesian minerals often produces a substance called **limonite,** yellowish to brownish in color and referred to in everyday language as just plain "rust." Limonite is not a true mineral because its composition is not fixed within narrow limits, but the term is universally applied to the iron oxides of uncertain composition that contain a variable amount of water. Limonite and some of the other iron oxides are responsible for the characteristic colors of most soils.

What happens to the magnesium produced by the weathering of the ferromagnesian minerals? Some of it may be removed in solution as a carbonate, but most of it tends to stay behind in newly formed minerals, particularly in the illite and montmorillonite clays.

Summary of weathering products If we know the mineral composition of an igneous rock, we can determine in a general way the products that the chemical weathering of that rock will probably yield. The chemical-weathering products of the common rock-forming minerals are listed in Table 4.1. These products include the minerals that make up most of our sedimentary rocks, and we shall discuss them again later (see Chapter 5).

4.3 RATES OF WEATHERING

Some rocks weather very rapidly and others only slowly. Rate of weathering is governed by the type of rock and a variety of other factors, from minerals and moisture, temperature and topography, to plant and animal activity.

RATE OF MINERAL WEATHERING

On the basis of field observations and laboratory experiments, the minerals commonly found in igneous rocks can be arranged according to the order in which they are chemically decomposed at the surface. We are not sure of all the details, and different investigators report different conclusions, but we can make the following general observations:

1 Quartz is highly resistant to chemical weathering.

2 The plagioclase feldspars weather more rapidly than orthoclase feldspar does.

3 Calcium plagioclase (anorthite) tends to weather more rapidly than sodium plagioclase (albite).

4 Olivine is less resistant than augite, and in many instances augite seems to weather more rapidly than hornblende.

5 Biotite mica weathers more slowly than do the other dark minerals, and muscovite mica is more resistant than biotite.

Notice that these points suggest a pattern (Figure 4.12) similar to that of Bowen's reaction series for crystallization from magma, discussed in Section 3.6 (illustrated in Figure 3.25). But there is one important difference: In weathering, the successive minerals formed do not react with one another as they do in a continuous reaction series. The relative resistance of these minerals to decomposition may reflect the difference between the surface conditions under which they weather and the conditions that existed when they were formed. Olivine, for example, forms at high temperatures and pressures, early in the crystallization of a melt. Consequently, as we might expect, it is extremely unstable under the low temperatures and pressures that prevail at the surface, and it weathers quite rapidly. On the other hand, quartz forms late in the reaction series, under considerably lower temperatures and pressures. Because these conditions are more similar to those at the surface, quartz is relatively stable and is very resistant to weathering.

Now we can qualify slightly the definition of weathering given at the beginning of this chapter. We have found that weathering disrupts the equilibrium that existed while the minerals were still buried in the earth's crust and that this disruption converts them into new minerals. Following Parry Reiche,[1] we may revise our definition as follows: **Weathering is the response of materials that were once in equilibrium within the earth's crust to new conditions at or near contact with air, water, and living matter.**

DEPTH AND RAPIDITY OF WEATHERING

Most weathering takes place in the upper few meters or tens of meters of the earth's crust, where rock is in closest contact with air, moisture, and organic matter. But some factors operate well below the surface and permit weathering to penetrate to great depths. For instance, when erosion strips away great quantities of material from the surface, the underlying rocks are free to expand. As a result, parting planes, or fractures—the joints that we spoke of earlier in the chapter—develop hundreds of meters below the surface.

Then, too, great quantities of water move through the soil and deep underground, transforming some of the materials there long before they are ever exposed at the surface. Rock salt that is located far below the surface in the form of a sedimentary rock often undergoes exactly this transformation. If a large quantity of underground water is circulating through, the salt is dissolved and carried off long before erosion can expose it.

[1] Parry Reiche, "A Survey of Weathering Processes and Products," *Univ. New Mexico Publ. Geol.*, no. 3, p. 5, 1950.

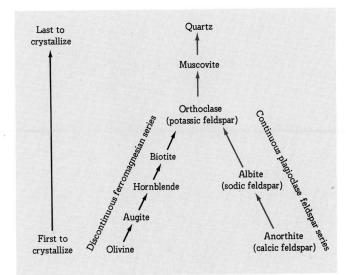

4.12 Relative rapidity of chemical weathering of the common igneous rock-forming minerals. The rate of weathering is most rapid at the bottom and decreases toward the top. Note that this table is in the same order as Bowen's reaction series (right). The discrepancy in the rate of chemical weathering between, for instance, olivine and quartz is explained by the fact that in the zone of weathering olivine is farther from its environment of formation than is quartz. It therefore reacts more rapidly than quartz does to its new environment and thus weathers more rapidly.

Weathering is sometimes so rapid that it can actually be recorded. The Krakatoa eruption of August 1883, described in Chapter 3, threw great quantities of volcanic ash into the air and deposited it to a depth exceeding 30 m on the nearby island of Long-Eiland. By 1928, 45 years later, a soil nearly 35 cm deep had developed on top of this deposit, and laboratory analyses showed that a significant change had taken place in the original materials. Chemical weathering had removed a measurable amount of the original potassium and sodium. Furthermore, mechanical weathering or chemical weathering or both had broken down the original particles so that they were generally smaller than the particles in the unweathered ash beneath.

In a more recent study, scientists from Ohio State University have demonstrated the nature and rate of soil development in unweathered material that has been exposed to the atmosphere with the retreat of Muir glacier in Glacier Bay National Park in southeastern Alaska. Geological and human records establish the successive positions of the retreating ice front since about 1700. Between this date and 1965, the front of the glacier retreated approximately 65 km. The age of soils at various points along this line of retreat can thus be demonstrated. Over a period of about 250 years, a soil about 35 cm thick has developed. Of this, the upper half is represented chiefly by the accumulation of organic material. The lower half, however, shows changes in the materials left by the retreating ice. After 250 years, virtually all the calcite and dolomite had been removed to a depth of at least 15 cm, and the soil acidity had increased markedly. The amount of iron oxide in the form of hematite had increased measurably, particularly in the lowest 10 cm.

Graveyards provide many fine examples of weathering within historic time. Calcite in the headstone pictured in Figure 4.13 has weathered so rapidly that the inscription to the memory of Moses L. Gould, carved in 1818, is only partially legible after more than a century and a half. (Examination of other marble slabs in the burying ground indicates that the earliest legible date is 1811.) Undoubtedly the rate of weathering has increased with time, for two reasons. First, continued weathering roughens the marble surface, exposing more and more of it to chemical attack and quickening the rate of decomposition. Second, as the number of factories and dwellings in Cambridge and neighboring towns has increased, the amount of carbon dioxide in the atmosphere also has increased. Consequently, rainwater in the twentieth century carries more carbonic acid than it did in the nineteenth, and it attacks calcite more rapidly.

In contrast to the strongly weathered marble is a headstone of slate erected in 1699 (Figure 4.14). Two hundred and sixty-nine years later, the inscription was still plainly visible. Slate is usually a metamorphosed shale, which, in turn, is composed largely of clay minerals formed by the weathering of feldspars. The clay minerals in the headstone were originally formed in the zone of weathering. Slight metamorphism has

4.13 Weathering of a marble headstone, burying ground of Christ Church, Cambridge, Massachusetts. The inscription, carved in 1818, is illegible. The monument illustrates the instability and rapid weathering of calcite (the predominant mineral in marble) in a humid climate. [Sheldon Judson, 1968.]

4.14 Weathering of a slate headstone, burying ground of Christ Church, Cambridge, Massachusetts. The inscription date, 1699, testifies to the durability of the slate. [Sheldon Judson, 1968.]

since changed many of the clay minerals to muscovite, a white mica, and the shale to slate (Section 6.3). The muscovite, however, is relatively stable in the zone of weathering, as indicated in Figure 4.12.

These examples show, then, that weathering often occurs rapidly enough to be measured during a lifetime. Let us now turn to the process that moves these products of weathering: erosion.

4.4 RATES OF EROSION

We found that chemical and mechanical weathering produces certain material from the rocks of the earth. It is these materials that the agents of erosion move from one place to another until they finally reach the settling basins of the world's oceans. Several agents are involved in this movement, including gravity, water, ice, and wind. These agents are individually treated in Chapters 11 to 16, but here it will be instructive to pause and consider how much material is being removed from the continents and at what rate.

Sometimes ancient ruins provide an index to the rates of erosion. Thus Figure 4.15 is a photograph of the remains of a cistern built 60 km north of Rome, Italy, in the second century. The footings exposed at the base of the finished wall indicate the amount of erosion that has occurred here since the structure was built. The rate of erosion from then to the present averages 30 cm/1,000 years.

This, however, is only a spot measurement at a specific place. What method can we use to measure the rates over large areas? One way is to measure the amount of material carried by a stream each year from its drainage basin. The amount averaged over the area of the basin gives an average figure of erosion. Now, obviously, the rate of erosion is not the same at every place in the river basin. In some places the material will be removed more rapidly than at others. Furthermore, there will be places at which deposition takes place, and the material is temporarily halted on its way out of the basin. Nevertheless, this method gives us an average figure for a unit area within the drainage basin.

A stream carries material in solution as well as solid material in the form of sediments. Most of the solid material is buoyed up by the flow of water and is said to be carried in **suspension**. A relatively smaller amount is pushed and bounced along the stream bottom in what is known as **bed load** or **traction load**. Suspended and dissolved matter can be measured without much difficulty. It is very difficult to measure traction load, but it is generally small and usually con-

4.15 Ruins of a cistern built in the second century for a villa about 60 km north of Rome. The exposed footings measure 1.3 m, as indicated by the tape. This is the amount of erosion since the cistern was built. [Sheldon Judson.]

sidered to be about 10 percent of the suspended load in the average stream.

With these facts in mind, we may ask how much material is carried annually by streams out of the drainage basin in which is located the ancient Roman cistern in Figure 4.15. The major stream here is the Tiber River. If we measure the load of the Tiber at Rome, we find that it is removing sediments on the average of about 7.5 million t of solid material each year from the area upstream. This converts to an erosion rate of 17 cm/1,000 years over the entire basin. If we had data on bed load and dissolved load for the Tiber, the figures would be higher but not by much.

Turning to the United States, we can examine Table 4.2, which describes the erosion going on in the major regions of the country. Not included here is the drainage to Hudson Bay and the Saint Lawrence River area. Also, the table does not include the area in the western states known as the Great Basin, where topography and rainfall are such that streams do not reach the sea. We see that the rates vary from region to region but average approximately 6 cm/1,000 years.

Data from the Amazon River, the world's largest, indicate that it is removing material from its basin at the rate of 4.7 cm/1,000 years. Another large tropical river, the Congo, is carrying enough material out of its

basin each year to reduce its drainage basin by approximately 2 cm/1,000 years.

A single large drainage basin integrates many factors affecting the rapidity of erosion. One we should consider carefully is the human influence. There is enough information from disciplines as diverse as archaeology and nuclear physics to indicate that when people occupy an area intensively and turn it to crop land, they increase the erosion rate 10 to 100 times over that of a naturally forested or grassed area.

How much material is transported each year to the oceans? We have already cited some figures for portions of the earth. These figures, however, cover only 10 percent of the land surface. Therefore, any estimate of the total amount of erosion of the earth's surface per unit time can be no better than that, a mere estimate. The information now at hand suggests, however, that something approaching 10 billion t/year was being delivered to the ocean before human beings became effective geological agents. This does not include windblown material, although it is thought to be negligible. Glacier ice is not considered but probably would not appreciably affect our figure. The rate of 10 billion t/year corresponds to a continental reduction of approximately 2.5 cm/1,000 years. If we take into consideration the human effect on erosion, it is estimated that at present approximately 24 billion t are moved annually by the rivers to the oceans. This is 2.5 times the amount of material moved before our intervention.

DIFFERENTIAL EROSION

Differential erosion is the process by which different rock masses or different sections of the same rock mass erode at different rates. Almost all rock masses of any size weather in this manner. The results vary from the boldly sculptured forms of the Grand Canyon to the slightly uneven surface of a marble tombstone. Unequal rates of erosion are chiefly caused by variations in the composition of the rock. The more resistant zones stand out as ridges, ribs, or pinnacles above the more rapidly weathered rock on either side (Figures 4.16 and 4.17).

A second cause of differential erosion is simply that the intensity of weathering varies from one section to another in the same rock. Figure 4.18 shows a memorial that has undergone mechanical weathering in certain spots. The rock is a coarse, red, homogeneous sandstone made up of quartz, feldspar, and mica, with

Rates of regional erosion in the United States[a]

Drainage region	Drainage area,[b] thousands of km²	Runoff, thousands of m³/s	Load, t/km²-yr			Erosion, cm/1,000 yr	Area sampled, %
			Dissolved	Solid	Total		
Colorado	629	0.6	23	417	440	17	56
Pacific Slopes, California	303	2.3	36	209	245	9	44
Western Gulf	829	1.6	41	101	142	5	9
Mississippi	3,238	17.5	39	94	133	5	99
South Atlantic and Eastern Gulf	736	9.2	61	48	109	4	19
North Atlantic	383	5.9	57	69	126	5	10
Columbia	679	9.8	57	44	101	4	39
Total (overall)	6,797	46.9	43	119	162	6	

[a] After Sheldon Judson and D. F. Ritter, "Rates of Regional Denudation in the United States," *J. Geophys. Res.*, vol. 69, p. 3399, 1964.
[b] Great Basin, Saint Lawrence, and Hudson Bay drainage not considered.

some red iron oxides. The inner sides of the pillars have disintegrated so badly that the original fluting has been entirely destroyed and the underside of the horizontal slab has scaled noticeably. Notice that these are the areas least accessible to the drying action of the sun. Consequently, moisture has tended to persist here, and the frost action has pried off flakes and loosened individual grains.

On a larger scale, differential erosion is shown in the cross section in Figure 4.19. The section is drawn across a portion of northeastern Pennsylvania near the Delaware Water Gap. Note that the highest ridge is underlain by a tough quartzite, which weathers very slowly. The major portion of the Great Valley is underlain by a shaley rock, which weathers easily. In this same valley are still lower spots underlain by a dolomitic limestone, which is susceptible to rapid chemical weathering in this humid climate. To the southeast lie the New Jersey Highlands, held up by resistant units of both igneous and metamorphic rocks.

4.5 SOILS

So far we have been discussing the ways in which weathering acts to break down existing rocks and to provide the material for new rocks, but weathering also plays a crucial role in the creation of the soils that cover the earth's surface and sustain all life. In fairly recent years the study of soils has developed into the science of **pedology** (from the Greek *pedon*, "soil," with *logos*, "reason," hence "soil science").[2]

SOIL CLASSIFICATION

In the early years of soil study in the United States, researchers thought that the parent material almost wholly determined the type of soil that would result from it. Thus, they reasoned, granite would weather to one type of soil and limestone to another.

It is true that a soil reflects to some degree the material from which it developed, and in some instances one can even map the distribution of rocks on the basis of the types of soil that lie above them. But as more and more information became available, it became apparent that the bedrock is not the only factor determining soil type. Russian soil scientists, following the pioneer work of V. V. Dokuchaev (1846–1903), have demonstrated that different soils develop over identical bedrock material in different areas when the climate varies from one area to another. The idea that climate exerts a major control over soil formation was introduced into this country in the 1920s by C. F. Marbut (1863–1935), for many years chief of the United States

[2]In the United States, the term **pedology** is sometimes confused with words based on *ped-* and *pedi-*, combining forms meaning "foot," or with words based on *ped-* and *pedo-*, combining forms meaning "boy," "child," as in **pediatrics**, the medical science that treats of the hygiene and diseases of children. Consequently, there is a tendency in this country to use **soil science** instead of **pedology**.

4.16 Differential weathering and erosion of Ledger Dolomite bedrock of Cambrian age in Lancaster, Pennsylvania. Solution has dissolved out less resistant portions, leaving behind pinnacles of more resistant rock, exposed when the soil cover was removed. [M. E. Kauffman.]

4.17 Differential weathering as displayed by the Double Arch in Kayenta Sandstone, Upper Muley Twist Canyon, Capitol Reef National Park, Utah. [U.S. National Park Service.]

4.5 SOILS

4.18 Differential weathering in a sandstone monument, burying ground of Christ Church, Cambridge, Massachusetts. Mechanical weathering has been most rapid on the shaded portions (far right). Here moisture remains longer, and consequently frost action has been more effective than on the less shaded portions. [Sheldon Judson, 1968.]

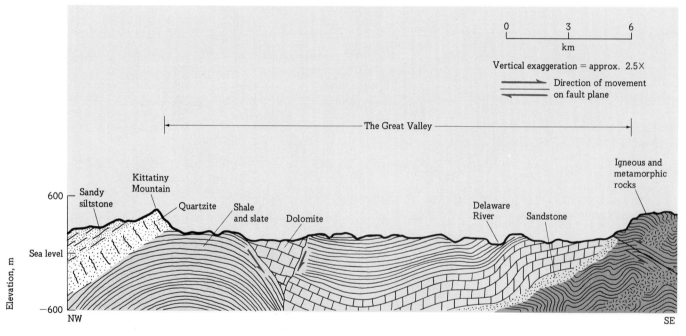

4.19 A generalized cross section in northeastern Pennsylvania and northwestern New Jersey (in the vicinity of the Delaware Water Gap) to show the relation of topography to the different rock types below the surface. See the text for discussion.

Soil Survey in the Department of Agriculture. Since that time, soil scientists have discovered that still other factors exercise important influences on soil development. For instance, the relief of the land surface plays a significant role. The soil on the crest of a hill is somewhat different from the soil on the slope, which in turn differs from the soil on the level ground at the foot of the hill; yet all three soils rest on identical bedrock. The

passage of time is another factor: A soil that has only begun to form differs from one that has been developing for thousands of years, although the climate, bedrock, and topography are the same in each instance. Finally, the vegetation in an area influences the type of soil that develops there: One type of soil will form beneath a pine forest, another beneath a forest of deciduous trees, and yet another on a grass-covered prairie.

Exactly what is a soil? It is a natural surficial material that supports plant life. Since each soil exhibits certain properties that are determined by climate and living organisms operating over periods of time on earth materials and on landscape of varying relief, and because all these factors are combined in various ways all over the land areas of the globe, the number of possible soil types is almost unlimited.

Soil scientists have developed a detailed classification system for soils of various types (see Table 4.3). Geologically, we are interested primarily in what materials are removed or added during the soil-forming process.

TABLE 4.3

Soil orders of the United States soil classification system[a]

Soil order	Soil characteristics
Alfisols	Subhumid climates, precipitation 15–125 cm, mostly under forest cover. Clay accumulation in *B* horizon. Leached upper gray-brown layer; red-brown lower layer, with aluminum and iron silicate concentrations
Aridisols	Dry climates with lime or gypsum accumulations, salt layers; accumulation of Ca, Mg, and K ions
Entisols	Recent soils with no recognizable horizons developed; many recent river floodplains, volcanic ash deposits, and recent sands
Histosols	Rich in organic content, including peats and mucks; accumulated plant remains in bogs and swamps
Inceptisols	Weakly developed horizons showing just the "inception" of soil development
Mollisols	Grassland soils; black, organic-rich near surface; high concentration of lime
Oxisols	Excessively weathered soils, often over 3 m deep; of low fertility; lots of iron and aluminum oxide clays
Spodosols	Sandy, leached soils of cool forests, mostly conifers; acid soil; lots of organic matter and/or Fe and Al clays
Ultisols	Strongly acid, well-weathered soils of tropical and subtropical climates; accumulation of residual silicates
Vertisols	High content of shrinking and expanding clays; very old weathered soils; deep wide vertical cracks when dry

[a] After R. L. Donahue, R. W. Miller, and J. C. Shickluna, *Soils: An Introduction to Soils and Plant Growth;* 4th Ed. Prentice-Hall, Inc., Englewood Cliffs, N. J., 1977.

Certain valid generalizations can be made about soils. We know, for example, that the composition of a soil varies with depth. A natural or artificial exposure of a soil reveals a series of zones, each recognizably different from the one above. Each of these zones is called a **soil horizon,** or, more simply, a **horizon.** The three major zones or horizons in a typical soil, shown in Figure 4.20, may be described as follows from the bottom upward.

C horizon The *C* horizon is a zone of partially disintegrated and decomposed rock material. Some of the original bedrock minerals are still present, but others have been transformed into new materials. The *C* horizon grades downward into the unweathered rock material.

B horizon The *B* horizon lies directly above the *C* horizon. Weathering here has proceeded still further than in the underlying zone, and only those minerals of the parent rock that are most resistant to the decomposition (quartz, for example) are still recognizable. The others have been converted into new minerals or into soluble salts. In moist climates the *B* horizon contains an accumulation of clayey material and iron oxides delivered by water percolating downward from the surface. In dry climates we generally find, in addition to the clay and iron oxides, deposits of more soluble minerals, such as calcite. This mineral too may have been brought down from above, but some is brought into the *B* horizon from below, as soil water is drawn upward by high evaporation rates. Because material is deposited in the *B* horizon, it is known as the "zone of accumulation" (see Figure 4.20).

4.20 The three major horizons of a soil. In many places it is possible to subdivide the zones themselves. Here the soil is shown as having developed from limestone.

A horizon The A horizon is the uppermost zone—the one into which we sink a spade when we dig a garden. This is the zone from which the iron oxides have been carried to the B horizon, and in dry climates it is the source of some soluble material that may be deposited in the B horizon. The process by which these materials have been moved downward by soil water is called **leaching**, and the A horizon is sometimes called the "zone of leaching" (Figure 4.20). Varying amounts of organic material tend to give the A horizon a gray-to-black color.

The three soil horizons have all developed from the underlying parent material. When this material is first exposed at the surface, the upper portion is subjected to intense weathering, and decomposition proceeds rapidly. As the decomposed material builds up, downward-percolating water begins to leach out some of the minerals and to deposit them lower down. Gradually the A horizon and the B horizon build up, weathering continues (at a slower rate now) on the underlying parent material, giving rise to the C horizon. With the passage of time, the C horizon reaches deeper and deeper into the unweathered material below, the B horizon keeps moving downward, and the A horizon in turn encroaches on the upper portion of the B horizon. Finally a "mature" soil is built up.

The thickness of the soil that forms depends on many factors, but in the northern United States and southern Canada, the material that was first exposed to weathering after the retreat of the last ice sheet some 10,000 years ago is now topped by a soil 60 to 80 cm thick. Farther south, where the surface was uncovered by the ice at an earlier time, the soils are thicker, and in some places the processes of weathering have extended themselves from 5 to 10 m below the present surface.

SOME SOIL TYPES

We can understand the farmer's interest in soil, but why is it important for the geologist to understand soils and the processes by which they are formed? There are several reasons: First, soils provide clues to the environment in which they were originally formed. By analyzing an ancient soil buried in the rock record, we may be able to determine the climate and physical conditions that prevailed when it was formed. Second, some soils are sources of valuable mineral deposits (see Chapter 18), and the weathering process often enriches otherwise low-grade mineral deposits, mak-

ing them profitable to mine. An understanding of soils and soil-forming processes, therefore, can serve as a guide in the search for ores. Third, because a soil reflects to some degree the nature of the rock material from which it has developed, we can sometimes determine the nature of the underlying rock by performing an analysis of the soil.

But most important of all, soils are the source of many of the sediments that are eventually converted into sedimentary rocks. And these in turn may be transformed into metamorphic rocks, or following another path in the rock cycle, may be converted into new soils. If we understand the processes and results of soil formation, we are in a better position to interpret the origin and evolution of many rock types.

The following pages will discuss three major types of soil. Two of them, the **pedalfers** and the **pedocals,** are typical of the middle latitudes. The third group, referred to as **laterites,** is found in tropical climates.

Pedalfers A **pedalfer** is a soil in which iron oxides, clays, or both have accumulated in the B horizon. The name is derived from *pedon* (Greek *pedon* "the ground") and the symbols Al and Fe for aluminum and iron. In general, soluble materials such as calcium carbonate or magnesium carbonate do not occur in the pedalfers. Pedalfers are commonly found in temperate, humid climates, usually beneath a forest vegetation. In the United States, most of the pedalfers lie east of a line that corresponds roughly to about 63 cm/year of rainfall. Northward in Canada, however, where the temperature is more important than the total rainfall in determining the distribution of pedalfers, the zone extends northwestward across Saskatchewan and Alberta coincident with a mean annual temperature of about 4.5°C or colder. Farther west, the pedalfers extend southward into the United States along the mountainous region of the Rockies, where the rainfall is somewhat higher and the temperatures lower than in the rest of western United States (see Figure 4.21).

In the formation of pedalfers, certain soluble compounds, particularly those that contain sodium, calcium, and magnesium, are rapidly removed from the A horizon by waters seeping into the soil from the surface. These soluble compounds proceed downward through the B horizon and are carried off by ground water. The less soluble iron oxides and clay are deposited in the B horizon, giving that zone a clayey character with a brownish to reddish color.

Using the information on the products of chemical weathering listed in Table 4.1, we can build up a picture of how a pedalfer develops from granite in temperate and humid areas. At some depth below the surface lies the unaltered granite. Directly above is the crumbly, partially disintegrated rock of the C horizon. Here we can still identify the minerals that made up the

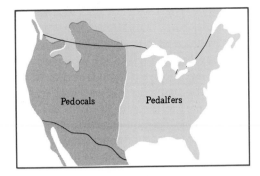

4.21 Generalized distribution of pedalfer and pedocal soils in the United States and southern Canada. In the United States, the pedalfers have developed in the more humid climates to the east of the line that marks approximately 63 cm/year of precipitation. To the west of this line, where precipitation is generally less than 63 cm/year, pedocal soils predominate. In Canada and in the northern Rocky Mountains of the United States, temperature is more critical than precipitation in determining the distribution of the two soils. There pedocals occur in areas where the average annual temperature is 4.5°C or less.

original granite, although the feldspars have started to decompose and have become cloudy, and the iron-bearing minerals have been partially oxidized.

Moving upward to the *B* horizon, the zone of accumulation, we find that here the feldspars have been converted to clay and that the material has a compact, clayey texture. Iron oxides or limonite stain the soil a reddish or brownish color. And because the grains of quartz released from the granite have undergone little change, we find some sand in this otherwise clayey zone.

The *A* horizon, a few centimeters thick, has a grayish to ashen color; the iron compounds have been leached from this zone and now color the *B* horizon below. Furthermore, the texture of this zone is sandier because most of the finer materials have also been moved downward to the *B* horizon, and the soluble salts have been largely dissolved and removed by water. The very top of the *A* horizon is a thin zone of dark humus material. The resulting pedalfer soil is shown in Figure 4.22.

Notice in Figure 4.23 how the original minerals of the granite are transformed as weathering progresses. On the left we have quartz, plagioclase, and orthoclase, which were released directly from the granite. Then, as weathering progresses, we find that the amount of kaolinite increases at the expense of the original minerals. The initial rise in the amount of quartz and orthoclase indicates simply that these minerals tend to accumulate in the soil because of their greater resistance to decomposition. Iron oxides also increase with weathering as the iron-bearing silicates decompose.

Forest

Forest litter
A horizon
Sandy gray loam

B horizon

Reddish-brown to yellowish sandy compact clay, grading into

C horizon

Rotted granite, still retaining original granitic minerals and texture, grading into

unaltered granite

1.25 m

4.22 A pedalfer soil that has developed on a granite. Note the transition from unaltered granite, upward through partially decomposed granite of the *C* horizon, into the *B* horizon, where no trace of the original granite structure remains, and finally into the *A* horizon, just below the surface.

There are several varieties of soils in the pedalfer group, including the red and yellow soils of the southeastern states as well as **podsol** (from the Russian for "ashy gray soil") and the gray-brown podsolic soils of the northeastern quarter of the United States and of southern and eastern Canada. Prairie soils are transitional varieties between the pedalfers of the East and the pedocals of the West.

Pedocals Pedocals are soils that contain an accumulation of calcium carbonate. Their name is derived from a combination of *pedon* with an abbreviation for "calcium." The soils of this major group are found in the temperate zones where the temperature is rela-

4.23 Change in mineral percentages as a granite is subjected to increased chemical weathering. As weathering progresses, the orthoclase and plagioclase of the granite decrease in abundance and give rise to an increasing amount of clay (kaolinite). At the same time, the percentage of iron oxides increases at the expense of the original iron silicate minerals (not shown). [Redrawn from S. S. Goldich, "A Study of Rock Weathering," *J. Geol.*, vol. 46, p. 33, 1938.]

tively high, the rainfall is low, and the vegetation is mostly grass or brush. In the United States, the pedocals are found generally to the west of the pedalfers; and in Canada, to the southwest.

In the formation of pedocals, calcium carbonate and, to a lesser extent, magnesium carbonate are deposited in the soil profile, particularly in the *B* horizon. This process occurs in areas where the temperature is high, the rainfall scant, and the upper level of the soil is hot and dry most of the time. Water evaporates before it can remove carbonates from the soil. Consequently these compounds are precipitated as **caliche,** a whitish accumulation made up largely of calcium carbonate. *Caliche* is a Spanish word, a derivative of the Latin *calix,* "lime." The occasional rain may carry the soluble material down from the *A* horizon into the *B* horizon; there it is later precipitated as the water evaporates. Soluble material may also move up into the soil from below. In this case, water beneath the soil or in its lower portion rises toward the surface through small capillary openings. Then, as the water in the upper portions evaporates, the dissolved materials will be precipitated out.

The tendency for pedocals to develop under a growth of brush and grass also helps to concentrate the soluble carbonates: The plants intercept them before they can be moved downward in the soil. When the plants die, the carbonates are either added to the soil, where they are used by other plants, or simply precipitated in the soil by high evaporation rates.

Because rainfall is light in the climates where pedocals form, chemical weathering proceeds only slowly, and clay is produced less rapidly than in more humid climates. For this reason pedocals contain a lower percentage of clay minerals than do the pedalfer soils.

The pedocal group includes the black and chestnut-colored soils of southern Alberta and Saskatchewan in Canada and the northern plains of the United States, the reddish soils farther south, and the red and gray desert soils of the drier western states.

Tropical soil (laterites)   The term **laterite** is applied to many tropical soils that are rich in hydrated aluminum and iron oxides. The name itself, from the Latin for "brick," suggests the characteristic color produced by the iron in these soils. The formation of laterites is not well understood. In fact, soil scientists are not even certain that the *A*, *B*, and *C* horizons characteristic of the pedalfers and the pedocals have their counterparts in the laterites, even though these soils exhibit recognizable zones.

In the development of the laterites, iron and aluminum accumulate in what is presumed to be the *B* horizon. The aluminum is in the form of $Al_2O_3 \cdot nH_2O$, which is generally called **bauxite,** an ore of aluminum. This ore appears to be developed when intense and prolonged weathering removes the silica from the clay minerals and leaves a residuum of hydrous aluminum oxide (that is, bauxite). In some laterites the concentration of iron oxides in the presumed *B* horizon is so great that it is profitable to mine them for iron, as is done in certain districts of Cuba.

The term **laterite** is most properly applied only to the zone in which iron and aluminum have accumulated. As we have seen, this is the zone that may be equivalent to the *B* horizon in more northerly soils. Overlying this zone, however, there is often a zone of crumbly loam, and below it is a light-colored, apparently leached zone that adjoins the parent material. Some soil scientists refer to these two zones as the *A* horizon and the *C* horizon, respectively.

PALEOSOLS

Not all soils occur at the surface. Some are buried beneath younger material and are known as **paleosols** (see Figure 4.24). Such buried soil zones are useful indicators of past periods of erosion and weathering.

4.24 Paleosol buried beneath younger material. The dark band in the center of this exposure is a buried soil and measures over 1 m in thickness. It is developed on water-laid pyroclastic material and is buried by beds of similar volcanic material. At the top of the younger beds, the present surface soil has been almost completely destroyed by erosion processes accelerated by human activity. Crater of Baccano along the Via Cassia, north of Rome, Italy. [Sheldon Judson.]

OUTLINE

Energy, the capacity for producing motion, is involved in every geological process.

    Potential energy is stored energy waiting to be used.

    Kinetic energy, possessed by every moving object, depends on the mass of the object and on the speed with which it is moving.

The **energy that drives the weathering process** comes from within the earth in terms of uplift in mountainous areas and from outside the earth—such as solar energy.

Weathering is the response of surface or near-surface material to contact with water, air, and living matter.

Types of weathering are mechanical and chemical.

    Mechanical weathering (disintegration) involves a reduction in the size of rock and mineral particles but no change in composition.

    Chemical weathering (decomposition) involves a change in the composition of the material weathered. The rate of chemical weathering increases with the decrease in the size of particles and with increase in temperature and moisture.

    Chemical weathering of igneous rocks, which include quartz, feldspar, and ferromagnesian minerals, gives clay, iron oxides, quartz, and soluble salts.

Rates of weathering vary with the type of material weathered and the environment. For example, olivine chemically weathers more rapidly than does quartz, and limestone weathers very rapidly in a moist climate but very slowly in a dry climate.

Rates of erosion indicate that before human beings began to use the landscape intensively, rivers annually transported about 10 billion t of material to the seas. Today it is two or three times that, chiefly because of human activities.

    Differential erosion is the process by which different rock masses or different sections of the same rock mass erode at different rates. For example, limestone is more resistant to erosion in a dry climate than is mudstone.

Soil is a naturally occurring surface material that supports life and generally is the product of weathering.

    Soil zones are the **A, B,** and **C horizons** from the surface downward.

    Soil types include pedalfers in moist temperate climates, pedocals in dry temperate climates, and laterites in moist tropical climates.

    Paleosols are buried soil zones indicative of past periods of erosion and weathering.

4.5 SOILS

89

SUPPLEMENTARY READINGS

Birkeland, P. W. *Pedology, Weathering and Geomorphological Research*, Oxford University Press, New York, 1974. *Study of soils in their natural setting—the field—with emphasis on the use of soils to date deposits on the basis of soil development. Attempts are made to reconstruct the environment that occurred during soil formation.*

Donahue, Roy L., R. W. Miller, and J. C. Shickluna *Soils: An Introduction to Soils and Plant Growth*, 4th ed., Prentice-Hall, Inc., Englewood Cliffs, N. J., 1977. *A basic introduction to the subject of soils and the interaction between soils and plants.*

Hodgson, J. M. *Soil Sampling and Soil Description*, Oxford University Press, New York, 1978. *Examines soil surveys around the world and reviews methods of soil and site descriptions.*

Judson, Sheldon "Erosion of the Land—or What's Happening to Our Continents?" *Am. Sci.*, vol 56, pp. 356–379, 1968. *An overview of worldwide erosion.*

Ritter, D. F. *Process Geomorphology*, William C. Brown Company, Publishers, Dubuque, Iowa, 1978. *A modern examination of the earth's surficial processes. Excellent coverage of physical and chemical weathering in the production of soils.*

QUESTIONS

1 Where do we find bedrock and why does it appear at the surface in some instances and not in others?

2 What is the difference between potential energy and kinetic energy?

3 What property of H_2O makes it such an important agent in weathering? How is this different from almost every other chemical substance?

4 Explain the process by which bedrock that has not been moved at all can come to be rounded, looking like so many boulders that have been rolled around to produce their rounded edges?

5 How is climate important in the chemical weathering of minerals and rocks?

6 How does the rate of chemical weathering of the minerals of an igneous rock compare with the pattern of crystallization from its original magma? What are the reasons for this relationship?

7 How can we determine *average* rates of erosion for whole continents?

8 What besides the bedrock itself will influence the kind of soil that develops in a given area?

9 Draw and label a typical "soil profile."

10 Of what use are buried soil zones or paleosols in interpreting the geologic history of an area?

11 With physical and chemical weathering taking place constantly, why do the weathering products not build up to staggering amounts on all continents? What happens to them?

ORIGIN AND OCCURRENCE OF SEDIMENTS AND SEDIMENTARY ROCKS

We have all dug our toes into a sandy beach, or picked our way over the gravels of a rushing stream, or perhaps slogged through the mud of a swamp. None of these—sand, gravel, or mud—immediately suggests hard, solid rock. Yet deposits of this sort or materials very similar to them are the stuff from which the great bulk of the rocks exposed at the earth's surface were formed. When we look down into the mile-deep Grand Canyon of the Colorado River in Arizona, we can see there layer upon layer of rocks that were once unconsolidated deposits of sand, gravel, and mud. Over the course of time, these loose sediments have been hardened into rocks that we call **sedimentary.**

It is extremely difficult to work out a concise, comprehensive definition of sedimentary rocks. The adjective, from the Latin *sedimentum*, means "settling." Therefore we might expect sedimentary rocks to be formed when individual particles settle out of a fluid, such as the water of a lake or an ocean. Fragments or minerals derived from the breakdown of rocks are swept into bodies of water, where they settle out as unconsolidated sediments. Later, they are hardened into true rocks. But other rocks, such as rock salt, are made up of minerals left behind by the evaporation of large bodies of water, and these rocks are as truly sedimentary rocks as those formed from particles that have settled on an ocean floor. Still other sedimentary rocks are made up largely of the shells and hard parts of animals, particularly invertebrate marine organisms.

Sedimentary rocks are often layered, or stratified. Unlike massive igneous rocks, such as granite, most sedimentary rocks are laid down in individual beds, one on another. The surface of each bed is essentially parallel to the horizon at the time of deposition, and a cross section exposes a series of layers like those of a giant cake. True, some igneous rocks, such as those formed from lava flows, are also layered. By and large however, **stratification is the single most characteristic feature of sedimentary rocks** (see Figure 5.1).

Some sediments, particularly sand and gravel, are consolidated into rock by a cementing process: Subsurface water trickling through the open spaces leaves behind a mineral deposit that serves to cement

91

5.1 Stratification or layering is the single most characteristic feature of sedimentary rocks and is shown here in the Grand Canyon of the Colorado. [Lee Krohn.]

the individual grains firmly together, giving the entire deposit the strength we associate with rock. Other sediments, such as fine deposits of mud, are transformed into rock by the weight of overlying deposits, which press, or compact, them into a smaller and smaller space. The sedimentary rock that results from either of these processes may eventually be exposed at the earth's surface. If the rock was formed beneath the bottom of the ocean, it may be exposed either by the slow withdrawal of the seas or by the upward motion of the sea floor, forming new areas of dry land.

5.1 AMOUNTS OF SEDIMENTS AND SEDIMENTARY ROCKS

About 75 percent of the rocks exposed at the earth's surface are sedimentary rocks or metamorphic rocks derived from them. Yet sedimentary rocks make up only about 5 percent by volume of the outer 10 km of the globe. The other 95 percent of the rocks in this zone are, or once were, igneous rocks (see Figure 5.2). The sedimentary cover is only as thick as a feather edge where it laps around the igneous rocks of the Adirondacks and the Rockies. In other places, it is thousands of meters thick. In Oklahoma, oil-drilling operations have cut into the crust nearly 9 km and have encountered nothing but sedimentary rocks. In the Ganges

River basin of India, the thickness of the sedimentary deposits has been estimated at between 13.5 and 18 km.

We have some estimates of the total mass of sedimentary rocks on the earth. Probably the best available estimate was made some time ago by Arie

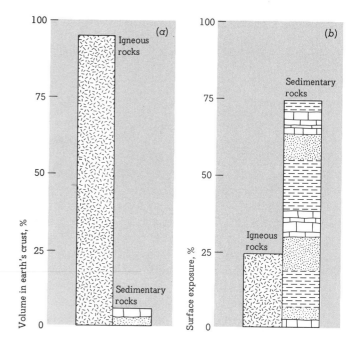

5.2 Graphs showing relative abundance of sedimentary rocks and igneous rocks. (a) The great bulk (95 percent) of the outer 15 km of the earth is made up of igneous rocks; only a small proportion (5 percent) is sedimentary. In contrast (b), the areal extent of sedimentary rocks at the earth's surface is three times that of igneous rocks. Metamorphic rocks are considered with either igneous or sedimentary rocks, depending on their origin.

Poldervaart[1] of Columbia University, who calculated their weight as $1,702 \times 10^{15}$ t. Of this total he estimated that 480×10^{15} t are presently on the continents and that the rest are in the oceans.

We have seen that sedimentary rocks are formed from the materials weathered from preexisting rocks. Eventually this material reaches the deep ocean basins. What estimate can we make about the amount of material delivered each year to the oceans? The figures presented in Table 5.1 are order-of-magnitude estimates. Therefore, a statement of the approximate amount of material delivered annually to the oceans —where most sedimentary rocks form—is approximately 10^{10} t. As Table 5.1 suggests and as we have implied in Section 4.4, the great bulk of this material is carried by rivers. That contributed by wind, by ice, or by extraterrestrial sources does not appreciably change the total amount of material deposited in the oceans.

TABLE 5.1

Order-of-magnitude estimates by source of materials delivered annually to the oceans[a]

Source	t/yr
Rivers	10 billion (10^{10})
Glaciers	100 million to 1 billion (10^8–10^9)
Wind	100 million (10^8)
Extraterrestrial	0.03 to 0.3 (3×10^{-2}–3×10^{-1})

[a]Estimated from various sources.

5.2 FORMATION OF SEDIMENTARY ROCKS

We found in Chapter 3 that igneous rocks harden from molten material that originates some place beneath the surface, under the high temperatures and pressures that prevail there. In contrast, sedimentary rocks form at the much lower temperatures and pressures that prevail at or near the earth's surface.

ORIGIN OF MATERIAL

The material from which sedimentary rocks are fashioned originates in two ways. First, the deposits may be accumulations of minerals and rocks derived either from the erosion of existing rock or from the weathered products of these rocks. Deposits of this type are called **detrital** (from the Latin for "worn down"), and sedimentary rocks formed from them are called **detrital sedimentary rocks.** Second, the deposits may be produced by chemical processes. We refer to these deposits as **nondetrital** or **chemical deposits** and to the rocks formed from them as **chemical sedimentary rocks.**

Gravel, sand, silt, and clay derived from the weathering and erosion of a land area are examples of detrital sediments. Let us take a specific example. The quartz grains freed by the weathering of a granite may be winnowed out by the running water of a stream and swept into the ocean. There they settle out as beds of sand, a detrital deposit. Later, when this deposit is cemented to form a hard rock, we have a sandstone, a detrital rock.

Chemically formed deposits are usually laid down by the precipitation of material dissolved in water. This process may take place either directly, through inorganic processes, or indirectly, through the intervention of plants and animals. The salt left behind after a salty body of water has evaporated is an example of a deposit laid down by inorganic chemical processes. On the other hand, certain organisms, such as the corals, extract calcium carbonate from seawater and use it to build up skeletons of calcite. When the animals die, their skeletons collect as a biochemical (from the Greek for "life") deposit, and the rock that subsequently forms is called a **biochemical rock**—in this case, limestone.

Although we distinguish between the two general groups of sedimentary rocks—detrital and chemical— most sedimentary rocks are mixtures of the two. We commonly find that a chemically formed rock contains a certain amount of detrital material. In similar fashion, predominantly detrital rocks include some material that has been chemically deposited.

Geologists use various terms to describe the environment in which a sediment originally accumulated. For example, if a limestone contains fossils of an animal that is known to have lived only in the sea, the rock is known as a **marine** limestone. **Fluvial**, from the Latin for "river," is applied to rocks formed by deposits laid down by a river. **Eolian**, derived from Aeolus, the Greek wind god, describes rock made up of wind-deposited material. Rocks formed from lake deposits are termed **lacustrine**, from the Latin word for "lake."

Detrital and chemical, however, are the main divisions of sedimentary rocks based on the origin of material, and, as we shall see later, they form the two major divisions in the classification of sedimentary rocks.

[1]Poldervaart, A. (ed.), "Crust of the Earth," *Geol. Soc. Am. Special Paper 62,* 1955.

The general process by which rock-forming material is laid down is called **sedimentation,** or deposition. The factors controlling sedimentation are easy to visualize: To have any deposition at all, there must obviously be something to deposit—another way of saying that a source of sediments must exist; we also need some process to transport this sediment; and finally there must be some place and some process for the deposition of the sedimentary material.

Methods of transportation Water—in streams and glaciers, underground, and in ocean currents—is the principal means of transporting material from one place to another. Landslides and other movements induced by gravity also play a role, as does the wind, but we shall look more closely at these processes in Chapters 11 through 16.

Processes of sedimentation Detrital material is **deposited** when its agent of transportation no longer has sufficient energy to move it farther. For example, a stream flowing along at a certain velocity possesses energy to move particles up to a certain maximum size. If the stream loses velocity, it also loses energy, and it is no longer able to transport all the material that it has been carrying at the higher velocity. The solid particles, beginning with the heaviest, start to settle to the bottom. The effect is much the same when a wind that has been driving sand across a desert suddenly dies. A loss of energy accompanies the loss in velocity.

Material that has been carried in solution is deposited in a different way, that is, by **precipitation,** a chemical process by which dissolved material is converted into a solid and separated from the liquid solvent. As already noted, precipitation may be either biochemical or inorganic.

Although at first glance the whole process of sedimentation seems quite simple, it is actually as complex as nature itself. Many factors are involved, and they can interact in a variety of ways. Consequently, the manner in which sedimentation takes place and the sediments that result from it differ greatly from one situation to another (see Figures 5.3 and 5.4). Think, for instance, of the different ways in which materials settle out of water. A swift, narrow mountain stream may deposit coarse sand and gravel along its bed, but farther downstream, as the valley widens, the same stream may overflow its banks and spread silt and

5.3 The environment of this quiet pond near Lexington, Massachusetts, favors the deposition of fine-grained sediments, largely mud. [Sheldon Judson.]

5.4 Exposed to the direct attack of ocean surf, the environment of this cliffed California coastline favors the deposition of coarse sand and gravel. [Sheldon Judson.]

mud over the surrounding country. A lake provides a different environment, varying from the delta of the inflowing stream to the deep lake bottom and the shallow, sandy shore zones. In the oceans, too, environment and sedimentation vary, from the brackish tidal lagoon to the zone of plunging surf and out to the broad, submerged shelves of the continents and to the ocean depths beyond.

MINERAL COMPOSITION OF SEDIMENTARY ROCKS

Sedimentary rocks, like igneous and metamorphic rocks, are accumulations of minerals. In sedimentary rocks, the three most common minerals are clay, quartz, and calcite, although, as we shall see, a few others are important in certain localities.

Rarely is a sedimentary rock made up of a single mineral, although one mineral may predominate. Limestone, for example, is composed mostly of calcite, but even the purest limestone contains small amounts of other minerals, such as clay or quartz. The grains of many sandstones are predominantly quartz, but the cementing material that holds these grains together may be calcite, dolomite, or iron oxide. In general, we may say that most sedimentary rocks are mixtures of two or more minerals.

Clay   In an earlier chapter we described how clay minerals develop from the weathering of the silicates, particularly the feldspars. These clays may be subsequently incorporated into sedimentary rocks; they may, for example, form an important constituent of mudstone and shale. Examination of recent and ancient marine deposits shows that the kaolinite and illite clays are the most common clays in sedimentary rocks and that the montmorillonite clays are relatively rare.

Quartz   Another important component of sedimentary rocks is silica, including the very common mineral quartz, as well as a number of much rarer forms, such as chert, flint, opal, and chalcedony.

The mechanical and chemical weathering of an igneous rock such as granite sets free individual grains of quartz that eventually may be incorporated into sediments. These quartz grains produce the detrital forms of silica and account for most of the volume of the sedimentary rock sandstone. But silica in solution or in particles of colloidal size is also produced by the weathering of an igneous rock. This silica may be precipitated or deposited in the form of quartz, particularly as a cementing agent in certain coarse-grained sedimentary rocks. Silica may also be precipitated in other forms such as **opal**, generally regarded as a hydrous silica, $SiO_2 \cdot nH_2O$. Opal is slightly softer than true quartz and has no true crystal structure.

Silica also occurs in sedimentary rocks in a form called **cryptocrystalline.** This term (from the Greek *kryptos*, "hidden," and "crystalline") indicates crystalline structure so fine that it cannot be seen under most ordinary microscopes. The microscope does reveal, however, that some cryptocrystalline silica has a granular pattern and that some has a fibrous pattern. To the naked eye, the surface of the granular form is somewhat duller than that of the fibrous form. Among the dull-surfaced, or granular, varieties is flint, usually dark in color. Flint is commonly found in certain limestone beds—the chalk beds of southern England, for example. Chert is similar to flint but tends to be lighter in color; and jasper is a red variety of granular cryptocrystalline form. The general term chalcedony is often applied to the fibrous types of cryptocrystalline silica, which have a higher, more waxy luster than do granular varieties. Sometimes the term is also used to describe a specific variety of brown translucent cryptocrystalline silica. Agate is a variegated form of silica, its bands of chalcedony alternating with bands of either opal or some variety of granular cryptocrystalline silica, such as jasper.

Calcite   The chief constituent of the sedimentary rock limestone, calcite, $CaCO_3$, is also the most common cementing material in the coarse-grained sedimentary rocks. The calcium is derived from igneous rocks that contain calcium-bearing minerals, such as calcium plagioclase and some of the ferromagnesian minerals. Calcium is carried from the zone of weathering as calcium bicarbonate, $Ca(HCO_3)_2$, and is eventually precipated as $CaCO_3$ through the intervention of plants, animals, or inorganic processes. The carbonate is ultimately derived from water and carbon dioxide.

Other materials in sedimentary rocks   Accumulations of clay, quartz and calcite, either alone or in combination, account for all but a very small percentage of the sedimentary rocks, but certain other materials occur in quantities large enough to form distinct strata. The mineral dolomite, $CaMg(CO_3)_2$, for example, is usually intimately associated with calcite; although it is far less abundant. (It is named after an eighteenth-century French geologist, Déodat de Dolomieu.) When the mineral is present in large amounts in a rock, the rock itself is also known as dolomite or dolostone. The mineral dolomite is easily confused with calcite; and because they often occur together, distinguishing them is important. Calcite effervesces freely in dilute hydrochloric acid; dolomite effervesces very slowly or not at all unless it is finely ground or powdered. The more rapid chemical activity results from the increase in surface area, an example of the general principle discussed in Section 4.2.

The feldspars and micas are abundant in some sedimentary rocks. We have found that chemical weathering converts these minerals into new minerals at a relatively rapid rate. Therefore, when we find mica and feldspar in a sedimentary rock, chances are that it was mechanical, rather than chemical, weathering that originally made them available for incorporation in the rock.

Iron produced by chemical weathering of the ferromagnesian minerals in igneous rocks may be caught up again in new minerals and be incorporated into sedimentary deposits. The iron-bearing minerals that occur most frequently in sedimentary rocks are hematite, goethite, and limonite. In some deposits these minerals predominate, but more commonly they act simply as coloring matter or as a cementing material.

Halite, $NaCl$, and gypsum, $CaSO_4 \cdot 2H_2O$, are minerals precipitated from solution by evaporation of the water in which they were dissolved. The salinity of the water—that is, the proportion of the dissolved material to the water—determines the type of mineral that will precipitate out. The gypsum begins to separate from seawater when the salinity (at 30°C) reaches a little over three times its normal value. Then, when the salinity of the seawater has increased to about ten times its normal value, halite begins to precipitate.

Pyroclastic rocks, mentioned in Chapter 3, are sedimentary rocks composed mostly of fragments blown from volcanoes. The fragments may be large pieces that have fallen close to the volcano or extremely fine ash that has been carried by the wind and deposited hundreds of kilometers from the volcanic eruption.

Finally, organic matter may be present in sedimentary rocks. In the sedimentary rock known as coal, plant materials are almost the only components. More commonly, however, organic matter is very sparsely disseminated throughout sedimentary deposits and the resulting rocks.

TEXTURE

Texture, as we have explained, refers to the general physical appearance of a rock—to the size, shape, and arrangements of the particles that make it up. There are two major types of texture in sedimentary rocks: clastic and nonclastic.

Clastic texture   The term clastic is derived from the Greek for "broken" or "fragmental," and rocks that

have been formed from deposits of mineral and rock fragments are said to have clastic texture. The size and shape of the original particles have a direct influence on the nature of the resulting texture. A rock formed from a bed of gravel and sand has a coarse, rubblelike texture that is very different from the sugary texture of a rock developed from a deposit of rounded, uniform sand grains. Furthermore, the process by which a sediment is deposited also affects the texture of the sedimentary rock that develops from it. Thus the debris dumped by a glacier is composed of a jumbled assortment of rock material ranging from particles of colloidal size to large boulders. A rock that develops from such a deposit has a very different texture from one that develops from a deposit of windblown sand, for instance, in which all the particles are approximately 0.15 to 0.30 mm in diameter.

Chemical sedimentary rocks may also show a clastic texture. A rock made up predominantly of shell fragments from a biochemical deposit has a clastic texture that is just as recognizable as the texture of a rock formed from sand deposits (see Figure 5.5).

One of the most useful factors in classifying sedimentary rocks is the size of the individual particles. In practice, we usually express the size of a particle in terms of its diameter rather than in terms of volume, weight, or surface area. When we speak of "diameter," we imply that the particle is a sphere; but it is very unlikely that any fragment in a sedimentary rock is a true sphere. In geological measurements, the term simply means the diameter that an irregularly shaped particle *would* have if it were a sphere of equivalent volume. Obviously, it would be a time-consuming, if not impossible, task to determine the volume of each sand grain or pebble in a rock and then to convert these measurements into appropriate diameters. So the diameters we use for particles are only approximations of their actual sizes. They are accurate enough, however, for our needs.

Several scales have been proposed to describe particles ranging in size from large boulders to minerals of microscopic dimensions. The Wentworth scale, presented in Table 5.2, is used widely, though not universally, by American and Canadian geologists. Notice that, although the term **clay** is used in the table to designate all particles below $1/256$ (0.0039) mm in diameter, the same term is used to describe certain minerals. To avoid confusion, we must always refer specifically to either "clay size" or "clay mineral" unless the context makes the meaning clear.

Because determining the size of particles calls for the use of special equipment, the procedure is normally carried out only in the laboratory. In the examination of specimens in the field, therefore, comparisons are made to samples of known size or to scales prepared for that purpose.

5.5 Viewed under the microscope, some limestones are seen to have clastic texture, as does this example from an ancient reef deposit in the Austrian Alps. The individual particles include the shells of one-celled marine animals (Foraminifera), as well as the fragments of other marine organisms and pellets of calcite. The particles are cemented together by clear calcite (dark). The black lines are also calcite that fills cracks in the rock. The area shown is 2.8 cm wide. [E. C. Bierwagen.]

TABLE 5.2

Wentworth scale of particle sizes
for clastic sediments[a]

Wentworth scale		For next larger size multiply by
Size, mm	Fragment	
	Boulder	
256	Cobble	4
64	Pebble	16
4	Granule	2
2	Sand	32
$\frac{1}{16}$ (0.0625)	Silt[b]	16
$\frac{1}{256}$ (0.0039)	Clay[b]	

[a] Modified after C. K. Wentworth, "A Scale of Grade and Class Terms for Clastic Sediments," *J. Geol.*, vol. 30, p. 381, 1922.
[b] Dust.

Nonclastic texture Some—but not all—sedimentary rocks formed by chemical processes have nonclastic texture, in which the grains are interlocked. These rocks have somewhat the same appearance as igneous rocks with crystalline texture. Actually, most of the sedimentary rocks with nonclastic texture have crystalline structure, although a few of them such as opal do not exhibit this structure.

The mineral crystals that precipitate from an aqueous solution are usually small. Because the fluid in which they form has a very low density, they usually settle out rapidly and accumulate on the bottom as mud. Eventually, under the weight of additional sediments, the mud is compacted more and more. Now the size of the individual crystals may begin to increase. Their growth may be induced by added pressure, which causes the favorably oriented grains to grow at the expense of less favorably oriented neighboring grains. Or crystals may grow as more and more mineral matter is added to them from the saturated solutions trapped in the original mud. In any event, the resulting rock is made up of interlocking crystals and has a texture similar to that of crystalline igneous rocks. Depending on the size of the crystals, we refer to these nonclastic textures as **fine-grained, medium-grained,** or **coarse-grained.** A coarse-grained texture has grains larger than 5 mm in diameter, and a fine-grained texture has grains less than 1 mm in diameter.

LITHIFICATION

The process of lithification converts unconsolidated rock-forming materials into consolidated, coherent rock. The term is derived from the Greek *lithos*, meaning "rock" and the Latin *facere* "to make." In the following subsections, we shall discuss the various ways in which sedimentary deposits are lithified.

Cementation In cementation the spaces between the individual particles of an unconsolidated deposit are filled up by some binding agent. Of the many minerals that serve as cementing agents, the most common are calcite, dolomite, and quartz. Others include iron oxide, opal, chalcedony, anhydrite, pyrite, and especially the clay minerals. Apparently the cementing material is carried in solution by water that percolates through the open spaces among the particles of the deposit. Then some factor in the new environment causes the mineral to be deposited, and the former unconsolidated deposit is cemented into a sedimentary rock.

In coarse-grained deposits there are relatively large interconnecting spaces among the particles. As we should expect, these deposits are very susceptible to cementation because the percolating water can move through them with great ease. Deposits of sand and gravel are transformed by cementation into the sedimentary rocks sandstone and conglomerate.

Compaction and desiccation In a fine-grained clastic deposit of silt-sized and clay-sized particles, the pore spaces are usually so small that water cannot freely circulate through them. Consequently, very little cementing material manages to find its way among the particles; but deposits of this sort are lithified by two other processes: compaction and desiccation.

In **compaction,** the pore space between adjacent grains is gradually reduced by the pressure of overlying sediments or by pressures resulting from earth movement. Coarse deposits of sand and gravel undergo some compaction, but fine-grained deposits of silt and clay respond much more readily. As the individual particles are pressed closer and closer together, the thickness of the deposit is reduced and its coherence is increased. It has been estimated that deposits of clay-sized particles buried to depths of 1,000 m have been compacted to about 60 percent of their original volume.

In **desiccation,** the water that originally filled the pore spaces of water-laid clay and silt deposits is forced out. Sometimes this is the direct result of compaction, but desiccation also takes place when a deposit is simply exposed to the air and the water evaporates.

Crystallization The crystallization of certain chemical deposits is in itself a form of lithification. Crystallization also serves to harden deposits that have been laid down by mechanical processes of sedimentation. For example, new minerals may crystallize within a deposit, or the crystals of existing minerals may increase in size. New minerals are sometimes produced by chemical reactions among amorphous, colloidal materials in fine-grained muds. Exactly how and when these reactions occur is not yet generally understood, but that new crystals have formed after the deposit was initially laid down becomes increasingly apparent as we make more and more detailed studies of sedimentary rocks. Furthermore, it seems clear that this crystallization promotes the process of lithification, particularly in the finer sediments.

Diagenesis After deposition, a sediment or sedimentary rock may undergo a series of physical, chemical, and biologic changes which we call **diagenesis.** These changes may be in the recombination or rearrangement of a mineral, resulting in the formation of a new mineral. Such changes are specifically limited to those which occur after the sediment has been deposited, during and after its lithification, but *before* any significant increases in temperature and pressure associated with metamorphism. This process also is limited to changes occurring prior to the time the sediment or sedimentary rock has been lifted up to come in contact with the atmosphere. Changes which occur under these conditions have already been discussed in the chapter on Weathering. Some examples of diagenesis include dewatering and decrease in porosity associated with compaction of sands, silts, and clay-muds; bacterial decomposition of organic matter; the distortion and destruction of bedding laminations by burrowing organisms; the removal of soluble materials such as calcium carbonate by dissolution in deep sea sediments; the formation of concretions; the formation of a variety of **authigenic minerals** (those **born** or **formed in place**), such as orthoclase feldspar, illite, and other clay materials, and even quartz (particularly quartz overgrowths on quartz sand particles). During diagenesis, carbonate rocks (limestones and dolostones) may undergo (1) **dissolution,** producing voids within the rock, (2) **cementation,** with the growth of new crystals into void space, or (3) **replacement,** involving nearly simultaneous dissolution of existing minerals and the precipitation of new minerals.

5.3 TYPES OF SEDIMENTARY ROCK

CLASSIFICATION

Having examined some of the factors involved in the formation of sedimentary rocks, we are in a better position to consider a classification for this rock family. The

Origin	Texture	Particle size or composition	Rock name
Detrital	Clastic	Granule or larger	Conglomerate (round grains) breccia (angular grains)
		Sand	Sandstone
		Silt	Siltstone
		Clay	Mudstone and shale
Chemical: Inorganic	Clastic or nonclastic	Calcite, $CaCO_3$	Limestone
		Dolomite, $CaMg(CO_3)_2$	Dolomite
		Halite, $NaCl$	Salt
		Gypsum, $CaSO_4 \cdot 2H_2O$	Gypsum
Biochemical	Clastic or nonclastic	$CaCO_3$ shells	Limestone, chalk, coquina
		SiO_2 diatoms	Diatomite
		Plant remains	Coal

classification presented in Table 5.3 represents only one of many possible schemes, but it will serve our purposes very adequately. Notice that there are two main groups—detrital and chemical—based on the origin of the rocks, and that the chemical category is further split into inorganic and biochemical. All the detrital rocks have clastic texture, whereas the chemical rocks have either clastic or nonclastic texture. We use particle size to subdivide the detrital rocks, and composition to subdivide the chemical rocks.

DETRITAL SEDIMENTARY ROCKS

Conglomerate A conglomerate is a detrital rock made up of more or less rounded fragments, an appreciable percentage of which are of granule size (2 to 4 mm in diameter) or larger. If the fragments are more angular than rounded, the rock is called a **breccia**. (Figure 5.6). Another type of conglomerate is **tillite**, a rock formed from deposits laid down directly by glacier ice (see Chapter 14). The large particles in a conglomerate are usually rock fragments, and the finer particles are usually minerals derived from preexisting rocks.

Sandstone A sandstone is formed by the consolidation of grains of sand size between $\frac{1}{16}$ (0.0625) and 2 mm in diameter. Sandstone is thus intermediate between coarse-grained conglomerate and fine-grained mudstone. Because the size of the grains varies from one sandstone to another, we speak of coarse-grained, medium-grained, and fine-grained sandstone.

Very often the grains of a sandstone are almost all quartz. When this is the case, the rock is called a **quartz arenite** (from the Latin for "sand") or **ortho-quartzite** (from the Greek here meaning "true," plus

"quartzite"). The name **quartzose sandstone** is also used. If the minerals are predominantly quartz and feldspar, the sandstone is called an **arkose**, a French word for the rock formed by the consolidation of debris derived from a mechanically weathered granite. Another variety of sandstone, called **graywacke**, or **lithic sandstone**, is characterized by its hardness and dark color and by angular grains of quartz, feldspar, and small fragments of rock set in a matrix of clay-sized particles.

Mudstone, pelite, and shale Fine-grained detrital rocks composed of clay and silt-sized particles (less

5.6 Breccia is a lithified deposit containing many angular fragments. This 12-cm-wide specimen is from eastern Nevada. [Willard Starks, Princeton University Museum of Natural History.]

than 0.06 mm in diameter) are termed **mudstones** or **pelites** or **shales.** Mudstones are fine-grained rocks with a massive or blocky aspect; pelites are the finest sedimentary rocks composed primarily of clay minerals; and shales are fine-grained rocks that split into platy slabs more or less parallel to the bedding. These terms are sometimes used interchangably for any very fine-grained detrital sedimentary rocks. The particles in these rocks are so small that it is difficult to examine them visually. To determine the precise mineral composition, we need to use X-ray diffraction and related techniques. We do know also that they contain not only clay minerals but also clay-sized and silt-sized particles of quartz, feldspar, calcite, and dolomite, to mention but a few.

CHEMICAL SEDIMENTARY ROCKS

Limestone   Limestone is a sedimentary rock that is made up chiefly of the mineral calcite that has been deposited by either inorganic or organic chemical processes. Many limestones have a clastic texture, but nonclastic, particularly crystalline, textures are common.

Inorganically formed limestone is made up of calcite that has been precipitated from solution by inorganic processes. Some calcite is precipitated from the fresh water of streams, springs, and caves, although the total amount of rock formed in this way is negligible. When calcium-bearing rocks undergo chemical weathering, calcium bicarbonate, $Ca(HCO_3)_2$, is produced in solution. If enough of the water evaporates, or if the temperature rises, or if the pressure falls, calcite is precipitated from this solution. For example, most **dripstone,** or **travertine,** is formed in caves by the evaporation of water that is carrying calcium bicarbonate in solution. And **tufa** (from the Italian for "soft rock") is a spongy, porous limestone formed by the precipitation of calcite from the water of streams and springs (see Figure 5.7).

Although geologists understand the inorganic processes by which limestone is formed by precipitation from fresh water, they are not quite sure how important these processes are in precipitation from seawater. Some observers have questioned whether they operate at all. On the floors of modern oceans and in rocks formed in ancient oceans, however, we find small spheroidal grains called **oölites,** the size of sand and often composed of calcite; these grains are thought to be formed by the inorganic precipitation of

calcium carbonate from seawater. (The term comes from the Greek for "egg" because an accumulation of oölites resembles a cluster of fish roe.) Cross sections show that many oölites, though not all, have grown up around a mineral grain or around a small fragment of shell that acts as a nucleus. Some limestones are made up largely of oölites. One, widely used for building, is the so-called Indiana or Spergen Limestone.

Dolomite   In discussing the mineral dolomite, $CaMg(CO_3)_2$, we mentioned that when it occurs in large concentrations, it forms a rock that is called **dolostone,** or **dolomite.** Extensive deposits of dolomite, the origin of which is not fully understood, appear to have been formed by replacement of preexisting deposits of calcite. There is now increasing agreement that most dolomite is in some way related to the local increase of the amount of magnesium in solution. Previously deposited calcite is modified by the movement through it of these magnesium-rich solutions. Field and laboratory observations show that in shallow-water intertidal zones, evaporation of seawater may cause precipitation of calcium-bearing deposits. The waters may be increased by an order of magnitude in their content of magnesium relative to calcite. Such high-magnesium-content waters may then circulate through underlying calcite deposits, replacing some of the calcium with magnesium and thus converting limestone to dolomite. In supratidal zones, just above high tide, thin crusts of primary dolomite crystals form in tropical and sub-tropical regions today. Such dolomites have been found in the Bahamas and in the Florida Keys.

Evaporites   An evaporite is a sedimentary rock composed of minerals that were precipitated from solution with the evaporation of the liquid in which they were dissolved. Rock salt (composed of the mineral halite) and gypsum are the most abundant evaporites. Anhydrite (from the Greek *anhydros*, "waterless") is an evaporite composed of the mineral of the same name, which is simply gypsum without its crystallization water, $CaSO_4$. Most evaporite deposits seem to have been precipitated from seawater according to a definite sequence. The less highly soluble minerals are the first to drop out of solution. Thus gypsum and anhydrite, both less soluble than halite, are deposited first. Then, as evaporation progresses, the more soluble halite is precipitated.

In the United States, the most extensive deposits of evaporites are found in Texas, Louisiana, and New Mexico. Here gypsum, anhydrite, and rock salt make up over 90 percent of the Castile Formation, which has a maximum thickness of nearly 1,200 m. In central New York State, there are thick deposits of rock salt;

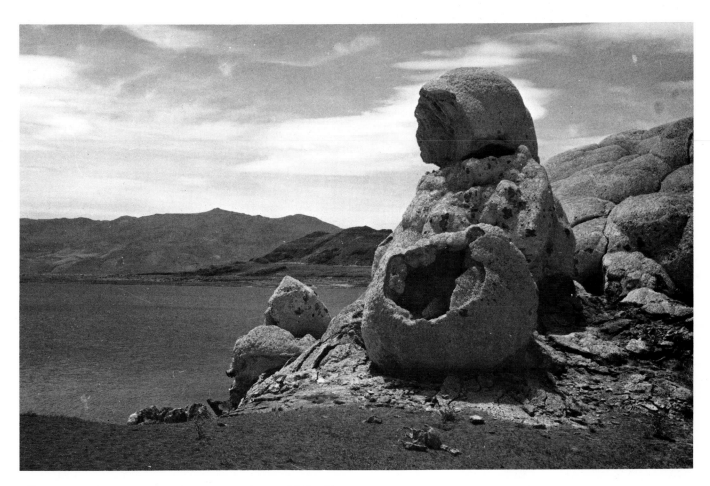

5.7 "Squaw and Basket." Tufa formations at Pyramid Lake, Nevada, the major remnant of the Late Pleistocene Lake Lahontan, which once covered a large part of the western Great Basin. [Adrian Atwater.]

and in central Michigan, there are layers of rock salt and gypsum. Some evaporite deposits are mined for their mineral content, and in certain areas, particularly in the Gulf Coast states, deposits of rock salt have pushed upward toward the surface to form salt domes containing commercially important reservoirs of petroleum (see Chapter 17).

ORGANIC ROCKS

A significant quantity of sedimentary rocks is the direct result of the accumulation of shells, plant fragments, and other remains of organisms. Together these are called **organic**, or **biochemical**, rocks. For example, diatoms are microscopic, single-celled plants that grow in marine or fresh water and secrete siliceous shells. When these shells accumulate in great numbers on the bottom of a basin, they may ultimately form a sedimentary rock called **diatomite**.

Biochemically formed limestones are created by the action of plants and animals that extract calcium carbonate from the water in which they live. The calcium carbonate may be either incorporated into the skeleton of the organism or precipitated directly. In any event, when the organism dies, it leaves behind a quantity of calcium carbonate, and over a long period of time, thick deposits of this material may be built up. Reefs, ancient and modern, are well-known examples of such accumulations. The most important builders of modern reefs are algae, mollusks, corals, and one-celled animals—the same animals whose ancestors built up the reefs of ancient seas—the reefs, now old and deeply buried, that are often valuable reservoirs of petroleum.

Chalk is made up in part of biochemically derived calcite in the form of the skeletons or skeletal fragments of microscopic oceanic plants and animals. These organic remains are found mixed with very fine-grained calcite deposits of either biochemical or inorganic chemical origin. A much coarser type of limestone composed of organic remains is known as **coquina** (from the Spanish for "shellfish" or "cockle") and is characterized by the accumulation of many large fragments of shells.

Coal   Coal is a rock composed of combustible matter derived from the partial decomposition of plants. We shall consider coal as a biochemically formed sedimentary rock, although some geologists prefer to think of it as a metamorphic rock because it passes through various stages.

The process of coal formation begins with an accumulation of plant remains in a swamp. This accumulation is known as **peat,** a soft, spongy, brownish deposit in which plant structures are easily recognizable. Time, coupled with the pressure produced by deep burial and sometimes by earth movement, gradually transforms the organic matter into coal. During this process, the percentage of carbon increases as the volatile hydrocarbons and water are forced out of the deposit. Coals are ranked according to the percentage of carbon they contain. Peat, with the least amount of carbon, is the lowest ranking; then come lignite, or brown coal; bituminous, or soft coal; and finally anthracite, or hard coal, the highest of all the coals in its percentage of carbon (see Section 17.2).

RELATIVE ABUNDANCE OF SEDIMENTARY ROCKS

Sandstone, mudstone and shale, and limestone constitute about 99 percent of all sedimentary rocks. Of these, mudstone and shale are the most abundant. On the basis of extrapolation from measurements made in the field, the estimates of the percentages of mudstone and shale approximate 50 percent of all sedimentary rocks. Similar calculations for limestone and sandstone suggest that the limestone forms about 22 percent of these rock types and that sandstone accounts for the remaining 28 percent. These percentages, however, do not agree with theoretical determinations of relative abundances. These are based on the determination of the products to be expected from the weathering of an average igneous rock. If these weathering products are assigned to the three major sedimentary rock types, then we find that shale should be considerably more important volumetrically than it appears to be on the basis of field measurements. On these theoretical grounds, mudstone and shale constitute approximately 75 percent of the three major sedimentary rock

types. Sandstone and limestone are approximately of equivalent volume and together constitute the other 25 percent. The discrepancy between the two estimates has not yet been resolved.

Estimates have been made of the average chemical composition of the world's sediments. One of these estimates is presented in Table 5.4.

TABLE 5.4

Average composition of all sediments (as oxides)[a]

Oxide	Wt. %
SiO_2	44.5
TiO_2	0.6
Al_2O_3	10.9
Fe_2O_3	4.0
FeO	0.9
MnO	0.3
MgO	2.6
CaO	19.7
Na_2O	1.1
K_2O	1.9
P_2O_5	0.1
CO_2	13.4
Total	100.0

[a]From Arie Poldervaart, "Chemistry of the Earth's Crust," *Geol. Soc. Am. Spec. Paper*, no. 62, p. 132, 1955.

5.4 FEATURES OF SEDIMENTARY ROCKS

We have mentioned that the stratification, or bedding, of sedimentary rocks is their single most characteristic feature. Now we shall look more closely at this feature, along with certain other characteristics of sedimentary rocks, including mud cracks and ripple marks, nodules, concretions, geodes, fossils, and color.

BEDDING

The beds, or layers, of sedimentary rocks are separated by **bedding planes,** along which the rocks tend to separate or break (see Figure 5.8). The varying thickness of the layers in a given sedimentary rock reflects the changing conditions that prevailed when each deposit was laid down. In general, each bedding plane marks the termination of one deposit and the beginning of another. For example, let us imagine a bay of an

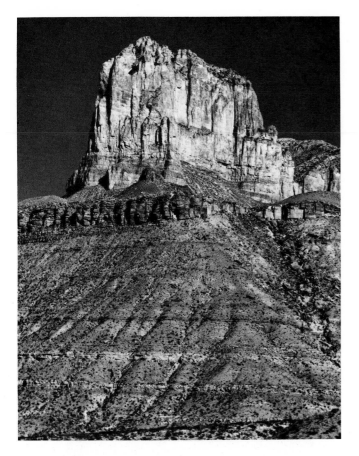

5.8 El Capitan Peak, Guadalupe Mountains National Park, showing nearly horizontal stratification, with thinly bedded shales and limestones forming gentler slopes and massively bedded limestones forming the cliffs. [U.S. National Park Service photo.]

ocean into which rivers normally carry fine silt from the nearby land. This silt settles out from the seawater to form a layer of mud. Now heavy rains or melting snows may cause the rivers suddenly to flood and thereby pick up coarser material, such as sand, from the river bed. This material will be carried along and dumped into the bay. There it settles to the bottom and blankets the silt that was deposited earlier. The plane of contact between the mud and the sand represents a bedding plane. If, later on, the silt and the sand are lithified into shale and sandstone, the bedding plane persists in the sedimentary rock. In fact, it marks a plane of weakness along which the rock tends to break.

Bedding planes are usually horizontal, and, when parallel, the bedding is called **parallel bedding.** Closely spaced parallel bedding planes form **laminated bedding.** But some beds are laid down at an angle to the horizontal, and such bedding is variously called **cross bedding** or **inclined bedding** (see Figure 5.9). Such nonhorizontal bedding can occur in several situations. For example, the bedding in sand

dunes may have high angles on the leeward side of the dune (see Chapter 16). The deposits laid down at the growing edge of a delta may be inclined from 5° to 30°, and such beds are usually given a special name, **foreset beds** (see Chapter 12). In flowing water, scouring of the stream floor by turbulent water may create small depressions in the channel deposits, which later will be refilled by inclined beds. Alternatively, cross bedding can occur as flowing water passes to somewhat quieter conditions and drops its load in a way similar to that in which a stream builds a foreset bed in a delta.

When the particles in a sedimentary bed vary from coarse at the bottom to fine at the top, the bedding is said to be **graded.** Such bedding is characteristic of rapid deposition from a water mass turbid with sedimentary particles of differing sizes. The largest particles settle most rapidly, and the finest most slowly. Some graded beds form **turbidites,** a term from the Latin *turbidus* for "disturbed," referring to the stirring up of sediments and water (see also Chapter 15).

Bedding may be distorted before the material becomes consolidated into firm rock. Thus unconsolidated sediments may sometimes slump or flow. In addition, animals may burrow or tunnel through unconsolidated layers to produce **disturbed,** or **mottled, bedding,** or in extreme cases destroy all vestiges of the original bedding.

RIPPLE MARKS, MUD CRACKS, AND SOLE MARKS

Ripple marks are the little waves of sand that commonly develop on the surface of a sand dune, along a beach, or on the bottom of a stream. **Mud cracks** are familiar on the dried surface of mud left exposed by the subsiding waters of a river or pond. **Sole marks** are the casts, or filling, of a primary sedimentary structure such as a groove or track found on the underside or lower surface of a siltstone or sandstone that had overlain such a structure in soft underlying sediment.

Ripple marks preserved in sedimentary rocks often furnish clues to the conditions that prevailed when a sediment was originally deposited. For instance, if the ripple marks are symmetric, with sharp or slightly rounded ridges separated by more gently rounded troughs, we are fairly safe in assuming that they were formed by the back-and-forth movement of water such as we find along a sea coast outside the surf zone. These marks are called ripple marks of **oscillation** (see

5.9 Cross bedding in Navajo Sandstone, Zion National Park, Utah. Middle inclined beds are approximately 5 m thick. These cross-bedded units represent eolian deposits (consolidated sand dunes). [Logan Campbell.]

Figure 5.10). If, on the other hand, the ripple marks are asymmetric, we can assume that they were formed by air or water moving more or less continuously in one direction. These marks are called **current** ripple marks.

Mud cracks make their appearance when a deposit of silt or clay dries out and shrinks; see Figures 5.11(a) and (b). The cracks outline roughly polygonal areas, making the surface of the deposit look like a section cut through a large honeycomb. Eventually, another deposit may come along to bury the first. If the deposits are later lithified, the outlines of the cracks may be accurately preserved for millions of years. Then, when the rock is split along the bedding plane between the two deposits, the cracks will be found much as they appeared when they were first formed, providing evidence that the original deposit underwent alternate flooding and drying.

Sole marks develop when a groove or trail is formed on the surface of soft sediment by a stick or other object carried along in the current. Under ideal conditions this depression will be preserved and filled by silt or sand. When lithified, this sequence of sedimentary beds can be used to decipher current directions, tops and bottoms of strata, and conditions of the environment.

5.10 Ripple marks of oscillation on a 35-cm-long slab of sandstone. The symmetrical nature of the ripples indicates that the current moved back and forth rather than continuously in one direction. [Willard Starks.]

NODULES, CONCRETIONS, AND GEODES

Many sedimentary rocks contain structures that were formed only *after* the original sediment was deposited. Among these are nodules, concretions, and geodes.

A nodule is an irregular, knobby-surfaced body of mineral matter that differs in composition from the sedimentary rock in which it has formed. It usually lies parallel to the bedding planes of the enclosing rock, and sometimes adjoining nodules coalesce to form a continuous bed. Nodules average about 30 cm in maximum dimension. Silica in the form of chert or flint is the major component of these bodies. They are most commonly found in limestone or dolomite. Most nodules are thought to have formed when silica replaced

5.11(a) Polygonal pattern of mud cracks resulting from dessication of modern fine-grained sediments in a playa lake in Nevada. Scale shown by pocket knife. [William C. Bradley.]

5.11(b) Mud cracks preserved in limy shale at Morgantown, Pennsylvania, in the Cambrian Conococheague Formation. [M. E. Kauffman.]

some of the materials of the original deposit; some, however, may consist of silica that was deposited at the same time as the main beds were laid down.

A **concretion** is a local concentration of the cementing material that has lithified a deposit into a sedimentary rock. Concretions range in size from a fraction of a centimeter to a meter or more in maximum dimen-

sion. Most are shaped like simple spheres or disks, although some have fantastic and complex forms. For some reason, when the cementing material entered the unconsolidated sediment, it tended to concentrate around a common center point or along a common center line. The particles of the resulting concretion are cemented together more firmly than the particles of the host rock that surrounds it. The cementing material usually consists of calcite, dolomite, iron oxide, or silica—in other words, the same cementing materials that we find in the sedimentary rocks themselves.

Geodes, more eye-catching than either concretions or nodules, are roughly spherical hollow structures up to 30 cm or more in diameter (see Figure 5.12). An outer layer of chalcedony is lined with crystals that project inward toward the hollow center. The crystals, often perfectly formed, are usually quartz, although crystals of calcite and dolomite have been found and, more rarely, crystals of other minerals. Geodes occur most commonly in limestone, but they also occur in shale.

How does a geode form? First, a water-filled pocket develops in a sedimentary deposit, probably as a result of the decay of some plant or animal that was buried in the sediments. As the deposit begins to consolidate into a sedimentary rock, a wall of silica with a jellylike consistency forms around the water, isolating it from the surrounding material. As time passes, fresh water may enter the sediments. The water inside the pocket has a higher salt concentration than does the water outside. To equalize the concentrations, there is

5.12 Two geodes broken open to show their internal structure. The dark outer layer of chalcedony is lined with milky-to-clear quartz crystals that project inward toward a hollow center. These structures are most commonly found in limestone, where they apparently form by the modification and enlargement of an original void. The smaller geode is about 8 cm in diameter. [Specimen from Harvard University Geological Museum, Walter R. Fleischer.]

5.4 FEATURES OF SEDIMENTARY ROCKS

a slow mixing of the two liquids through the silica wall or membrane that separates them. This process of mixing is called **osmosis.** As long as the osmotic action continues, pressure is exerted outward toward the surrounding rock. The original pocket expands bit by bit until the salt concentrations of the liquids inside and outside are equalized. At this point, osmosis stops, the outward pressure ceases, and the pocket stops growing. Now the silica wall dries, crystallizes to form chalcedony, contracts, and cracks. If, at some later time, mineral-bearing water finds its way into the deposit, it may seep in through the cracks in the wall of chalcedony. There the minerals are precipitated, and crystals begin to grow inward, toward the center, from the interior walls. Finally, we have a crystal-lined geode imbedded in the surrounding rock. Notice that the crystals in a geode grow inward; in a concretion they grow outward.

Other geodes are known to have occurred in a very different way. Replacement of evaporite (anhydrite) nodules by silica produces silicified geodes. The evaporite nodules can occur under conditions of extreme imbalance between precipitation and evaporation. After they have formed, silica-rich fluids may come in contact with them and result in the silicification on a molecule-by-molecule replacement basis.

FOSSILS

The word **fossil** (derived from the Latin *fodere*, "to dig up") originally referred to anything that was dug from the ground, particularly a mineral or some inexplicable form. It is still used in that sense occasionally, as in the term **fossil fuel** (see Chapter 17). But today, fossil generally means any direct evidence of past life—for example, the bones of a dinosaur, the shell of an ancient clam, the footprints of a long-extinct animal, or the delicate impression of a leaf (see Figures 5.13 and 5.14).

Fossils are usually found in sedimentary rocks, although they sometimes turn up in igneous and metamorphic rocks. They are most abundant in mudstone, shale, and limestone but are also found in sandstone, dolomite, and conglomerate. Fossils account for almost the entire volume of certain rocks, such as the coquina and limestones that have been formed from ancient reefs.

The remains of plants and animals are completely destroyed if they are left exposed on the earth's surface; but if they are somehow protected from destructive forces, they may become incorporated in a sedimentary deposit, where they will be preserved for millions of years. In the quiet water of the ocean, for example, the remains of starfish, snails, and fish may be buried by sediments as they settle slowly to the bottom. If these sediments are subsequently lithified, the remains are preserved as fossils that tell us about the sort of life that existed at the time when the sediments were laid down.

Fossils are also preserved in deposits that have settled out of fresh water. Countless remains of land animals, large and small, have been dug from the beds of such once-watery environments as extinct lakes, flood plains, and swamps.

The detailed story of the development of life as recorded by fossils is properly a part of historical geology, and although we do not have time to trace it here, in Chapter 7 we note that fossils are extremely useful in subdividing geologic time and constructing the geologic column.

COLOR OF SEDIMENTARY ROCKS

Throughout the western and southwestern areas of the United States, bare cliffs and steep-walled canyons provide a brilliant display of the great variety of colors exhibited by sedimentary rocks. The Grand Canyon of the Colorado River in Arizona cuts through rocks that vary in color from gray, through purple and red, to brown, buff, and green. Bryce Canyon in southern Utah is fashioned of rocks tinted a delicate pink, and the Painted Desert, farther south in Arizona, exhibits a wide range of colors, including red, gray, purple, and pink.

The most important sources of color in sedimentary rocks are the iron oxides. Hematite, Fe_2O_3, for example, gives rocks a red or pink color, and limonite or goethite produces tones of yellow and brown. Some of the green, purple, and black colors may be caused by iron but in exactly what form is not completely understood. Only a very small amount of iron oxide is needed to color a rock. In fact, few sedimentary rocks contain more than 6 percent of iron, and most contain very much less.

Organic matter, when present, may also contribute to the coloring of sedimentary rocks, usually making them gray to black. Generally, but not always, the higher the organic content, the darker the rock.

The size of the individual particles in a rock also influences the color or, at least, the intensity of the color. For example, fine-grained clastic rocks are usually somewhat darker than coarse-grained rocks of the same mineral composition.

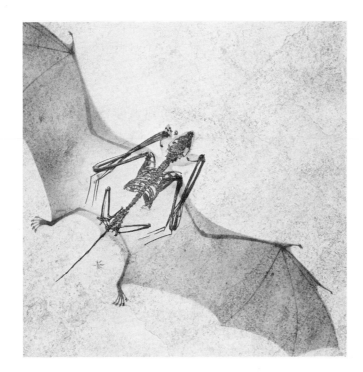

5.13 This fossil bat, *Icaronycteris index,* oldest known flying mammal, was found in the Green River Formation of the early Eocene age in southwestern Wyoming. It measures 12 cm in length, and its wing span (here restored) is 30 cm. Detailed study suggests that the bat was a young male that died on a summer evening while foraging for insects or small fish close to the surface of the lake in which the Green River Formation was laid down. Just to the right of the tail, a small flower has been preserved. [Willard Starks, Princeton University Museum of Natural History.]

5.14 Casts of burrows made by worms that lived in Devonian muds are shown on the underside of this slab from a locality near Ithaca, New York. [Sheldon Judson.]

SEDIMENTARY FACIES

If we examine the environments of deposition that exist at any one time over a wide area, we find that they differ from place to place. Thus the freshwater environment of a river changes to a brackish-water environ-

ment as the river nears the ocean. In the ocean itself, marine conditions prevail. But even here the marine environment changes—from shallow water to deep water, for example. And as the environment changes, the nature of the sediments that are laid down also changes. The deposits in one environment show characteristics that are different from the characteristics of deposits laid down at the same time in another environment. This change in the "look" of the sediments is called a change in **sedimentary facies,** the latter word derived from the Latin word for the notion of "aspect" or "form."

We may define sedimentary facies as an accumulation of deposits that exhibits specific characteristics, reflecting a particular depositional environment, and grades laterally into other sedimentary accumulations formed at the same time but exhibiting different characteristics. The concept of facies is widely used in studying sedimentary rocks and the conditions that gave rise to metamorphic rocks (see Chapter 6). The concept is generally not used in referring to igneous rocks.

Let us consider a specific example of facies. Figure 5.15 shows a coastline where rivers from the land empty into a lagoon. The lagoon is separated from the open ocean by a sandbar. The fine silts and clays dumped into the quiet waters of the lagoon settle to the bottom as a layer of mud. At the same time, waves are eroding coarse sand from a nearby headland outside the lagoon. This sand is transported by currents and waves and deposited as a sandy layer seaward of the sandbar. Different environments exist inside and outside the lagoon; therefore different deposits are being laid down simultaneously. Notice that the mud and the sand grade into each other along the sandbar. Now imagine that these deposits have been consolidated into rock and then exposed to view at the earth's surface. We should find a shale layer grading into sandstone; that is, one sedimentary facies is grading into another.

But the picture is not always so simple. Recent sediments range from sand, through a sandy mud, to mud, and in a few areas there are limy deposits. Where the sea floor is rocky, little or no recent sedimentation has taken place. Should the soft sediments become lithified, a sedimentary rock ranging from sandstone, through sandy shale, to shale and limestone would result.

Ancient sedimentary deposits show exactly this kind of variation in facies. Figure 5.16 pictures the actual pattern of rock types that have been identified in the middle portion of the Maquoketa Formation in Illinois. Notice that the types include limestone, shale, limy shale, and a limy shale containing some sand.

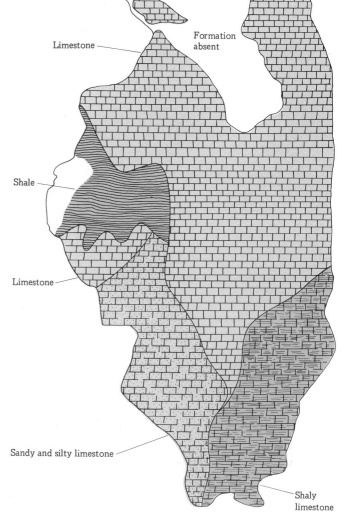

5.16 Variation in facies of ancient sediments comprising the Maquoketa Formation in Illinois. [After E. P. Du-Bois, "Subsurface Relations of the Maquoketa and 'Trenton' Formations in Illinois," *Ill. State Geol. Surv. Rep. Invest.*, no. 105, pp. 7–33, 1945.]

5.15 Diagram to illustrate a change in sedimentary facies. Here the fine-grained muds are deposited in a lagoon close to shore. A sandbar separates them from sand deposits farther away from shore. The sand in this instance has been derived from a sea cliff and transported by waves and currents.

5.5 PLATE TECTONICS AND SEDIMENTATION

As noted in Chapter 1, we are just beginning to understand the crustal motions of the colossal portions of the lithosphere known as **plates**. Areas exist on the earth's crust where sediments accumulate to great thicknesses. Many of these regions are ultimately pushed up into mountains, developing most commonly along the margins of the plates.

Thick accumulations of sediments may occur either where there is a deep hole into which the sediments are dumped or where the crust subsides as more and more sediments are deposited in relatively shallow water. We have evidence of both in the geologic record, but the overwhelming bulk of sedimentary rocks found in mountain ranges apparently formed originally as shallow-water sediments, as proved by indicators such as certain kinds of ripple marks, mud cracks, and shallow-water organisms.

GEOSYNCLINES ✓

Very large syncline

Following the work of James Hall, a nineteenth-century pioneer in American geology, Marshall Kay, and other twentieth-century proponents of the idea, the concept of geosynclines has developed into a dominant theme in the interpretation of many mountain belts. Folded rocks in mountain ranges of the world have apparently developed from thick deposits of sediments that have accumulated in these geosynclines—literally, "earth syncline," or large-scale downwarp in the sedimentary rocks found in the earth's crust. A geosyncline is thus a large sediment-filled elongate basin. We shall examine the problem of deformation of these sedimentary rocks and the actual formation of mountains in Chapter 8.

Divergent Boundary

Parallel adjacent belts characterize geosynclines. These have been called **miogeosyncline** (or simply **miogeocline** by Robert Dietz and others), where sediments are deposited on continental basement and where volcanic rocks and deeper-water sediments are lacking; and **eugeosyncline** (or **eugeocline**), where sediments are deposited on noncontinental basement

Convergent Boundary

and where there is an abundance of interstratified volcanic rocks and deeper water sediments.

PLATE BOUNDARIES AND SEDIMENTATION

As discussed in Chapter 1, there are three basic types of plate boundaries: **convergent**, where old lithosphere is carried downward along subduction zones and magma rises to form volcanic regions, as the west coast of South America or the Aleutians; **divergent**, where old lithosphere pulls apart and new lithosphere forms along midoceanic spreading centers, as the Atlantic ridge; and **transform**, or **parallel**, where two plates slide past one another with no new lithosphere forming and no old lithosphere being destroyed, as along the San Andreas fault.

In relating sedimentation and geosynclinal concepts to those of plate-tectonic theory, we can interpret the miogeosynclines as representing thick sediment wedges forming on divergent continental margins and the eugeosynclines as representing sediments deposited farther offshore or in areas marked by volcanic island arcs, perhaps at a convergent plate boundary. Transform margins are marked by numerous basins that are later filled with thick sequences of sediments similar to those found along the San Andreas fault zone.

The relative motion of the plates on which or between which the sediments have accumulated will determine the ultimate structure of the mountain systems that later form from these sediments and sedimentary rocks (see Chapter 8).

OUTLINE

Sedimentary rocks cover about 75 percent of the earth's surface and make up about 5 percent by volume of the outer 10 km of the solid earth.

Formation of sedimentary rocks takes place at or near the earth's surface.

Detrital material worn from the landmasses and **chemical** deposits precipitated from solution are the two chief types of sediments.

Sedimentation is the process by which rock-forming materials are laid down, and the resulting deposits vary with the source of material, the methods of transportation, the processes of deposition, and the environment of deposition.

Clay, quartz, and **calcite** are the most common minerals in sedimentary rocks. Other minerals include dolomite, goethite, hematite, limonite, feldspar, mica, halite, and gypsum.

Texture depends on the size, shape, and arrangement of the particles. Texture may be clastic or nonclastic.

Lithification converts unconsolidated sediments to firm rock by cementation, compaction, desiccation, crystallization, and diagenesis.

Types of sedimentary rocks include detrital, chemical, and organic forms.

Detrital rocks include conglomerate, sandstone, mudstone, and shale.

Chemical rocks include limestone, dolomite, and evaporites.

Organic rocks include chalk, coquina, diatomite, and coal.

Most abundant are shale and mudstone, sandstone, and limestone, in that order. They form 99 percent of the sedimentary rock family.

Features of sedimentary rocks include bedding, mud cracks, nodules, concretions, geodes, and fossils.

Color of sedimentary rocks is due largely to small amounts of the iron oxide minerals and less importantly to organic matter.

Sedimentary facies refers to an accumulation of deposits that exhibits specific characteristics and that grades laterally into other accumulations formed at the same time but showing different characteristics.

Sedimentation is closely related to **plate boundaries,** with thick accumulations along some margins.

Geosynclines are large-scale downwarps in the earth's crust in which large accumulations of sediments occur.

Miogeoclines are thick accumulations of strata on margins of continental blocks.

Eugeoclines represent rocks developed farther offshore or in areas marked by volcanic island arcs.

SUPPLEMENTARY READINGS

Blatt, H., G. Middleton, and R. Murray *Origin of Sedimentary Rocks*, 2nd ed., Prentice-Hall, Inc., Englewood Cliffs, N. J., 1980. *A modern text that emphasizes physical and chemical mechanisms and processes of sedimentation; good treatment of facies and environments of deposition.*

Greensmith, J. T. *Petrology of the Sedimentary Rocks*, George Allen & Unwin, Ltd., London, 1980. *A modern treatment of sedimentary rocks in hand samples, in thin-sections, and as representatives of specific origins and environments of deposition.*

LaPorte, Leo F. *Ancient Environments*, Prentice-Hall, Inc., Englewood Cliffs, N. J., 1968. *An excellent treatment of the way sedimentary materials can be used to determine environments.*

Pettijohn, F. J. *Sedimentary Rocks*, 3rd ed., Harper & Row, Publishers, New York, 1975. *An excellent standard text on the whole field of sedimentary rocks.*

Reading, H. G. (ed.) *Sedimentary Environments and Facies*, Elsevier North-Holland, Inc., New York, 1978. *A single comprehensive coverage of modern and ancient environments. It shows how ancient environments may be reconstructed by interpreting modern processes that give rise to certain facies.*

Reineck, H. E., and I. B. Singh *Depositional Sedimentary Environments*, 2nd ed., Springer-Verlag New York, Inc., New York, 1980. *Profusely illustrated textbook on modern environmental studies, with reference to examples of ancient environments as found in the stratigraphic record.*

QUESTIONS

1 Stratification is the single most characteristic feature of sedimentary rocks. How does it form?

2 How do detrital sedimentary rocks differ from chemical rocks?

3 What are the most common mineral components of the majority of all sedimentary rocks?

4 How does iron affect the color of the resulting sedimentary rock?

5 Why do some very ancient sediments never undergo lithification, whereas some very recent sediments are lithified almost overnight?

6   Where are limestones forming today?

7   Where are many evaporites forming today?

8   Why do we consider coal to be a sedimentary rock? What is the principal component of that rock?

9   When we examine the sedimentary rocks at the earth's surface, we find three principal kinds. What are these three rocks and why do you suppose they are so abundant?

10   Under what conditions will sediments accumulate in a condition other than horizontal?

11   What is meant by the term sedimentary facies? Give one example, other than those given in the text, where we might get several sedimentary facies adjacent to one another.

12   How is it possible to get great thicknesses of sediments, all of which bear evidence of being accumulated in very shallow water? Explain.

6

METAMORPHISM AND METAMORPHIC ROCKS

Many rocks exposed at the earth's surface today show evidence that they have undergone changes since they were first deposited as sedimentary rocks or were crystallized as igneous rocks. At first glance, some of these rocks resemble crystalline igneous rocks, but then we discover that their mineral grains are arranged in a peculiar planar manner (see Figure 6.1). Other rocks may be highly contorted. Some may have the same composition as limestone, but they seem to have developed much larger mineral grains. Still others are strikingly different from both igneous and sedimentary rocks. All these are **metamorphic** ("changed-form") rocks.

6.1 METAMORPHISM

Some sedimentary and igneous rocks have been changed while in the solid state, in response to pronounced changes in their environment. These changes may bring about modifications *within* the rocks themselves through the process called **metamorphism.** Metamorphism occurs within the earth's crust, below the zone of weathering and cementation and outside the zone of remelting. In this environment, rocks undergo chemical and structural changes to adjust to conditions different from those under which they were originally formed.

Since metamorphism normally takes place in the crust, it is not readily available for direct study as are

6.1 Metamorphic rock, showing alignment of previously unoriented minerals. The light-colored bands are mainly orthoclase and quartz; the dark streaks are biotite and other ferromagnesian minerals. The bulk composition is that of granite, but in contrast to the random mixing in granite, the minerals here are distributed in relatively systematic patterns. [Robert Navias.]

weathering, sedimentation, and igneous activity. Nevertheless, laboratory simulation of temperature and pressure conditions permits us to examine changes that occur during the processes of metamorphism.

AGENTS OF METAMORPHISM

Metamorphism is limited to changes that take place in the texture or composition of solid rocks because after their melting point has been reached, a magma is formed and we are in the realm of igneous activity. (Rock is also considered to be solid when it is in the plastic state.) The agents of metamorphism are heat, pressure, and chemically active fluids.

Heat Heat may be the most essential agent of metamorphism. Some geologists question whether pressure alone could produce changes in rocks without a simultaneous increase in temperature. In fact, they say that metamorphism appears as if it were invariably controlled by temperature.

Pressure Under the influence of pressure, changes take place to reduce the space occupied by the mineral components of a rock mass. Pressure may produce a closer atomic packing of the elements in a mineral, recrystallization of the mineral, or formation of new minerals.

When rocks are buried to depths of several kilometers, they gradually become plastic and responsive to the heat and deforming forces that are active in the

earth's crust and mantle (see Figure 6.2). When plastic, they deform by intergranular motion, by the formation of minute shear planes within the rock, by changes in texture, by reorientation of grains, and by crystal growth. Francis J. Turner[1] says that depth of burial combined with a thermal gradient of from 10 to 15°C/km explains "burial metamorphism." There is some question, however, whether burial alone is sufficient to produce metamorphism. The type of the original rock, again, has an important effect on the results achieved by burial and deformation.

Chemically active fluids Hydrothermal solutions released in the solidification of magma often percolate beyond the margins of the magma reservoir and react on surrounding rocks. Sometimes these solutions remove ions and substitute others. Or they may add ions to the rock minerals to produce new minerals. When chemical reactions within the rock or the introduction of ions from an external source causes one mineral to grow or change into another of different composition, the process is called **metasomatism**. The term de-

[1]*Metamorphic Petrology*, McGraw-Hill Book Company, New York, 1968.

6.2 Metamorphosed conglomerate with stretched pebbles, north shore of Lake Superior. [M. E. Kauffman.]

6.1 METAMORPHISM

113

scribes all ionic transfers, not just those that involve gases or solutions from a magma. Fine-grained rocks are more readily changed than are others because they expose greater areas of grain surface to chemically active fluids.

Some of the chemically active fluid of metamorphism is the liquid already present in the pores of a rock. It is believed that such pore liquid may often act as a catalyst; that is, it expedites changes without itself undergoing change. Sometimes the changes of metamorphism consist essentially of progressive dehydration.

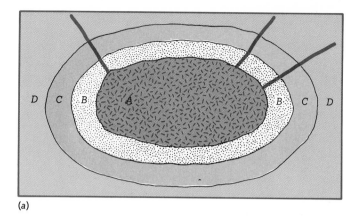

(a)

6.2 TYPES OF METAMORPHISM

Several types of metamorphism occur, but we shall concern ourselves here with the two basic ones: contact metamorphism and regional metamorphism. A third type, **cataclastic metamorphism,** results from the grinding related to faulting and extreme folding, where minerals and rocks are broken and even pulverized, producing coarse-grained **fault breccias,** or finer-grained varieties called **mylonites.** These rocks show the intensity of the pressures and temperatures associated with faulting and folding.

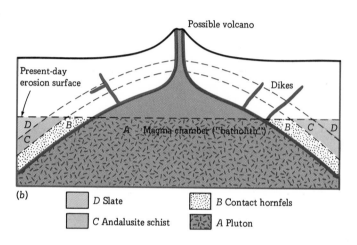

(b)

	D Slate		B Contact hornfels
	C Andalusite schist		A Pluton

6.3 (a) Geologic map, showing contact metamorphic aureoles around pluton. (b) Inferred cross section at time of intrusion of pluton. [Modified from various sources.]

CONTACT METAMORPHISM

When magma is intruded into the earth's crust, it alters the surrounding rock. The alteration of rocks at or near their contact with a body of magma is called <u>contact metamorphism.</u> Minerals formed by this process are called **contact metamorphic minerals.** The type of reaction depends on the temperature, the composition of the intruding mass, and the properties of the intruded rock. At the actual surface of contact, all the elements of a rock may be changed or replaced by other elements introduced by the hydrothermal solutions escaping from the magma. Farther away, the replacement may be only partial.

Contact metamorphism occurs in zones called <u>aureoles</u> ("halos"), which seldom measure more than a few hundred meters in width and may be only a fraction of a centimeter wide. Aureoles are found bordering plutons (see Figure 6.3): sills, dikes, laccoliths, stocks, lopoliths, and batholiths. During contact meta-

morphism, temperatures may range from 300 to 800°C, and load pressures may range from 100 to 3,000 atm (atmospheres) or bars.[2]

Two kinds of contact metamorphic mineral are recognized: those produced by heating up the intruded rock and those produced by the hydrothermal solutions reacting with the intruded rock. Hydrothermal metamorphism develops at relatively shallow depths. It is only late in the cooling of a body of magma, when it nears the surface, that large quantities of hydrothermal solutions are released.

Contact metamorphic minerals Many new minerals are formed during contact metamorphism. For example, when an impure limestone is subjected to thermal contact metamorphism, its dolomite, clay, or quartz may be changed to new minerals. Calcite and quartz may combine to form wollastonite plus carbon dioxide:

$$CaCO_3 + SiO_2 \rightarrow CaSiO_3 + CO_2$$

[2]One atmosphere (or approximately one bar) is the "unit of pressure equal to the pressure exerted per unit area at sea level by the column of air from sea level to the top of the earth's atmosphere."

(handwritten margin notes: "Small Area", "metamorphic halo")

Dolomite may react with the quartz to form diopside plus carbon dioxide:

$$CaMg(CO_3)_2 + 2SiO_2 \rightarrow CaMg(Si_2O_6) + 2CO_2$$

Aluminum, in the clay, will react to form corundum, spinel, or garnet. If carbonaceous materials are present, they may be converted to graphite. Many other minerals are produced by hydrothermal contact metamorphism. Solutions given off by the magma react to produce new minerals containing elements not present in the limestone. In this type of metamorphism, oxide and sulfide minerals are frequently formed and may constitute ore deposits of economic importance (see Chapter 18).

REGIONAL METAMORPHISM

Regional metamorphism is developed over extensive areas, often involving thousands of square kilometers of rock thousands of meters thick. It is believed that regional metamorphism is related to the building of mountain ranges (see Chapter 8). Metamorphic rocks are found in the root regions of old mountains and in the Precambrian continental shields. Thousands of meters of rock have had to be eroded in order to expose these metamorphic rocks.

Regional metamorphic minerals During regional metamorphism, new minerals are developed as rocks respond to increases in temperature and pressure. These include some silicate minerals not found in igneous and sedimentary rocks, such as sillimanite, kyanite, andalusite, staurolite, almandite, brown biotite, epidote, and chlorite.

The first three of these new minerals are aluminosilicates and have the formula Al_2SiO_5. Their independent SiO_4 tetrahedra are bound together by positive ions of aluminum. Sillimanite develops in long, slender crystals that are brown, green, or white. Kyanite forms bladelike blue crystals. Andalusite forms coarse, nearly square prisms.

Staurolite is a silicate composed of independent tetrahedra bound together by positive ions of iron and aluminum. It has a unique crystal habit, which is striking and easy to recognize: It develops six-sided prisms that intersect either at 90°, forming a cross, or at 60°, forming an "X" (see Figure 6.4).

Garnets are a group of metamorphic silicate minerals. All have the same atomic structure of independent SiO_4 tetrahedra, but a wide variety of chemical compositions is produced by the many positive ions that bind the tetrahedra together. These ions may be iron, magnesium, aluminum, calcium, manganese, or chromium. But whatever the chemical composition,

6.4 Crystal of staurolite, 2 cm wide, from Georgia. [Benjamin M. Shaub.]

garnets appear as distinctive 12-sided or 24-sided fully developed crystals. Actually, it is difficult to distinguish one kind of garnet from another without resorting to chemical analysis. A common deep-red garnet containing iron and aluminum is called almandite.

Epidote is a complex silicate of calcium, aluminum, and iron, in which the tetrahedra are in pairs independent of each other. This mineral is pistachio green or yellowish to greenish black.

Chlorite is a sheet silicate of calcium, magnesium, aluminum, and iron. The characteristic green color of chlorite was the basis for its name, from the Greek *chlōros*, "green." Chlorite exhibits a cleavage similar to that of mica, but the small scales produced by the cleavage are not elastic, like those of mica. Chlorite occurs either as aggregates of minute scales or as individual scales that will be scattered throughout a rock.

Regional metamorphic zones The temperature and pressure conditions that develop under regional metamorphism may result in rocks characteristic of different zones: high-grade, middle-grade, and low-grade. High-grade metamorphism occurs in rocks that have been subjected to the most intense pressure and temperature. Low-grade metamorphic rocks have undergone the least change and blend along their outer margins into unchanged rocks. Middle-grade metamorphism falls between these two extremes.

Metamorphic zones are identified by certain diagnostic metamorphic minerals called index minerals. Zones of regional metamorphism reflect the varied mineralogical response of chemically similar rocks to different physical conditions. Each index mineral gives an indication of the conditions at the time of its formation. For minerals composed of Al_2SiO_5, for example, the temperature and pressure conditions will determine whether the resulting mineral will be kyanite (low temperature with a wide range of pressures), andalusite (moderate temperature with low pressures), or sillimanite (high temperature with a wide range of pressures) as shown in Figure 6.5.

The first appearance of chlorite tells us that we are at the beginning of a low-grade metamorphic zone. The first appearance of almandite garnet is evidence of the beginning of a middle-grade metamorphic zone. And the first appearance of sillimanite marks a high-grade zone. Other minerals sometimes occur in association with each of these index minerals (see Table 6.1), but they are usually of little help in determining the degree of metamorphism of a given zone.

By noting the appearance of the minerals that are characteristic of each metamorphic zone, it is possible to draw a map of the regional metamorphism of an entire area. Of course, the rocks must have the proper chemical composition to allow these minerals to form. The boundary between two metamorphic zones is commonly shown on geologic maps as an **isograd**, a line of equal grade of metamorphism drawn to distinguish one zone of index minerals from another. These lines can be interpreted as connecting similar temperature-pressure values.

Regional metamorphic facies In Chapter 5, we defined facies as it applies to sedimentary rocks. Actually, it has a broader application: **A facies is an assemblage of mineral, rock, or fossil features reflecting the environment in which the rock was formed.**

A **metamorphic facies** is an assemblage of minerals that reached equilibrium during metamorphism under a specific set of conditions. Each facies is named after a common metamorphic rock that belongs to it, and every metamorphic rock is assigned to a facies according to the conditions that attended its formation, not according to its composition. Although it is not always possible to assign a rock to a particular metamorphic facies on the basis of a single hand specimen, it is usually possible to do so after examining various rocks in the region.

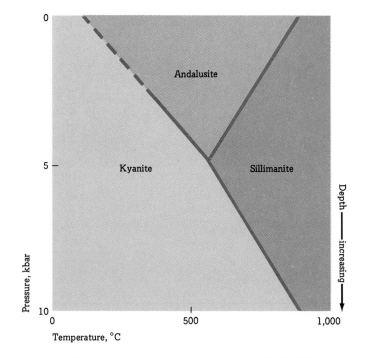

6.5 Temperatures and pressures at which Al_2SiO_5 forms different minerals. [Modified from various sources.]

Widely recognized regional metamorphic facies are shown in Table 6.2. They are the zeolite, greenschist, amphibolite, granulite, blueschist (glaucophane schist), and eclogite facies. Typical mineral assemblages are shown for metamorphism of a shale and of a basalt.

The zeolite facies rocks are transitional from diagenesis[3] (see Section 5.2) to metamorphism, and form under pressures of 2,000 to 3,000 bars at temperatures of 200° to 300°C, with extreme limits of 4,000 bars and 400°C.

Greenschists are metamorphic rocks that form from preexisting rocks under conditions of low and intermediate temperatures (300°–500°C), where pressures may vary from low to high (3,000 to 8,000 bars). They are characterized by many green-colored minerals such as chlorite, epidote, and actinolite.

The amphibolite facies includes metamorphic rocks formed under the same range of pressures (3,000 to 8,000 bars) and fairly high temperatures (450° to 700°C).

The granulite facies metamorphic rocks form at high temperatures (above 650°C) and a broad range of pressures (3,000 to 12,000 bars).

Blueschist or glaucophane schist facies include metamorphic rocks representing pressures greater than 5,000 bars and temperatures of 200° to 450°C. It is considered to be the high-pressure glaucophanitic

---

[3]**Diagenesis** is the recombination or rearrangement of a mineral, resulting in a new mineral, specifically those chemical, physical, and biologic changes undergone by a sediment after it has been deposited, during and after its lithification, but specifically before any metamorphic changes.

TABLE 6.1

Regional metamorphic
minerals

Zone	Grade of metamorphism	Minerals	
(Increasing ↓ metamorphism) CHLORITE	Low	Chlorite, muscovite, quartz	*green schist*
BIOTITE	Low	Biotite, muscovite, chlorite, quartz	
GARNET	Middle	Garnet (almandite), muscovite, biotite, quartz	
STAUROLITE	Middle	Staurolite, garnet, biotite, muscovite, quartz	
KYANITE	Middle	Kyanite, garnet, biotite, muscovite, quartz	
SILLIMANITE	High	Sillimanite, quartz, garnet, muscovite, biotite, oligoclase, orthoclase	*Blueschist*

TABLE 6.2

Regional metamorphic facies
and minerals typically produced
in the metamorphism of a
shale and a basalt[a]

Metamorphic facies	Mineral assemblages developed from:	
	Shale	Basalt
ZEOLITE	Illite, chlorite, quartz	Zeolites
GREENSCHIST	Muscovite, chlorite, quartz	Albite, epidote, chlorite, actinolite
AMPHIBOLITE	Muscovite, biotite, garnet, quartz, plagioclase	Amphibole, plagioclase, garnet
GRANULITE	Garnet, sillimanite, plagioclase, quartz, biotite, pyroxene	Calcic pyroxene, plagioclase
BLUESCHIST (GLAUCOPHANE SCHIST)	Muscovite, chlorite, quartz, manganese-aluminum garnet	Glaucophane, quartz, lawsonite (high-pressure alteration of plagioclase)
ECLOGITE	Not known	Sodic pyroxene, magnesian garnet

[a] Modified from various sources.

(hence the blue color) phase of the greenschist facies. Blueschists commonly occur along oceanic trenches where sedimentary rocks have been carried down along a subduction zone.

Eclogites form over a large range of pressures (7,500 to more than 10,000 bars) and temperatures (350° to 750°C), intruding upon other facies. Some metamorphic petrologists would abandon the eclogite facies concept. It does, however, represent those metamorphic rocks that formed most deeply within the earth.

Pressure-temperature fields for all metamorphic facies are shown in Figure 6.6. Also shown in this figure are those temperature and pressure conditions below which metamorphic changes do not commonly occur. This is the realm of diagenesis. Note also the lower pressure region in which moderate-to-high temperature contact metamorphism produces hornfels.

Table 6.3 shows the general relationships among zones and resulting metamorphic rocks in the progressive metamorphism of several original rock types.

LOW-GRADE REGIONAL METAMORPHISM IN PROGRESS

Studies on the cuttings and cores of three wells in sedimentary deposits of the Imperial Valley of California have indicated that low-grade metamorphism is in progress on a regional scale and at relatively shallow depth.[4] Two of the wells are located 400 m apart in a geothermal field at the southeastern end of the Salton Sea. They are part of a group of ten wells that have been drilled since 1960 to extract the elements contained in solution in the hot brines of the region and to tap the heat energy. The third was a dry oil test well drilled in 1963. It is located 35 km south-southeast of the geothermal field, in the Salton Trough (see Figure 6.7). Table 6.4 lists the three wells, gives their depths, bottom-hole temperatures, and geothermal gradients.

In this region the sedimentary deposits are geologically young. The oldest deposits in the Trough were deposited in the early Pliocene age, and those in the geothermal field during the Pleistocene age. They consist of poorly sorted sandstones and siltstones of the Colorado River delta. The original dominant minerals were quartz, calcite, subordinate dolomite, plagio-

[4] L. J. Patrick Muffler and Donald E. White, "Active Metamorphism of Upper Cenozoic Sediments in the Salton Sea Geothermal Field and the Salton Trough, Southeastern California," Bull. Geol. Soc. Am., vol. 80, pp. 157–182, 1969.

6.6 Generalized pressure-temperature fields for metamorphic facies. Boundaries are not sharp and distinct lines but, rather, vary over considerable zones. [Modified from various sources.]

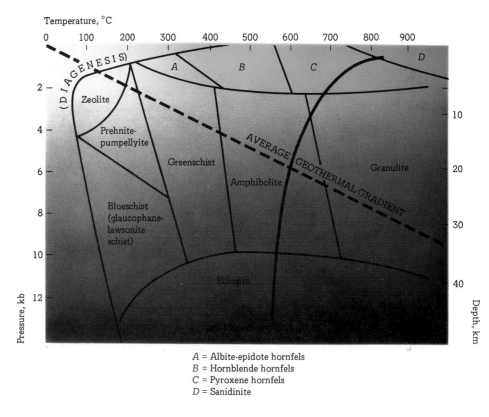

A = Albite-epidote hornfels
B = Hornblende hornfels
C = Pyroxene hornfels
D = Sanidinite

TABLE 6.3

Products of regional metamorphism[a]

	Metamorphic rocks resulting from increasing regional metamorphism				
Original rock	Chlorite zone	Biotite zone	Almandite (garnet) zone	Staurolite zone	Sillimanite zone
Shale	Slate	Biotite phyllite	Biotite-garnet phyllite	Biotite-garnet-staurolite schist	Sillimanite schist or gneiss
Clayey sandstone	Micaceous sandstone	Quartz-mica schist	Quartz-mica-garnet schist or gneiss	Quarta-mica-garnet schist or gneiss	Quarta-mica-garnet schist or gneiss
Quartz sandstone	Quartzite	Quartzite	Quartzite	Quartzite	Quartzite
Limestone and dolomite	Limestone and dolomite	Marble	Marble	Marble	Marble
Basalt	Chlorite-epidote-albite schist (greenschists)	Chlorite-epidote-albite schist (greenschists)	Albite-epidote amphibolite	Amphibolite	Amphibolite
Granite	Granite	Granite gneiss	Granite gneiss	Granite gneiss	Granite gneiss
Rhyolite	Rhyolite	Biotite schist or gneiss	Biotite schist or gneiss	Biotite schist or gneiss	Biotite schist or gneiss

[a] After Marland P. Billings, *The Geology of New Hampshire. Part II: Bedrock Geology*, p. 139, Granite State Press, Inc., Manchester, 1956.

clase, potassium feldspar, montmorillonite, illite, and kaolinite.

Samples from the three wells show increasing hardness and a regular sequence of mineral changes with depth as the temperature increases. Some detrital minerals drop out, and other minerals are formed. The

TABLE 6.4

Wells providing data on metamorphism

Location	Depth of well, m	Bottom-hole temperature, °C	Gradient, °C/100 m
Sportsman 1 (geothermal field)	1,419	310	20
I.I.D. 1 (geothermal field)	1,570	328	20
Wilson 1 (Salton Trough)	4,033	260	7

rocks look like ordinary sedimentary rocks but all have undergone mineralogical change at depth since deposition. This change is brought about by progressive metamorphism.

Geothermal field The wells in the geothermal field produce a brine containing over 250,000 ppm (parts per million) of dissolved solids, primarily Cl, Na, Ca, K, and Fe, plus a host of minor elements. Mineral transformations are due to hydrothermal metamorphism in the high-temperature environment (see Figure 6.8). There is continuous transition from sediments through indurated sedimentary rocks to low-grade metamorphic rocks of the greenschist facies. The resultant rocks are more porous than are most metamorphic rocks, probably because of the low pressures and short duration of metamorphism. No schistosity is developed. The rocks are analogous to products of burial metamorphism.

6.2 TYPES OF METAMORPHISM

119

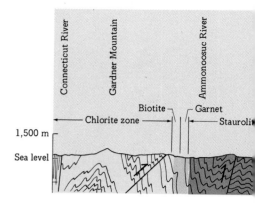

6.8 Mineral occurrence as a function of temperature in the Salton Sea geothermal field. [Data from L. J. Patrick Muffler and Donald E. White, "Active Metamorphism of Upper Cenozoic Sediments in the Salton Sea Geothermal Field and the Salton Trough, Southeastern California," *Bull. Geol. Soc. Am.*, vol. 80, p. 165, 1969.]

Outside the geothermal field Wilson 1 was a dry, oil test well drilled in 1963. The temperature at the bottom of the well, at 4,033 m, was 260°C, which represents an average gradient of 60°C/km. Mineralogical transformations occur over the same temperature ranges in this well as in the geothermal field, but because the thermal gradient is less, new minerals form at greater depths and extend over a much wider depth range. The temperatures encountered at the bottom of the well were not high enough for the development of high-temperature mineral forms.

Because the same mineralogical changes occurred at the same temperature in all three of the wells, it seems clear that temperature is the controlling factor. (See geothermal discussion in Chapter 17.)

6.3 METAMORPHIC ROCKS

Contact and regional metamorphic rocks are found in mountain ranges, at roots of mountain ranges, and on continental shields. The regional metamorphic rocks are by far the most widespread. They vary greatly in appearance, texture, and composition.

Even when different regional metamorphic rocks have all been formed by changes of a single uniform rock type, such as a shale, they are sometimes so drastically changed that they are thought to be unrelated.

These and other characteristics of regionally metamorphosed shale are well illustrated in New Hampshire. Along the Connecticut River, west of the White Mountains, rocks that were originally shales are found in a low-grade metamorphic zone as slate, with chlorite. Southeast of these rocks, the original shale is now found as phyllite, grading into schist. New metamorphic minerals appear one after the other toward the southeast: almandite, staurolite, and then sillimanite. These metamorphosed shales occur in a belt surrounding the White Mountain batholith (see Figure 6.9). The closer the area is to the batholith, the higher is its grade of regional metamorphism. This correlation is indicated by the presence of the index minerals found in the metamorphic rock.

TEXTURE OF METAMORPHIC ROCKS

In most rocks that have been subjected to heat and deforming pressures during regional metamorphism, the minerals tend to be arranged in parallel layers of flat or elongated grains (see Figure 6.10). This arrangement gives the rocks a property called **foliation** (from the Latin *foliatus*, "leaved" or "leafy," hence consisting of thin sheets).

The textures most commonly used to classify metamorphic rocks are **unfoliated** (either aphanitic or granular) and **foliated.** Let us look first at the unfoliated textures. In rocks with aphanitic texture, the individual grains cannot be distinguished by the unaided eye, and these rocks do not exhibit rock cleavage. You will remember that we have used the term **cleavage** to describe the relative ease with which a mineral breaks along parallel planes. But notice here that we are using the modifier "rock" to distinguish rock cleavage from mineral cleavage. In rocks with granular texture, the individual grains are clearly visible, but again no rock cleavage is evident.

6.9 Cross section of New Hampshire, showing metamorphic zones around the White Mountain batholith of the Older Appalachians. The length of the section is approximately 80 km. [After Marland P. Billings, *The Geology of New Hampshire. Part II: Bedrock Geology*, p. 139, Granite State Press, Inc., Manchester, 1956.]

Rocks with foliated texture, however, invariably exhibit rock cleavage. There are four degrees of rock cleavage:

1 **Slaty** (from the old French *esclat*, "fragment," "splinter"), in which the cleavage occurs along planes separated by distances of microscopic dimensions.

2 **Phyllitic** (from the Greek *phyllon*, "leaf"), in which the cleavage produces flakes barely visible to the unaided eye. Phyllitic cleavage produces fragments thicker than those of slaty cleavage.

3 **Schistose** (from the Greek *schistos*, "divided" or "divisible"), in which the cleavage produces flakes that are clearly visible. Here cleavage surfaces are rougher than in slaty or phyllitic cleavage.

4 **Gneissic** (from the Greek *gneis*, "spark," for the luster of certain of the components), in which the surfaces of breaking are from a few millimeters to a centimeter or so apart.

TYPES OF METAMORPHIC ROCK

The many types of metamorphic rock stem from the great variety of original rocks and the varying kinds and degrees of metamorphism. Metamorphic rocks may be derived from any of the sedimentary or igneous rocks and other metamorphic rocks of lower grade.

Metamorphic rocks are usually named on the basis of texture. A few may be further classified by including the name of a mineral, such as chlorite schist, mica schist, and hornblende schist.

Slate Slate is a metamorphic rock that has been produced from the low-grade metamorphism of shale or pyroclastic igneous rock (see Figure 6.11). It is aphanitic with a slaty cleavage caused by the alignment of platy minerals by the stress field. Some of the clay min-

erals in the original shale have been transformed by heat into chlorite and mica. In fact, slate is composed predominantly of small colorless mica flakes and some chlorite. It occurs in a wide variety of colors. Dark-colored slate owes its color to the presence of carbonaceous material or iron sulfides.

Phyllite Phyllite is a metamorphic rock with much the same composition as slate, but its minerals exist in larger units. Phyllite is actually slate that has undergone further metamorphism. When slate is subjected to heat greater than 250 to 300°C, the chlorite and mica minerals of which it is composed develop large flakes, giving the resulting rock its characteristic phyllitic cleavage and a silky sheen on freshly broken surfaces. The predominant minerals in phyllite are chlorite and muscovite. This rock usually contains the same impurities as slate, but sometimes a new metamorphic mineral, such as tourmaline or magnesium garnet, makes its appearance.

Schist Of the metamorphic rocks formed by regional metamorphism, schist is the most abundant. There are many varieties of schist; for it can be derived from many igneous, sedimentary, or lower-grade metamorphic rocks; but all schists are dominated by clearly visible flakes of some platy mineral, such as mica, talc, chlorite, or hematite. Fibrous minerals are commonly present as well. Schist tends to break between the platy or fibrous minerals, giving the rock its characteristic schistose cleavage.

Schists often contain large quantities of quartz and feldspar as well as lesser amounts of minerals such as augite, hornblende, garnet, epidote, and magnetite. A green schistose rock produced by low-grade metamorphism, sometimes called a **greenschist**, owes its color to the presence of the minerals chlorite and epidote.

(→) metamorphism

shale → slate → phyllite → schist

6.3 METAMORPHIC ROCKS

121

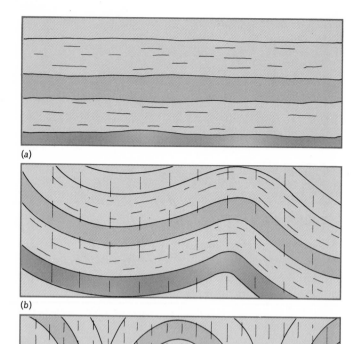

(a)

(b)

(c)

6.10 Development of schistosity in argillaceous sedimentary rocks: (a) Nearly horizontal bedding soon after deposition. (b) Moderately folded rocks with beginning of schistosity produced by realignment and growth of clay minerals perpendicular to regional stresses (coming from the sides). Note angle between bedding and schistosity at various places around folds. (c) Strongly folded rocks with highly developed schistosity nearly parallel to axial planes of folds and nearly parallel to most of the beds.

Table 6.5 lists some of the more common varieties of schist, together with the names of the rocks from which they were derived.

Amphibolite Amphibolite is composed mainly of hornblende and plagioclase. There is some foliation or lineation due to alignment of hornblende grains, but it is less conspicuous than in schists. Amphibolites may

6.11 Vermont slate quarry, showing how rock cleavage controls breaking. [Harvard University, Gardner Collection.]

TABLE 6.5

Common schists

Variety	Rock from which derived
Chlorite schist	Shale
Mica schist	Shale
Hornblende schist	Basalt or gabbro
Biotite schist	Basalt or gabbro
Quartz schist	Impure sandstone
Calc-schist	Impure limestone

be green, gray, or black and sometimes contain such minerals as epidote, green augite, biotite, and almandite. They are products of the medium-grade to high-grade regional metamorphism of ferromagnesian igneous rocks and of some impure calcareous sediments.

Gneiss A granular metamorphic rock, gneiss is most commonly formed during high-grade regional metamorphism. A banded appearance makes it easy to recognize in the field. Although gneiss exhibits rock cleavage, this is far less pronounced than in the schists.

In gneiss derived from igneous rocks such as granite, gabbro, or diorite, the component minerals are arranged in parallel layers: The quartz and the feldspars alternate with the ferromagnesians. In gneiss formed from the metamorphism of clayey sedimentary rocks such as graywackes, bands of quartz or feldspar usually alternate with layers of platy or fibrous minerals such as chlorite, mica, graphite, hornblende, kyanite, staurolite, sillimanite, and wollastonite.

Marble A familiar metamorphic rock, marble is essentially composed of calcite or dolomite, is granular, and was derived during the contact or regional metamorphism of limestone or dolomite. It does not exhibit rock cleavage. Marble differs from the original rock in having larger mineral grains. In most marble, the crystallographic direction of its calcite is nearly parallel; this is in response to the metamorphic pressures to which it was subjected. The rock shows no foliation, however, because the grains have the same color, and lineation does not show up.

Although the purest variety of marble is snow white, many marbles contain small percentages of other minerals that were formed during metamorphism from impurities in the original sedimentary rock.

6.12 Marble quarry, showing how blocks are sawed out for dimension stone to be used in buildings. [Benjamin M. Shaub.]

These impurities account for the wide variety of color in marble. Black marbles are colored by bituminous matter; green marbles, by diopside, hornblende, serpentine, or talc; red marbles, by the iron oxide hematite; and brown marbles by the iron oxide limonite. Garnets have often been found in marble, as have rubies on rare occasions.

Marble occurs most commonly in areas of regional metamorphism (see Figure 6.12), where it is often found in layers between mica schists or phyllites.

Quartzite The metamorphism of quartz-rich sandstone forms the rock quartzite. The quartz in the original sandstone has become firmly bonded by the entry of silica into the pore spaces. Quartzite may also be formed by percolating water under the temperatures and pressures of ordinary sedimentary processes working near the surface of the earth. Many quartzites, however, are true metamorphic rocks and may have been formed by metamorphism of any grade.

Quartzite is unfoliated and is distinguishable from sandstone in two ways: (1) there are fewer pore spaces in the quartzite, and (2) the rock breaks right through the sand grains that make it up, rather than around them. The structure of quartzite cannot be recognized without a microscope; but when we cut it into thin sections, we can identify both the original rounded sand grains and the silica that has filled the old pore spaces.

Pure quartzite is white, but iron or other impurities sometimes give the rock a reddish or dark color. Among the minor minerals that often occur in quartzite are feldspar, muscovite, chlorite, zircon, tourmaline, garnet, biotite, epidote, hornblende, and sillimanite.

6.4 METAMORPHISM AND PLATE BOUNDARIES

Different types of metamorphism and metamorphic products can be expected to develop along the margins of lithospheric plates, depending on the kind of boundary and its geologic setting.

CONVERGENT-BOUNDARY METAMORPHISM

Crustal plates are commonly subducted at trenches under convergent island-arc regions (such as the island arcs of the western Pacific) and at convergent ocean-continent boundaries (such as the Andean coast of South America). In these regions, subducting slabs of lithosphere interact with sediments that have accumulated on the sea floor. These sediments tend to consist of a **mélange** (heterogeneous mixture) of fine-grained deep-sea facies combined with exotic blocks of varying ages up to several kilometers in size. A good example of such an assemblage is the Franciscan Formation (Jurassic to Eocene in age) of California, which consists of a mélange of deep-sea sediments, oceanic-crust fragments, and various metamorphic assemblages (greenschist facies).

A second convergent-boundary setting involves the downwarping of thick piles of sediments that have accumulated between an island arc and an adjacent continent or on the ocean edge of a continental block. Low-grade regional metamorphism commonly occurs at the bottom of such a wedge of sediments. Plutonic activity also develops so that contact metamorphism affects the rocks adjacent to these intrusions.

The third principal convergent boundary involves collision of two continents (such as the Alpine belt). The dominant metamorphic assemblage here is the blue-schist facies involving extensive thrusting of rocks caught between the converging continental plates.

In all these settings, descending slabs of lithosphere undergo metamorphism caused by increasing temperature and pressure. Such effects can be calculated for slabs of a given composition, for varying thicknesses and sizes of slabs, and for different rates of movement. These measurements become very complicated but are possible within certain limiting conditions for these parameters.

The lithospheric slabs apparently retain their mechanical competence and are relatively cool in contrast to their surroundings until thermal equilibrium is reached (Figure 6.13). One analysis estimated that a slab would reach thermal equilibrium with the surrounding mantle at a depth of about 650 km for a lithospheric plate being subducted at a rate of 8 cm/year.[5]

Metamorphism is caused by heat generated along this subduction zone from friction of the descending lithosphere slipping past other rocks. At a given depth, melting can occur, producing magmas that rise to supply surface volcanoes (Figure 6.14).

Different pressure-temperature conditions prevail at different parts of this descending plate: (1) A zone of low heat flow has been observed over what is considered to be the upper portion of this descending slab. Here low-temperature metamorphic minerals are created. (2) An intermediate zone of relatively higher temperature and pressure occurs along the slab at intermediate depths. (3) A zone of high heat flow occurs over the deepest part of the downward-moving lithosphere, where melting occurs and heat is transferred upward by rising magmas. High-temperature–high-pressure metamorphic minerals are created here (Figure 6.14).

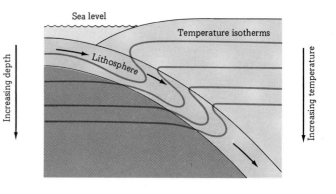

6.13 Relative temperature increases at the convergent boundary. The lithospheric plate undergoing subduction retains its cooler temperature, deflecting the isotherms downward. [From various sources.]

DIVERGENT-BOUNDARY METAMORPHISM

Metamorphism (and attendant mineralization) occurs along divergent plate boundaries, notably at mid-ocean spreading centers such as the Mid-Atlantic Ridge. Outflowings of basaltic lavas heat adjacent rocks to produce contact metamorphic minerals.

Seawater over midocean ridges forms hydrothermal solutions, penetrating fissures, dissolving minerals from rocks under the ridge crests, and precipitating those minerals in highly concentrated deposits often of economic importance (see Chapter 18 and Figure 6.15).

Metallic sulfides have been found in rocks dredged from midoceanic ridges at a number of places, including the Mid-Atlantic and Indian Ridges.

The island of Cyprus is famous for a variety of rich mineral deposits, first mined by the Phoenicians. The origin of this island is closely related to sea-floor spreading from a midoceanic ridge that existed in the earliest development of the eastern Mediterranean.

The Red Sea is perhaps the very early product of a divergent plate boundary currently forming between the African and Arabian plates. Very salty hot brines have been found associated with sediments containing significant quantities of sulfide minerals, including iron, zinc, and copper sulfides, plus some silver and

[5]M. N. Toksoz, J. W. Minea, and B. R. Julian: "Temperature Field and Geophysical Effects of a Downgoing Slab," *J. Geophys. Res.*, vol. 76, pp. 1114–1138, 1971.

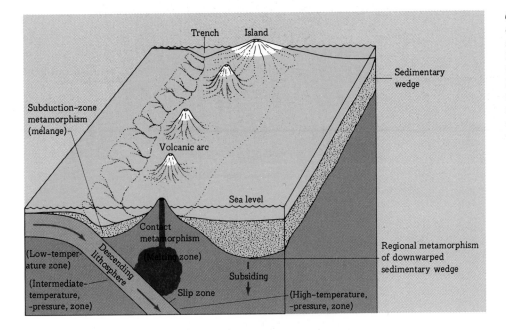

6.14 Thermal and metamorphic zones associated with descending lithospheric plate under a converging island-arc region. The outer zone near the trench is a zone of relatively low heat flow. The zone associated with the deepest portion of the descending plate is a zone of high heat flow, with melting and the production of magma, contact metamorphism, and overlying volcanic island chain. Types of metamorphism are indicated for various parts of the system. [Modified from various sources.]

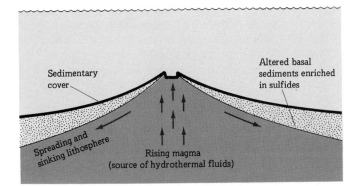

6.15 Cross section of divergent boundary at spreading center, showing hydrothermal fluids and their contact-metamorphosed sediments.

gold. The brines are thought to be hydrothermal solutions from which these minerals precipitate or into which minerals are introduced from volcanic sources beneath the Red Sea or from sediments with high mineral content found adjacent to the basins.

TRANSFORM- (PARALLEL-) BOUNDARY METAMORPHISM

Fractures occur along transform boundaries as plates slide past one another. These fractures and faults provide zones of weakness along which magma can rise. Volcanoes occur along such zones in oceanic regions. Traces of plates moving past one another can also be observed on land, for example, along the San Andreas fault system. Heat and pressure generated by these rock masses moving past one another produce meta-

morphic minerals. Mineralization of economic significance occurs in some important mining districts along major fault zones, especially in California and in similar geologic settings around the world.

6.5 ORIGIN OF GRANITE

The eighteenth-century geologist James Hutton once stated that granite was produced by the crystallization of minerals from a molten mass, and ever since, most geologists have accepted the magmatic origin of granite. However, several investigators have questioned this conclusion, suggesting instead that granite is a metamorphic rock produced from preexisting rocks by a process called **granitization.**

In discussing batholiths, we mentioned that one of the reasons for questioning the magmatic origin of granite was the mystery of what happened to the great mass of rock that must have been displaced by the intrusion of the granite batholiths. This so-called space problem has led some geologists to conclude that batholiths actually represent preexisting rocks transformed into granite by metasomatic processes. Certain rock formations support this theory: These sedimentary rocks were originally formed in a continuous layer but now grade into schists and then into **migmatites** ("mixed rocks"), apparently formed when magma squeezed in between the layers of schist.

The concept of migmatite as a rock group was introduced by J. J. Sederholm in 1907.[6] It is a generic term that covers a large number of petrological combinations and is to be applied to rock formed from two principal ingredients: the host of early-formed material and the material introduced by permeation, metasomatism, or injection of liquids. The average composition of a migmatite is granitic. Its diagnostic characteristics are structure, texture, and broad regional relationships.

The migmatites, in turn, grade into rocks that contain the large, abundant feldspars characteristic of granite but that also seem to show shadowy remnants of schistose structure. Finally, these rocks grade into pure granite. The proponents of the granitization theory say that the granite is the result of extreme metasomatism and that the schists, migmatites, and granitelike rocks with schistose structure are way stations in transforming sedimentary rocks into granite.

What mechanism could have brought about granitization? Perhaps ions migrated through the original solid rock, building up the elements characteristic of granite, such as sodium and potassium, and removing superfluous elements, such as calcium, iron, and magnesium. The limit to which the migrating ions are supposed to have deposited the sodium and potassium is called the **granitic front.** The limit to which the migrating ions are supposed to have carried the calcium, iron, and magnesium is called the **simatic front.**

In the middle of the twentieth century, geologists were carrying on an enthusiastic debate over the origin of granite, but they had reached agreement on one fundamental point; that is, various rocks with the composition and structure of granite may have different histories. In other words, some may be igneous and others metasomatic. So the debate between "magmatists" and "granitizationists" has been reduced to the question of what percentage of the world's granite is metasomatic and what percentage is magmatic. Those who favor magmatic origin admit that perhaps as much as a quarter of the granite exposed at the earth's surface is metasomatic. But the granitizationists reverse the percentages and insist that about three-quarters or more is metasomatic and only one-quarter or less is of magmatic origin.

[6]J. J. Sederholm, "Om Granit och Gneiss," *Comm. Geol. Finland Bull.*, vol. 23, p. 110, 1907.

Field relationships in some cases suggest that great granite bodies have formed as second-generation igneous rocks in the cores of mountain ranges. This would involve remelting of deep portions of geosynclines and would leave no space problem or need for widespread metasomatism.

6.6 Geothermometers

Geologists have long sought out ways to determine what conditions have existed at various times in the past—conditions of pressure, climate, wind direction, stream flow direction, and especially **temperature.** Any mechanism that can be used to determine the earth's temperature at a specific time and place is known as a **geothermometer.** These geothermometers may be animal, plant, or mineral.[7]

The changes that rocks and minerals undergo during metamorphism are caused in part by increasing temperatures. These same temperature increases can cause changes in the enclosed plants and animals, changes that in some cases can be instructive in telling us just what temperatures were encountered during metamorphism.

We usually think of metamorphism as occurring at temperatures in excess of about 200°C. We discussed those minerals that develop at higher temperature ranges of metamorphism when we examined metamorphic facies (see Section 6.2). At temperatures less than 200°C, other changes can occur in what we have called the "diagenetic" range or the realm of diagenesis. Such mineralogic and chemical changes affect organic constituents most noticeably. Let us look at a few such changes induced by temperature increases.

TEMPERATURE OF COAL FORMATION

Coal was defined in Chapter 5 as a sedimentary rock formed from accumulating plant material. In fact, higher grade coals are really metamorphic rocks. They have undergone changes due to increased temperature and pressure, coal being very sensitive to such changes. Some differences in coal types result from differences in the character of the original plant material and the environment in which the coal accumulates. Such differences, however, are far overshadowed by changes produced by variations in the maximum temperature to which the coal has been exposed during its history. The effect of pressure also is minor when compared to temperature effects.

[7]Some of the discussion that follows resulted from personal communications with A. G. Harris, U.S. Geological Survey.

TABLE 6.6

Geothermometers; showing changes that occur as temperature increases

Temperature range (°C)[a]	Metamorphic grade	Pelitic rocks	Carbonate rocks	Coals	Conodonts	Hydrocarbons
20°	D I A G E N E S I S	Shales contain clay minerals	Calcium and magnesium carbonates	Change from peat to brown coal to lignite	First color change from pale yellow to very pale brown	Beginning stages of oil generation
50°		Submicroscopic changes in clay minerals	Unchanged	Subbituminous to bituminous	Very pale brown to light brown	Peak stages of oil generation; early generation of gas
100°		Microscopic changes in clay minerals	Unchanged	Bituminous to semianthracite	Brown to dark brown	Phase out of oil generation; peak of gas generation
200°	Very low-grade metamorphism	Change to slates	Recrystallized	Anthracite	Very dark grayish brown to dark reddish brown	Hydrocarbons are metamorphosed and destroyed
300°		Change to phyllites and chlorite schists	Coarsely crystalline marbles	Meta-anthracite	Black	
400°	Low-grade metamorphism	Biotite to garnet schist	Marbles; impure carbonates contain talc and tremolite	Graphite	Very black	
500°	Medium-grade metamorphism	Andalusite to kyanite schist	Marbles; impure carbonates contain tremolite and actinolite	Graphite	Colors fade	
600°	High-grade metamorphism	Sillimanite schist	Replaced by calcium silicate (e.g., wollastonite)	Graphite to diamonds (if sufficient pressure)	Clear	

[a] Arbitrary boundaries; summarized from various sources, especially A. G. Harris, U.S. Geological Survey, personal communication.

Coal passes through several "ranks" during the development of the coal we know as **anthracite,** or "hard coal." These ranks include such stages as peat, brown coal, lignite, subbituminous, bituminous, anthracite, meta-anthracite, and finally graphite if the thermal effects reach high enough stages.

With increasing depth of burial and increasing temperature, the coal-producing material changes in important ways: (1) color and luster change from earthy brown to bright black to metallic silver, (2) light reflectivity or reflectance increases, (3) porosity decreases and the coal becomes more compact, (4) recognizable plant particles disappear, (5) carbon content increases, (6) hydrogen, moisture, and other volatiles decrease, and (7) the heat-producing character of the coal increases.

The conversion from peat to brown coal to lignite usually occurs within the upper 400 m of burial and in the temperature range of 20°C to 50°C (see Table 6.6). Lignite changes to subbituminous and then bituminous coal between 50° and 200°C at depths of 400 to 5,000 m of burial. Anthracite, meta-anthracite, and graphite require even higher temperatures but may also require directed pressures, such as those found in areas undergoing mountain building. There are exceptions to this general need for higher pressures to produce high-rank coals. For example, some high-rank coals appear to be related to large igneous intrusions. There are other examples where normal bituminous material has been altered to anthracite adjacent to dikes and sills.

CONODONT COLOR CHANGES

Recent studies of microfossils, 0.1 to 1 mm in size, of a poorly known group called **conodonts,** have led to the conclusion that these organisms change color from pale yellow to black with increasing temperatures associated with increasing depth of burial. A. G. Epstein, J. B. Epstein, and L. D. Harris[8] concluded that color changes begin at temperatures of 50° to 90°C, with the black color occurring at temperatures in excess of 300°C. These temperature guides can be very useful geothermometers, especially in marine carbonate rocks ranging in age from Cambrian through Triassic, when conodonts appear to have had worldwide distribution.

[8]A. G. Epstein, J. B. Epstein, and L. D. Harris, *Conodont Color Alteration—An Index to Organic Metamorphism,* U.S. Geological Survey Prof. Paper 995, 1977.

OUTLINE

Metamorphism produces metamorphic rocks by changing igneous and sedimentary rocks while they are in the solid state.

Agents of metamorphism are heat, pressure, and chemically active fluids.

Heat may be the essential agent.

Pressure may be great enough to induce plastic deformation.

Chemically active fluids, particularly those released late in the solidification of magma, react on surrounding rocks.

Types of metamorphism are contact and regional.

Contact metamorphism occurs at or near an intrusive body of magma.

Contact metamorphic minerals include wollastonite, diopside, and some oxides and sulfides constituting ore minerals.

Regional metamorphism is developed over extensive areas and is related to the formation of some mountain ranges.

Regional metamorphic minerals include sillimanite, kyanite, andalusite, staurolite, almandite garnet, brown biotite, epidote, and chlorite.

Regional metamorphic zones are identified by diagnostic index minerals.

Regional metamorphic facies is an assemblage of minerals that reached equilibrium during metamorphism under a specific set of conditions.

Metamorphic rocks are found in mountain ranges, at mountain roots, and on continental shields.

Textures of metamorphic rocks are unfoliated and foliated.

Unfoliated rocks do not exhibit rock cleavage.

Foliated rocks exhibit rock cleavage as slaty, phyllitic, schistose, or gneissic.

Types of metamorphic rocks are many because of the variety of original rocks and the varying kinds of metamorphism.

Metamorphism and plate boundaries are closely related.

Convergent boundaries show increased temperature and pressure effects on subducting slabs of lithosphere with sediments intruded by batholiths, downwarped thick sedimentary sequences, and plates that have collided.

Divergent boundaries have high heat flow at spreading centers with mineralization including sulfides and evaporites.

Transform or parallel boundaries have less igneous activity but include some metamorphism and some economic mineralization.

Origin of granite may be igneous or metamorphic or by melting of sediments.

Geothermometers are animal, plant, or mineral indicators of changes in the earth's temperature.

Mason, Roger *Petrology of the Metamorphic Rocks*, George Allen & Unwin, Ltd., London, 1978. *The main theme is the recognition of assemblages of minerals coexisting during metamorphism.*

Miyashiro, A. *Metamorphism and Metamorphic Belts*, George Allen & Unwin, Ltd., London, 1973. *Shows temperature-pressure fields for different metamorphic facies and typical mineral assemblages of basic igneous rocks and pelitic rocks for each facies.*

Read, H. H. *The Granite Controversy*, Thomas Murby and Company, London, 1957. *Eight addresses, delivered between 1939 and 1954, concerned with the origin of granitic and associated rocks.*

Turner, Francis J. *Metamorphic Petrology*, McGraw-Hill Book Company, New York, 1968. *A general text for advanced students. Embodies a survey of mineralogical aspects of metamorphism.*

Winkler, H. G. F. *Petrogenesis of Metamorphic Rocks*, 4th ed., Springer-Verlag New York, Inc., New York, 1976. *Stresses the mineralogical-chemical and physicochemical aspects of metamorphism, especially as related to specific rock groups of common composition.*

SUPPLEMENTARY READINGS

QUESTIONS

1   What limits are placed on conditions under which the process of metamorphism is considered to occur?

2   Why is heat considered to be the most essential agent of metamorphism?

3   How does pressure produce changes during metamorphism?

4   How do contact aureoles develop around igneous intrusions?

5   Regional metamorphism produces zones characterized by which index minerals?

6   What is a regional metamorphic facies? How is it similar to a sedimentary facies? How does it differ?

7   Why is diagenesis considered by some to be the "final stage" before metamorphism?

8   Where and under what conditions do we find low-grade regional metamorphism taking place today?

9   What is the typical progression in the increasing regional metamorphism of a shaly rock? Rearrange these metamorphic rocks in their proper order: phyllite, schist, shale, slate.

10   Quartzite is unfoliated and can be distinguished from a sandstone in what two ways?

11   Why do temperature isograds make sharp bends along subducted plates as shown in Figure 6.13?

12   How is it possible to produce a granite by extension of the processes that have produced metamorphism?

13   What is meant by a geothermometer? Give an example of a plant, a mineral, and an animal geothermometer.

7

GEOLOGIC TIME

In Chapter 1, we stressed the vastness of geologic time and emphasized its importance to an understanding of the processes of geology. In this chapter, we shall look at some of the ways of measuring geologic time and at how the geologic-time scale has developed.

7.1 RELATIVE TIME

Before geologists knew how to determine absolute time, they had discovered events in earth history that convinced them of the great length of geologic time. In putting these events in chronological order, they found themselves subdividing geologic time on a relative basis and using certain labels to indicate relative time. You have probably picked up a newspaper or magazine (or indeed this textbook) and have read of the dis-

covery of a dinosaur that lived 100 million years ago during the Cretaceous Period or of the development of a coal deposit in strata formed 280 million years ago in the Pennsylvanian Period. The names of the periods are terms used by geologists to designate certain units of **relative geologic time.** In this section we shall look at how such units have been set up and how absolute dates have been suggested for them.

Relative geologic time has been determined largely by the relative positions of sedimentary rocks. Remember that a given layer of sedimentary rock represents a certain amount of time—the time it took for the original deposit to accumulate. By arranging various sedimentary rocks in their proper chronological sequence, we are in effect arranging units of time in *their* proper order. Our first task in constructing a relative-time scale, then, is to arrange the sedimentary rocks in their proper order.

THE LAW OF SUPERPOSITION

The principle used to determine whether one sedimentary rock is older than another is very simple, and it is known as the **law of superposition.** Here is an exam-

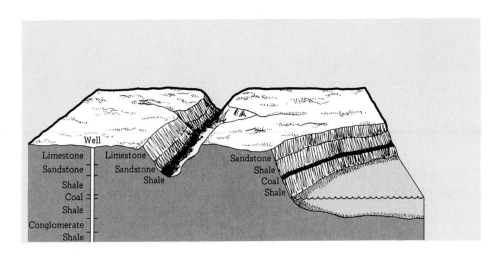

ple. A deposit of mud laid down this year in, say, the Gulf of Mexico will rest on top of a layer that was deposited last year. Last year's deposit, in turn, rests on successively older deposits that extend backward into time for as long as deposition has been going on in the gulf. If we could slice through these deposits, we should expose a chronological record with the oldest deposit on the bottom and the youngest on top. This sequence would illustrate the law of superposition: **If a series of sedimentary rocks has not been overturned, the topmost layer is always the youngest and the lowermost layer is always the oldest.**

The law of superposition was first derived and used by Nicolaus Steno (1638–1687), a physician and naturalist who later turned theologian. His geologic studies were carried on in Tuscany, in northern Italy, and were published in 1669. On first glance the principle worked out by Steno is absurdly simple. For instance, in a cliff of sedimentary rocks, with one layer lying on top of another, it is perfectly obvious that the oldest is on the bottom and the youngest on top. We can quickly determine the relative age of any one layer in the cliff in relation to any other layer. The difficulty, however, lies in the fact that unknown tens of thousands of meters of sedimentary rock have been deposited during geologic time and that there is no one cliff, no one area, where *all* these rocks are exposed to view. The rocks in one place may be older or younger or of the same age as those in some other place. The task is to find out how the rocks around the world fit into some kind of relative time scale.

CORRELATION OF SEDIMENTARY ROCKS

Because we cannot find sedimentary rocks representing all of earth time neatly arranged in one convenient area, we must piece together the rock sequence from locality to locality. This process of tying one rock sequence in one place to another in some other place is

known as **correlation,** from the Latin for "together" plus "relate."

Correlation by physical features When sedimentary rocks show rather constant and distinctive features over a wide geographic area, we can sometimes connect sequences of rock layers from different localities. Figures 7.1 and 7.2 illustrate how this is done. Here is a series of sedimentary rocks exposed in a sea cliff. The topmost, and hence the youngest, is a sandstone. Beneath the sandstone we first find a shale, then a seam of coal, and then more shale extending down to the level of the modern beach. We can trace these layers for some distance along the cliff face, but how are they related to other rocks farther inland?

Along the rim of a canyon that lies inland from the cliff, we find that limestone rocks have been exposed. Are they older or younger than the sandstone in the cliff face? Scrambling down the canyon walls, we come to a ledge of sandstone that looks very much like the sandstone in the cliff. If it *is* the same, then the limestone must be younger because it lies above it. The only trouble is that we cannot be certain that the two sandstone beds *are* the same; so we continue down to the bottom of the canyon and there find some shale beds very similar to the shale beds exposed in the sea cliff beneath the sandstone. We feel fairly confident that the sandstone and the shale in the canyon are the same beds as the sandstone and upper shale in the sea cliff, but we must admit the possibility that we are dealing with different rocks.

In searching for further data, we find a well drilled still farther inland, its drill bit cutting through the same limestone that we saw in the canyon walls. As the bit cuts deeper and deeper, it encounters sandstone,

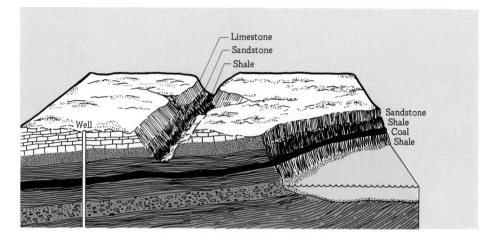

7.2 Similar lithologies and sequences of beds in the three different localities of Figure 7.1 suggest the correlation of rock layers shown in this diagram.

Limestone
Sandstone
Shale

Well

Sandstone
Shale
Coal
Shale

shale, a coal seam, more shale, and then a bed of conglomerate before the drilling finally stops in another shale bed. This sequence duplicates the one we observed in the sea cliff and reveals a limestone and an underlying conglomerate and shale that we have not seen before. We may now feel justified in correlating the sandstone in the sea cliff with that in the canyon and that in the well hole. The limestone is the youngest rock in the area; the conglomerate and lowest shale are the oldest rocks. This correlation is shown in Figure 7.2.

Many sedimentary formations are correlated in just this way, especially when physical features are our only keys to rock correlation. But as we extend the range of our correlation over a wider and wider area, physical features become more and more difficult to use. Clearly it is impossible to use physical characteristics to determine the relative age of two sequences of layered rocks in, say, England and the eastern United States. Fortunately we have another method of correlation, a method that involves the use of fossils.

Correlation by fossils Around the turn of the nineteenth century, an English surveyor and civil engineer named William Smith (1769–1839) became impressed with the relationship of rock strata to the success of various engineering projects, particularly the building of canals. As he investigated rock strata from place to place, he noticed that many of them contained fossils. Furthermore, he observed that (no matter where found) some rock layers contained identical fossils, whereas the fossils in rock layers above or below were different. Eventually Smith became so skillful that when confronted with a fossil, he could name the rock from which it had come.

At about the same time, two French geologists, Georges Cuvier (1769–1832) and Alexandre Brongniart (1770–1847), were studying and mapping the fossil-bearing strata that surround Paris. They too found that certain fossils were restricted to certain rock layers, and they had also used the law of superposition to arrange the rocks of the Paris area in chronological order, just as Smith had done in England. Then Cuvier and Brongniart arranged their collection of fossils in the same order as the rocks from which the fossils had been dug. They discovered that the fossil assemblages varied in a systematic way with the chronological positions of the rocks. In comparing the fossil forms with modern forms of life, Cuvier and Brongniart discovered that the fossils from the higher rock layers bore a closer resemblance to modern forms than did the fossils from rocks lower down.

From all these observations it became evident that the **relative age** of a layer of sedimentary rock could be determined by the nature of the fossils that it contained. This fact has been verified time and again by other workers throughout the world. It has become an axiom in geology that **fossils are a key to correlating rocks** and that **rocks containing the same fossil assemblages are similar in age.**

But when we apply this axiom to actual situations involving the use of fossils for correlation, certain complications arise. For instance, it is obvious that in our modern world the distribution of living forms varies with the environment. This fact was presumably as true in the past as it is now. We found, in discussing facies of sedimentary rocks, that different sediments were laid down in different places at the same time. Plants and animals also reflect changes in environment, particularly if they happen to live on the sea bottom. Organisms living in an area where mud is being deposited are different from those living in an area where sand is being laid down. Thus we find fossils in a bed of shale (formed from mud) somewhat different from those in a bed of sandstone (formed from sand).

If both the physical features and the fossils are different, how can we correlate the rocks in two different areas? There are two possible ways: first, we may actually be able to see that two different rock types, with their differing fossils, grade into each other laterally as we follow the beds along a cliff face; second, we may find that a few fossils occur in both environments. In an oceanic environment, for instance, some forms float or swim over a wide geographic area that takes in more than one condition of deposition on the sea floor. The remains of these swimming or floating forms may settle to the bottom, to be incorporated in different kinds of sediment forming on the ocean floor.

7.3 The dark areas in this specimen from Grafton, New Hampshire, are made up of the uranium-bearing mineral gummite. Compare with Figure 7.4. [Specimen in the Princeton University Museum of Natural History, Willard Starks.]

7.2 ABSOLUTE TIME

We earthlings use two basic units of time: (1) the day, the interval required for our planet (in the present epoch) to complete one rotation on its axis; and (2) the year, the interval required for the earth to complete one revolution around the sun. In geology, however, the problem is to determine how many of these units of time elapsed in the dim past when nobody was around to count and record them. Our most valuable clues in solving this problem are provided by the decay rates of radioactive elements.

RADIOACTIVITY

The nuclei of certain elements spontaneously emit particles, changing in size and/or form and producing new elements. This process is known as **radioactivity.** Shortly after the turn of the twentieth century, researchers suggested that minerals containing radioactive elements could be used to determine the age of other minerals in terms of absolute time (see Figures 7.3 and 7.4 and Chapter 1).

Let us take a single radioactive element as an example of how this method works. Regardless of what element we use, we must know what products result from its radioactive decay. Let us choose uranium 238, $^{238}_{92}$U, which is known to yield helium and lead, $^{206}_{82}$Pb, as end products. We know too the rate at which uranium 238 decays. As far as we can determine, this rate is constant and is unaffected by any known chemical or physical agency. The rate at which a radioactive element decays is expressed in terms of what we call its **half-life**—the time required for half of the nuclei in a sample of that element to decay. The half-life of uranium 238 is 4.51×10^9 years, which means that if we

7.4 A photograph of the same specimen as that in Figure 7.3. In this photograph the specimen was placed on a photographic plate. As the uranium and some of its radioactive daughter products decayed, the plate was exposed to the emission of particles. The white areas mark the location of the uranium-bearing minerals. [Willard Starks.]

start with 1 g of uranium 238, there will be only 0.5 g left after 4,510 million years (Figure 7.5).

The history of 1 g of uranium 238 may be recorded as shown in Table 7.1. At any instant during this process, there is a unique ratio of lead 206 to uranium 238. This ratio depends on the length of time decay has been going on. Theoretically, then, we may find the age of a uranium mineral by determining how much lead 206 is present and how much uranium 238 is present. The ratio of lead to uranium then serves as an index to the age of the mineral.

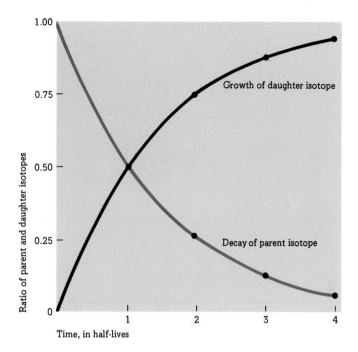

7.5 Graph showing the change in ratio of parent and daughter isotopes plotted against half-lives.

TABLE 7.1

History of 1 g of uranium 238[a]

Age, millions of yr	$^{206}_{82}$Pb formed, g	$^{238}_{92}$U remaining, g
100	0.013	0.985
1,000	0.116	0.825
2,000	0.219	0.747
3,000	0.306	0.646
4,500 (1 half-life)	0.433	0.500
9,000 (2 half-lives)	0.650	0.250
13,500 (3 half-lives)	0.758	0.125

[a]From various sources.

One of the assumptions we have made in applying radioactivity to age determination is that the laws governing the rate of decay have remained constant over incredibly long periods of time; we are justified in wondering whether this assumption can be valid.

Geology supplies one piece of confirming evidence in the form of **pleochroic** ("many-colored") **halos.** These are minute, concentric, spherical zones of dark or colored material no more than 7.5×10^{-3} cm in diameter—the thickness of an ordinary sheet of paper—that form around inclusions of radioactive materials in biotite and in a few other minerals. If such a sphere is sliced through the center, the resulting sections show the pleochroic halos as rings. Each ring is a region in which alpha particles, shot out of the decaying radioactive mineral, came to rest and ionized the host material. The effect resembles that of light on a photographic film.

The energy possessed by an alpha particle depends on what element released it and on the stage of decay at which the particle was released. The pleochroic halos have radii that correspond to the energy of present-day alpha particles. Apparently the energy of an alpha particle today is the same as it was hundreds of millions of years ago. This fact implies that the fundamental constants of nuclear physics that govern the travel of alpha particles have not changed.

This evidence, of course, does not prove explicitly that the rate of decay has always been the same. But if the laws governing the energy of the particles have not changed, it is reasonable to assume that related laws governing the rate of decay have also continued unchanged.

Uranium 238 is not the only element found useful in the age determination of rock material; there are others. But the same basic idea prevails, whatever element we choose. If we discover a mineral that contains one or more radioactive elements, we may be able (after proper chemical analysis) to determine how many years ago the mineral was formed. If the mineral was formed at the same time as the rock in which it is enclosed, the age of the mineral will also give us the age of the rock.

A radioactive element decays in one of several ways. The nucleus may lose an alpha particle—the nucleus of the helium atom, which, as we saw in Chapter 2, consists of two protons and two neutrons. This process is called **alpha decay,** and the mass of the element decreases by four (two protons plus two neutrons), and the atomic number decreases by two (two protons). An element undergoing **beta decay** loses an electron (a beta particle) from one of the neutrons of the nucleus. The neutron thereby becomes a proton, and the atomic number is increased by one. The mass—the sum of all protons and neutrons—remains the same. In a third type of decay, **electron-capture decay,** the nucleus changes by picking up an electron from its orbital electrons. This electron is added to a proton within the nucleus, converting the proton to a neutron and decreasing the atomic number by one.

Many elements, in decaying from a radioactive to a nonradioactive state, go through a series of transformations until one or more stable end products are reached. Thus uranium 238 begins to decay by alpha emissions before lead 206 is produced.

Some elements may follow two or more different paths in their decay. Potassium 40 is an example. It

TABLE 7.2

Some of the more useful elements for radioactive dating

Parent element	Half-life, yr	Daughter element	Type of decay
Carbon 14	5,730	Nitrogen 14	Beta
Potassium 40	1,300 million	Argon 40	Electron capture
Rubidium 87	47,000 million	Strontium 87	Beta
Uranium 235	713 million	Lead 207	Seven alpha and four beta
Uranium 238	4,510 million	Lead 206	Eight alpha and six beta

may decay either by beta emission to form calcium 40 or by electron capture to form argon 40.

Methods of determining age by radioactivity have produced tens of thousands of dates for events in earth history, and new ones are constantly being reported. The most reliable dates utilize more than one isotopic pair and yield consistent results.

Rocks from southwestern Minnesota and Greenland have been dated as approximately 3.8 billion years old. Field relations show that other, still older rocks exist. A granite about 3.2 billion years old from Pretoria, South Africa, contains inclusions of a quartzite, positive indication that older sedimentary rocks existed before the intrusion of the granite. The exact age of the earth itself is still undetermined, but several lines of evidence converge to suggest an age of about 4.5 billion years.

Today's discoveries, then, have fully vindicated the assumptions made over a century and a half ago that geologic time *is* vast and that within earth history there *is* abundant time for slow processes to accomplish prodigious feats.

There are scores of radioactive isotopes, but only a few are useful in geologic dating (see Table 7.2). For older rocks, potassium 40, rubidium 87, uranium 235, and uranium 238 have proved most important. Potassium 40 has also proved useful in dating some of the more recent events of earth history, events younger than 2 million years. For very recent events, however, the radioactive isotope of carbon, carbon 14, $^{14}_{6}C$, has proved most versatile. It is usable only on organic material that is around 50,000 years old or less than that age.

The carbon 14 method, first developed at the University of Chicago by Willard F. Libby, works as follows. When neutrons from outer space, sometimes called **cosmic rays,** bombard nitrogen in the outer atmosphere, they knock a proton out of the nitrogen nucleus, thereby forming **carbon 14:**

Neutron

$\downarrow$

$^{14}_{7}N$

Proton $\swarrow$ $\searrow$ $^{14}_{6}C$

The carbon 14 combines with oxygen to form a special carbon dioxide, $^{14}CO_2$, which circulates in the atmos-

phere and eventually reaches the earth's surface, where it is absorbed by living matter. It has been found that the distribution of carbon 14 around the world is almost constant. Its abundance is independent of longitude, latitude, altitude, and the type of habitat of living matter.

The bulk of carbon in living material is the stable isotope carbon 12. Nevertheless, there is a certain small amount of carbon 14 in all living matter. And when the organism—whether it is a plant or an animal—dies, its supply of carbon 14 is, of course, no longer replenished by life processes. Instead, the carbon 14, with a half-life of about 5,730 years, begins spontaneously to change back to $^{14}_{7}N$ by beta decay. The longer the time that has elapsed since the death of the organism, the less the amount of carbon 14 that remains. So when we find carbon 14 in a buried piece of wood or in a charred bone, by comparing the amount present with the universal modern $^{14}C/^{12}C$ ratio, we can calculate the amount of time that has elapsed since the material ceased to take in carbon 14, that is, since the organism died.

We have seen that there are assumptions on which we base our evaluation of the validity of radioactive dating. These assumptions are reasonable and have some support from observation. Nevertheless, the more we use radioactive methods, the more sophisticated our understanding of them—which is another way of saying that we become aware of the problems involved. Let us take carbon 14 as an example of how increased knowledge reveals increased complexity.

When Libby did his initial work on carbon 14, its half-life was determined to be 5,570 years. More refined measurements now indicate it to be about 5,730 years. By common agreement, however, 5,570 is still used as the "accepted value," and therefore all published dates should be adjusted upward 3 percent. Beyond this, Hans Suess, while working with the United States Geological Survey, demonstrated that since about 1850, the amount of nonradioactive, or "dead," carbon 12 poured into the atmosphere by the combustion of coal and petroleum has diluted the amount of carbon 14 to 98 percent of its original concentration before human beings began tampering with the atmosphere. Working in the opposite direction, we have created a great amount of new carbon 14 since 1945, when we began atmospheric nuclear explosions. In fact this source of contamination has doubled the carbon 14 activity of the atmosphere.

Of more historical interest, perhaps, is the recent discovery that radiocarbon dates obtained on materials from about the time of Christ and going back for another 5,000 years, at least, are younger than we should expect when we compare them with carbon 14 dates of samples of known age dated independently of carbon 14. These samples have included historically dated material from the Mediterranean area, particularly from Egypt, and tree rings of the bristlecone pine from the western United States. This tree, which includes the world's oldest known living material, has provided us with a set of tree rings that goes back over 7,000 years and is securely tied to our modern solar calendar. When we subject tree rings older than 2,000 years to carbon 14 dating, the radiocarbon date is younger than the actual date. The farther back we go, the greater is the divergence. For example, by the time the tree ring count reaches 3200 B.C., the radiocarbon age is only 2700 B.C. (see Figure 7.6). We do not yet know the cause of this divergence or how far back into time it may continue. But it suggests that we are well-advised to use the term **radiocarbon years** to indicate that radiocarbon dates are not absolutely synonymous with dates in our solar calendar.

Despite these difficulties, the thousands of radiocarbon dates now available have revolutionized many aspects of archaeology as well as the study of the last 50,000 years of earth history.

Fission-track dating One of the newest methods of absolute age dating results from the spontaneous fission of certain heavy elements, particularly the relatively abundant isotope uranium 238. As this isotope undergoes fission, the resulting parts of its nucleus leave characteristic tracks in the crystals of rocks where they occur. The number of such tracks gives us a clue to the absolute ages of those rocks, ages that can be measured from a few tens of years ago to the beginning of the solar system. This tremendous range makes the fission-track dating method extremely useful.

RADIOACTIVITY AND SEDIMENTARY ROCKS

Until recently, radioactive minerals suitable for dating geologic events were sought chiefly in the igneous rocks. These rocks were usually the uranium- and thorium-bearing minerals. Even today the great bulk of the radioactively dated rocks are igneous in origin. The development of new techniques, however, particularly the use of radioactive potassium, has extended radioactive dating to some of the sedimentary rocks.

Some sandstones and, more rarely, shales contain glauconite, a silicate mineral similar to biotite. Glauconite, formed in certain marine environments when the sedimentary layers are deposited, contains radioactive potassium, another geologic hourglass that reveals the age of the mineral and hence the age of the rock. The end products of the decay of radioactive potassium are argon and calcium. Age determinations are based on the ratio of potassium to argon.

Some pyroclastic rocks are composed mostly or completely of volcanic ash. Biotite in these rocks includes radioactive potassium and so offers a way of dating the biotite and sometimes the rock itself.

SEDIMENTATION AND ABSOLUTE TIME

Another way of establishing absolute dates for sedimentary strata is to determine the rate of their deposition.

Certain sedimentary rocks show a succession of thinly laminated beds. Various lines of evidence suggest that, in some instances at least, each one of these beds represents a single year of deposition. Therefore, by counting the beds, we can determine the total time it took for the rock to be deposited.

Unfortunately we have been able to link this kind of information to our modern calendar in only a very few places, such as the Scandinavian countries. Here the Swedish geologist Baron de Geer counted the annual deposits, or laminations, that formed in extinct glacial lakes. These laminations, called **varves** (see Figure 7.7 and Chapter 14), enable us to piece together some of the geologic events of the last 20,000 years or so in the countries ringing the Baltic Sea.

Much longer sequences of laminated sediments have been found in other places, but they tell us only the **total length of time** during which sedimentation took place, **not how long ago** it happened in absolute

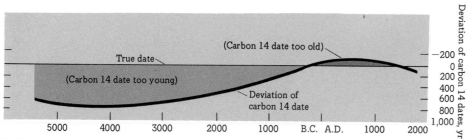

Dendrochronological dates

Deviation of carbon 14 dates, yr

7.7 Varves from Puget Lowland, Washington. Strips of paper are used to record thickness and succession of the various layers. Each pair of light (summer) and dark (winter) layers is thought to represent a single year of sedimentation. (Note pocket knife holding uppermost paper.) [J. Hoover Mackin.]

7.8 The Green River Shale in Wyoming is composed of minute annual layers. Because each step in this block is 100 layers high, the entire block records 700 years of sedimentation. A portion of a fossil fish is seen on the large step on the left-hand side of the block. Counting of the layers indicates that the fish died 471 years after the formation of the lowest layer. [Specimen from Princeton University Museum of Natural History, Willard Starks.]

7.3 THE GEOLOGIC COLUMN

Using the law of superposition and the concept that fossils are an index to time, geologists have made chronological arrangements of sedimentary rocks from all over the world. They have pictured the rocks as forming a great column, with the oldest at the bottom and the youngest at the top. The pioneer work in developing this pattern was carried out in the British Isles and in western Europe, where modern geology had its birth. There geologists recognized that the change in the fossil record between one layer and the next was

time. One excellent example of absolute-time sequence is recorded in the Green River shales of Wyoming (see Figure 7.8). Here, each bed, interpreted as an annual deposit, is less than 0.02 cm thick, and the total thickness of the layers is about 980 m. These shales then represent approximately 5 million years of time.

TABLE 7.3

The geologic column

Era	System	Series	Some major events	Dominant life form[a]		
						Man
Cenozoic	Quaternary[b]	Holocene (Recent)				
		Pleistocene				
	Tertiary[b]	Pliocene		Mammals		
		Miocene	Grasses become abundant			Flowering plants
		Oligocene				
		Eocene	Horses first appear			
		Paleocene				
Mesozoic	Cretaceous	—[c]	Extinction of dinosaurs		Conifer and cycad plants	
	Jurassic		Birds first appear	Reptiles		
	Triassic		Dinosaurs first appear			
Paleozoic	Permian	—[c]				Amphibia
	Pennsylvanian		Coal-forming swamps	Spore-bearing land plants		
	Mississippian					
	Devonian				Fish	
	Silurian		Vertebrates first appear (fish)			Marine invertebrates
	Ordovician					
	Cambrian		First abundant fossil record (marine invertebrates)			Marine plants
Precambrian	—[d]			Primitive marine plants and invertebrates One-celled organisms		

[a]This column does not give the complete time range of the forms listed. For example, fish are known from pre-Silurian rocks and obviously exist today; but when the Silurian and Devonian rocks were being formed, fish represented the most advanced form of animal life.
[b]Some geologists prefer to use the term Cenozoic for the Quaternary and the Tertiary.
[c]Many Mesozoic and Paleozoic series are distinguished but are not necessary here.
[d]Precambrian rocks are abundant, but worldwide subdivisions are not generally agreed upon.

not gradual but sudden. It seemed as if whole segments recording the slow change of plants and animals had been left out of the rock sequence. These breaks, or gaps, in the fossil record served as boundaries between adjoining strata. The names assigned to the various groups of sedimentary rocks are given in the **geologic column** of Table 7.3. (See also front endpaper.)

Notice that the oldest rocks are called **Precambrian,** a general term applied to all the rocks that lie beneath the Cambrian rocks. Although the Precambrian rocks represent the great bulk of geologic time, we have yet to work out satisfactory subdivisions for them because there is an almost complete absence of fossil remains, so plentiful in Cambrian and younger rocks. Without fossils to aid them, geologists have been forced to base their correlations on the physical features of the rocks and on dates obtained from radioactive minerals. Physical features have been useful for establishing local sequences of Precambrian rocks, but these sequences cannot be extended to worldwide subdivisions. On the other hand, radioactive dates are

not yet numerous enough to permit subdivisions of the Precambrian rocks, although such dates will become increasingly important as more are assembled.

We have said that the geologic column was originally separated into different groups of rocks on the basis of apparent gaps in the fossil record. But as geologic research progressed and as the area of investigation spread from Europe to other continents, new discoveries narrowed the gaps in the fossil record. It is now apparent that the change in fossil forms has been continuous and that the original gaps can be filled in with data from other localities. This increase in information has made it more and more difficult to draw clear boundaries between groups of rocks, and yet, despite the increasing number of "boundary" problems, the broad framework of the geologic column is still valid.

7.4 THE GEOLOGIC-TIME SCALE

The names in the geologic column refer to rock units that have been arranged in a chronological sequence from oldest to youngest. Because each of the units was formed during a definite interval of time, they provide a basis for setting up time divisions in geologic history.

In effect, the terms we apply to time units are the terms that were originally used to distinguish rock units. Thus we speak either of **Cambrian time** or of **Cambrian rocks.** When we speak of **time units,** we are referring to the **geologic-time scale.** When we speak of **rock units,** we are referring to the **geologic column.**

The geologic-time scale is given in Table 7.4 as well as in the front endpaper. Notice the terms **Era, Period,** and **Epoch** across the top of the table. These are general time terms. Thus we can speak of the Paleozoic Era, or the Permian Period, or the Pleistocene Epoch. In Table 7.3, the terms **System** and **Series,** used as general terms for rock units, correspond to the time units of "period" and "epoch," respectively. The term **Erathem** is becoming the generally accepted rock term equivalent to the "era" of the geologic-time scale.

7.9 This diagram illustrates the law of crosscutting relationships, which states that a rock is younger than any rock that it cuts. The rock units are arranged in order of decreasing age from 1 to 5. The manner in which radioactive ages of the igneous rock (2 and 4 of the diagram) are used to give approximate ages in terms of years for the sedimentary rocks (1, 3, and 5 of the diagram) is discussed in the text and in Table 7.5.

ABSOLUTE DATES IN THE GEOLOGIC-TIME SCALE

We have found that the geologic-time scale is made up of units of relative time and that these units can be arranged in proper order without the use of any designations of absolute time. This relative-time scale has been constructed on the basis of sedimentary rocks. As noted earlier, geologists have only recently been able to date some sedimentary rocks by radioactive methods, and most dates have come from the igneous rocks. How can the dates obtained from igneous rocks be fitted in with the relative-time units from sedimentary rocks?

To insert the absolute age of igneous rocks into the geologic-time scale, we must know the relative-time relationships between the sedimentary and igneous rocks. The basic rule here is called the **law of crosscutting relationships,** which states that **a rock is younger than any rock that it cuts across.**

Let us consider the example given in Figure 7.9, which shows a hypothetical section of the earth's crust with both igneous and sedimentary rocks exposed. The sedimentary rocks are arranged in three assemblages, numbered 1, 3, and 5, from oldest to youngest. The igneous rocks are 2 and 4, also from older to younger.

The sedimentary rocks labeled 1 are the oldest rocks in the diagram. First they were folded by earth forces; then a dike of igneous rock was injected into them. Because the sedimentary rocks had to be present before the dike could cut across them, they must be older than the dike. After the first igneous intrusion, erosion beveled both the sedimentary rocks and the dike, and across this surface were deposited the sedimentary rocks labeled 3. At some later time, the batholith, labeled 4, cut across all the older rocks. In time,

this batholith and the sedimentary rocks 3 were also beveled by erosion, and the sedimentary rocks labeled 5 were laid across this surface. We now have established the relative ages of the rocks, from oldest to youngest, as 1, 2, 3, 4, and 5.

Now, if we can date the igneous rocks by means of radioactive minerals, we can fit these dates into the relative time sequence. If we establish that the batholith is 230 million years old and that the dike is 310 million years old, the ages of the sedimentary rocks may be expressed in relation to the known dates; the final arrangement will be as shown in Table 7.5.

By this general method, approximate dates have been assigned to the relative-time units of the geologic-time scale, as shown in Table 7.4. These dates may be revised and refined as new techniques for dating develop. One of the most exciting developments in this field lies in the direct application of dating methods to the radioactive minerals of sedimentary rocks, as suggested earlier.

MAGNETOSTRATIGRAPHY

The direction of the earth's magnetic field in past geologic ages can be determined by examining lava flows on land, deep-sea cores of sediments, and igneous rocks from many parts of the earth's surface (see Chapter 10). These rocks contain iron minerals (magnetite, ilmenite, hematite) that become magnetized at the time

TABLE 7.4

The geologic-time scale[a]

Era	Period	Epoch	Millions of yr	
			Duration	Before present
Cenozoic	Quaternary	Holocene	0.01	
		Pleistocene	1.5–2	
	Tertiary	Pliocene	5–5.5	
Mesozoic		Miocene	19	
		Oligocene	11–12	
Paleozoic		Eocene	15–17	
		Paleocene	11–12	
				65
	Cretaceous		71	
	Jurassic		54–59	
	Triassic		30–35	
				225
	Permian		55	
	Pennsylvanian		45	
Precambrian	Mississippian		20	
	Devonian		50	
	Silurian		35–45	
	Ordovician		60–70	
	Cambrian		70	
				570
	Precambrian time[b]			

[a] From W. B. Harland, A. Gilbert Smith, and B. Wilcock (eds.), "The Phanerozoic Time-Scale," *Quart. J. Geol. Soc. London*, vol. 120S (suppl.), 1964.
[b] The oldest rocks are about 3.8 billion years old; the earth itself is 4 to 5 billion years old.

of their formation, thus creating a "magnetic finger-print" in the rock for a certain instant of geologic time. These rock records indicate that the earth's magnetic field has changed directions many times, with the present field representing **normal** directions for north and south. During intervals of **reversed polarity,** the poles apparently switch from south to north and vice versa. These changes have taken place fairly rapidly (geologically speaking) and have affected newly forming magnetic minerals. The process of reversal takes about 2,000 years according to studies conducted at the University of Pittsburgh.

Reversals have apparently occurred during at least the last billion years. Studies of magnetism in rocks from various parts of the world can be used with the geologic time scale to work out a magnetic time chart. A pattern emerges for the polarity reversals through geologic time, with **polarity epochs**[1] lasting roughly 1 million years. Within these epochs there occur several **polarity events**[2] in which the earth's polarity switches from one direction to another and back to the first, with durations on the order of 100,000 years for each polarity. Such a pattern has existed for the last 70 million years (see Figure 7.10). Prior to that time, there appeared long intervals in which the polarity directions remained fairly constant. This latter uniform interval was preceded by a "mixed interval," when

[1] Called polarity **zones** by the International Subcommittee on Stratigraphic Classification (ISSC), *Geology,* vol. 7, pp. 578–583, December 1979.
[2] Called polarity **subzones** by ISSC; ibid.

TABLE 7.5

Relative, absolute, and approximate ages of rocks[a]

Event	Age		
	Relative	Absolute, millions of yr	Approximate, millions of yr
Sedimentary rocks	5[b]		<230
Erosion			<230
Batholith	4	230	
Sedimentary rocks	3		>230, <310
Erosion			>230, <310
Dike	2	310	
Folding			>310
Sedimentary rocks	1		>310

[a] See Figure 18.10 and the text.
[b] Youngest.

changes occurred frequently. Such polarity episodes of rapid change alternating with long intervals of little change occurred as far back as a billion years, according to present understanding. As we go farther back in geologic time, of course, the difficulties of age dating and measuring magnetic directions increase enormously. Nevertheless a distinct pattern is emerging for the use of magnetized rocks in determining a **magnetic stratigraphy (magnetostratigraphy)** for the earth.

Geologic time can be expressed in either relative or absolute terms.

Relative time has been determined largely by the relative position of sedimentary rocks to each other.

The rocks are arranged in proper chronological position according to the **law of superposition.**

Correlation of sedimentary rocks from one area to another allows the extension of a chronology of relative time from region to region and continent to continent. Correlation is based on **physical features** but most importantly on **fossils.**

Absolute time is expressed in terms of years and is measured by the rate of decay of radioactive elements. Among the most useful such elements are **carbon 14, potassium 40, rubidium 87, uranium 235,** and **uranium 238.**

The geologic column is a chronological sequence of rocks, from oldest to youngest. The units are rock units.

The geologic-time scale is the chronological sequence of units of earth time represented by the rock units of the world geologic column. Absolute dates in the geologic-time scale are based on the relation of radioactively dated igneous, sedimentary, and metamorphic rocks of the geologic column. Position in the time scale of events dated from radioactive rocks is determined by the relative position of those rocks and the fossils they contain.

Magnetic reversals have occurred at intervals during the geologic past and can be used to subdivide earth history into **epochs** of **magnetostratigraphy.**

7.10 Chart of magnetostratigraphy, showing magnetic-anomaly numbering system (based on marine magnetic-anomaly analysis), polarity epochs (based on potassium-argon dating and paleomagnetic studies of outcropping Pliocene and Pleistocene igneous rocks), and the polarity-epoch numbering system (based on micropaleontological and paleomagnetic analysis of piston cores). See also Figure 10.8. [Modified from N. D. Watkins, *Geotimes*, April 1976.]

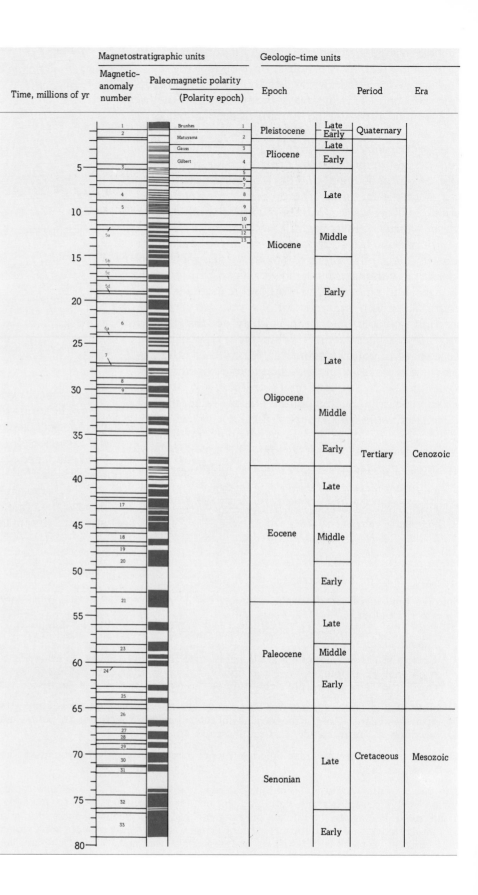

Berry, William B. N. *Growth of a Prehistoric Time Scale*, W. H. Freeman & Company Publishers, San Francisco, 1968. *Historical account of the development of the geologic time scale, with emphasis on the role of organic evolution.*

Libby, W. F. "Radiocarbon Dating," *Am. Sci.*, vol. 44, no. 1, pp. 98–112, 1956. *A summary of the method written by the man who discovered it.*

Ojakangas, R. W., and D. G. Darby *The Earth, Past and Present*, McGraw-Hill Book Company, New York, 1976. *A condensed review of the earth through time, including biologic and physical developments.*

Ralph, E. K., and H. N. Michael "Twenty-Five Years of Radiocarbon Dating," *Am. Sci.*, vol. 62, no. 5, pp. 553–560, 1974. *A review of errors inherent in radiocarbon dating and the use of long-lived bristlecone pines to correct radiocarbon dates.*

Watkins, N. D. "Polarity Subcommission sets up some Guidelines," *Geotimes*, vol. 21, no. 4, pp. 18–20, 1976. *A short history of its development and presentation of data and chart for magnetostratigraphy.*

1 Reword and simplify the law of superposition. What does it mean in simple terms?

2 How are rocks correlated from one place to another by physical features? By fossils?

3 Explain briefly the carbon 14 age-dating method.

4 How are radioactive age-dating techniques possible on sedimentary rocks?

5 What proportion of the total age of the earth is represented by Precambrian time? By Phanerozoic (that is, post-Precambrian) time?

6 Explain how age dating of igneous intrusions aids in the establishment of a geologic-time scale.

7 What is meant by normally polarized rocks? By reversed polarity?

8 What causes reversals in the earth's polarity?

8

DEFORMATION

Many rocks that we see at the earth's surface were formed from materials transported and deposited by wind, water, ice, or gravity. Nearly all such rocks were originally deposited in a horizontal position. Today, we find many of these same rocks tilted, folded, or even highly contorted (see Figure 8.1).

The crust of the earth is continuously changing. We have already discussed changes produced by weathering and erosion, but the crust is also being changed in other more fundamental ways. It is being deformed by forces acting within it. These forces help to maintain the surface of the land above sea level and work to offset, or compensate for, the destructive effects of erosion.

8.1 DEFORMATION OF THE EARTH'S CRUST

We see bedrock at the earth's surface in natural stream-cut and beach exposures, in quarries, and in road or railroad cuts formed by human activities. Frequently these rocks show distorted layering. We speak of these rocks as having been **deformed** from their original nearly flat-lying condition.

Evidence of deformation can be seen everywhere. Sediments that were deposited on the bottom of the ocean are now found hardened into rocks in mountainous areas high above sea level. These rocks contain fossils that attest to their sedimentary origin. Far below sea level, miners often find pieces of trees or other plants embedded in layers of coal. This shows that these beds, now deeply buried, were at one time lying on the earth's surface.

Deforming forces elevate and depress large landmasses, and they also work to distort the earth's crust.

8.1 Rocks tilted at an angle to the horizontal. Originally these beds were flat-lying. Outcrop of Devonian Catskill Formation along the Susquehanna River, 20 km north of Harrisburg, Pennsylvania. [M. E. Kauffman.]

8.2 Photograph of deformed rocks along Interstate Highay 81 in north-central Pennsylvania. [J. H. Moss.]

This deformation we call **tectonism.** Rocks that are rigid at the surface become plastic at depth and respond to deforming forces by folding (see Figure 8.2). These shapes could not have been produced in a rigid rock.

ABRUPT MOVEMENTS

Crustal deformation has occurred during some large earthquakes. This has resulted in measurable displacements in historic times; surveys have revealed that large portions of the earth's crust have been warped, tilted, or moved horizontally. Earthquakes

have also altered the earth's surface by triggering landslides.

Three major earthquakes in 1811 and 1812 were centered near New Madrid, Missouri, on December 16, January 23, and February 7. These were accompanied by changes in topography over an area of 8,000 to 13,000 km². The crust sank 1.5 to 6 m in places, forming

8.3 Relative horizontal movement of earth's crust near Prince William Sound as determined by 1964 surveys of United States Coast and Geodetic Survey.

swamps and lakes and drowning forests. The largest lake to form was Reelfoot Lake in Tennessee.

The Alaskan Prince William Sound earthquake of 1964 was accompanied by considerable vertical and horizontal displacements. These occurred over more than 120,000 km². The Coast and Geodetic Survey found that there was a general south-to-north rotation of a portion of the earth's crust centered at the northern coast of Prince William Sound. A maximum uplift of 13 m occurred off the southwestern tip of Montague Island. The largest subsidence of 2 m occurred 43 km north of Glennallen. Maximum horizontal movements were found between Montague and Latouche Islands, which, as we shall see, were 5 to 7 m closer together after the earthquake than before. Surveys showed that the southwest side of Montague Island moved northwest with respect to Latouche Island. Figure 8.3 shows horizontal movements that occurred and the areas involved. One of the largest changes took place at Val-

dez when the earthquake triggered movement that deepened the harbor as much as 100 m in one place.

Vertical displacements Vertical displacements during one earthquake cannot account for the amount of elevation of the earth's crust associated with mountains or for the large depressions of sedimentary layers found in deep coal mines. However, if displacements occur often enough throughout geologic time, the total can become significant. For instance, on the shore of Sagami Bay, Japan, not far from Yokohama, a cliff reared up during an earthquake on September 1, 1923. The amount of movement was measured by using some marine bivalves called *Lithophaga* ("rock eaters") as a reference. These little animals scoop out small caves for their 5-cm shells at mean sea level and spend their lives waiting for the sea to bring food at each rise of the tide. After the 1923 earthquake, rows of *Lithophaga* were found starved to death 5 m above the

8.4 Drawing showing elevated cliff on Sagami Bay, Japan. Inset, *Lithophaga* shells.

waters that used to feed them. Other rows of *Litho-phaga* holes in this same cliff were found and correlated with quakes in A.D. 1703, 818, and 33 (see Figure 8.4). The total elevation over that 2,000-year interval was 15 m. At this rate the elevation over a geologic time interval of 2 million years would be 15,000 m.

Horizontal displacements The San Andreas fault is a large scar in the earth's crust approximately 1,000 km long extending in a line from Cape Mendocino southeastward to the Gulf of California (see Figure 8.5). Displacements have occurred along portions of the fault at the times of some earthquakes but not over its entire length during any one quake (see Figure 8.6). At the time of the earthquake of April 18, 1906, there were horizontal displacements along 400 km of the northerly

8.5 (above) A section of the San Andreas fault in the Indio Hills, California; view looks northwest. An abrupt movement of 6 m along part of this fault occurred at the time of the San Francisco earthquake of 1906. [Spence Air Photos.]

8.6 (right) Section of the San Andreas fault that slipped at the time of the earthquake in 1906. [After John H. Hodgson, *Earthquakes and Earth Structure*, p. 15, Prentice-Hall, Inc., Englewood Cliffs, N. J., 1964.]

8.1   DEFORMATION OF THE EARTH'S CRUST

end from Point Arena to San Juan Bautista. The maximum slippage was 7 m on the shore near Tomales Bay. This displacement died out in both directions.

Rocks on both sides of the San Andreas fault have been studied, and some geologists believe formations can be correlated that were joined across the fault as a single unit 150 million years ago. Estimates of the total displacement run in the scores of kilometers up to as many as 500 km.

SLOW MOVEMENTS

Not all crustal movements are accompanied by earthquakes. Slow movements of the crust are going on today. These are being measured along faults in California by government geologists and university investigators in a broad, continuing program. They use cross-fault strain meters, which show that deformation occurs in zones up to 10 m across. Creep is in progress along the Hayward fault from Richmond to south of Fremont, along the Calaveras fault in the Hollister area, and along the San Andreas fault from San Juan Bautista to Cholame (see Figure 8.7). The time and amount of movement are shown in Figure 8.8. It is interesting to note that at San Juan Bautista and at the Cienega Winery, the creep occurs as distinct events, whereas the pattern of movement is different at Stone Canyon Observatory. Creep does not occur simultaneously at San Juan Bautista and Cienega Winery but may be several days apart. The difference in time at which creep starts suggests that it propagates from San Juan Bautista to the Cienega Winery at a rate of approximately 10 km/day.[1] The average movement along a 100-km section of the San Andreas in the vicinity of Parkfield is 5 cm/year, with the Pacific side of the fault moving northwestward.[2]

Another example is that of Mount Suribachi on Iwo Jima—in 1945, the scene of a famous flag raising by the

[1] Robert D. Nason, "Preliminary Instrumental Measurements of Fault Creep Slippage on the San Andreas Fault, California," *Earthquake Notes*, vol. 40, pp. 6–10, 1969.
[2] Stewart W. Smith and Max Wyss, "Displacement on the San Andreas Fault Subsequent to the 1966 Parkfield Earthquake," *Bull. Seis. Soc. Am.*, vol. 58, pp. 1955–1973, 1968.

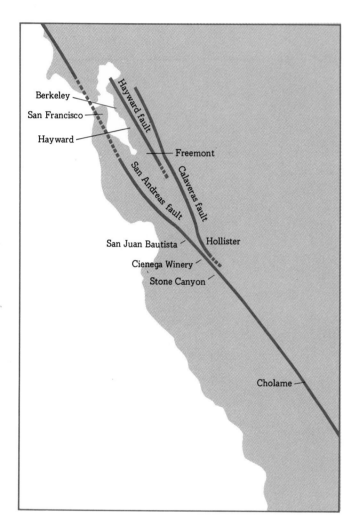

8.7 Sections of the San Andreas, Hayward, and Calaveras faults, showing places of creep measurement. Heavier lines indicate sections of faults undergoing active movement.

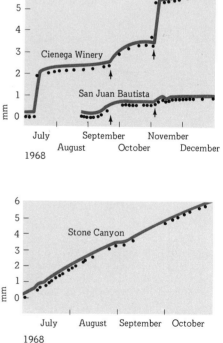

8.8 Fault movement at Cienega Winery, San Juan Bautista, and Stone Canyon (18 km southeast of Cienega Winery.)

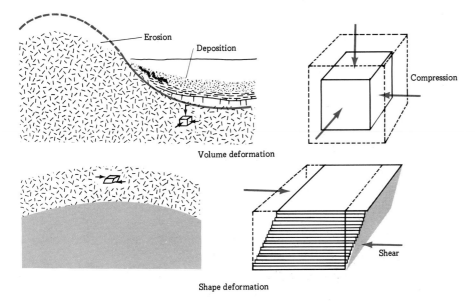

Deformation may produce change in volume without change in shape, or change in shape without change in volume, or a combination of the two.

Volume deformation

Shape deformation

United States marines. According to geologist Takeyo Kosaka, pressure from underlying magma caused the surface to rise 7 m in 23 years from that date.

8.2 STRESS, STRAIN, AND STRENGTH OF ROCKS

Deformation of the earth's crust can be described in terms of change of volume, change of shape, or a combination of both. The kinds of deformation are illustrated in Figure 8.9. If a material changes shape without a change of volume, the deformation is called **shear.** An analogy for pure shear is supplied by staggering a deck of cards. If the bottom card is held fixed and the others are slid forward by amounts proportional to their distances from the bottom, the shape of the space occupied by the deck is altered, but its volume (the total amount of cardboard) remains the same. A force applied to material that tends to change that material's dimensions is called **stress.** This is commonly known as **unit stress,** defined as the total force divided by the area over which the force is applied.

STRESS AND STRAIN

The effects of some deforming forces on earth materials have been studied in the laboratory by measuring the strain produced by application of stress. For example, if a cylindrical specimen of rock is stretched, the amount of its extension can be plotted against the force producing it. If its original length is l_0 and the force used to stretch it is P, the increase in length Δl leaves it at a new length l. The **strain,** sometimes called **unit**

strain, is the change in length per unit length, or $\Delta l/l_0$. The stress, or unit stress, is the total force P divided by the area of the cross section of the specimen, or P/A (see Figure 8.10). Graphs of stress versus strain are usually set up with unit stress and unit strain as the axes.

In most materials, as stress increases, for a time the strain is proportional to stress. If stress is removed, the strain goes back to zero. This is the range of elastic deformation. Then, when the stress reaches the yield point, the strength of the material has been overcome. Deformation becomes plastic; that is, with no additional stress, strain continues to increase. When stress is removed, the plastic portion is not recovered (Figure 8.11).

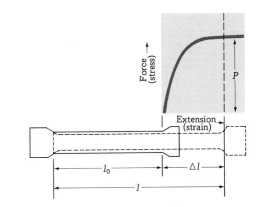

8.10 Deformation of a cylindrical specimen of rock by extension. See the text for discussion.

8.11 Behavior of material under stress. (a) As stress increases, for a while strain is proportional to stress. This is the range of elastic deformation. Beyond the yield point, deformation becomes plastic for a while until the material again resists, and stress must be increased to produce more strain. (b) After deformation to point *B*, if the stress is removed, (1), the elastic part of the strain (deformation) is recovered (to point *C*), but the rest is not. A retardation of the recovery occurs. If stress is again applied, (2), deformation picks up where it left off at *B* and continues as though there had been no interruption.

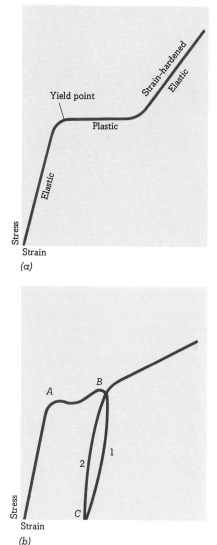

Retardation of recovery Rocks are said to be deformed elastically if the deformation disappears when the stress is removed. However, they do not ordinarily regain their former shape the instant the stress is removed. There is a time lag over which recovery is made.

It has been suggested that a demonstration of this retardation is provided on a grand scale by downwarping of the earth's crust under loads of glacial ice and subsequent recovery upon melting of the ice. The meltings and upwarpings from the last ice advance began about 15,000 years ago, but recovery is still progressing in areas where the ice completely disappeared as much as 10,000 years ago. Table 8.1 lists some areas where upwarping has occurred and the amounts recovered so far.

TABLE 8.1

Upwarping in regions of complete or partial deglaciation[a]

Area	Maximum observed uplift, m
Scotland	30
Iceland	120
Greenland	146 +
Novaya Zemlya	100
New Zealand (South Island)	100 ±
Antarctica	100
Norway and Sweden	275 +
Eastern Canada–Labrador	270 +
Newfoundland	137

[a]R. A. Daly, *Strength and Structure of the Earth,* Prentice-Hall, Inc., Englewood Cliffs, N.J., 1940.

Elastic deformation Elastic deformation of rock is recoverable. A solid is deformed elastically if it can recover its size and shape after the deforming force has been removed. An elastic solid resists deformation up to a certain point. Its resistance to a change in volume without a change of shape is defined as the **ratio of stress to strain:**

$$\frac{\text{Increase in force per unit area}}{\text{Change in volume per unit volume}}$$

A material's resistance to elastic shear is called **rigidity.**

Plastic deformation Plastic deformation is permanent. It involves a property of rock called **viscosity.** A material that is deforming plastically does so by the propagation and movement of dislocations (or small structural defects in the material). If the rate of the flow is proportional to the stress causing it, the material is said to be **viscous;** and if we plot the rate of flow against the stress, the slope of the graph is the viscosity (Figure 8.12). **Viscosity equals stress per rate of flow.** Viscosity is an important property in some geological processes. It is the property that governs the ability of magma to flow during igneous activity and enables the mantle to adjust to crustal loads. It is measured experimentally by observing rate of flow, the force needed to turn a paddle wheel at constant speed, or the rate of sinking of standard spheres (see Figure 8.13).

We do not know the absolute viscosity of materials in the earth's interior. This is difficult to determine ex-

8.12 Demonstration of viscosity. If an oil film of thickness d rests on a surface OX under a block P and P is displaced sidewise with a velocity V, the change in the angle α that OP makes with the vertical can be used to measure the shear (change in shape) of the oil. If the rate of shear is proportional to the force causing it, the deformation is said to be *viscous*.

8.13 Methods of measuring viscosity: (*a*) rate of flow, (*b*) force necessary to turn paddle wheel at known speed, and (*c*) rate at which spheres of the same material but different radii fall through the fluid. For c, $x = 0.22gr^2(d - d')/v$, where x is the velocity of the sphere when the motion is steady (terminal velocity), g is the acceleration of gravity, d is the density of the sphere, d' is the density of the fluid, r is the radius of the sphere, and v is the viscosity of the fluid. In a granite magma with the viscosity of hard pitch (10^{13} times that of water, or 10^{10} times that of glycerine) a sphere of gneiss with a diameter of 2 m would sink about 10 cm/day. A 4-m sphere would sink four times as fast.

perimentally. However, the effect of viscosity may be illustrated by considering a granite magma with a viscosity of hard pitch, 10^{13} times that of water. A sphere of gneiss 2 m in diameter would sink in this magma about 10 cm/day. A sphere 4 m in diameter would sink four times as fast.

Time factor Within the earth, materials show a duality of response. They transmit earthquake waves elasti-

cally but move plastically during the formation of mountains and other surface features. Here an important factor that cannot be reduced to a graph is **time**. Material may be elastic for short-time stresses in transmitting earthquake waves but plastic over a longer time.

For a short time, a tuning fork formed from pitch can ring as though it were purest steel. Yet, given time, the fork will flow into a formless blob from its own weight. A steel bar supported at its ends and loaded in the middle will appear to have strength entirely adequate to support the load but will in time slowly bend. And some metamorphic rocks testify to the deformation of rocks under prolonged periods of stress and increased temperature. On geologic-time scales, the rocks of the crust have shown themselves to be weak.

Types of stress There are different types of stress: **tension, compression,** and **shear.** Tension is a stretching stress and can increase the volume of a material. Compression tends to decrease the volume. A shear stress, as described, produces changes in shape. A stress beyond a material's strength can cause rupture.

STRENGTH OF ROCKS

Rocks possess strength. **Strength** is the limiting stress that a solid can withstand without failing by **rupture** or continuous plastic flow.

Rocks possess several types of strength, as they respond differently to different stresses. Laboratory test results for rocks generally give **compression strength, shear strength,** and **tension strength.** These tests have shown that a rock's strength in tension is smaller than its strength in compression. When the difference between these two is large, we say the material is **brittle.** When it is small, we say the material is **ductile.** As the confining pressure (in the earth's crust the all-sided confining pressure is **lithostatic pressure** resulting from the load of overlying rocks) increases, a brittle rock tends to become ductile (Figure 8.14). It does not rupture in the ordinary sense of disintegration but flows along an indefinite number of shear planes. Rupture in tension may be important near the earth's surface, but it will not occur at great depth. The strength in compression for brittle rocks is about 30 times larger than the strength in tension.

How a rock mass responds to stress depends on temperature and depth of burial. At depth, rocks will respond to stress by flowing before they rupture,

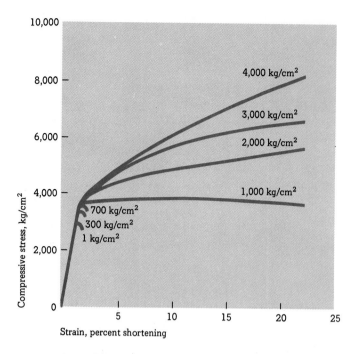

8.14 Effect of confining pressure on behavior of limestone under compression. [After E. C. Robertson, "Experimental study of the strength of rocks," *Bull. Geol. Soc. Am.* vol. 66, pp. 1294–1314, 1955.]

whereas at the earth's surface they will break. The stresses at the earth's surface are quite different from those at depth, where temperatures and pressures increase considerably. Whether a rock breaks or flows under a given stress depends on the prevailing temperatures. Rocks are generally weaker at high temperatures than they are at low ones. If a rock is surrounded by equal pressure on all sides, it tends to increase in strength. Such pressure strengthens a rock, and increased temperature weakens it.

Different rocks have different compressive strengths. Hard, brittle, competent rocks such as sandstone and limestone have a different strength than does clay or shale.

ISOSTASY AND EQUILIBRIUM

A mass standing high above its surroundings will exert a gravitational attraction that can be computed and measured. One device for measuring this attraction makes use of a plumb bob suspended on a plumb line. Like every other object on the globe, the suspended bob is pulled by gravity. On the surface of a perfect sphere with uniform density, the bob would be pulled straight down, and the plumb line would point directly toward the center of the sphere. But if there is any variation from these ideal conditions—that is, if there are surface irregularities on the sphere—the plumb line will be deflected as the bob is attracted by their concentrations of mass.

In 1749, Pierre Bouger found that plumb bobs were deflected by the Chimborazo Mountain in the Andes, but by amounts much less than calculated.[3] In 1849, in the Pyrenees, British surveyors actually found that the plumb bob appeared to be deflected away from the mountains. Similar discrepancies between calculated and measured values of gravitational attraction of mountains were also observed in the middle of the nineteenth century by British surveyors in India. They were using the plumb line to sight stars, in an attempt to fix the latitude of Kaliana, near Delhi in northern India, and of Kalianpur, about 600 km due south. They observed that the difference in latitude between these two stations was 5°23'37.06". Then they checked the difference directly by standard surveying methods. The difference computed from these measurements was 5°23'42.29". There was a discrepancy of 5.23", or about 150 m, between measurements. That may not seem very much over a distance of 600 km, but it was too large to be explained by errors of observation. Scientists then concluded that the plumb line at Kaliana had been deflected more by the attraction of the Himalaya Range than it had been at Kalianpur farther south. Actually, however, the discrepancy should have been three times as large on the assumption that the mountains were of the same average density as the surrounding terrain and that they were resting as a dead load on the earth's crust.

Two quite different explanations were proposed to explain the discrepancies between computed and observed values of gravity. The first was made by G. B. Airy in 1855. He regarded the earth's crust as having the same density everywhere and suggested that differences in elevation resulted from differences in the thickness of the outer layer. Hydrostatic equilibrium was achieved by lighter material's floating in a denser substratum. The depth of compensation is variable. Continents, mountains, and other topographic features are in equilibrium because they have roots like icebergs and are floating in a denser material.

J. H. Pratt, on the other hand, proposed that all portions of the earth's crust have the same total mass above a certain uniform level, called the **level of compensation**. Any section with an elevation higher than its surroundings would have a proportionately lower density. These ideas of Airy and Pratt are illustrated in Figure 8.15.

In 1889, C. E. Dutton suggested that different por-

[3]Pierre Bouger, *La Figure de la Terre*, p. 391, Paris, 1749.

Airy hypothesis

Pratt hypothesis

8.15 Two explanations for why mountains stand high.

tions of the earth's crust should balance out, depending on differences in their volume and specific gravity (see Figure 8.16). He called this **isostasy.** In his words:

If the earth were composed of homogeneous matter its normal figure of equilibrium without strain would be a true spheroid of revolution; but if heterogeneous, if some parts were denser or lighter than others, its normal figure would no longer be spheroidal. Where the lighter matter was accumulated there would be a tendency to bulge, and where the denser matter existed there would be a tendency to flatten or depress the surface. For this condition of equilibrium of figure, to which gravitation tends to reduce a planetary body, irrespective of whether it be homogeneous or not, I propose the name *isostasy* (from the Greek *isostasios,* meaning "in equipoise with"; compare *isos,* equal, and *statikos,* stable). I would have preferred the word *isobary,* but it is preoccupied. We may also use the corresponding adjective *isostatic.* An

isostatic earth, composed of homogeneous matter and without rotation, would be truly spherical. . . .[4]

GRAVITY

Gravity is a universal force. On earth it is the force that makes things fall to earth. In fact, it was long thought that this force did not apply to the heavenly bodies because they did not fall to earth. In 1666, when he was 24 years old, Isaac Newton proposed that gravity is a universal force: Every aggregation of matter attracts every other aggregation of matter by a force that is directly proportional to their respective masses and inversely proportional to the square of the distance between them. Newton reasoned that the moon does not fall to earth because its speed in orbit just balances the pull of the earth's and its own gravity. The universal law of gravity can be expressed as:

$$F = \frac{mM}{r^2} G$$

where F is force, m and M are the masses, r is the distance between them, and G is the universal constant of gravitation. Newton's law provides an excellent summary of the observed motions in our solar system.

On the earth, masses can best be compared by a process called **weighing.** This actually compares the earth's attraction for one mass with its attraction for another. The **weight** of a body on earth is the force that gravity exerts upon it. This **force is proportional to its mass,** that is, to the quantity of matter it contains. This can be written:

$$\frac{F}{m} = \frac{GM}{R^2}$$

where R is the earth's radius and M is its mass. G is

[4]"On Some of the Greater Problems of Physical Geology," *Bull. Phil. Soc. Wash.,* vol. 11, p. 51, 1889. Reprinted in *J. Wash. Acad. Sci.,* vol. 15, p. 359, 1925; also in *Bull. Natl. Res. Council (U.S.),* vol. 78, p. 203, 1931.

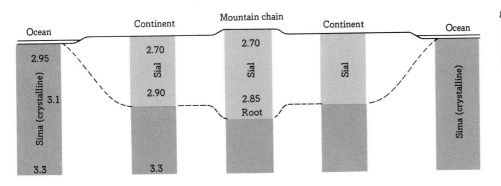

8.16 Hypothesis explaining isostasy, with variations in density of crustal material. [After R. C. Daly, *Strength and Structure of the Earth,* p. 61, Prentice-Hall, Inc., Englewood Cliffs, N. J., 1940.]

8.2 STRESS, STRAIN, AND STRENGTH OF ROCKS

again the universal gravitational constant. The quantities R, G, and M do not change.

Acceleration of gravity Gravity governs the fall of a body toward the earth and causes it to fall faster during each second. This is known as the **acceleration due to gravity,** represented by g. Falling freely in a vacuum, a body increases its speed by 980 cm/s during each second of fall. This is written:

$$g = 980 \text{ cm/s}^2$$

which is read "g equals 980 centimeters per second per second."

A body falling in the earth's atmosphere does not increase in speed without limit, however. Friction of the atmosphere causes it to reach a terminal velocity and go no faster after that. For a person falling freely, this is about 200 km/h.

The force we call **weight** is mass times the acceleration of gravity:

$$W = mg$$

On the moon, the acceleration of gravity is one-sixth of what it is on the earth; so a person would weigh one-sixth as much. But because the moon has no atmosphere, a body falling toward it would not have a terminal velocity as it does on earth.

Gravity anomalies—evidence for inequilibrium If the earth were a smooth, homogeneous, nonrotating sphere, gravity on its surface would be the same everywhere. It is not the same, however, because of differences in elevation, which change the distance from the earth's center; effects of rotation, which are different at different latitudes; and differences in the density of materials in the earth's crust.

Because $W = mg$ and m for a given body remains the same, changes in g, the acceleration due to gravity, will be reflected in changes in the weight of a body. An instrument for making such measurements is called a **gravity meter,** or **gravimeter.**

An acceleration of 1 cm/s² has been defined as a unit called the **Gal,** after the renaissance physicist Galileo. The acceleration of gravity at the earth's surface is around 980 Gal, but changes as small as a ten-millionth of this may be significant. Consequently, the common, or practical, unit in which gravity differences are expressed is the **milligal** (mGal), or 0.001 Gal.

Measurements of the force of gravity have to be corrected for differences in altitude, latitude, and geological structure. Altitude of the point of measurement

is taken into account by subtracting 0.3086 mGal for each meter of height above sea level or adding the same for each meter below sea level. Adjustment of an observed value to what it would be at sea level is known as the **free-air correction.**

The sea-level value of g varies from 978.049 Gal at the equator to 985.221 Gal at the poles because the earth's radius is *less* through the poles and because the centrifugal force of the earth's rotation tends to counteract the pull of gravity toward the center. Two kilometers away from the equator, the gravity is reduced by 1 mGal.

Corrections also need to be made for hills and valleys in the vicinity of the measuring point, as indicated in Figure 8.17.

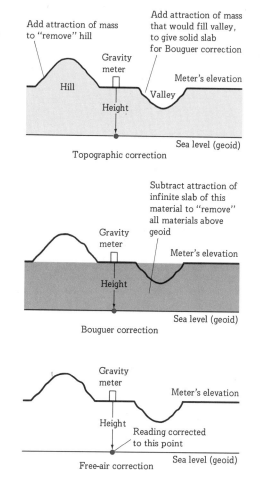

8.17 Correcting gravity-meter readings to a common level of reference at sea level. Because the hill rises above the meter, its attraction is upward and the meter would show a greater reading if the hill were not there. So the hill is "removed" by adding a figure to represent its attraction. In effect, the topographic correction reduces the surface to a plane. The Bouguer correction leaves a reading that the meter should show if it were suspended with nothing but free air between it and sea level. The free-air correction leaves the reading at a value that it would have if the meter were at sea level vertically below the point at which it actually stands.

After these factors have been evaluated, a number is obtained that represents the expected value of gravity at the point. If the measured value is different, this difference is called a **gravity anomaly.**

A sphere of salt 600 m in radius, with its center 1,200 m below the surface and surrounded by sedimentary formations about 10 percent more dense, causes gravity to be about 1 mGal less directly above it than at surface points beyond its range of influence. A buried granitic ridge under Kansas, with 1,500 m of relief, has a gravity anomaly of only 2 mGal. The effect of a volcanic pipe 3 km in diameter and of considerable depth may be 30 mGal.

There are significant negative gravity anomalies over ocean deeps, such as the Mindanao Deep off the Philippines, the Nero Deep off Guam, and the deep fronting Java and Sumatra in the East Indies. These have been cited as possibly caused by subducting lithospheric plates involved in ocean-floor spreading (see Chapter 10).

WHY DO WE STILL HAVE MOUNTAINS?

If gravity were the only force acting on the earth's surface, all masses of surface rocks would be standing today at heights governed by their thickness and by the ratio of their specific gravity to that of the rocks supporting them. Mountains, however, are great masses of rocks once deeply buried but now standing above adjacent rocks. One of the problems of geology is to explain why such masses stand high after millions and even billions of years of weathering and erosion, and three possible explanations for this phenomenon have been suggested.

One is that the crust underlying mountains is strong enough to support them as dead loads. Laboratory experiments, however, have shown that no rocks are sufficiently strong to support the weight of even comparatively low hills. Consequently we must conclude that the crust beneath mountains is not by itself capable of supporting their dead weight.

Another possible explanation is that mountains retain their elevations because the forces that originally formed them are still active. Certainly in areas where mountain-building forces still make their presence known through earthquakes and changes of sea level it is quite reasonable to assume that these forces are contributing to the continued elevation of mountains. In fact, it would be difficult to point with assurance to any region on the globe that lacks at least some kind of active internal force. Plate boundaries are regions of especially active mountain building (see further particulars in Chapter 10).

A third explanation for the continued elevation of mountains is that they are isostatic in relation to surrounding portions of the crust. That is, mountains are

merely the tops of great masses of rock floating in a substratum, as icebergs float in water; see Figure 8.18(a). Such a situation requires a substratum of rock, at not too great a depth, that will flow to adjust itself to an excess load. This rock, however, need not be a liquid in the ordinary sense of the word. It could be rock in a state not unlike that of silicone putty, which can be shaped into a ball that bounces but which under a very slight load—even under its own weight—gradually loses its shape entirely. Hence this rock would be rigid enough for stresses of short duration but would have very little real strength over a long period of time. A small piece of erasing gum nine-tenths as dense as silicone putty would, in time, sink into the putty until it was floating with only one-tenth of its volume above the surface. Likewise, a mountain range

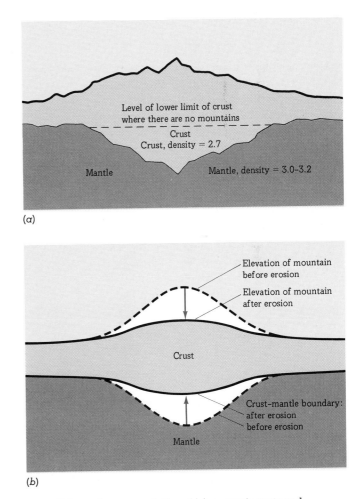

8.18 Schematic representation of (a) mountain roots and (b) isostasy.

8.2 STRESS, STRAIN, AND STRENGTH OF ROCKS

with an average specific gravity of 2.7 (that of granite) can sink into a layer of plastic rock with a specific gravity of 3.0 to 3.2 (like the mantle) until the range is floating with a root of about nine-tenths and a mountain of one-tenth its total volume. When some of the top of the mountain is removed by weathering and erosion, the mass will float upward to a new position of equilibrium; see Figure 8.18(b), where the root will no longer be quite so deep, but the same ratio of mass (roughly one-tenth) will form the mountain and nine-tenths of this new mass will serve as the root. In this way a mountain range can continue to exist by floating isostatic equilibrium after millions of years of erosion. In most parts of the earth isostatic equilibrium does prevail. Only in those areas where modern tectonism is actively deforming the crust is it likely that isostatic equilibrium is lacking.

8.3 STRUCTURAL FEATURES

Structural geology deals primarily with deformed masses of rock, their shapes, and stresses that caused the deformation. In describing structural features, the concept of **relief** is sometimes helpful. This is familiar in connection with the ground surface, where it denotes the difference in elevation between the highest and lowest points in a specified area. For example, the topographical relief of California is 4,400 m. This is the difference between the elevation of Mount Whitney and that of Death Valley.

Structural relief is the difference in elevation of parts of a deformed bed and is a measure of the extent of deformation. Sometimes the shape of a rock mass is determined by drilling and sometimes by hypothetical reconstruction of eroded portions.

Sedimentary rocks and some igneous rocks form in approximately horizontal layers. When these are found tilted, folded, or broken, they provide evidence of deformation of the earth's crust. In some areas, erosion has stripped away as much as thousands of meters of uplifted portions of the crust, revealing structures that were once buried deeply. Deformation features include folds, joints, faults, and unconformities.

In describing the attitudes of structural features, geologists have found it convenient to use two special terms: **dip** and **strike**. These are more easily described with reference to layered rocks. If a rock layer is not horizontal, the amount of its slope is called its **dip** (see Figure 8.19), measured by specifying the acute angle that the layer makes with the horizontal. The dip is measured in the direction of the greatest amount of in-

8.19 Measuring dip with a Brunton compass.

8.20 Jurassic sandstone bed cropping out as a hogback west of Denver, Colorado, showing dip and strike measurements. [M. E. Kauffman.]

8.21 Syncline and adjacent anticline in sandstone beds of the Moccasin Formation along the river road to Goodwin's Ferry, Giles County, Virginia. [T. M. Gathright II.]

clination. Its **strike** is the course, or bearing, of the outcrop of an inclined bed or structure on a level surface. If the rock layer is tilted so that it disappears below the surface but protrudes somewhat because of resistance to weathering, as in Figure 8.20, strike is the direction in which the resulting ridge runs. A bed that dips either east or west has a north-south strike, usually designated simply as north. A bed that dips either to the north or to the south has an east-west strike.

Folds are a common feature of rock deformation. They are produced by compressive stress (see Figure 8.21). They range in size from microscopic wrinkles in a piece

8.3 STRUCTURAL FEATURES

157

of metamorphic rock to huge folds involving thousands of meters of thickness for distances of hundreds of kilometers. Folds are seldom isolated structures but generally occur in closely related groups. Sometimes they are very broad and sometimes tight and narrow. Sometimes they are tilted to one side and sometimes to one end.

In some areas, sedimentary layers are only slightly bent, whereas in others, usually associated with mountain structures, they are intensely deformed. The difference in the kind of folding is mainly dependent on the amount of deforming stress in relation to the strength of the rocks and on the temperature and confining pressure during the deformation.

Mechanism of folding Folding can be classified as **concentric** or **flow**. **Concentric folding** is basically an elastic bending of an originally horizontal sheet. It occurs in surface layers of the crust. The size of a concentric fold is determined by the thickness of the beds involved and by their elastic properties. Beds are buckled during concentric folding, but the thickness of the beds and the volume remain the same. Large-amplitude folds form from thick strata and small-amplitude folds from thin strata. The distance between crests is also controlled by the thickness of the beds.

Flow is a type of deformation that occurs in rocks when they are in the plastic state. It involves internal movement that may take any direction. It occurs in weak rocks near the surface and in strong rocks at depth. Flow is due to conditions of confining pressures and high temperatures. It is the only true plastic deformation of rocks.

Monoclines Monoclines are relatively simple examples of deformation, involving an elastic bending of sedimentary layers. A **monocline** is a double flexure connecting strata at one level with the same strata at another level. Extensive horizontal layers are bent down and pass beneath younger horizontal strata. The flexed layers did not accumulate with the bend in them. As bending progressed, erosion worked to strip away the higher portion.

There are many monoclines in the Colorado Plateau. Some involve displacements as great as 4,000 m and lengths of 250 km. The Waterpocket fold in southern Utah is one of the better known (see Figure 8.22). It involves displacements of close to 2,000 m. Estimates of

8.22 Aerial view of Waterpocket fold, Capitol Reef National Park, Utah. Beds are nearly horizontal on the left and right extremes of the photo but are dipping steeply toward the right near the center. [U.S. National Park Service.]

8.23 Massive limestone and dolomite beds "draped" over west side of Bighorn Mountains near Five Springs, Wyoming. [M. E. Kauffman.]

the rate of its deformation range from a few centimeters to possibly a few meters per century.

Anticlines and synclines Local arching of layered rocks is the most common form of a structure called an **anticline** (Figures 8.23 to 8.26). The two sides of the fold, called its **limbs**, dip away from each other. The **axis** of an anticline is the direction of an imaginary line drawn on the surface of a single layer parallel to its intersection with the axial plane. The axes of most geological folds are inclined. The angle of dip of its axis is the **plunge** of the fold. Parts of a fold are illustrated in Figure 8.24.

Anticlines can be spectacular when their anatomy has been exposed by erosion (see Figure 8.26). The

oldest rocks are at the core and the youngest on the outer flanks.

A **syncline** is a downward fold, the opposite of an anticline (Figure 8.27). The limbs dip toward the central axis. Synclines are best seen in mountainous areas where there has been uplift along with the folding. The youngest rocks are in the center of the fold, with successively older rocks farther out (Figure 8.28).

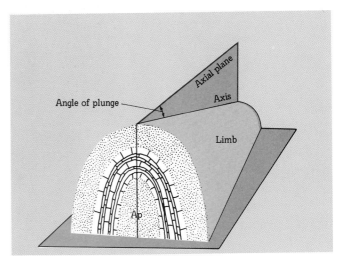

8.24 The parts of a fold, shown on a plunging anticline but applicable to a syncline also. The axis is the line joining the places of sharpest folding. The axial plane (Ap) includes the axis and divides the fold as symmetrically as possible. If the axis is not horizontal, the fold is plunging, and the angle between the axis and the horizontal is the angle of plunge. The sides of a fold are the limbs.

8.25 Broad anticlinal fold in Devonian rocks along the southwestern coast of England (opening approximately 1 m high and 4 m wide). [M. E. Kauffman.]

(a)

8.27 Tight syncline in Devonian rocks along the southwestern coast of England. Thickest bed is 0.5 m thick. [M. E. Kauffman.]

(b)

8.26 (a) Aerial photograph showing resistant beds of an eroded plunging anticline. (b) Sketch illustrating the relationship of outcrops to fold. Western Sahara, Africa. [U.S. Army Air Corps.]

Anticlines and synclines commonly occur together in elongate groups like belts of wrinkles in the earth's outer skin. Anticlines and synclines are said to be **symmetrical** if their opposite sides have approximately equal dips, and **asymmetrical** if one limb is steeper than the other. Types of fold are listed in Table 8.2 and illustrated in Figures 8.29 and 8.30.

8.28 View of plunging syncline from top of cliff overlooking tightly folded Devonian rocks along the southwestern coast of England. See figures on beach for scale. [M. E. Kauffman.]

TABLE 8.2

Types of fold

Name	Description
Anticline	Upfold, or arch
Syncline	Downfold, or trough
Monocline	Local steepening of an otherwise uniform dip
Dome	Anticline roughly as wide as it is long, with dips in all directions from a high point
Basin	Doubly plunging syncline
Asymmetrical	Strata of one "limb of the fold" dip more steeply than those of the other
Overturned	Limbs tilted beyond the vertical; both dip in the same direction, although perhaps not the same amount
Recumbent anticline	Beds on the lower limb upside down
Recumbent syncline	Upper limb inverted
Isoclinal	Beds on both limbs nearly parallel, whether fold upright, overturned, or recumbent

Structural domes Structural domes are the result of pressure acting upward from below to produce an uplifted portion of the crust with beds dipping outward on all sides. Where erosion has removed the top of a dome, concentric ridges are formed by one or more relatively resistant beds (see Figure 8.31). The troughs are formed by erosion of weaker rocks. It is possible to project the dip and strike of the beds upward over the dome and reconstruct its whole shape before erosion removed any parts.

Some structural domes are formed by the upwelling of plastic material such as salt or magma. It is impossible to know whether slight folding started the upwelling or whether hydrostatic adjustment of loads on the crust was the primary cause. Where there is a local thickening or thinning of the crust, the plastic material starts to rise and continues as a result of the unequal static load. Most domes show marked upward bending of the surrounding beds against the stock, often accompanied by faulting. The beds above are domed and stretched by the push from below and often exhibit an intricate pattern of normal tension faults.

8.29 Types of folds. (Ap = axial plane.)

Dome

Monocline

Asymmetrical folds

Recumbent folds

Basin
Youngest rock
Anticline Syncline Anticline
Ap Ap Ap
Oldest rock
Symmetrical folds
Normal limb Inverted limb
Ap
Ap Ap Ap
Overturned folds

Ap Ap Ap Ap
Isoclinal folds

8.30 Recumbent fold in Rheems Quarry, Pennsylvania. Rocks are limestones and dolomites of the Beekmantown Formation of Ordovician age. [M. E. Kauffman.]

8.31 Section across the Black Hills in South Dakota and Wyoming, illustrating a structural dome exposed by erosion. [After N. H. Darton.]

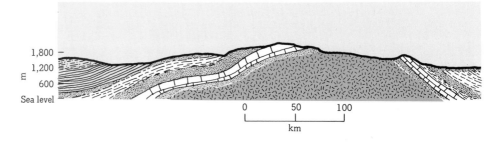

JOINTS

Masses of rock may be broken, with or without movement along the cracks. If there has been no slippage along the fracture surface, the crack is called a **joint.** If there has been displacement, the structural feature is called a **fault.**

Joints are found in all kinds of rock and are the most common feature of rocks exposed at the surface. They usually occur in sets, the spacing between them ranging from just a few centimeters to a few meters. As a rule, the joints in any given set are almost parallel to one another, but the whole set may run in any direction—vertically, horizontally, or at some angle (see Figure 8.32). Most rock masses are traversed by more than one set of joints, often with two sets intersecting at

approximately right angles. Such a combination of intersecting joint sets is called a **joint system.** A regional pattern of joint systems may occur over areas of hundreds of square kilometers in a given type of rock exposed at the surface. As a rule, the kinds of rock have a marked influence on what joint systems develop. A massive sandstone and a sandy shale, for example, each show characteristic jointing directions and spacing.

It is rarely possible to determine the age or origin of a joint set. It is generally accepted, however, that different sets in a joint system were probably made at different times and under different conditions. In sedimentary rocks, one set could develop while the rock was being compressed by the weight of overlying rocks during burial and consolidation. Another set could be produced when pressure is released during unloading by erosion. In other cases, sets may be developed during deformation by tension. Some joint sets are known to be formed during the cooling of a mass of igneous rock, and others during movements of magma.

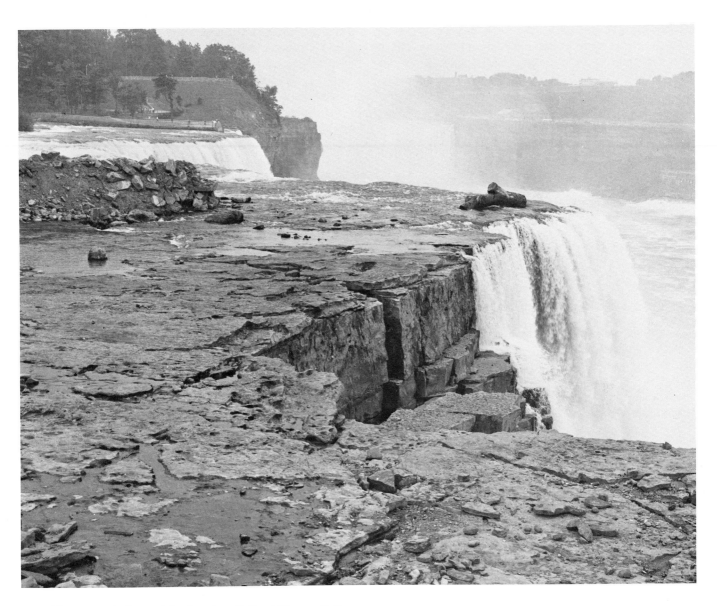

8.32 The sawtooth shape of the American Falls of Niagara Falls results from the prevailing joint pattern. (Photographed during unwatering of June, 1960, for Niagara power project.) [State of New York Power Authority.]

Columnar jointing Columnar jointing is a special pattern of jointing found in some masses of fine-grained igneous rocks. It consists of sets of cracks produced by the mechanism of cooling. In certain lava flows and shallow intrusive bodies, joints are formed as the rock shrinks in cooling. Dikes or sills cool by fairly uniform loss of heat over the entire surface that is in contact with cold wall rock. This allows shrinkage about many equally spaced centers, not unlike that which produces mud cracks. Such contraction means tension between the centers, which results in a system of short straight cracks separating each center of contraction from its neighbors. These form a polygonal pattern, ideally hexagonal but sometimes four- or five-sided because of irregularities in cooling. As cooling progresses into the body, contraction advances with it, and each column may grow until it reaches the opposite side.

Columnar jointing is not clearly developed in all masses of basalt. It seems to be most characteristic of tabular masses, where columns form across the narrow dimension of the mass. The columns are often broken into sections of varying lengths by transverse joints (see Figure 8.33). One of the best-known areas in the world where columnar jointing can be seen is the Giant's Causeway, near Portrush, Antrim, Northern Ireland. Columnar jointing is also well developed in Devil's Post Pile National Monument, Sierra Nevada, California, along the Snake River in Washington, and in Devil's Tower, Wyoming.

Sheeting   A pattern of essentially horizontal joints is called **sheeting** (see Figure 8.34). Here the joints occur fairly close together near the surface, but less and less frequently the deeper we follow them down until they seem to disappear altogether a few tens of meters below the surface. But even at considerable depth, the rock shows a tendency to break along surfaces parallel to the surfaces of sheeting above. This type of jointing is especially common in masses of granite, and engineers often put it to good use in planning blasting operations.

Joints and dikes formed during movements of magma
Some joint systems are formed during movements of magma within the crust. Moving magma exerts pressure on adjacent rocks, developing cracks. Magma is then injected into the cracks as they open, enlarging them. This produces such structural features as dikes and sills, which are thought to form before the magma reaches the surface and is extruded. After extrusion begins, pressure is reduced. Because extrusion takes less pressure than intrusion, the magma wells out instead of intruding into adjacent rocks.

Dikes are exposed at the earth's surface after softer rocks have been eroded away. They vary in width from a few centimeters to kilometers across and may be many kilometers in length. The largest dike known is in Rhodesia. It is 480 km long and up to 8 km wide.

Dikes occur in different patterns: ring dikes, cone sheets, and swarms. **Ring dikes** and **cone sheets** consist of a concentric arrangement of dikes formed by a definite center of stress. They occupy fractures around a center where magma has pushed up a section of the earth's crust. Dikes are filled by upwelling magma (see Figure 8.35). Ring dikes have diameters ranging from 2 to 25 km, with individual dikes sometimes as much as 500 to 1,000 m wide. Ring dikes dip away from the magmatic body that cracked the crust to provide avenues of weakness along which they moved into place. Cone sheets outline inverted cones with sides dipping toward the magmatic body. From their angle of dip, it is possible to estimate the depth below the surface to the body of magma that supplied them.

In some places, dikes occur in roughly parallel groups called **dike swarms.** Some dikes in a swarm are up to 400 km in length. Dike swarms differ from ring dikes in their pattern which suggests that they originate in a different kind of stress field.

Some dike swarms are caused by an elastic release of pressure. This has been documented in several places in Scotland. The Caledonian orogenic belt contains dike swarms running parallel to the main folding axis, which trends northwest-southeast (see Figure 8.36). The trend of the dikes suggests that they were formed by an elastic release after compression. Dike swarms are also associated with strike-slip faults in Scotland. These faults include the Great Glen fault, the Highland Boundary fault, and the Southern Upland fault, all of which trend northeast-southwest. Their direction suggests that they were formed by a north-south compression. The dike swarms trend east-west, perpendicular to the compressive stress. This again suggests that the cracks in the earth's crust resulted from an elastic release of compression.

Some dike swarms are associated with the folding process but do not involve an elastic release of compression. They are the result of tension. Such a case has been reported in the bend of a monocline on the east coast of Greenland.[5] The monocline consists of a sheet of basalt blanketing metamorphic rocks stretching many hundreds of kilometers and varying in dip. The dike swarm is located on the convex side of the flexure and is absent in the concave part (see Fig-

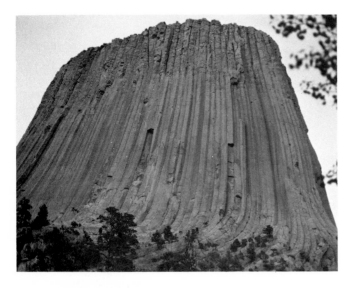

8.33 Devil's Tower, Wyoming, showing well-developed columnar jointing. Note the abrupt changes in column orientation. Nearly vertical sides are approximately 200 m high. [M. E. Kauffman.]

[5] L. R. Wager and W. A. Deer, "A Dike Swarm and Crustal Flexure in East Greenland," *Geol. Mag.*, vol. 75, pp. 39–46, 1938.

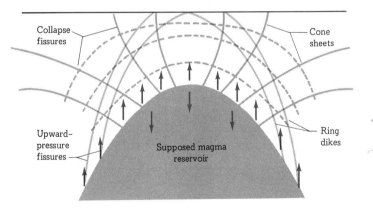

8.35 Ring dikes and cone sheets. [After E. M. Anderson, *Proc. Roy. Soc. Edinburgh*, vol. 56, pp. 128–157, 1936.]

ure 8.37). Where the flexure dips 55°, there are more than 70 dikes per kilometer across the structure. Where the dip is 12°, there are only about 15 dikes per kilometer; and where the dip is 7°, there are only a few dikes. The dikes dip at about right angles to the basalt sheet, and their position indicates that they fill fissures devel-

oped by tension as the crust bent. The monocline has a maximum relief of at least 8 km.

FAULTS

Faults are deformation by rupture in which the sections on each side of the break move relative to each other. Theoretically they may be of any size, but here we shall discuss only the types that are large shearing surfaces in the earth's crust.

A useful basis for the classification of faults is the nature of the relative movement of the rock masses on opposite sides of the fault. If displacement is in the direction of dip, the fault is a **dip-slip fault.** The sections separated by the fault are named as they were by miners who encountered faults underground. The block that is overhead is called the **hanging wall,** and the

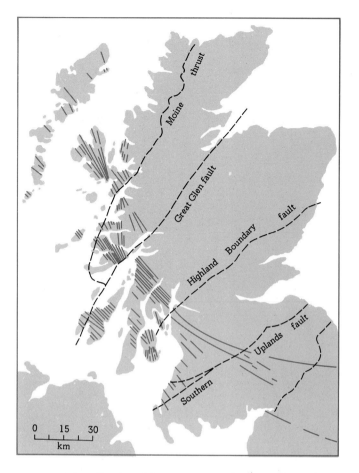

8.36 Tertiary dike swarms in Scotland. [After J. E. Richey, *Trans. Geol. Soc. Edinburgh*, vol. 13, p. 393, 1939.]

8.37 Dike swarms on the eastern Greenland coast. [After L. R. Wager, *Med. om Grønland*, vol. 134, no. 5, pp. 1–62, 1947.]

one beneath is the **foot wall**. Dip-slip faults are classified according to the relative movement of these blocks. If the hanging wall seems to have moved downward in relation to the foot wall, the fault is called **normal**. Normal faults occur when the vertical stress is greater than the horizontal stress.

Crustal movements which cause the surface to stretch may produce a series of related normal faults. A **graben** is an elongated, trenchlike structural form bounded by parallel normal faults created when the block that forms the trench floor has moved downward relative to the blocks that form the sides. The term comes from the German for "trough" or "ditch" and is the same in both singular and plural. A **horst** is an elongated block bounded by parallel normal faults in such a way that it stands above the blocks on both sides. The term comes from the German and is used figuratively for a crag or height. The type of movement involved in the formation of a graben and a horst is illustrated in Figure 8.38.

A dip-slip fault in which the hanging wall appears to have moved upward in relation to the foot wall is a **thrust fault** (or **reverse fault** if high angle, generally greater than 45°). It is the result of largely horizontal compressive stresses.

A fault along which the movement has been predominantly horizontal is called a **strike-slip fault** because the slipping has been parallel to the strike of the fault. Such faults have also been called **transcurrent faults**. Strike-slip faults are further designated as **right-lateral** or **left-lateral**, depending on how the ground opposite you appears to have moved when you stand facing the fault. Types of fault are listed in Table 8.3 and are shown in Figure 8.39.

The directions of movement involved in faulting are entirely relative. It is convenient to indicate one block as having moved upward or downward or as moving left or right. However, the absolute direction of movement usually cannot be determined, and the best that can be done is to indicate relative movements. During the upward warping of a large region, for ex-

TABLE 8.3

Types of fault

Type	Predominant relative movement
Dip-slip	Parallel to the dip
Normal	Hanging wall down
Thrust	Hanging wall up
Strike-slip (transcurrent)	Parallel to the strike
Right-lateral	Offset to the right
Left-lateral	Offset to the left
Oblique-slip	Components along both strike and dip
Hinge	Displacement dies out perceptibly along strike and ends at definite point

Footwall block — Hanging-wall block

Graben

Hanging-wall block — Footwall block

Horst

8.38 A graben and a horst.

ample, all blocks could have moved upward, but some may have moved less than others. These lagging blocks could be said to have dropped relative to their neighbors even though all parts at the finish were at higher elevations than when they started.

Normal faults In the earth's crust there are large zones that have repeatedly been disturbed by large-scale normal faulting. The faults form horsts and rift valleys. The most famous, the African rift zone, extends over 6,000 km in a north-south direction. In some areas volcanoes are found along the zone where magma has worked its way up along fault surfaces. In Europe there are several zones of normal faults, including the rift zone of the Upper Rhine valley. In the western part of the United States we find an extensive zone of normal faulting in the Basin and Range Physiographic Province. Innumerable smaller normal fault zones can be found all over the world.

The Basin and Range Physiographic Province consists of tilted fault blocks. Its eastern limit is the Wasatch fault. The main faults in the vicinity of Wasatch Mountain dip 50° to 55°. Net slip on major fault planes in the Wasatch Range is from 2,200 to 2,600 m. Farther west are the faulted Oquirrh Range, the Stansbury

Range, and the Cedar Range. Each is a tilted fault block some 30 km in breadth. The western limit is the great normal fault that limits the tilted Sierra Nevada block on the east.

Thrust faults Thrust faults are usually closely associated with the folding process. Big, steep thrust faults are characteristic of the margins of mountain chains and also occur in their uplifted central blocks of crystalline rocks. Rocks above the fault plane seem to have been pushed up and over those below. Along the fault plane, rocks are considerably crushed and sheared. Thrust faulting is more clearly seen in eroded regions where metamorphic rocks that were once deeply buried now overlie sedimentary rocks that were formed near the surface. Old rocks are generally thrust over younger ones.

The total movement along a thrust fault can be only estimated in most cases, but some faults involve total vertical displacements as much as 1,000 m, combined with horizontal movements of tens of kilometers. The dips of thrust faults vary from less than 10° up to 60°. **Overthrusts** are far-travelled, low-angle thrusts.

Some well-known regions of thrust faulting are in the Alps, the southern Appalachians, the central and northern Rocky Mountains, and southern Nevada. Besides thrust faults of considerable displacement, every deformed region contains numerous small thrust faults (Figure 8.40). These may have displacements as small as 2 m but do not usually appear on geologic maps unless they occur in a mining area.

Thrust faults are often associated with asymmetric anticlines where rupture occurs in the highest part of the fold. A thrust fault of this type is a superficial structure because it extends down only to the basal plane and occurs in brittle rocks near the earth's surface. Some are found in the Valley and Ridge Physiographic Province of the Appalachians, with the best known being the Cumberland overthrust in Virginia (Figure 8.41).

Strike-slip (transcurrent) faults Strike-slip faults are found all over the world. A well-known example of left-lateral strike-slip faulting is the Great Glen fault, which intersects Scotland from coast to coast. There is a string of lakes, including Loch Ness, along its eroded trace, which is marked by a belt of crushed, sheared rock up to 1.5 km in width. Horizontal displacements along the fault of as much as 100 km have been suggested on the basis of correlating geological structures on both sides. It dates from the Late Paleozoic.

8.39 Types of fault.

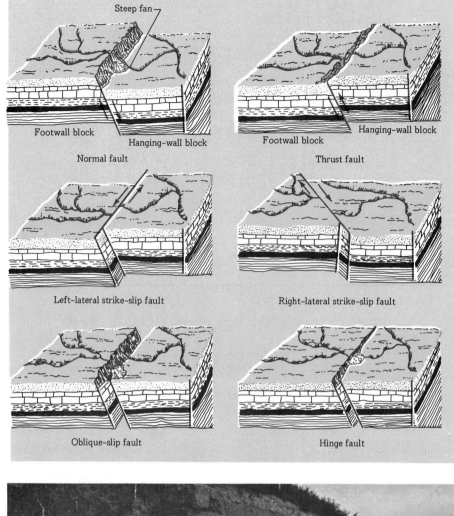

Steep fan

Footwall block

Hanging-wall block

Normal fault

Footwall block

Hanging-wall block

Thrust fault

Left-lateral strike-slip fault

Right-lateral strike-slip fault

Oblique-slip fault

Hinge fault

8.40 Thrust fault on the Pan American Highway 40 km northwest of San Salvador. [Thomas F. Thompson.]

Hanging Wall

Footwall

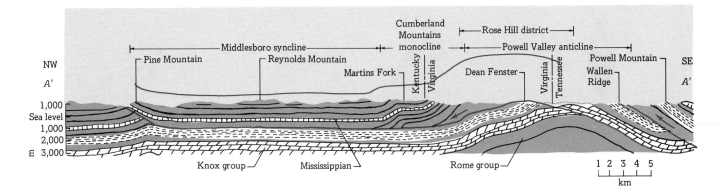

8.41 Cumberland overthrust, Valley and Ridge Physiographic Province of the Appalachians. [After R. L. Miller and J. O. Fuller, "Geology and Oil Resources of the Rose Hill District," *Bull. Va. Geol. Surv.*, vol. 71, p. 383, 1955.]

Several left-lateral strike-slip faults rip across folds of the Jura Mountains, not far from Lake Geneva and Lake Neuchâtel. They clearly originated during the folding process.

The San Andreas fault in California is a right-lateral strike-slip fault that has been traced for nearly 1,000 km. Terrace deposits cut by the fault show offsets of as much as 10 km, and stream channels show shifts of 25 km within relatively short geological intervals. The total strike slip has been debated. Evidence has been interpreted as showing a displacement of 16 km since the Pleistocene, of 370 km since late Eocene time, and 580 km since pre-Cretaceous time. An offshoot from the San Andreas, the Garlock fault separates two clusters of dike swarms that seem pretty clearly to have been intruded as a single event. They are now separated by 80 km in a sense of movement that makes the Garlock a left-lateral strike-slip fault.

Transform faults   Another basis for the classification of faults is their relationship to other structural features.[6] It has been suggested that deformation of the earth's crust produces major structural forms: mountains, deep-sea trenches, midocean ridges, and strike-slip faults with large horizontal displacements. J. Tuzo Wilson proposed that these features are interrelated, that they do not come to dead ends but link in continuous networks girdling the globe, and that any major feature at its apparent termination may change into one of the other types. A junction where this takes place is called a **transform**. It has been proposed that the name **transform fault** be applied to a class of faults connecting other major structural features. Figure 8.42 suggests the relationships that exist along a transform fault connecting two segments of a fragmented oceanic ridge and compares them with the relationships along

[6] J. Tuzo Wilson, "A New Class of Faults and Their Bearing on Continental Drift," *Nature*, vol. 207, pp. 343–347, 1965.

a transversely faulted ridge. Analysis of earthquakes along the faults connecting segments of ridges indicates the motion to be as shown in Figure 8.42(a). This is exactly opposite to that which would result from a simple offsetting of the ridge by transcurrent faults, as shown in Figure 8.42(b). The original offset of the ridges along transform faults may well have been due to transcurrent faulting. But with the establishment of transform motion, the ridge crests do not change position with time. In the case of continued transcurrent faulting, the offset increases with time. Many of the faults across the ocean ridges have been interpreted as transform faults: for example, those of the equatorial Mid-Atlantic Ridge shown in Figure 8.43.

UNCONFORMITIES

There is no known place on earth where sedimentation has been continuous throughout geologic time. During the formation of a continent, large sections of the crust have been raised out of the shallow seas in which its rocks were formed, subjected to erosion, and then lowered again to levels where deposition of sediments is renewed. Activity of this sort produces a buried erosion surface with younger rocks overlying older rocks. Some surfaces of this kind can be seen today because of another cycle of uplift. A buried erosion surface separating two rock masses of which the older was exposed to erosion for a long interval of time before deposition of the younger is called an **unconformity**.

The time represented by an unconformity is important geologic evidence of the history of a region, marking an interval when the surface was above the sea and sediments were not being deposited. Some unconformities represent gaps of a few thousand years; others, as many as 400 million years.

8.42 Comparison between transform and transcurrent faults. The perspective sketch and plan in (a) show a ridge-to-ridge transform fault. New crustal material is postulated as being continuously added at the crest and transferred laterally away from the crest. There is no net displacement of the forms of the ridge segments with time. Transcurrent faults are shown in perspective and plan in (b). The ridge fragments are increasingly offset with time.

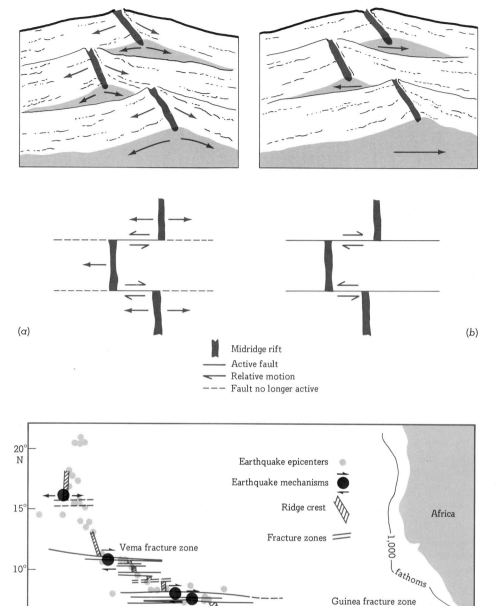

(a)   (b)

▮ Midridge rift
── Active fault
← Relative motion
--- Fault no longer active

8.43 Equatorial section of the Mid-Atlantic Ridge showing ridge crest and fracture zones. [After L. R. Sykes, "Seismological Evidence for Transform Faults, Sea Floor Spreading and Continental Drift," in R. A. Phinney (ed.), *The History of the Earth's Crust*, Princeton University Press, Princeton, 1968. © 1968 by Princeton University Press. Reprinted by permission.]

There are three principal types of unconformity: angular unconformity, disconformity, and nonconformity.

Angular unconformity An unconformity in which the older strata dip at an angle different from that of the younger strata is called an **angular unconformity**. On a wall of Box Canyon, near Ouray, Colorado, can be seen some layers of Precambrian sedimentary rocks dipping at nearly 90°. Above them are some nearly horizontal Devonian sedimentary rocks. The older rocks were deposited under the waters of the ocean and were then folded and uplifted above the water. While exposed at the surface, the tilted beds were beveled by erosion. Then, as time passed, these tilted and eroded rocks sank again beneath the ocean, where they were covered by new layers of sediments. Both were later elevated and exposed to view (see Figure 8.44). On the basis of fossil evidence we know that the second sedimentation began during the Devonian Period, indicating that more than 180 million years elapsed between the two periods of sedimentation. The Precambrian rocks meet the younger Devonian rocks at an angle; so this unconformity is an angular unconformity.

Disconformity An unconformity with parallel strata on opposite sides is called a **disconformity**. It is formed when layered rocks are elevated, exposed to erosion, and then lowered to undergo further deposition without being folded. Careful study and long experience are required to recognize a disconformity because the younger beds are parallel to the older ones. Geologists rely heavily on fossils to correlate the times of deposition of beds above and below a disconformity. Fossils are a useful tool because certain ones lived during a definite geologic time range.

Nonconformity An unconformity between profoundly different rocks is called a **nonconformity**. Nonconformities are formed where intrusive igneous rocks or metamorphic rocks are exposed to erosion and then are downwarped to be covered by sedimentary rocks. A structure of this sort is illustrated in Figure 8.45, which shows a sandstone deposit lying on top of an eroded surface of granite. Field studies have revealed that pieces of weathered granite occur in the bottom layers of the sandstone and that cracks in the granite are filled with sandstone. This evidence supports the view that the granite did not come into its present position as an intrusion after the sandstone was formed.

Unconformities in the Grand Canyon The Grand Canyon of the Colorado exposes several unconformities (see Figure 8.46). This region has undergone three major sequences of uplift and erosion, subsidence and deposition, and crustal deformation and crustal stability. These have produced three major rock units: Vishnu Schist, Grand Canyon Series, and Paleozoic Series.

The Vishnu Schist is the base rock of the canyon. It is composed of highly metamorphosed sedimentary rock and some volcanics. It has yielded no fossils and may be 2 billion years old.

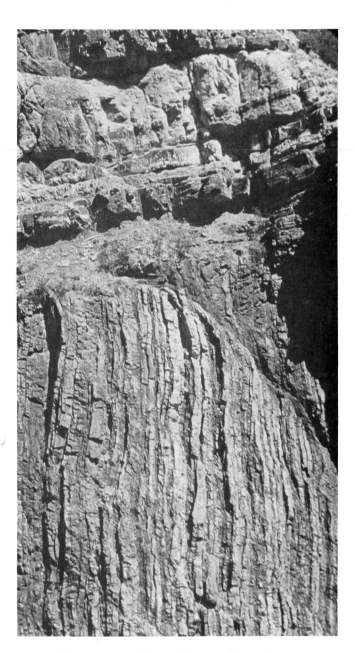

8.44 A striking unconformity between Precambrian sedimentary rocks that were twisted from their original horizontal position before deposition of overlying Devonian beds. In Box Canyon, near Ouray, Colorado. [Kirtley F. Mather.]

Resting unconformably on the Vishnu Schist in some places are tilted beds of the Precambrian Grand Canyon Series. The beds consist of 75 m of Bass Limestone, 180 to 240 m of sandy shale, and 300 m of relatively resistant beds of Shinumo Sandstone, topped off by 600 m of shaly sandstone. It has been estimated that

8.45 A nonconformity between light-colored granite and dark overlying Table Mountain Sandstone on the Cape of Good Hope, South Africa. The cave, about 6 m above sea level, was cut by waves when sea level was higher than it is now. [R. A. Daly.]

8.46 Diagrammatic representation of one wall of the Grand Canyon along the Kaibab Trail, showing unconformities. [After W. J. Breed and E. Roat (eds.), *Geology of the Grand Canyon*, Museum of Northern Arizona—Grand Canyon Natural History Association, Flagstaff and Grand Canyon, Ariz., 1976.]

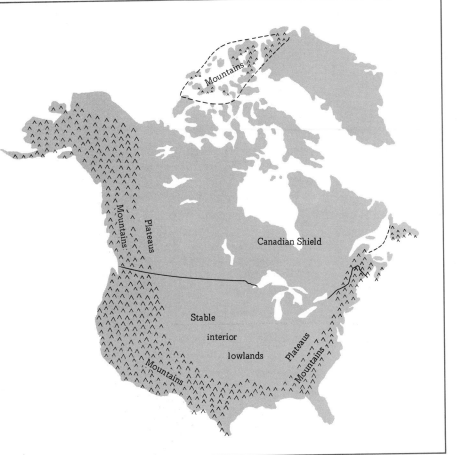

an additional 3,000 m of Grand Canyon sediments were deposited on the beds before the region was elevated, deformed, and then eroded to the surface on which Paleozoic sediments were deposited. An uplift of at least 3,600 m had to take place to bring the lowest Grand Canyon layers above sea level. Erosion then removed all but the lower sections of some tilted blocks. It is not possible at the present time to say how long it took to complete the downwarping of the Vishnu Schist, deposition, sedimentation, deformation, elevation, and erosion.

When the Grand Canyon Series was eroded to a surface of low relief and then submerged, Paleozoic sediments were deposited. These sediments are found today in horizontal layers along the upper part of the canyon walls, indicating uplifting of the region without deformation. The Paleozoic Series contain at least four discontinuities. The longest gap is between the Muav Limestone and the Temple Butte Formation. The Muav Limestone was deposited in Cambrian time. The Temple Butte Formation immediately overlying these and essentially parallel to them contains fossils of primitive armored fish that occur only in Devonian rocks. Apparently, there was no deposition during Ordovician or Silurian time, a lapse of some 80 million years.

In areas where all the Grand Canyon Series had been eroded away before deposition of Paleozoic sedi-

ments, we find the "great unconformity." Its time gap is estimated as several hundred million years.

BROAD SURFACE FEATURES

Every continent has a nucleus of Precambrian rocks. These are called **shields,** and the continents have apparently grown around them. The oldest rocks were formed about 3 billion years ago along chains of volcanoes. By about 2 billion years ago, platforms of the earliest rocks had developed on which sediments accumulated. The full range of mountain-building cycles has been operating since then. Deformation has produced the continental land features we find today—the plains, the plateaus, and the mountains.

North America's shield, the Canadian shield (Figure 8.47), is exposed over an appreciable part of the northeastern portion of the continent. It covers more than 5 million km² and includes all of Labrador, much of Quebec, Ontario, and the Northwest Territories, the northeastern parts of Manitoba and Saskatchewan, and part of the Arctic islands. Exposure at the surface could not have taken place without elevation—in this case brought about by a broad warping of the basement complex.

Bordering the shield are the interior lowlands, extending to eastern Ohio on the east, the Ouachita Mountains on the south, and the Arctic Ocean on the north. They are composed of sedimentary rocks, the oldest having been deposited in seas lapping up onto the shield during Cambrian time. On 14 different occasions, portions of this continental lowland sank beneath the ocean and later rose above the water. So it was a restless area during the building of this continent, and the rocks bear the traces of unsettled times. Gently tilted in places, folded into small domes and basins in others, the thickness of these sedimentary rocks varies as much as 3,000 m in areas of exceptional subsidence and deposition; yet the beds lie relatively low and flat and form plains.

The sedimentary rocks surrounding the interior lowland plains have been broadly upwarped into plateaus, although the rocks still lie nearly flat. Still farther from the shield the plateaus merge into mountain ranges, which represent deformation at its greatest. As much as 15,000 m of sediments accumulated here because of subsidence in the basement complex; they then consolidated into sedimentary rock. Intense folding, rupture, and elevation of the entire region resulted in mountain ranges such as the Appalachians and the Rockies.

Plains are regions of low relief composed of nearly horizontal sedimentary rocks. Plateaus are higher regions of nearly horizontal strata with a relief generally over 150 m. Plateaus may originate by uplift or by deposition of lava flows. Mountains encompass a wide range of geological structures and relief, including fault-block mountains, thrust mountains, and mountain ranges. Youthful mountains stand today thousands of meters high, whereas the older mountain chains that have been attacked by many more years of erosion have low relief.

Broad warps Some large sections of the continents are composed of horizontal strata that have been gently bent upward or downward. In North America, these strata are found in the lowlands between the Rocky Mountains and the Appalachians. Regional downwarping forms basins; upwarping forms arches or plateaus.

The Great Basin, 470,000 km² between the mountains, is centered in Nevada and includes adjacent parts of California, Oregon, and Utah. Its relief is close to 3,900 m.

OUTLINE

Evidence of deformation of the earth's crust is best preserved in sedimentary and metamorphic rocks.

Abrupt movements of the earth's crust have occurred at the times of some large earthquakes.

Vertical displacements have been such as to amount to 15,000 m if continued at the same rate for 2 million years.

Horizontal displacements have been estimated as great as 200 km in 150 million years.

Slow movements are being measured instrumentally in several places today, and on the San Andreas fault near Parkfield they average 5 cm/year.

Kinds of deformation are described in terms of change of volume, change of shape, or a combination of both.

Stress-strain relationships are used to define a material's response to stress.

Elastic deformation is recoverable; plastic deformation is not.

Plastic deformation is permanent and involves a property of rock called **viscosity**.

Time factor sometimes results in a material's being elastic for a short time but plastic over a longer time.

Types of stress are **tension**, **compression**, and **shear**.

Strength of rocks is the stress at which they begin to be permanently deformed.

Gravitational attraction of mountains can be computed and measured; **isotasy** is the principle that different portions of the earth's crust should be in balance.

Gravity is a universal force by which every aggregation of matter attracts every other aggregation of matter.

Acceleration of gravity is the amount by which a freely falling body increases its speed of fall each second.

Gravity anomalies are differences between expected and measured values of gravity.

Mountain elevation may be maintained because the forces that originally formed mountains are still active or because they are isostatic.

Structural features are the shapes of deformed masses of rock, including folds, joints, faults, and unconformities.

Folds are a common feature of rock deformation and are produced by stresses.

Mechanism of folding is elastic bending or plastic flow.

Monoclines are double flexures.

Anticlines and **synclines** occur together in elongate groups like belts of wrinkles in the earth's outer skin.

Structural domes result from pressures acting upward.

Joints usually occur in sets.

Columnar jointing is sets of cracks produced by the mechanism of cooling certain tabular plutons to produce columns.

Sheeting is a pattern of essentially horizontal joints.

Joints formed during movements of magma include dikes and sills.

Faults are deformation by rupture in which the sections on each side move relative to each other.

Normal faults have the hanging wall apparently dropped relative to the foot wall.

Thrust faults have the hanging wall apparently moving higher than the foot wall.

Strike-slip faults have movement predominantly horizontal, parallel to the strike of the fault.

Transform faults are strike-slip faults that terminate in a mountain range, mid-ocean ridge, or deep-sea trench.

Unconformities are buried erosion surfaces separating rock masses where the older was exposed to erosion for a long interval of time before deposition of the younger.

Angular unconformity is one in which the older strata dip at an angle different from that of the younger.

Disconformity is an unconformity with parallel strata on opposite sides.

Nonconformity is an unconformity between profoundly different rocks.

Unconformities in the Grand Canyon show three major sequences of uplift and erosion, subsidence and deposition, and crustal deformation and crustal stability.

Broad surface features are basins, plains, plateaus, mountains, and Precambrian shields with continents built around them.

Broad warps have produced basins and arches or plateaus.

Billings, Marland P. *Structural Geology*, 3rd ed., Prentice-Hall, Inc., Englewood Cliffs, N. J., 1972. *A textbook designed to follow an introductory course in physical geology.*

Dennis, J. G. *Structural Geology*, The Ronald Press Company, New York, 1972. *A very understandably written introduction to the subject.*

De Sitter, L. U. *Structural Geology*, 2nd ed., McGraw-Hill Book Company, New York, 1964. *A textbook presupposing a certain familiarity with the elements of structural geology and its terminology.*

Hobbs, Bruce E., W. D. Means, and P. F. Williams *An Outline of Structural Geology*, John Wiley & Sons, Inc., New York, 1976. *A textbook also assuming some background in basic concepts of structural geology; covers most modern concepts in the field.*

Umbgrove, J. H. F. *The Pulse of the Earth*, 2nd ed., Martinus Nijhoff, The Hague, 1947. *A treatise that must be rated a classic for its impact on the broad philosophical aspects of structural geology.*

SUPPLEMENTARY READINGS

QUESTIONS

1 How do we know that most folding of rocks takes place far below the earth's surface rather than near the surface?

2 Which displacements are greatest during earthquakes—vertical or horizontal? Explain your answer.

3 What differences are noted at the earth's surface between crustal movements that are slow as compared to those that are rapid?

4 Distinguish between stress and strain; between elastic and plastic deformation.

5 Define and illustrate strength as related to rocks.

6 How do gravity anomalies occur? What information about the earth can be learned from them?

7 Why do we still have mountains when weathering and erosion have been at work on them for eons of time?

8 What are the surface patterns of outcrops for plunging anticlines and synclines?

9 How are normal and reverse faults different? How does each form?

10 Differentiate between transcurrent and transform faults.

11 How does a disconformity differ from a nonconformity? How might one merge into the other?

12 How are basins and arches produced on the plateaus of many continents?

EARTHQUAKES AND THE EARTH'S INTERIOR

Earthquakes have profound effects on many people and on their works. They also cause changes in the earth's surface and give us controlling data on the structure and nature of the earth's interior. For example, one of the most significant measurements in physical geology is that some earthquakes have sources as deep as 700 km but that **no earthquakes are deeper.** Of significance also is their distribution (Figures 9.1 and 9.2).

9.1 SEISMOLOGY

The scientific study of earthquakes is called **seismology,** from the Greek *seismos,* "earthquake," and *logos,* "reason" or "speech." At the turn of the twentieth century, there were approximately a dozen scientists in the world who would have been classified as professional seismologists. The subject matured, however, until by midcentury there were close to 400 seismograph stations that had seismologists recording and studying earthquakes and other ground vibrations. Data from seismology have become an integral part of physical geology in its growth from a descriptive natural history to a science that includes geophysics, a category that cuts across the groupings of knowledge labeled "physics" and "geology."

9.2 EFFECTS OF EARTHQUAKES

Earthquakes are interesting to most people because of their effects on human beings and structures. Their geological effects may also be profound. However, of all the earthquakes that occur every year, only one or two are likely to produce spectacular geological effects such as landslides or the elevation or depression of

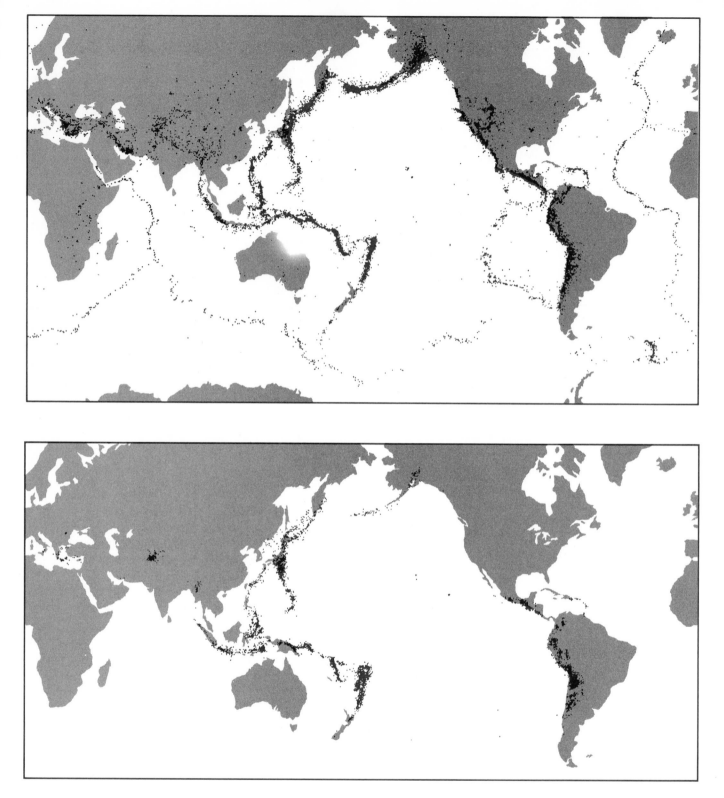

9.1 (top) Locations of earthquakes during a 7-year period, with focal depths of 0 to 100 km. [Data from ESSA, U.S. Coast and Geodetic Survey.]

9.2 (bottom) Locations of earthquakes during a 7-year period, with focal depths of 100 to 700 km. [Data from ESSA, U.S. Coast and Geodetic Survey.]

large landmasses. A hundred or so may be strong enough near their sources to destroy human life and property. The rest are too small to have serious effects.

FIRE

When an earthquake occurs near a modern city, fire can be a greater hazard than the shaking of the ground. In fact, fire has caused an estimated 95 percent of the total loss caused by some earthquakes.

One reason for the rapid spread of fire after an earthquake is that the vibrations often disrupt the gas and water systems in the area. In San Francisco, for example, some 23,000 service pipes were broken by the great earthquake of 1906 (often called the "San Francisco Fire" rather than "earthquake"). Water pressure throughout the city fell so sharply that when the hoses were attached to fire hydrants, only a small stream of water trickled out. Since that time, a system of valves has been installed to isolate any affected area and keep water pressure high in unbroken pipes in the rest of the city.

DAMAGE TO STRUCTURES

Modern, well-designed buildings of steel-frame construction have withstood the shaking of even some of the most severe earthquakes. Chimneys have often been particularly sensitive to earthquake vibrations, however, because they tend to shake in one direction while the building shakes in another. Consequently chimneys often break off at the roof line. In contrast, tunnels and other underground structures are little affected by the vibrations of even the largest earthquakes because they move as a unit with the surrounding ground.

In some earthquakes the extent to which buildings are affected by vibrations depends in part on the type of ground on which they stand. For example, in the San Francisco earthquake of 1906, buildings set on water-soaked sand, gravel, or clay suffered up to ten times as much damage as similar structures built on solid rock nearby. (See Liquefaction, p. 184.)

It has been found that the duration as well as the intensity of ground motion is a factor in causing damage to buildings.[1] Continued motion may cause damage even if it does not do so at the start. Reinforced concrete buildings start to show hairline shear cracks at the beginning of damaging motion. With continuing motion the cracks become enlarged, and eventually disintegration results if shaking continues long

enough. Repeated strong movements will bring destruction even to steel buildings.

In the Alaskan earthquake of March 27, 1964, the duration of damaging motion was approximately 3 min, or three times as long as the damaging shaking in San Francisco in 1906. Many buildings withstood the early vibrations, only to collapse in the later stages. Had the buildings in Alaska been subjected to vibrations for only 1 min, many probably would not have collapsed. Some of the buildings in Alaska that suffered severe damage had been built to conform to the earthquake provisions of the Uniform Building Code recommended by the Structural Engineers Association of California (see Figure 9.3). These provisions were intended to safeguard against major structural failures during earthquakes. However, they were formulated on building damage resulting from California earthquakes, and no California earthquake has ever had damaging shaking exceeding 1 min in duration. It will therefore be necessary to establish new building codes that will take into account the stresses involved in long-lasting damaging shaking.

The duration of strong shaking is also blamed for the extensive landslides that destroyed many homes in Anchorage.

SEISMIC SEA WAVES

If you are ever fortunate enough to be basking in the sun on Waikiki Beach and the water suddenly pulls away from the shore and disappears over the horizon, do not start picking up seashells or digging clams—a seismic sea wave is coming, and the withdrawal of the water is the first warning of its approach.

Some submarine earthquakes abruptly elevate or lower portions of the sea bottom, setting up great sea waves in the water. The same effect may also be produced by submarine landslides at the time of a quake. These giant waves are called **seismic sea waves,** or **tsunami,** a Japanese term, which has the same form for both singular and plural. Seismic sea waves may be generated by volcanic eruptions, as in 1883, when Krakatoa erupted. These waves killed 36,500 in the East Indies. Seismic sea waves have devastated oceanic islands and continental coastlines from time to time throughout history. The Hawaiian Islands have been hit 37 times since their discovery by Captain Cook in 1778.

On April 1, 1946, a severe earthquake occurred at 53.5°N, 163°W, 130 km southeast of Unimak Island,

[1] Fergus J. Wood (ed.), *The Prince William Sound, Alaska, Earthquake of 1964 and Aftershocks*, vol. II, Publication 10-3, U.S. Coast and Geodetic Survey, Washington, D.C., 1967.

Contrast in building response to the Prince William Sound earthquake, Alaska, 1964. [U.S. Coast and Geodetic Survey.]

Alaska, in the Aleutian Trench. Here the ocean is 4,000 m deep. Minutes after the earthquake occurred, waves more than 33 m high smashed the lighthouse at Scotch Cap, Unimak Island, killing five persons. About 4.5 h later, the first seismic sea wave from this quake reached Oahu, Hawaii, after traveling 3,600 km at 800 km/h. At the time, marine geologist Francis P. Shepard and his wife were living in a seashore cottage on northern Oahu and were awakened by a loud hissing noise. They dashed to the window just in time to see waters of the ocean boiling up over a high ridge and heading toward their house. Shepard grabbed his camera, but when he got to the door, much to his disappointment, he saw the water retreating rapidly oceanward. The sea's level quickly dropped 10 m. It was then that he realized he had seen a seismic sea wave. Several more times after this, the water surged up over high-tide levels and then ran back into its basin.

Starting in the Alaskan waters where the earthquake occurred, these seismic sea waves had spread out much as waves do when a rock is thrown into a pond. But these waves were tremendous in length. Their crests were about 160 km apart. They swept out into the Pacific Ocean, moving at terrific speeds of over 800 km/h. As they passed ships, however, these waves went unnoticed. They were about 1 m high in water 3,000 m deep, with 160 km between crests; so their effect was similar to the ground level rising 1 m as you walk 160 km. But when they reached Oahu and other Pacific shores, the effect was dramatic. The energy that moved thousands of meters of water in the open ocean became concentrated on moving a few meters of water at a shallow shore. There the water curled into giant crests that increased in height until they washed up over shores meters above high tide: 12 m on Oahu and 18 m on Hawaii. These same seismic sea waves swept on down the Pacific. They reached Valparaiso, Chile, 18 h after their launching, when they still had enough energy to cause 2 m rises of the water after traveling 13,000 km. Some even returned to hit the other side of Hawaii 18 h later. In fact, tide gauges showed that seismic sea waves sloshed around the Pacific basin for days after the earthquake was over.

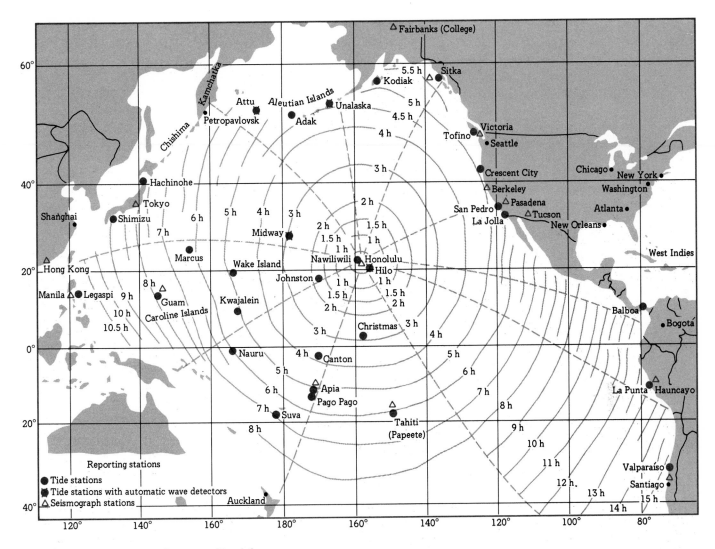

9.4 Seismic-sea-wave travel times to Honolulu.

This 1946 seismic sea wave was one of the most destructive to hit Hawaii. This one came without warning and was rated the worst natural disaster in Hawaii's history, but it was the last destructive seismic sea wave to surprise the Hawaiian Islands.

Seismic-sea-wave warning system In 1948, the United States Coast and Geodetic Survey established a seismic-sea-wave warning system (SSWWS) for Hawaii (see Figure 9.4). It operates continuously, recording visible seismograms at its seismological observatory in Honolulu. These are equipped with an automatic alarm system that sounds whenever an unusually strong earthquake is recorded. Tide stations have also been set up to detect waves with characteristics of seismic sea waves. These gauges filter out the normal tides and the wind waves. When the alarm rings, requests are made for immediate readings from other seismograph stations around the Pacific. Within approximately 1 h, the earthquake's location is deter-

mined. If it is found that the earthquake was in the Pacific Ocean or near its perimeter, tide gauges are then checked to see whether they show the existence of a seismic sea wave. If they do, an estimate of its arrival time is made, an alert is sounded, and people are warned to evacuate coastal areas.

The seismic-sea-wave warning system proved very effective on November 4, 1962, when there was an earthquake under the sea off Kamchatka Peninsula at 17 h 7 min Greenwich time. Within about 1 h, with the help of reports from seismograph stations in Alaska, Arizona, and California, Honolulu located the earthquake at 51°N, 158°E. Reports from tide stations at Attu and Dutch Harbor (Aleutian Islands) indicated that seismic sea waves had been started by the quake. Honolulu thereupon computed the time it would take for the first wave to reach Oahu and Hawaii, as well as the other Hawaiian Islands. It was due at Honolulu at 23 h 30 min Greenwich time, 6 h and 23 min after the quake occurred off Kamchatka. In a little over 3 h, Midway reported that it was covered by 3 m of water as the first wave raced over it. At Honolulu and Hilo, the waves were not so large as those of 1946, although damage was great. But not a single life was lost, thanks in great measure to the warning system.

The Chilean seismic sea wave of May 22, 1960, was the most destructive in recent history, causing deaths and extensive damage in Chile, Hawaii, the Philippines, Okinawa, and Japan. As usual, Honolulu had computed the location and time of the earthquake; so when word was flashed that seismic sea waves were racing out over the Pacific, the Honolulu observatory issued warnings of the danger, urging evacuation of coastal areas and correctly predicting the arrival time of the first waves. These arrived on schedule, 6 h after the warnings were broadcast, 15 h after they started off South America. Through failure of many people to heed the warnings, however, 61 lives were lost in Hilo. Two hundred and twenty-nine dwellings were destroyed or severely damaged (see Figure 9.5). In Japan, no general seismic-sea-wave warning was issued; for it was not known that a seismic sea wave of such distant origin could be so destructive. About 8 h after hitting Hawaii, more than 22 h after the earthquake that started them, the seismic sea waves roared

9.5 A mail truck lies half-buried in the shambles of a residential area in Hilo, Hawaii, after seismic sea waves from the Chilean earthquake of May 22, 1960, devastated the community. [Wide World.]

up the coasts of Honshu and Hokkaido. There, more than 17,000 km from where they started, they brought death to 180 people and caused extensive damage. The waves also did considerable damage in New Zealand. All Chilean coastal towns between the thirty-sixth and forty-fourth parallels were destroyed or severely damaged.

Fortunately, only an extremely small fraction of all submarine or coastal earthquakes cause seismic sea waves.

LANDSLIDES

In regions where there are many hills with steep slopes or areas with special soil conditions sensitive to vibration, earthquakes are often accompanied by landslides. These slides occur within a zone seldom exceeding 40 to 50 km in radius, although the very largest earthquakes have affected areas as far away as 150 km. The Alaskan quake of 1964 did this.

One of the worst earthquake-caused landslides on record occurred on June 7, 1692. More than 20,000 lives were lost and much property destroyed in a large section of the then-bustling town of Port Royal, Jamaica. The whole waterfront was launched into the sea, together with several streets of two- and four-story brick houses. The houses and other buildings had been built on loose sands, gravel, and filled land. Shaken loose by the quake, the underlying gravel and sand gave way and slid into the sea; two-thirds of the town, consisting of the government buildings, wharf, streets, homes, and people, went along.

In the province of Kansu, China, in deposits of loess (wind-deposited silt) an earthquake on December 16, 1920, caused some of the most spectacular landslides on record. The death toll was 100,000. Great masses of surface material moved nearly 2 km, and some of the blocks carried along damaged roads, trees, and houses.

Several landslides occurred in the vicinity of Hebgen Reservoir, Montana, when the area was shaken by an earthquake just before midnight on August 17, 1959. The largest slide dammed the Madison River to form Earthquake Lake, as shown in Figures 9.6, 11.8, and 11.9.

CRACKS IN THE GROUND

One of the most persistent fears about earthquakes is that the earth is likely to open up and swallow everyone and everything in the vicinity. Such fears have been nourished by a good many tall tales, as well as by pictures like Figure 9.7, which shows a fault scarp resulting from an earthquake.

One account of the Lisbon, Portugal, earthquake of November 1, 1755, claimed that about 40 km from

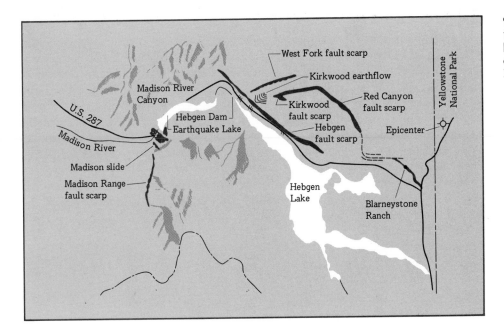

9.6 Region of Earthquake Lake, Montana: earthquake of August 17, 1959. [After John H. Hodgson, *Earthquakes and Earth Structure*, Prentice-Hall, Inc., Englewood Cliffs, N. J., 1964.]

9.7 Fault scarp resulting from earthquake near Fairview Peak, Nevada, some 160 km east of Reno. Quake measured 7.1 on the Richter scale. Note extension of scarp on left of photo. Rupture zone was 50 km long. [Adrian Atwater.]

Lisbon, the earth opened up and swallowed a village's 10,000 inhabitants with all their cattle and belongings and then closed again. The story probably got its start when a landslide buried some village in the area.

In Japan, widespread ideas about the earth's ability to swallow people may have sprung from an allusion in one of the Buddhist scriptures: When Devadatta, one of Sakya Muni's disciples, turned against his master and even made attempts on his life,

Heaven punished him by consigning him to Hades, whereupon the ground opened and immediately swallowed him up.

It is true that landslides do bury people and buildings and under special conditions they may even open

9.2 EFFECTS OF EARTHQUAKES

183

small, shallow cracks. In California, in 1906, a cow did fall into such a crack and was buried with only her tail protruding. But there is no authenticated case in which solid rock has yawned open and swallowed anything.

LAND MOVEMENTS

Some earthquakes are accompanied by significant vertical and horizontal movement of the land surface. The surface sinks in some places, rises in others, and is often tilted, and portions can even be moved laterally great distances.

Fault creep is slow, periodic movement of the land on opposite sides of a fault (see Chapter 8). Measurable movement along the San Andreas is on the order of 1 or 2 cm/year. During the past 20 million years, this has resulted in offsets of several hundred kilometers.

Vertical displacements in excess of 10 m have been recorded in a single earthquake, and areas of thousands of square kilometers may be affected. Details of some examples are discussed in Chapter 8.

LIQUEFACTION

Damage to structures as a result of eathquakes may be caused more by foundation failure and sliding due to liquefaction of saturated soil or loose sediments than by the actual shaking of the ground. Liquefaction often occurs several minutes to tens of minutes after a quake strikes a region, when underground sewers, storage tanks, pipes, and piles driven into the ground may be observed to float up to the surface. Nearby structures may settle several meters into the ground. Water and sand may be ejected a meter or more into the air for several minutes after the quake.

All these phenomena apparently result from turning what had been relatively stable deposits near the surface into liquefied material incapable of supporting structures. This most commonly occurs where the deposits are thoroughly saturated, such as along seacoasts, the shores of lakes or ponds, or anywhere that subsurface water has filled the fractures, voids, and pores in the ground. So it appears that soils and loose deposits that are free-draining are less likely to undergo liquefaction during earthquakes and are therefore better foundations for structures.

SOUND

When an earthquake occurs, the vibrations in the ground often disturb the air and produce sound waves that are within the range of the human ear. These are known as **earthquake sounds**. They have been variously described, usually as low, booming noises. Very near the source of an earthquake, sharp snaps are sometimes audible, suggesting the tearing apart of great blocks of rock. Farther away, the sounds have been likened to heavy vehicles passing rapidly over hard ground, the dragging of heavy boxes or furniture over the floor, a loud, distinct clap of thunder, an explosion, the boom of a distant cannon, or the fall of heavy bodies or great loads of stone. The true earthquake sound, of course, is quite distinct from the rumble and roar of shaking buildings, but in some cases the sounds are probably confused.

9.3 INTERPRETING EARTHQUAKES

By examining the location and strength of earthquakes, we can learn much about how earthquakes do their damage.

WORLDWIDE STANDARD SEISMOGRAPH NETWORK

In 1961, the United States Coast and Geodetic Survey began establishing a worldwide network of 125 standard seismograph stations (see Figure 9.8). Since then, seismological knowledge has been expanding rapidly because of the quantity and quality of earthquake measurements. Also, computers have helped to speed the processing of data for determining earthquake locations, depths of foci, and size. The United States Coast and Geodetic Survey and the United States Geological Survey now locate approximately 6,000 quakes per year, whereas 15 years before, the number had been one-tenth that. This increase does not mean that the seismicity of the earth has increased. It only appears to have done so because the improved equipment in the increased number of recording stations gathers more and better data.

From this worldwide seismic network and from other seismograph stations come measurements that give us a new perspective on the sequences of earthquake activity associated with large shocks and information on the size of the area involved. For example, after the Rat Islands earthquake of February 4, 1965, 1,300 aftershocks were recorded during the first 45 days. These occurred over an area roughly 600 km long by 300 km wide. It has been found that the largest earthquakes and their aftershocks involve areas up to 200,000 km².

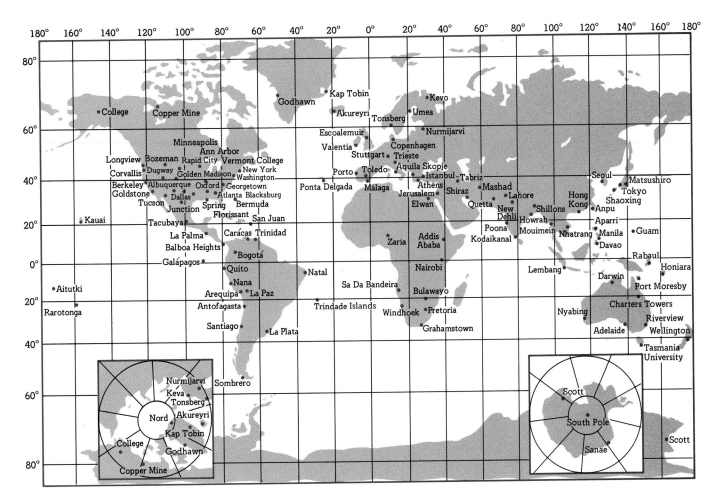

9.8 Stations of the worldwide standard seismograph network.

EARTHQUAKE FOCUS

In seismology the term **focus** designates the source of a given set of earthquake waves. Just what is this source? As we know, the waves that constitute an earthquake are generated by the rupture of earth materials. When these waves are recorded by instruments at distant points, their pattern indicates that they originated within a limited region. Most sources have dimensions probably closer to 50 km in length and breadth than 5 or 500 km. A few of the largest earthquakes may involve up to 1,000 km. But trying to fix these dimensions more accurately offers a real problem, one that has not yet been solved.

In any event, the focus of an earthquake is usually at some depth below the surface of the earth. An area on the surface vertically above the focus is called the **epicentral area**, or **epicenter**, from the Greek *epi*, "above," and "center" (Figure 9.9).

Foci have been located at all depths down to 700 km, a little more than one-tenth of the earth's radius (see Table 9.1). On some continental margins, foci have clustered along a plane dipping toward the inte-

9.9 Diagram showing the positions of the focus and epicenter of an earthquake.

rior of the continent, as in Figure 9.10. The deepest have been limited to the Tonga-Fiji area and the Andes as shown in Figure 9.11. (These are at the bottom of active subduction zones. See Section 10.4.)

TABLE 9.1

Focal depths for a 7-year period[a]

Depth, km	Foci
< 75	27,650
100 ± 25	3,533
150 ± 25	1,942
200 ± 25	974
250 ± 25	447
300 ± 25	223
350 ± 25	196
400 ± 25	200
450 ± 25	205
500 ± 25	292
550 ± 25	488
600 ± 25	448
650 ± 25	154
>650 ± 25	8
Total	36,760

[a]Summarized from United States Coast and Geodetic Survey's preliminary determination of epicenter data (personal communication from Leonard M. Murphy, Chief, Seismology Division).

EARTHQUAKE INTENSITY

How to specify the size of an earthquake has always posed a problem. Before the development of instrumental seismology, some of the early investigations of earthquakes led to various attempts to describe the intensity of the shaking at a specific place. For instance, a missionary in some remote region would keep a diary of earthquakes rated as weak, strong, or very strong. This was at best a personal scale.

In 1883, an intensity scale was developed that combined the efforts of M. S. de Rossi of Rome and F. A. Forel of Lausanne, Switzerland. For half a cen-

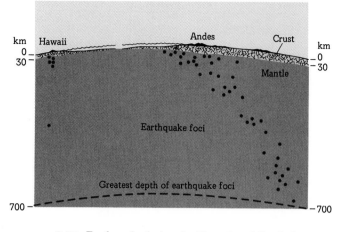

9.10 Earthquake foci under Hawaii and South America.

tury the Rossi-Forel intensity scale was widely used throughout the world. According to this scale, earthquake effects were classified in terms of 10 degrees of intensity. It had definite limitations, however, from a scientific standpoint. For example, the definition of the sixth degree of intensity included "general awakening of those asleep; general ringing of bells; oscillation of chandeliers; stopping of clocks; some startled persons leave their dwellings." But an earthquake that produced those effects in Italy or Switzerland might not even wake the baby in Japan, or it might cause a general stampede in Boston.

Objections to the Rossi-Forel intensity scale led L. Mercalli of Italy to set up a new scale in 1902. This was modified in 1931 by H. O. Wood and Frank Neumann and is the scale currently in use by the United States Geological Survey to evaluate earthquake intensity. It has 12 degrees of intensity and takes into account varying types of construction:

I Not felt except by a very few under specially favorable circumstances. (I on the Rossi-Forel scale.)

9.11 Earthquake foci between 600 and 700 km in depth.

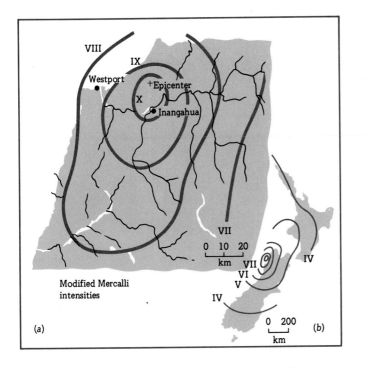

9.12 (a) Isoseismic map of epicentral region of Inangahua earthquake, New Zealand, May 24, 1968; (b) isoseismic map of New Zealand for the same earthquake.

II Felt only by a few persons at rest, especially on upper floors of buildings. Delicately suspended objects may swing. (I to II, Rossi-Forel scale.)

III Felt quite noticeably indoors, especially on upper floors of buildings, but many people do not recognize it as an earthquake. Standing motorcars may rock slightly. Vibration like passing of truck. Duration estimated. (III, Rossi-Forel scale.)

IV During the day, felt indoors by many, outdoors by few. At night, some awakened. Dishes, windows, doors disturbed; walls make creaking sound. Sensation like heavy truck striking building. Standing motorcars rocked noticeably. (IV to V, Rossi-Forel scale.)

V Felt by nearly everyone, many awakened. Some dishes, windows, and the like broken; a few instances of cracked plaster; unstable objects overturned. Disturbances of trees, poles, and other tall objects sometimes noticed. Pendulum clocks may stop. (V to VI, Rossi-Forel scale.)

VI Felt by all; many frightened and run outdoors. Some heavy furniture moved; a few instances of fallen plaster or damaged chimneys. Damage slight. (VI to VII, Rossi-Forel scale.)

VII Everybody runs outdoors. Damage negligible in buildings of good design and construction; slight to moderate in well-built ordinary structures; considerable in poorly built or badly designed structures; some chimneys broken. Noticed by persons driving motorcars. (VIII, Rossi-Forel scale.)

VIII Damage slight in specially designed structures; considerable in ordinary, substantial buildings, with partial collapse; great in poorly built structures. Panel walls thrown out of frame structures. Fall of chimneys, factory stacks, columns, monuments, walls. Heavy furniture overturned. Sand and mud ejected in small amounts. Changes in well water. Persons driving motorcars disturbed. (VIII+ to IX−, Rossi-Forel scale.)

IX Damage considerable in specially designed structures; well-designed frame structures thrown out of plumb; great in substantial buildings, with partial collapse. Buildings shifted off foundations. Ground cracked conspicuously. Underground pipes broken. (IX+, Rossi-Forel scale.)

X Some well-built wooden structures destroyed; most masonry and frame structures destroyed with their foundations; ground badly cracked. Rails bent. Landslides considerable from river banks and steep slopes. Shifted sand and mud. Water splashed (slopped) over banks. (X, Rossi-Forel scale.)

XI Few, if any, (masonry) structures remain standing. Bridges destroyed. Broad fissures in ground. Underground pipelines completely out of service. Earth slumps and land slips in soft ground. Rails bent greatly.

XII Damage total. Waves seen on ground surfaces. Line of sight and level distorted. Objects thrown upward into air.

By means of postcards, letters, and interviews, investigators for the United States Geological Survey make surveys of the effects caused by each earthquake. They then determine what places have been shaken equally. These places are plotted on a map, and the points are connected by **isoseismic lines,** or lines of equal shaking. Figure 9.12 shows an isoseismic map for the New Zealand earthquake that occurred in May 1968.

EARTHQUAKE MAGNITUDE AND ENERGY

Having to rely on the impressions of people is a very unsatisfactory way of compiling accurate information on the actual size of an earthquake. Therefore a scale was devised in 1935, based on instrumental records.[2] By computation based on the amount of motion in certain waves, it ascribed to each earthquake a number called the earthquake's **magnitude,** an index of the quake's energy at its source. This was refined in 1956 by B. Gutenberg and C. F. Richter and in 1967 by an International Committee on Magnitude. According to this scale, shallow earthquakes near inhabited areas can have the following effects: One of magnitude 2.5 would be just large enough to be felt nearby; one of magnitude 4.5 or over is capable of causing some very local damage; one of 6 or over is potentially destructive. A magnitude 7 or over represents a major earthquake.

It should be noted that the magnitude calculation for any given earthquake is corrected for distance from

[2]C. F. Richter, "An Instrumental Earthquake Magnitude Scale," *Bull. Seis. Soc. Am,* vol. 25, pp. 1–32, 1935.

the epicenter, so that seismic stations all over the world calculate the same magnitude for the same event.

The **Richter scale** is measured on a logarithmic basis: magnitude varies logarithmically with the wave amplitude of the quake as recorded on the seismograph. Each higher whole number on the scale represents an increase of 10 times in the measured wave amplitude of an earthquake. Therefore, the amplitude of a magnitude 8.0 earthquake is not twice as large as one of magnitude 4.0 but 10,000 times as large.

We can also estimate the **amount of energy** released during a quake by use of the Richter scale magnitudes. An increase of one unit on the magnitude scale represents a 31-fold increase in energy released. Therefore, a quake of magnitude 6 releases 31 times as much energy as one of magnitude 5, and over 900 times as much as one of magnitude 4. In the example from the previous paragraph, a magnitude 8.0 earthquake releases almost one million times more energy than one of magnitude 4.0 on the Richter scale.

An earthquake of magnitude 5 releases approximately the same amount of energy as did the first atomic bomb when it was tested on the New Mexico desert, July 16, 1945. A megaton nuclear device releases the same amount of energy as an earthquake of magnitude 6. Of course, the energy is applied in quite different ways—highly concentrated in the device, widely dispersed in the earthquake—and the results are correspondingly different.

Some earthquakes that are major catastrophes in a human and property sense are not of the greatest magnitude. Near the coast of northern Peru, at 9.2°S, 78.8°W, on May 31, 1970, a great quake killed an estimated 50,000 persons and left 800,000 homeless while causing $230 million in damage. Hualas Canyon towns were flooded by burst dams and were buried under landslides and mudslides. The quake was felt along 1,000 km of Peru; yet its magnitude was 7.8, less than that of any of the quakes that have been listed in Table 9.2.

The largest magnitude assigned to an earthquake to date has been 8.6. As Table 9.2 shows, four of that size have occurred during the modern observational period: in Alaska, September 10, 1899; in Colombia, January 31, 1906; in the Tibet-Burma region, August 15, 1950; and in Alaska, March 27, 1964. The only earthquake in history that might have been larger—if we judge from the reported effects—was the Lisbon quake of 1755. Possibly its magnitude was between 8.7 and 9.0. An earthquake with a magnitude of over 10 should theoretically be perceptible in scattered areas over the

TABLE 9.2

Great earthquakes

Date	Location	Magnitude
September 10, 1899	Alaska, 60°N, 140°E	8.6
January 31, 1906	Colombia, 1°N, 82°W	8.6
August 17, 1906	Chile, 33°S, 72°W	8.4
January 3, 1911	Soviet Union-China, 44°N, 78°E	8.4
December 16, 1920	China, 36°N, 106°E	8.5
March 2, 1933	Japan, 39°N, 145°E	8.5
August 15, 1950	Pakistan-Tibet-Burma, 29°N, 97°E	8.6
May 22, 1960	Chile, 38°S, 73.5°W	8.4
March 27, 1964	Alaska, 61.1°N, 147.7°W	8.6

entire earth, but such an occurrence has never been reported.

It should be clear that the location of earthquake epicenters, with respect to population density, can account for lower magnitude events doing more damage than some larger events. In addition, local rock conditions and style of construction are major determinants in loss of life and property damage.

9.4 DISTRIBUTION OF EARTHQUAKES

From year to year there are wide variations in the total energy released, as well as in the number of individual shocks. For example, 1906 showed 6 times the average energy released between 1904 and 1952 and 40 times the minimum; 1950 was another very active year, second only to 1906; and 1976 was also an extremely active year for earthquakes. (See Section 9.6.)

Most of the energy released is concentrated in a relatively small number of very large earthquakes. A single earthquake of magnitude 8.4 releases just about as much energy as was released, on the average, each year during the first half of the twentieth century. It is not unusual for the energy of one great earthquake to exceed that of all the others in a given year or of several years put together. The nine great earthquakes listed in Table 9.2 represent nearly a quarter of the total energy released from 1899 through 1964.

The maximum energy released by earthquakes becomes progressively less as depth of focus increases. The nine earthquakes in the table were relatively shallow and had the magnitudes shown; the five largest intermediate-depth shocks over the same interval had magnitudes of 8.1, 8.2, 8.1, 7.9, and 8.0; and the three largest deep shocks had magnitudes of only 8.0, 7.75, and 7.75.

TABLE 9.3

*Average annual number and
magnitude of earthquakes[a]*

	Magnitude	Av. number	Release of energy, approx. explosive equiv.
Actually observed:			
Great	7.7–8.6	2	50,000 1-megaton (hydrogen) bombs
Major	7.0–7.6	12	
Potentially destructive	6.0–6.9	108	
Estimates based on sampling special regions	5.0–5.9	800	1-megaton bomb
			1 small atom bomb (20,000 t TNT)
	4.0–4.9	6,200	
	3.0–3.9	49,000	
	2.5–2.9	100,000	
			0.5 kg TNT
	2.5	700,000	

[a] Modified from B. Gutenberg and C. F. Richter, *Seismicity of the Earth*, Princeton University Press, Princeton, 1949.

The average annual number of earthquakes from 1904 through 1946 was estimated by Gutenberg and Richter (Table 9.3), who placed the number of earthquakes large enough to be felt by someone nearby at more than 150,000 per year. Gutenberg and Richter estimated that the total number of earthquakes "may well be of the order of a million each year." Seismograph stations throughout the world are now supplying data leading to the location of close to 6,000 earthquakes per year, although locations have been noted from seismographic records since 1899. Figures 9.1 and 9.2 show the distribution of quakes for 1961 through 1967, the first years of standardized networks and computer-programming techniques. These active zones were not significantly different in the past.

Earthquakes tend to occur in belts, or zones, also marked by active volcanoes. The earthquakes that occur in the most active zone, around the borders of the Pacific Ocean, account for a little over 80 percent of the total energy released throughout the world. The greatest activity is near Japan, western Mexico, Melanesia, and the Philippines. The loop of islands bordering the Pacific has a high proportion of great shocks at all focal depths. (See also the discussion of earthquakes resulting from plate movements, Section 9.6.) Another 15 percent of the total energy released by all earthquakes is in a zone that extends from Burma through the Himalaya Range, into Baluchistan, across Iran, and westerly through the Alpine structures of Mediterranean Europe. This is sometimes called the "Mediterranean and trans-Asiatic" zone. Earthquakes in this zone have foci aligned along mountain chains. That leaves 5 percent of the energy to be released throughout the rest of the world. Narrow belts of activity are found to follow the oceanic ridge systems (see Figure 9.1).

With the widespread acceptance of the theory of plate tectonics (Chapters 1 and 10), it now becomes evident that most earthquakes occur along the margins of plates, where one plate comes into contact with another and stresses develop. The release of energy associated with these stress conditions is in the form of earthquakes.

Maps showing zones of earthquake expectancy in the United States have been issued by the United States government (Figures 9.13 and 9.14). The zones in Figure 9.13 were outlined after a 2-year study of 28,000 earthquakes and show where earthquakes may be expected in the next 100 years. Figure 9.14 gives a quick method for evaluating relative earthquake hazards: Levels of ground shaking are shown by contour lines that indicate the percentages of the force of gravity likely to occur at least once in a 50-year period. For example, a contour of 40 percent of gravity means one can be relatively certain that that region will not experience ground shaking more than 40 percent of the force of gravity. (All percentages are given with 90-percent relative certainty, or at the 90-percent probability level.)

SEISMICITY OF ALASKA AND ALEUTIAN ISLANDS

Before instruments were developed, reports of earthquakes were dependent on population distribution. Alaska was settled in 1783 by Russians on Kodiak Island, who first reported an earthquake occurring south

9.13 Seismic risk map for the United States, issued January 1969, shows earthquake damage zones of reasonable expectancy in the next hundred years. [Developed by U.S. Coast and Geodetic Survey, *ESSA Rel. ES-1*, January 14, 1969.]

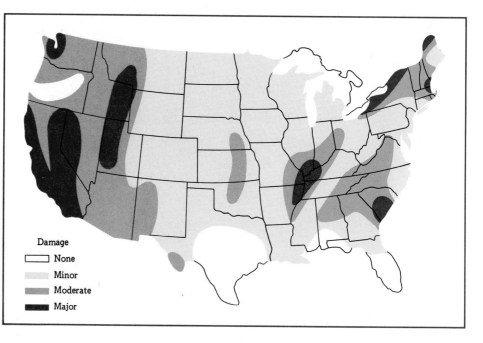

Damage

☐ None
☐ Minor
▨ Moderate
■ Major

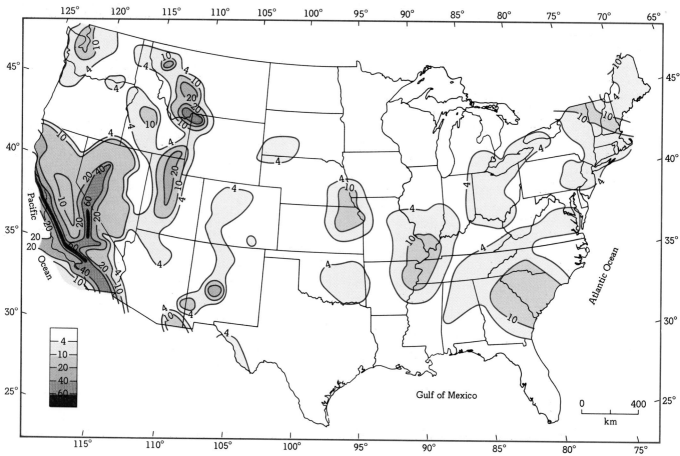

9.14 Map of expectable levels of earthquake shaking hazards. Levels of ground shaking for different regions are shown by contour lines, which express in percentages of the force of gravity the maximum amount of shaking likely to occur at least once in a 50-year period. [Modified from S. T. Algermissen, and D. M. Perkins; "A Probabilistic Estimate of Maximum Acceleration in the Contiguous United States," *U.S. Geol. Surv. Open-File Rep. 76–416*, July 1976.]

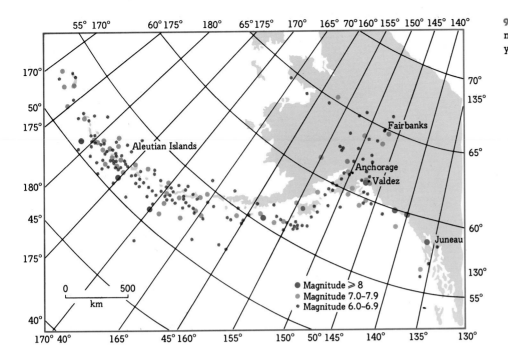

9.15 Earthquakes in Alaska with a magnitude of 6 or greater during a 65-year period.

of the Alaskan peninsula on July 27, 1788. From that time, more than 1,500 earthquakes have been felt by persons living in the area. Information on early earthquakes was obtained from historical records, newspapers, and ships' logs. From 1899 on, there have been more than 300 instrumentally recorded earthquakes of which the magnitude has been 6 or greater (see Figure 9.15).

The Prince William Sound earthquake of 1964   On March 27, 1964, at 5:36 P.M. Alaskan standard time (or 3 h 36 min 13 s Greenwich mean time, March 28, 1964), an earthquake of magnitude 8.6 rocked the Prince William Sound area in south-central Alaska. It released twice as much energy as the great San Francisco earthquake of 1906 and was the most destructive earthquake ever recorded on the North American continent—one of our greatest natural disasters.

Its center was located at 61.1°N, 147.7°W, at a depth of 33 km (see Figure 9.16). The earthquake was felt over 1.3 million km², and major damage to buildings, ports, and transportation occurred over an area of 130,000 km². Although only 10 percent of the land area of Alaska was involved in the effects of the earthquake, this was occupied by 50 percent of Alaska's total population and included all the major seaports, highways, and railroads. Property damage was estimated at $86 million. In spite of all the damage, there were only nine lives lost in Anchorage and six elsewhere from building collapse and landslides. If it had been earlier in the day, the casualties could have been much higher. A seismic sea wave generated by the earthquake accounted for 98 deaths in Alaska; 11 in Crescent City, California; and 1 in Seaside, Oregon.

9.16 Epicenter of Alaskan earthquake of March 27, 1964. Elevation changes and the position of a tidal wave 1 h after the quake are also shown.

9.17 Uplift and subsidence associated with Alaskan quake of March 27, 1964.

As soon as news of the disaster reached the United States Coast and Geodetic Survey and the United States Geological Survey in Washington, they took immediate action. They organized the largest project ever undertaken to obtain data about an earthquake. Investigations were conducted simultaneously from the land, sea, and air. Portable seismographs were sent into the area to record aftershocks. Survey teams determined the amounts and location of surface movements and alteration of the sea bottom. Engineers studied building damage. Much was learned from these

investigations about the effects resulting from the unusually prolonged and violent shaking.

Many of the towns in the affected region had been built on unconsolidated deposits. As a result, landslides occurred, causing extensive damage to buildings and other structures. Displacements of the land and sea bottom modified shorelines and changed seafloor topography. Because Alaska is dependent to a large extent on marine commerce for its survival, the Coast and Geodetic Survey acted swiftly to determine which sea lanes were navigable and to chart new lanes where necessary.

Surveys showed that permanent vertical and horizontal displacements occurred across considerable distances (see Figure 9.17). At the entrance to Prince William Sound, 150 km from the epicenter, uplifts of

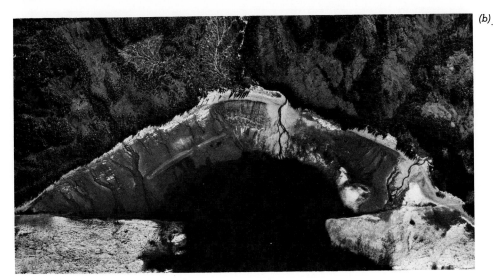

9.18 Hanning Bay Fault Cove, northeast of Macleod Harbor, Montague Island, Alaska. (a) Photograph taken by the United States Forest Service on June 8, 1959. The sharp break in the vegetation in the extreme lower right marks an old fault line. (b) Photograph taken on July 7, 1964, by the United States Coast and Geodetic Survey. Note the fault line extending across the harbor entrance and the amount of uplift. Comparison of the photographs shows the uplift and its effect in damming the outlet to the lake on the lower right. Note also the channels being cut through deltaic deposits of the two major streams.

(a)

(b)

the sea bottom exceeding 13 m were discovered (see Figure 9.18). The entire land area from Anchorage to Seward, a distance of 150 km, subsided 1 to 2 m. Horizontal displacements resulted in changes of distances between points. These generally ranged from 1 to 2 m. The absolute amount and direction of movement could not be accurately determined as reference points were also disturbed. Surveys across Montague Strait indicated that Montague and Latouche Islands were 5 to 7 m closer together after the quake than they had been in 1933. The general pattern of horizontal and vertical movements suggests turbulence of the earth's crust rather than simple displacement.

The earthquake was of sufficient magnitude to trigger the SSWWS at College, Alaska, and 8 min later, the alarm sounded in the Honolulu Observatory. Within 1 h, Honolulu had determined a preliminary epicenter, and coastal cities were warned of an impending inundation.

The seismic sea wave was the first major one associated with an earthquake whose epicenter was on land. However, it was started when submarine landslides and vertical displacements disturbed a large area of the sea bottom. It was highly destructive around the Gulf of Alaska, causing extensive damage to many of Alaska's principal harbors: Anchorage, Cordova, Kodiak, Seward, Valdez, and Whittier. It also caused damage to the west coast of Canada and Crescent City, California. It accounted for the greatest loss of life.

In Anchorage, 120 km from the center of the quake, a substantial portion of earthquake damage was attributable to landslides. The largest and most damaging slide occurred in the Turnagain Heights section (see Figure 9.19). This is a residential area built on a bluff overlooking an arm of Cook Inlet. The vibrations triggered multiple slumps in the bluff. Downward and

9.4 DISTRIBUTION OF EARTHQUAKES

9.19 Turnagain Heights, Anchorage, Alaska. (a) Photograph taken on August 12, 1961; (b) Photograph on July 26, 1964. Bootlegger Cove clay, which has little bearing strength when wet, underlies most of Anchorage. Earthquake-produced ground vibrations triggered multiple slides in the bluff area adjacent to Knik Arm. The slide covered an area about 2,400 m long and up to 400 m wide. The slide stopped when enough material collected at the toe to prevent further sliding. Clearing of debris and leveling of the ground has been started [lower left in (b)]. Note also the roadways through the slide area. [ESSA, U.S. Coast and Geodetic Survey.]

(a)

(b)

outward movement occurred over a 2,400-m section of coastline extending from east to west. The slide extended inland approximately 200 m at the east end and 400 m at the west. The edge of the bluff slid more than 200 m into the inlet. Some blocks moved intact as much as 170 m. Eyewitness accounts indicated that the sliding retrogressed from the edge of the bluff, with movements beginning approximately 2 min after the start of the earthquake. More than 75 houses were destroyed in the Turnagain Heights section.

Valdez, 80 km from the epicenter, suffered extensive damage. The town was built on an alluvial fan that had grown out into the head of a deep, steep-sided fiord. The earthquake caused a major landslide, which brought total destruction to the entire waterfront. There were extensive shoreline changes, and

nearly 6 months later the land was found to be still shifting. The town was relocated approximately 6.5 km northwest on more stable ground.

Seward, 150 km from the epicenter, also suffered extensive damage. The waterfront for a length of 1,500 m slid into Resurrection Bay. This slide, coupled with the destructive effect of an 8-m seismic sea wave, left no usable docks or piers. About 90 percent of Seward's industry was obliterated, and the rest was damaged. Investigations revealed that the northwest section of the bay was 27 to 30 m deeper than previously charted (see Figures 9.20 and 9.21).

Kodiak, on Kodiak Island, 450 km from the epicenter, had most of its business section and port facilities destroyed by the seismic sea wave.

Aftershocks More than 7,500 shocks were instrumentally recorded in the months that followed. These occurred in a belt 300 km wide, stretching 800 km from the main shock southwestward to a region off the

9.20 Seward, Alaska. The tsunami of March 27, 1964, hit the Seward port section with sufficient force to hurl this fishing vessel across the road. Note also the displaced buoy and other debris. [ESSA, U.S. Coast and Geodetic Survey.]

(a)

9.21 Seward, Alaska (a) in September, 1948; (b) in August, 1964. Extensive shoreline changes and railroad damage resulted from the 1964 earthquake (1,500 m of waterfront slumping into Resurrection Bay), tsunami, and fire. The steep and unstable newly formed shoreline developed in front of many tension cracks. [ESSA, U.S. Coast and Geodetic Survey.]

(b)

southwest coast of Kodiak Island. Their focal depths ranged from 20 to 60 km.

SEISMIC AND VOLCANIC ACTIVITY

The locations of volcanoes relative to major earthquake belts have been coming into focus as one of the results of increased data on number and quality of earthquakes.

At the scene of Chile's 1960 earthquake series, there is a chain of 16 active volcanoes paralleling the coast. One of these, Puyehue, erupted the day after the largest earthquake of the series, and 10 were variously active during the following year. Figure 9.22 shows the affected region and the volcanoes.

It has been estimated that approximately 6 percent of the world's large earthquakes occur in the Aleutian

9.4 DISTRIBUTION OF EARTHQUAKES

195

9.22 Location of volcanoes of Chile in the region affected by the earthquakes of 1960.

Islands and continental Alaska. The Aleutian Islands arc is more than 3,200 km in length and contains 76 volcanoes, 36 active since 1760. Two large earthquakes occurred off the adjacent coast after the 1912 eruption of Katmai.

It has been fashionable for decades to think of "volcanic earthquakes" as a special feature of explosive eruptions, small in energy and seldom damaging. But there is now emerging a realization that volcanoes and earthquakes may have a common ultimate cause in deep movements of mantle materials, and the coincidence of belts of major earthquake activity with belts that include active volcanoes supports the idea. The most obvious common cause of seismic and volcanic activity relates to plate interactions (see Chapters 1 and 10) where fracture zones afford access for volcanic materials welling up from the lower crust and mantle. These boundaries are also areas where earth-

quakes would naturally occur by interaction of the plates in zones of convergence, or divergence, or where two plates slide past one another along transform boundaries. (See Section 9.6.) On the other hand, there are **aseismic ridges** (see Section 15.4), such as the Hawaiian Island-Emperor Seamount Ridge (Figures 3.17 and 15.9), along which there may be little or no seismic activity.

9.5 CAUSE OF EARTHQUAKES

Centuries ago (and in some places today), people believed that the mysterious shakings of the earth were caused by the restlessness of a monster that was supposed to be supporting the globe. In Japan, it was first a great spider, and then a giant catfish; in some parts of South America, it was a whale; and some of the North American Indians decided that the earth rested on the back of a giant tortoise. The lamas of Mongolia had another idea. They assured their devout followers that, after God had made the earth, he placed it on the back of an immense frog; every time the frog shook his head or stretched one of his feet, an earthquake occurred immediately above the moving part (see Figure 9.23).

The Greek philosopher Aristotle (384–322 B.C.) held that all earthquakes were caused by air or gases struggling to escape from subterranean cavities. Because the wind must first have been forced into the cavities, he explained that just before an earthquake the atmosphere became close and stifling. As time passed, people began to refer to "earthquake weather," and to this day some insist that the air turns

9.23 Cause of earthquakes, according to the lamas of Mongolia. It was believed that when the frog lifted a foot, there was an earthquake immediately above the part that had moved.

humid and stuffy when an earthquake is about to occur. A Japanese scientist checked on this belief by investigating the conditions that had preceded 18 catastrophic earthquakes in Japan between 1361 and 1891. He found that the weather was fair or clear on 12 occasions, cloudy on 2, rainy or snowy on 3, and rainy and windy once, but that humid and sultry earthquake weather had never just preceded an earthquake.

Aristotle's idea that earthquakes were caused by imprisoned gases has long been abandoned as we have learned more about the behavior of gases, the structure of the earth's crust, and finally the depths at which earthquakes occur. What, then, causes earthquakes? Most are caused by deforming forces in the earth; and the immediate cause is the **sudden rupture** of earth materials distorted beyond the limit of their strength.

ELASTIC-REBOUND HYPOTHESIS

After the California earthquake of 1906, it was generally accepted that the immediate cause of earthquakes was faulting due to rupture of rocks of the earth's crust. The displacements along the San Andreas fault provided an unusual opportunity to study the mechanics of faulting and led to the formulation of Reid's **elastic-rebound hypothesis.** Surveys of part of the fault that broke in 1906 had been made during the years preceding the earthquake. H. F. Reid, of Johns Hopkins University, analyzed these measurements in three groups: 1851 to 1865, 1874 to 1892, and 1906 to 1907. The first two groups showed that the ground was twisting in the area of the fault. The third showed displacements that occurred at the time of the earthquake. From these Reid reconstructed a history of the movement.

Although there was no direct evidence, Reid assumed that the rocks in the earth's crust in the vicinity of the fault had been storing elastic energy at a uniform rate over the entire interval and that the region had started from an unstrained condition approximately a century before the earthquake. As the years passed, a line that in 1800 had cut straight across the fault at right angles was assumed to have become progressively more and more warped. When the relative movement on either side of the fault became as great

as 6 m in places, the strength of the rock was exceeded, and rupture occurred. The blocks snapped back toward an unstrained position, driven by the stored elastic energy (see Figure 9.24).

The San Andreas fault runs roughly from northwest to southeast. Land on the western, or Pacific Ocean, side of the fault has moved northwest relative to land on the eastern side. In 1906, a section of the fault 450 km long broke. The strains and adjustments were greatest within a zone extending 10 km on each side of the fault. Imagine a straight line 20 km long crossing this zone at right angles to the fault, which was in the center of the zone. After the earthquake, this line was broken at the fault and was shifted into two curves. The broken ends were separated by 6 m at the break, but the other ends were unmoved. Fences, roads, and rows of vegetation provided short sections of lines by which the fault displacement could be gauged (Figure 9.25).

Elastic rebound is the result of faulting and associated loss of confinement. This idea was based on observed surface movements. It was also based on our knowledge of how rocks behave when they have been subjected to deforming forces. In the laboratory, rocks have been subjected to pressures that are equivalent to the pressures in the earth's crust. Under these pressures the rocks gradually change shape; but they resist more and more as the pressure builds up until they finally reach the breaking point. Then they tear apart and snap back into unrestrained positions. This snapping back is called **elastic rebound** and was considered by Reid as the mechanism associated with the generation of earthquakes.

EARTHQUAKES NOT NECESSARILY ASSOCIATED WITH ELASTIC REBOUND

Until recently, it was believed that all earthquakes were caused by faulting due to elastic rebound. However, elastic rebound assumes that earthquakes origi-

9.25 Looking along a fence that crossed the San Andreas fault at approximately right angles. Before the earthquake of April 18, 1906, this fence was a continuous straight line. It is shown here as it appeared after the earthquake. [J. C. Branner.]

nate from distortion and rupture of elastic rocks capable of storing strain energy until it is released abruptly at the time of rupture. From laboratory measurements,

it has been found that this is a reasonable assumption for rocks similar to those at the earth's surface. But with the increased pressures of burial under 20 or 30 km of overburden, all rocks deform plastically; and plastic rocks neither store strain elastically nor are able to rebound. Therefore the elastic-rebound hypothesis in its original form would apply only to earthquakes with surface foci or with foci that are at extremely shallow depths.

In the 1920s, it became apparent that at least some foci were at depths where plastic deformation takes place. The first thoughts were that deep-focus earthquakes might represent a different mechanism of rupture than shallow-focus earthquakes do. However, there were so many features of earthquake records that indicated no really fundamental difference in the wave-generating mechanisms. So an alternative suggestion was made involving the property of matter called **viscosity**, the ratio of deforming force to **rate** at which a substance changes shape in response to the force. A material may have such high viscosity that although it deforms plastically, it does this slowly enough to have at the same time a buildup of force sufficient to allow rupture before complete adjustment has been accomplished through flow.

During the mid-twentieth century, improvements in seismological recording and interpretation have accumulated an impressive number of measurements showing that a large number of earthquakes have their foci in the mantle and regions of the lower crust where earth materials behave as plastics.

In the 1950s, Percy Bridgman at Harvard University documented by laboratory measurements the mechanism of rupture of earth materials under pressures equivalent to depths of hundreds of kilometers. He described the deformation as taking place steadily by plastic flow, but this is interrupted at times when the material ruptures, or in his words "lets go to get a fresh hold." Thus in an environment of continuous plastic deformation, there would be sudden "jerks," or rupture, capable of causing earthquakes. These sudden movements could be considered faulting which allows rebound to occur. Even deep plastic flow, therefore, could still be related to a form of elastic-rebound.

9.6 EARTHQUAKES RESULTING FROM PLATE MOVEMENTS

We suggested in Chapter 1 and earlier in this chapter that most earthquakes occur in narrow belts that also mark the edges of crustal plates. As these plates move about the earth like rafts over the plastic material of the

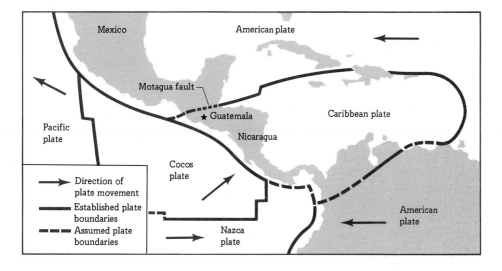

9.26 Guatemala earthquake of February, 1976. The quake was the result of movement along the Motagua fault zone at the boundary between the American plate and the Caribbean plate.

asthenosphere, they collide, separate, or simply slide past each other. Earthquakes frequently accompany these plate interactions.

The American and Pacific plates are converging at the rate of about 5 cm per year. As they slide past each other, enormous amounts of stress build up within the plates. This stress is commonly released in sudden slipping action producing the numerous earthquakes along this fault zone (the San Andreas being the most famous of such faults).

There are, however, sections of this zone that have become locked across the fault. This produces a **seismic gap** or region of the fault zone along which little earthquake activity has occurred for many years. Such gaps have been located at a number of localities along the contact between the American and Pacific plates. One occurs near San Francisco and has led to concern that stresses have built to the point where a major earthquake may be in the offing.

Another gap occurs in southern Alaska near Cape Yakataga. There have been no major earthquakes in this area during this century. It is feared that an earthquake of magnitude 8 or more could occur in the next few decades. An earthquake of magnitude 7.7 occurred in February, 1979, at St. Elias about 100 km to the east of this gap; and the great 1964 earthquake of magnitude 8.5 occurred as close as 150 km to the west. These earthquakes relieved the stresses that had built up in adjacent areas, but it is believed the Cape Yakataga gap may experience an earthquake of magnitude 8 or so in the near future.

CLUSTERS OF SEVERE EARTHQUAKES

There appear to be episodes of more frequent occurrences of severe earthquakes. The year 1976 was a time of several rather severe earthquakes in widely separated parts of the earth, all occurring in active earthquake-prone regions along plate boundaries.

Guatemala A severe quake jolted much of Central America on February 4, 1976, measuring 7.5 on the Richter scale. More than 23,000 persons died. This earthquake was apparently caused by movement of plates in the Caribbean region (Figure 9.26): The American plate moves westward relative to the Caribbean plate; the northern portion of the American plate carries Canada, the United States, and Mexico, while the southern portion carries most of South America in this westerly direction. A small plate off the southwest coast of Central America, the Cocos plate, moves northeasterly and slides beneath Central America, forming a deep trench along a subduction zone just off the west coast.

Movement of these plates has averaged about 4 to 6 cm/year, building up enormous stress over thousands of years. There had been no major movement along these plates in Guatemala for at least 200 years. Suddenly the stresses could no longer be constrained, and rupture occurred. The American plate slipped westward as much as a meter along a zone extending more than 150 km. This break occurred along the Motagua fault zone, where a number of faults had been mapped, but where the exact location of the boundary between the American plate and the Caribbean plate had never before been confirmed.

China On July 27,1976, a series of very severe earthquakes hit parts of northeast China in the province of Hopei, approximately 125 to 150 km east-southeast of

Peking. This series of shocks reportedly devastated Tangshan, an industrial city of a million and a half, and caused casualties and damage in cities as far away as Peking. Estimates of resulting deaths reached staggering proportions, with fears of more than half a million. The initial shock registered 8.2 on the Richter scale. This was followed by many aftershocks, some registering as high as 7.9. Although there were no apparent precautionary measures taken prior to this particular series of quakes, accurate prediction of earlier quakes that had hit other provinces in China in February, 1975, and May, 1976, had permitted authorities to take measures to minimize casualties. China has been a leader in the study of the prediction of earthquakes (see Section 9.11).

Turkey A quake of magnitude 7.6 struck near the Turkish-Soviet border on November 24, 1976, destroying over 100 villages and killing almost 4,000 persons. This quake was the strongest in Turkey since tremors in 1939 that registered 7.9 and killed 30,000. The province hit by the 1976 quake lies along the Anatolia fault, stretching inland from Turkey's Aegean coast. It is part of the major Mediterranean boundary zone between the African and Eurasian plates.

9.7 EARLY INSTRUMENTAL OBSERVATIONS

Information about the earth's interior comes from studying the records of earth waves. Earlier in this chapter we learned that some large earthquakes involve violent displacements of the earth's crust. These movements use up some of the energy released at the time of the earthquake. The rest is carried to great distances by earth waves. These may travel through the whole earth and around its entire surface if the earthquake is large enough. At great distances, sensitive instruments are needed to detect their passage. By studying the manner in which earth waves travel out from earthquakes, supplemented by the study of waves generated by dynamite and nuclear explosions, geologists have assembled a wealth of information about the structure of the globe from surface to center. But before we look at the results of these studies, let us review some of the methods used to obtain information.

The earliest instrument used to detect an earthquake was built around A.D. 136.[3] Its design is credited to a Chinese philosopher named Chang Hêng. His instrument was said to resemble a wine jar about 2 m in diameter. Evenly spaced around this were eight dragon heads under each of which was a toad with head back and mouth open. In each dragon's mouth was a ball (see Figure 9.27). When there was an earthquake, one of the balls was supposed to fall into a toad's mouth. There is no known record of what was inside the jar. Speculations over the centuries have assumed a pendulum of some sort, which would swing as the ground moved, knocking the ball out of a dragon's mouth in line with the swing. Chang Hêng had the notion that if a frog on the south side caught a ball, the earthquake had happened to the north of the instrument. His instrument was intended to record the occurrence of an earthquake and show the direction of its origin from the observer. It was a seismoscope, as it had no provision for a written record of the motion.

In 1703, J. de la Haute Feuille had the same general idea. He proposed a seismoscope design that he felt would respond to the tilting of the earth's surface at the time of an earthquake. His instrument consisted of a bowl of mercury so placed that when the earth's surface tilted, mercury would spill over the rim and fall into one of eight channels. Each channel represented a principal direction of the compass; the spilled mercury was to show the direction of the earthquake, and the amount of spillage the earthquake's intensity. However, there is no record that this seismoscope was ever built.

The first European to use a mechanical device for studying earthquakes was Nicholas Cirillo. He used his device to observe the motion of the ground from a series of earthquakes in Naples in 1731. His instrument

[3]James Dewey and Perry Byerly, "The Early History of Seismometry (to 1900)," *Bull. Seis. Soc. Am.*, vol. 59, pp. 183–227, 1969.

9.27 Chang Hêng's seismoscope, as visualized by Wang Chen-To.

was a simple pendulum whose swing indicated the amplitude of ground motion. He made observations at different distances from earthquakes to see how the motion died out with distance. He found that the amplitude decreased with the inverse square of the distance.

In 1783 there occurred a series of earthquakes in Calabria, Italy, that resulted in 50,000 deaths. These stimulated the development of other mechanical devices for studying such events. It also led to the appointment of the first earthquake commission. Shortly after the first large quake, a clockmaker and mechanic named D. Domenico Salsano had an instrument operating in Naples approximately 320 km from Calabria. It consisted of a long common pendulum with a brush attached that was supposed to trace out the motion with slow-drying ink on an ivory slab, the first attempt to preserve a written record. His instrument was also equipped with a bell that was to ring when motions were large enough. It has been reported that the bell rang on several occasions. A similar instrument was reported in use in Cincinnati around the time of the New Madrid, Missouri, earthquakes (1811 and 1812).

A series of small earthquakes near Comrie, Scotland, led to the development of an inverted-pendulum seismometer by James Forbes in 1844. His instrument consisted of a vertical metal rod with a movable mass on it. This was supported by a steel wire. The stiffness of the wire and the position of the mass could be adjusted to alter the free period of the pendulum. Forbes was trying to give his pendulum a long period so that it would remain stationary when the ground moved during an earthquake. At one end of the metal rod he attached a pencil that came into contact with a stationary paper-lined dome above it, to preserve a written record of the movement (see Figure 9.28). By moving the pencil some distance from the mass, he was able to magnify the motion of the pendulum two or three times. This instrument, a crude seismograph, was not successful because friction between the writing pencil and the recording surface reduced the sensitivity of the instrument, ruining recording of felt local quakes.

The first record of a distant earthquake was obtained accidentally by Ernst von Rebeur-Paschwitz at Potsdam, Germany, on April 17, 1889.[4] The earthquake was felt in Tokyo about 1 h before it was recorded in Germany. The motion was too small and was on a time scale that was too compressed to give much information. However, it showed that the energy from an earthquake could travel halfway around the world. And because the record gave the time of day, it indicated how long the waves took to travel that far, how long the motion lasted, and that the motion of the ground had a horizontal component.

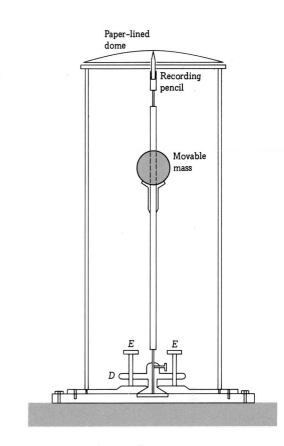

9.28 Forbes' seismometer. Screws, *E*, acting on a support, *D*, were to help set the pendulum in an upright position.

9.8 MODERN SEISMOLOGY

With the discovery that earth waves could travel to such great distances from an earthquake, modern seismology was born. The first records of earthquake motion registered on a tilt instrument were too fuzzy to be analyzed accurately. They were too small in amplitude, and the events of 1 h were compressed into the space of 6 mm. To obtain meaningful data about earthquakes, it was necessary to develop seismographs with increased recording speed and sensitivity so that the fuzzy lines would be spread out and enlarged.

A group of British professors teaching in Japan in the late nineteenth century initiated the activities that developed into all aspects of seismology as we find it today. Chief among these in his influence on the entire

[4]E. von Rebeur-Paschwitz, "The Earthquake of Tokyo, April 18, 1889," *Nature*, vol. 40, pp. 294–295, 1889.

field was John Milne. He was the one mainly responsible for developing a seismograph capable of picking up and recording intelligible signals from distant earthquakes. Milne was also responsible for promoting the first worldwide network of seismograph stations for systematic, continuous registration and interpretation of their records. By 1900, Milne's seismographs were operating on all the inhabited continents (see Figure 9.29). Sixteen stations were regularly recording and sending their data to Milne.

MODERN INSTRUMENTS

A modern seismograph consists of three basic parts: an inertia member, a transducer, and a recorder.

The inertia member is a weight suspended by a wire, or spring, so that it acts like a pendulum but is so constructed that it can move in only one direction. The inertia member tends to remain at rest as earth waves pass by. It has to be damped so that the mass will not swing freely.

The transducer is a device that picks up the relative motion between the mass and the ground. It converts this into a recordable form. It may be a mechanical lever or an electrodynamic system. In an electrodynamic system, a coil of wire moves back and forth in a magnetic field. This movement creates an electric current that passes through a galvanometer to be recorded on a sheet of paper.

To record motion in all directions, it is necessary to have three seismographs. One records vertical motion, and two record horizontal motion in directions at right angles to each other. A well-equipped seismograph station will have a set of three components: vertical, north-south, and east-west.

A mass on a spring will oscillate if displaced and then released. The time required to complete one oscillation is called the **period.** If the ground under this system oscillates with a shorter period, the mass hangs still in space, or nearly so. It then serves as a point of reference from which to measure the earth's motion. The inertia members are built to stand still during the passage of waves of a certain selected period range. Generally, two sets are used, one for short-period waves of 5 s or less and one for long-period waves of 5 to 60 s. Because the motion of the ground at distant stations is microscopic, the motion must be magnified to make a visible record. Therefore the transducer includes a system that will serve to magnify the ground motion.

9.29 The Milne horizontal seismograph. Light from *L* was reflected by *M* onto photographic paper through the intersection of two crossed slits. The lower illustration is a top view of the instrument with its outer case removed. *T* is a flexible wire holding up the boom. The weight, *W,* was pivoted on the boom.

A standard recorder is a cylindrical drum on which is wrapped a sheet of recording paper. It rotates at a constant speed and by a helical drive moves sidewise as it turns. This produces a continuous record, a series of parallel lines. These lines appear straight when there is no ground motion; but if there is motion, the lines move in response. A clock-controlled device jogs the recording line briefly once per minute and for a longer time each hour. These jogs on the record identify the time of day. Each sheet of paper includes the capacity for 24 h of continuous recording (see Figure 9.30).

9.9 EARTH WAVES

When earth materials rupture and cause an earthquake, some of the energy released travels away by means of earth waves. The manner in which earth waves transmit energy can be illustrated by the behavior of waves on the surface of water. A pebble dropped into a quiet pool creates ripples that travel outward over the water's surface in concentric circles. These ripples carry away part of the energy that the pebble possessed as it struck the water. A listening device at some distant point beneath the surface can detect the noise of impact. The noise is transmitted through the body of the water by sound waves, far different from surface waves and not visible by ordinary means.

Just as with water-borne waves, there are two general classes of earth waves: **body waves,** which travel through the interior of the mass in which they are generated, and **surface waves,** which travel only along the surface.

BODY WAVES

Body waves are of two general types: **push-pull** and **shake.** Each is defined by its manner of moving particles as it travels along.

Push-pull waves, more commonly known as **sound waves,** can travel through any material—solid, liquid, or gas. They move the particles forward and backward; consequently the materials in the path of these waves are alternately compressed and rarefied. For example, when we strike a tuning fork sharply, the prongs vibrate back and forth, first pushing and then pulling the molecules of air with which they come in contact. Each molecule bumps the next one, and a wave of pressure is set in motion through the air. If the molecules next to your eardrum are compressed at the rate of 440 times/s, you hear a tone that is called "middle A."

Shake waves can travel only through materials that resist a change in shape. These waves shake the particles in their path at right angles to the direction of their advance. Imagine that you are holding one end of a rope fastened to a wall. If you move your hand up and down regularly, a series of waves will travel along the rope to the wall. As each wave moves along, the particles in the rope move up and down, just as the particles in your hand did. In other words, the particles move at **right angles** to the direction of the wave's advance. The same is true when you move your hand from side to side instead of "vibrating" it up and down.

SURFACE WAVES

Surface waves can travel along the surface of any material. Let us look again at the manner in which waves transmit energy along the surface of water. If you stand on the shore and throw a pebble into a quiet pool, setting up surface waves, some of the water seems to be moving toward you. Actually, though, what is coming toward you is **energy** in the form of waves. The particles of water move in a definite pattern as each wave advances: up, forward, down, and back, in a small circle. We can observe this pattern by dropping a small cork on the surface into the path of the waves (see Figure 9.31).

When surface waves are generated in rock, one common type of particle motion is just the reverse of the water-particle motion—that is: forward, up, back, and down.

9.10 RECORDS OF EARTHQUAKE WAVES

As we have seen, the first records of earthquake motion were difficult to analyze accurately. But when the drum speed was increased to spread the events of 1 h over a space of 1 m, a sharp pattern emerged.

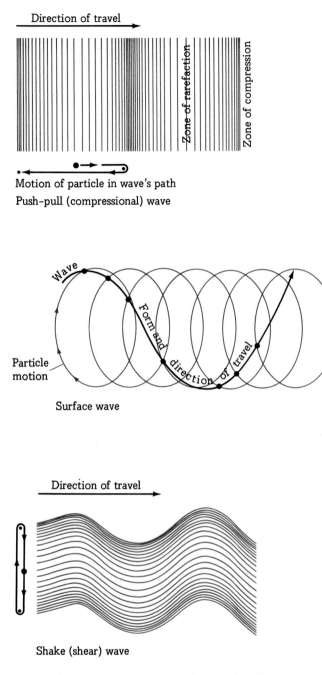

Direction of travel

Zone of rarefaction

Zone of compression

Motion of particle in wave's path

Push-pull (compressional) wave

Wave

Form and direction of travel

Particle motion

Surface wave

Direction of travel

Shake (shear) wave

9.31 Motion produced by three earthquake wave types.

This pattern consisted of three sets of earth waves. The first waves to arrive at the recording station were named **primary**; the second to arrive were named **secondary**; and the last to arrive were named **large waves**. The symbols P, S, and L are commonly used for these three types. Closer study revealed that the P

waves are push-pull waves and travel with speeds that vary with the density of the material. It also revealed that the S waves are shake waves and likewise travel at speeds that vary with density.

The P and S waves travel from the focus of an earthquake through the interior of the earth to the recording station. The L waves are surface waves that travel along the surface to the recording station from the area directly above the focus.

The P waves arrive at a station before the S waves because although they follow the same general paths of travel, they go at different speeds. The push-pull mechanism by which P waves travel generates more rapid speed than does the shake mechanism of the S waves. The S waves travel at about two-fifths the speed of P waves in any given earth material. The L waves are last to arrive because they travel at slower speeds and over longer routes (see Figure 9.32).

TIME-DISTANCE GRAPHS

As data accumulated, Milne began plotting the arrival time of the first P, S, and L waves from felt earthquakes. He set up time-distance graphs for each wave. He found that all three waves took longer and longer to travel to greater and greater distances. He also found that the time between the arrival of each wave type increased as the distance from the source increased.

Travel times If earth waves traveled through a material of uniform elasticity, their paths would be straight lines as they went at constant speed to greater and greater distances. However, if there is a progressive increase in elasticity with depth, they will gradually increase their speed, and their paths will become smooth curves. When there is a sharp increase in elasticity, the wave directions are changed, and this is reflected by sharp changes of direction in the line on a time-distance graph.

Distance is sometimes expressed as the number of kilometers for the length of a great-circle arc between two surface points or as the number of degrees for the angle at the earth's center subtended by that arc. For example, one-quarter of the way around the earth is 10,000 km, or 90°.

Less than 11,000 km From thousands of measurements the world over, it has been learned that P, S, and L waves have regular travel schedules for distances up to 11,000 km. From an earthquake in San Francisco for example, we can predict that P will reach El Paso, 1,600 km away, in 3 min 22 s and S in 6 min 3 s; P will reach Indianapolis, 3,220 km away, in 5 min

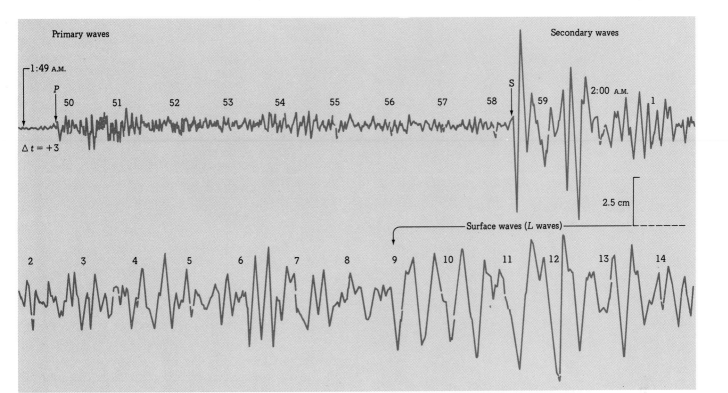

9.32 A record of an earthquake on a seismogram at Harvard, Massachusetts. All waves started in Rumania at the same instant. They arrived as indicated above because of different speeds and paths. The distance was 7,445 km.

TABLE 9.4

Sample timetables for P and S

Distance from source, km	Travel time				Interval between P and S (S − P)	
	P		S			
	min	s	min	s	min	s
2,000	4	06	7	25	3	19
4,000	6	58	12	36	5	38
6,000	9	21	16	56	7	35
8,000	11	23	20	45	9	22
10,000	12	57	23	56	10	59
11,000	13	39	25	18	11	39

Distance km	Time (S-P)
500	50 sec
	40
300	30
200	20
100	10
75	8
50	6
25	4
20	2
10	
5	
0	0

56 s and S in 10 min 48 s; P will reach Costa Rica, 4,800 km away, in 8 min 1 s and S in 14 min 28 s.

The travel schedules move along systematically out to a distance of 11,000 km, as shown in Table 9.4.

Beyond 11,000 km Beyond 11,000 km, however, something happens to the schedule, and the P waves are delayed. By 16,000 km they are 3 min late. When we consider that up to 11,000 km we could predict their arrival by seconds, a 3-min delay is significant. The fate of the S waves is even more spectacular:

they disappear altogether, never to be heard from again.

When the strange case of the late P waves and the missing S waves was first recognized, seismologists became excited; for they realized that they were not just recording earthquakes but were developing a picture of the interior of the earth. (See Section 9.12.)

9.10 RECORDS OF EARTHQUAKE WAVES

205

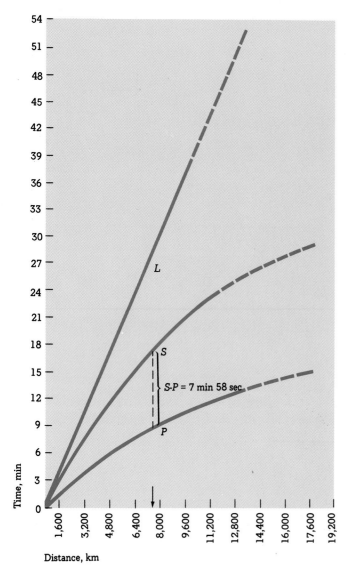

9.33 Travel-time curves for earth waves. An example is shown for an earthquake whose epicenter is approximately 7,500 km from the seismograph station. The delay between arrival of the primary wave and the secondary wave is demonstrated by the dashed line (*S-P*).

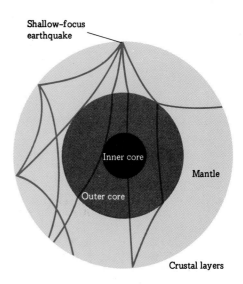

9.34 The paths of some of the more common earthquake waves.

server first determines the interval between the *P* and the *S* waves, as shown in Table 9.5. Next, this interval is plotted on the travel-time curve, always keeping the lower mark on the *P* line. At one place—and only one—the distance representing this time interval (*S-P*) will occur on the graph. This will happen at the place that corresponds to the quake's distance, in this case 7,440 km, or 67°. The graph also shows that *P* travels 7,440 km in 10 min 52 s. The seismologist subtracts this travel time from the arrival time of *P* and has the time of the earthquake at its source:

P arrived	15 h 09 min 06 s
P had traveled	10 min 52 s
P started at	14 h 58 min 14 s

This process is carried out for several seismograph stations that have recorded the quake. Then an arc is drawn on a globe to represent the computed distance from each station. The point where all the arcs intersect indicates the center of the disturbance.

LOCATING EARTHQUAKES

We now have timetables for earth waves for all possible distances from an earthquake. These are represented in the graph of Figure 9.33 and 9.34. Data of this sort are the essential tools of the seismologist.

When the records of a station give clear evidence of the *P*, *S*, and *L* waves from an earthquake, the ob-

TABLE 9.5

Interval between arriving waves

Time, beginning with *P*			Interval, time after *P*	
h	min	s	min	s
15	09	06		
15	17	04	7	58

The example in Table 9.5 and Figure 9.33 was based on a record taken at the station in Alert,

Canada.[5] Palisades, New York, recorded it at 3,580 km, and Pasadena, California, at 2,675 km. The United States Coast and Geodetic Survey located it in south-central Mexico, where they reported extensive property damage, 3 deaths, and 16 injured.

Although this whole procedure is essentially very simple, some have found its accuracy hard to believe. On December 16, 1920, seismologists all over the world found the record of an exceptionally severe earthquake on their seismographs. Each computed the distance of the quake and sent the information along to the central bureaus, where the various reports were assembled. The next day, the location of the earthquake was announced to the press, unlike many lesser shocks that fail to make the news. The announcement stated that a very severe earthquake had occurred at 5 min 43 s after 12:00, Greenwich time, December 16, 1920, in the vicinity of 35.6°N, 105.7°E. That placed it in the Chinese province of Kansu, about 1,600 km inland from Shanghai, on the border of Tibet. This area is densely populated but quite isolated. No reports of damage came in, however, and the matter was soon forgotten by the general public. But it was not forgotten by the members of the press, who were sure that they had been misinformed. Then, 3 months later, a survivor staggered into the range of modern communications with a story of a catastrophe in Kansu on the day and at the time announced, a catastrophe that had killed an estimated 100,000 persons and had created untold havoc by causing the great landslides described earlier in this chapter.

In contrast, however, on another occasion the news traveled faster than the waves themselves. On August 20, 1937, at 6:59 P.M., an earthquake occurred at Manila, in the Philippines, and the story was transmitted with unusual promptness. It was flashed to North America and found its way to the Boston office of a news agency. An operator there, 1 h after the earthquake happened, picked up a phone and called the Harvard Seismograph Station at Harvard, Massachusetts, 13,440 km from Manila, to inquire whether the disturbance had been recorded. The conversation took place 10 min before the earthquake's surface waves reached the station.

9.11 EARTHQUAKE PREDICTION AND CONTROL

From time to time observers have tried to relate the occurrence of earthquakes to sunspots, tides, positions of heavenly bodies, and other phenomena; but they

[5]John H. Hodgson, *Earthquakes and Earth Structure*, Prentice-Hall, Inc., Englewood Cliffs, N. J., 1964.

have all ignored the facts in one way or another. Occasionally someone steps forward to predict earthquakes on the basis of some completely fancied correlation, usually keeping the supporting data secret. One such prophet stated that on each of certain days during three months of one year, there would be "an earthquake in the southwest Pacific." Because he did not specify magnitude, time, or place in this highly active zone, statistically speaking, it was not possible for him to miss.

During the 1970s, however, great strides were made toward scientifically accurate measurement, interpretation, and potentially precise prediction of earthquakes. This has been an elusive goal of seismologists and astrologers alike for centuries. As a result of recent advances in the earth sciences, we appear to be on the verge of making this dream a practical reality.

EARTHQUAKE PRECURSORS

Many different kinds of both short-term and long-term effects have been observed to precede earthquakes. These **precursors** include crustal movements, unexpected changes in surface tilting, and changes in fluid pressure and in electrical and magnetic fields, as well as local small quakes.

It has long been alleged that dogs, cattle, and other animals behave strangely immediately prior to the occurrence of an earthquake. It now appears that these animals are not soothsayers but are merely sensing minute changes in the earth and its activities that tend to escape human senses.

DILATANCY MODEL

Observations for a number of earthquakes in Russia, Japan, and the Adirondack region of the United States have shown that prior to each earthquake, there is a marked change in the velocity of seismic waves passing through the rocks near the quake area.

Laboratory experiments have long ago demonstrated that rocks undergo **dilatancy,** or increase in volume, before failure. This is apparently produced by the formation and gradual outward movement of cracks within the rock. This dilatancy begins to occur at stresses as low as half the breaking strength of the rock. Water is released as the cracks expand. The rocks become more fragile and ultimately are too weak to withstand the stress; rupture results.

A bowing up of the earth's crust commonly occurs in the region of the earthquake epicenter. This changes the tilt of the surface, which can often be measured with sensitive instruments.

All these changes are thought to be related to dilatancy. If such changes could be observed in sufficient time, it would be possible to predict the earthquake.

A formula has been developed by J. H. Whitcomb, J. D. Garmany, and D. L. Anderson of the California Institute of Technology, which relates the duration of the precursor events with the magnitude of the predicted earthquake. A quake with magnitude 5 on the Richter scale should have precursor events up to 4 months prior to the quake. One with magnitude 7 would be preceded by about 14 years of such precursor activity.

Until now, only about ten earthquakes have been successfully predicted prior to their actual occurrence. With continued close observation of earthquakes and their precursor events, earthquake prediction could soon be a reality. It has become apparent, however, that the dilatancy model is not sufficient to serve as the only explanation for precursor events. Indeed it has been established that earthquakes may be preceded by quite different physical events and changes in different geologic regions, and earthquake-prediction methods may thus vary according to different sets of criteria for different regions.

CONTROL OF EARTHQUAKES

A fortuitous discovery made some years ago has led to the distinct possibility of controlling or modifying at least some earthquakes.

Waste fluids had been injected into a deep well near Denver, Colorado, over a period of several years. There soon developed a series of minor earthquakes, which were subsequently demonstrated to coincide with periods of fluid injection. During lulls in the injections, fewer or no quakes occurred. Since then, laboratory and field investigations have shown that movement along a fault zone can be initiated by reducing frictional resistance across the fault. This can be accomplished by injecting fluids along the fault. Conversely, faults can be locked when fluids are withdrawn. In the Rangely oil fields in Colorado, some United States Geological Survey workers demonstrated this ability to turn the seismic activity in an area on and off.

Although it is not likely that a major active fault zone will be treated in this manner in the near future, this method of control over fault movements may well be commonplace for future generations.

9.12 STRUCTURE OF THE EARTH

Studies of the travel habits of body waves through the earth and of surface waves around the earth have given us information about its structure from surface to center. These studies have been made possible by our knowledge of the speed of these earth waves and of their behavior in different materials (for example, body waves travel at greater speeds through denser materials than through less dense materials).

When body waves move from one kind of material to another, they are refracted (bent) and reflected (see Figure 9.35). These waves have revealed several places within the earth where there are changes in physical properties. Such changes could be due to changes in composition, atomic structure, or atomic state. The boundary where change takes place is called a **discontinuity**.

For body waves to reach greater and greater distances on the surface, they must penetrate deeper and deeper into the earth's interior. Thus in traveling from an earthquake in San Francisco to a station at Dallas, a surface distance of 2,400 km, the body waves penetrate to 480 km below the surface. This holds true for any other 2,400-km surface distance. To reach a station 11,200 km away, the body waves dip into the interior to a maximum depth of 2,900 km and bring us information from that depth.

On the basis of data assembled from studies of the travel habits of earth waves, the earth has been divided into three major zones: **crust, mantle,** and **core.**

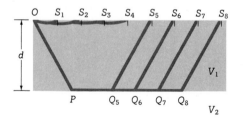

9.35 Paths of refracted waves used to determine the thickness of a single layer. If velocity in V_2 material at depth d is greater than V_1, waves from O arrive before any others at stations S_1, S_2, S_3, and S_4 by way of surface paths. Waves from O through P to Q_5S_5, Q_6S_6, Q_7S_7, and Q_8S_8 take the lead after S_5 and are first to arrive. By plotting time-distance relationships here, seismologists measure V_1, V_2, and d.

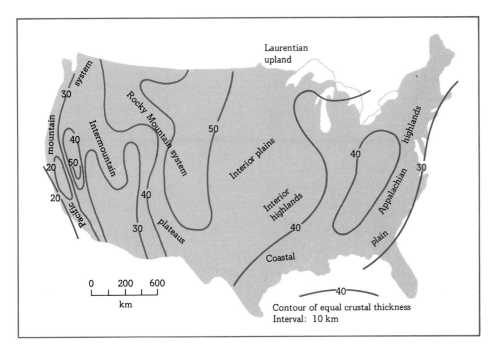

9.36 Crustal thickness determined from seismic data. [After L. C. Pakiser and I. Zietz.]

Laurentian upland

Rocky Mountain system

mountain system

Intermountain

Pacific

plateaus

Coastal

Interior plains

Interior highlands

Appalachian highlands

plain

0   200   600
km

40

Contour of equal crustal thickness
Interval: 10 km

CRUST

Information on the earth's crust comes primarily through observation of the velocities of *P* and *S* waves from local earthquakes (within 1,000 km) and from dynamite and nuclear blasts. One of the first things revealed is that the earth's crust is solid rock. Early in the history of crustal studies, a seismologist in Yugoslavia, A. Mohorovičic, made a study of records of the earth waves from an earthquake on October 8, 1909, in the Kulpa Valley, Croatia. He observed two *P* waves and two *S* waves and concluded that the first *P* and *S* had encountered something that caused some of their energy to be reflected to the surface. He also concluded that velocities of *P* and *S* waves increased abruptly below a depth of about 50 km. This abrupt change in the speed of *P* and *S* waves indicated a change in material and became known as the **Mohorovičic discontinuity.** For convenience it is now referred to as the "Moho," or the **M discontinuity.** The Moho marks the base of the earth's crust, separating it from the mantle.

Crust of the continents The depth to the Moho varies in different parts of the continents. However, it is generally greater under mountain regions than it is under lowlands. The crust is also variable in composition.

In the United States, the continental crust has an average thickness of 33 km but varies between 20 and 60 km (see Figure 9.36). Under the intermountain plateaus and the Great Basin Ranges, it averages less than 30 km; under the Great Valley of California,

20 km; under the Sierra Nevada, more than 50 km. It is thickest under the eastern front of the Rocky Mountain ranges. The crust under New England is about 36 km thick.

It has been difficult to get precise data on the earth's crust from the waves of earthquakes. Waves from dynamite blasts, however, with precisely known locations and times of detonation, have filled in some of the details. In 1941, the Harvard Seismograph Station determined the structure of the continental crust in New England. Analysis of many blast records revealed that in New England the continental crust has three layers, each one with different elastic properties, indicating different rock types. The layered structure of this crust may be representative of other sections of the continents. Table 9.6 summarizes the data.

The average velocity of waves traveling through the upper crustal layer is similar to the velocity expected for granite, granodiorite, or gneiss. Because these rocks are rich in silica and aluminum, we say that the material of the upper crust is **sialic** in composition; **sial** (Si for silicon and Al for aluminum) is generally used in speaking of this layer, which is found only on the continental areas of the earth. The second and third layers are more and more mafic in composition. The third is believed to be basalt. These darker, heavier rocks are sometimes designated collectively as **sima,** a name coined from "Si" for silicon and "ma"

9.12   STRUCTURE OF THE EARTH

TABLE 9.6

Earth's crust under New England[a]

| Thickness, km (and zone) | Velocity, km/s | | Rock type |
	P	S	
16 (layer 1)	6.1	3.5	Sialic
13 (layer 2)	6.8	3.9	Intermediate
7 (layer 3)	7.2	4.3	Simatic
(Moho)			
(Top of mantle)	8.4	4.6	

[a]L. D. Leet, "Trial Travel Times for Northeastern America," *Bull. Seis. Soc. Am.*, vol. 31, pp. 325–334, 1941.

for magnesium. A simatic layer encircles the earth, presumably underlying the sial of the continents, and is believed to be the outermost rock layer under deep permanent ocean basins.

Crust under oceans Our knowledge of the structure of the crust beneath the oceans is based on observations of rocks exposed on volcanic islands and on studies of the velocities of L waves from earthquakes, supplemented by dynamite-wave profiles and magnetic anomalies (Chapters 1 and 10).

The types of rock found on islands help to determine the edge of the Pacific basin. The **andesite line** (see Figure 3.3) has, on its ocean side, rocks composed primarily of basalt, whereas on the other side they are principally andesite. This has been viewed as the dividing line between the oceanic and the continental crusts.

On the basis of seismic-wave velocities, it appears that the crust under the Pacific basin is not layered and is appreciably thinner than the crust of the continents. Its thickness also varies but averages about 5 km. It is thinner under the Gulf of California and in the northeast Pacific. It is apparently composed of simatic rocks.

Crust under Hawaii The crust under the island of Hawaii was surveyed by seismic-refraction methods in 1964.[6] It was concluded that it is 10 km thick under the summit of Kilauea and 12 to 15 km under other parts of the island.

The crust under Hawaii appears to be divided into two layers: the upper layer, 4 to 8 km thick, is probably

[6]David P. Hill, "Crustal Structure of the Island of Hawaii from Seismic-Refraction Measurements," *Bull. Seis. Soc. Am.*, vol. 59, pp. 101–130, 1969.

an accumulation of lava flows that form the bulk of the island; the lower layer, also 4 to 8 km thick, is probably the original basaltic crust modified by a complex intrusive system associated with the central vents and rift zones of the island (see also Table 3.1).

MANTLE

Below the earth's crust is a second major zone, the **mantle,** which extends to a depth of approximately 2,900 km into the interior of the earth. Our knowledge of the mantle is based in part on evidence supplied by the behavior of P and S waves recorded between 1,100 and 11,000 km (Figure 9.37).

At the Moho, the speeds of P and S waves increase sharply, an indication that the composition of the material suddenly changes. We have no direct evidence of the new material's nature, but the change in speed suggests that it may contain more ferromagnesian minerals than does the crust. Scientists have accepted the idea that the mantle is solid because it is capable of transmitting S waves. To explain mountain-building processes, observers have emphasized that the mantle material undergoes slow flow as it adapts to changing conditions on the surface.

There are also variations in the composition of the upper mantle. We know this from variations in the velocity of P_n, the wave that travels through the upper portion of the mantle. Speeds of P_n vary from 7.5 to 8.5 km/s. This difference may be due in part to variation in elastic properties of the mantle material.

9.37 Structure of the mantle and core.

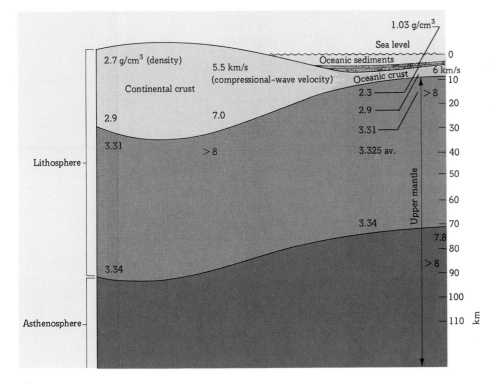

Within figure:

1.03 g/cm³

Sea level

2.7 g/cm³ (density)

5.5 km/s (compressional–wave velocity)

Continental crust

Oceanic sediments

Oceanic crust

6 km/s

2.9   7.0

2.3

2.9

3.31

3.31   >8

3.325 av.

Lithosphere

>8

3.34

3.34

Asthenosphere

7.8

>8

Upper mantle

0, 10, 20, 30, 40, 50, 60, 70, 80, 90, 100, 110 km

LITHOSPHERE AND ASTHENOSPHERE

The **lithosphere** is the rigid, outer layer of the earth, including the crust and the upper part of the mantle; the **asthenosphere** is a zone within the earth's mantle where plastic movements occur; it lies below the lithosphere, extending from about 50 to 100 km below the earth's surface downward to 500 km. There is a seismic discontinuity in the mantle. It occurs at the so-called low-velocity zone, which varies from about 50 to 80 km under oceans to about 100 km under continents (Figure 9.38). This apparently marks the boundary between the upper, colder portion of the mantle and the lower, hotter mantle. It also coincides, therefore, with the boundary between the lithosphere and the asthenosphere. When we speak of plates we are essentially referring to this lithosphere, which moves over the asthenosphere. It consists of crust *and* upper mantle, although we often refer to it as **crustal plates.** This is the part of the earth that moves slowly but persistently. In essence, the upper mantle is in motion, carrying the oceanic and continental crusts on its top in piggyback fashion.

CORE

We come now to the core, a zone that extends from the 2,900-km inner limit of the mantle to the center of the earth at a depth of 6,370 km. An analysis of seismographic records from earthquakes 11,200 km or more distant reveals that the core has two parts, an outer zone 2,200 km thick and an inner zone with a radius of 1,270 km.

In traveling between two points on the surface 11,200 km apart, *P* and *S* waves penetrate 2,900 km into the interior. But after they go deeper than that, they enter a material that delays *P* and eliminates *S* altogether (Figure 9.39). There have been various suggestions to explain these observations. The most widely favored theory postulates that the outer core is liquid. By definition, *S* waves are capable of traveling only through substances that possess rigidity; and since rigidity is a property of the solid state, it is assumed that the outer core, not transmitting *S* waves, is in a liquid state or at least behaving as a liquid.

P waves speed up again at a depth of about 5,100 km, indicating that the inner core is solid.

As the earth swings around the sun, it behaves like a sphere with a specific gravity of 5.5, but geologists have found that the average specific gravity of rocks exposed at the surface is less than 3.0. And even if rocks with this same specific gravity were squeezed under 2,900 km of similar rocks, their specific gravity would increase to only 5.7. Geophysicists have computed that the specific gravity of the core must be about 15.0 to give the whole globe an average of 5.5. To meet this requirement, it has been suggested that the core may be composed primarily of iron, possibly mixed with about 8 percent of nickel and some cobalt, in the same proportions found in metallic meteorites.

9.39 Interior of earth as deduced from interpretation of seismic waves. Shadow zone occurs 103° to 143° away from each earthquake. No *P* or *S* waves penetrate this region. The last ray to miss the outer core arrives just below 103°. The next ray catches the outer core: its *S* wave is not transmitted through the liquid outer core; its *P* wave is refracted to a position just beyond 143° from the quake. Only *P* waves penetrate the outer core to arrive at the surface on the opposite side of the earth.

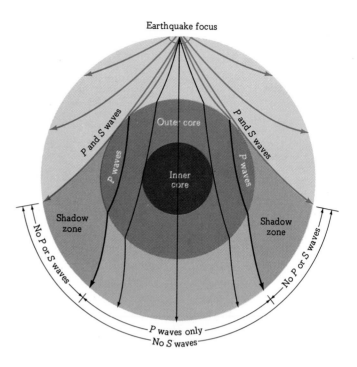

Earthquake focus

Outer core

Inner core

P and S waves

P waves

Shadow zone

Shadow zone

No *P* or *S* waves

No *P* or *S* waves

P waves only
No *S* waves

OUTLINE

Earthquakes are vibrations caused when earth materials have been distorted until they rupture.

Seismology is the scientific study of earthquakes.

Effects of earthquakes include fire, damage to structures, seismic sea waves, landslides, cracks in the ground, changes in land level, and sound.

    Fire has caused an estimated 95 percent of the total loss from some earthquakes.

    Damage to structures depends on intensity and duration of the shaking.

    Seismic sea waves are now being detected, and a **seismic-sea-wave warning system** operated by the United States Coast and Geodetic Survey alerts threatened communities.

    Landslides often accompany the largest earthquakes.

    Cracks in the ground, in the sense of yawning chasms in rock, have not happened.

    Land movements result in slow creep or rapid offsets both horizontally and vertically.

    Liquefaction of soil and loose sediments often follows earthquakes and commonly causes more damage than the actual shaking by the quake.

    Sounds associated with earthquakes vary from inaudible to booming and thunderous.

    Worldwide standard seismograph network is supervised by the Coast and Geodetic Survey.

Earthquakes are interpreted and evaluated from different facts and scales.

    Earthquake focus is a term used to designate the source of a given set of earthquake waves.

    Earthquake-intensity scales are used to estimate the amount of shaking at different places.

    Earthquake magnitude evaluates the amount of energy at the earthquake's source.

Distribution of earthquakes, in terms of energy released, is concentrated, with about 80 percent occurring around the borders of the Pacific Ocean.

Seismicity of Alaska and Aleutian Islands is estimated at 6 percent of the world's total.

The Prince William Sound earthquake of 1964 was the most destructive earthquake ever recorded on the North American continent.

Seismic and volcanic activity may have a common ultimate cause in deep movements of mantle materials, especially along crustal-plate margins.

Cause of earthquakes is the sudden rupture of earth materials distorted beyond the limit of their strength.

Elastic-rebound hypothesis proposed snapping back of distorted and ruptured rocks as the cause of earthquakes.

Plate movements generate a large proportion of all earthquakes. **Rates of movement** average 4 to 6 cm/year.

Interior of the earth is known to us from studying the records of earth waves.

Early instrumental observations were mostly by seismoscopes; the first record of a distant earthquake was made in 1889.

Modern seismology had its beginnings with a group of British professors teaching in Japan in the late nineteenth century.

Modern instruments consist of an inertia member, a transducer, and a recorder.

Earth waves carry away some of the energy released when earth materials rupture.

Body waves travel through the interior of the earth.

Surface waves travel along the surface.

Records of earthquake waves are characterized by three sets of earth waves: *P*, *S*, and *L* waves.

Time-distance graphs are bases for analyzing the history of earth waves.

Travel times are governed by the materials through which waves have traveled.

Locating earthquakes requires data from several stations.

Earthquake prediction may be possible primarily by observing and interpreting **precursor** events, including **crustal movements, tilt of earth's surface,** and **changes in fluid pressure** and **in electrical and magnetic fields.**

The dilatancy model is based on volumetric increases in rocks immediately prior to their rupture.

Control of earthquakes may be achieved by injection and removal of fluids from potentially active fault zones.

Structure of the earth is made up of crust, mantle, and core.

The earth's **crust** is separated from the mantle at a **discontinuity** called the **Mohorovičic,** or the **Moho,** or simply the **M discontinuity.**

The **crust of the continental** United States varies between 20 and 60 km in thickness.

The **crust under the oceans** averages about 5 km in thickness.

The **mantle** extends to 2,900 km in depth.

The **lithosphere** contains the crust and the upper, colder part of the mantle.

The **asthenosphere** extends downward from its border with the lithosphere to about 500 km.

The **core** is thought to have a liquid outer zone and a solid center; the outer zone behaves as a liquid in not allowing **S** waves to penetrate.

SUPPLEMENTARY READINGS

Allen, C. R. (chm.) *Predicting Earthquakes, A Scientific and Technical Evaluation— With Implications for Society*, Panel on Earthquake Prediction of the Committee on Seismology, National Academy of Sciences, Washington, D.C., 1976. *Evaluates the status of earthquake prediction and previews future trends. Presents urgent scientific and technical recommendations for further study and implementation of earthquake-prediction capabilities.*

Clark, S. P. *Structure of the Earth*, Prentice-Hall, Inc., Englewood Cliffs, N. J., 1970. *An excellent treatment of the subject.*

Dewey, James, and Perry Byerly "The Early History of Seismometry (to 1900)," *Bull. Seis. Soc. Am.*, vol. 59, pp. 183–227, 1969. *Describes the first instruments tried for recording ground motion from earthquakes, the first successful record of a distant earthquake, and early research that was the forerunner of all aspects of modern seismology.*

Hodgson, John H. *Earthquakes and Earth Structure*, Prentice-Hall, Inc., Englewood Cliffs, N. J., 1964. *A good account written for the nonspecialists.*

Press, Frank "Earthquake Prediction," *Sci. Am.*, vol. 232, no. 5, pp. 14–23, 1975. *A summary of recent technical advances, which have brought the United States and several other countries into a position of nearly achieving reliable long-term and short-term forecasts within a decade.*

Wood, Fergus J. (ed.) *The Prince William Sound, Alaska, Earthquake of 1964 and After-shocks*, vols. I–III, Publication 10-3, U.S. Coast and Geodetic Survey, Washington, D.C., 1966, 1967, 1969. *Definitive reports on the greatest earthquake ever recorded on the North American Continent, making it also the most thoroughly studied.*

QUESTIONS

1 What is the greatest hazard associated with earthquakes? Why is this so?

2 How does the type of ground affect the probability of a building's undergoing severe damage during an earthquake?

3 Why are oceanfront structures thousands of kilometers from an earthquake likely to be in danger?

4 What produces the impetus for large landslides associated with earthquakes?

5 What difference is there between an earthquake's focus and its epicenter? Make a sketch to show the relationship.

6 How do we distinguish between an earthquake's intensity and its magnitude? Which is more objective? Why?

7 How does the distribution of earthquakes compare to the distribution of volcanoes? What is the reason for this relationship?

8 What is the elastic-rebound hypothesis for the cause of earthquakes?

9 Why are there "seismic gaps" along some major fault zones, such as the San Andreas?

10 What are the principal earthquake waves? How are they similar? How do they differ from one another?

11 Explain briefly how the epicenter of an earthquake can be located accurately from three or more recording seismograph stations.

12 What are some of the principal kinds of earthquake precursors? What may be the reason or reasons for these to occur?

13 How close are we to being able to predict earthquakes? Explain.

14 How close are we to being able to control earthquakes? Explain.

15 What have earthquake studies taught us about the structure of the earth's interior? Make a simplified sketch to show the various parts of the earth.

10

CONTINENTS, OCEANS, PLATES, AND DRIFT

Recent observations from drill-hole depths of several kilometers inside the earth and measurements from satellites hundreds to thousands of kilometers above the earth permit us to gain a unique view of this planet and its crust. This picture was only indirectly observable prior to the last decade. A major result of the data obtained has been the development of a central theme called the "new global geology," relating paleomagnetism, sea-floor spreading, plate tectonics, and continental drift.

The immense amount of time and energy required to build mountains, erode continents, and shift shorelines defies our imagination. Yet we have come to accept these changes because they can be documented from evidence we see in our everyday experiences. But we must consider change on an even larger scale, challenging not only our sense of direction but our concept of geographic permanency as well. We now have evidence to suggest that the north and south poles have shifted position through time. We want to know why and how. We also now are led to believe that our landmasses have been torn from their moorings and have wandered over the earth's surface, a concept known as **continental drift.** One of the possible mech-

anisms for continental drift is the opening up of ocean basins through the process of **sea-floor spreading.** The concept of **plate tectonics** pictures the earth's crust as a jigsaw puzzle of rigid plates jostling one another, growing in some places and decaying in others. In this chapter we will show how attempts have been made to determine the movement rates and to detect the signs of the most recent movements, and will also discuss the implications of such global tectonics.

10.1 ORIGIN AND PERMANENCE OF CONTINENTS AND OCEAN BASINS

How and when did the continents and ocean basins form? Were they always distinct and separate entities, or have they only recently developed their characteris-

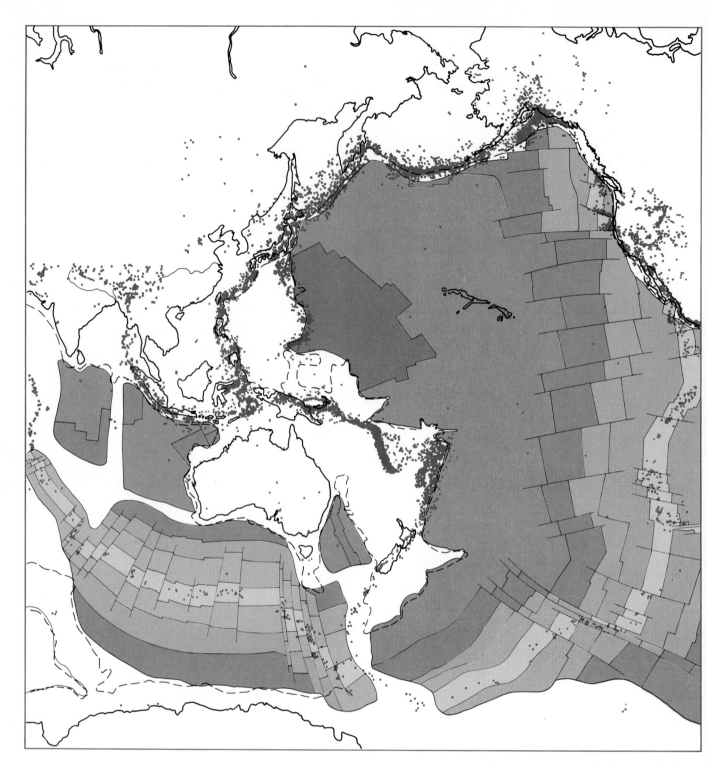

10.1 The age of the ocean floor is shown as bands of different colors, with the youngest ocean floor near midoceanic spreading centers and the oldest farthest away. The edge of the deep ocean floor is marked by the dashed line. [After Sheldon Judson, Kenneth S. Deffeyes, and Robert B. Hargraves,

Age of sea floor, millions of yr

5
23
38
65
135
190

Earthquake .

Physical Geology, Prentice-Hall, Inc., Englewood Cliffs, N. J., 1976; and from the map by W. C. Pitmann III, R. L. Larson, and E. M. Herron, "Age of the Ocean Basins Determined from Magnetic Anomaly Lineations," The Geological Society of America, Inc., Boulder, CO, 1974.]

tic identities? These are questions of primary importance to geology.

The oldest rocks on earth have been dated by rubidium-strontium methods as 3.8 billion years old (see Chapter 7). These ages came from highly metamorphosed rocks from widely separated parts of the earth —the Baltic Shield, Greenland, and Minnesota. By interpreting the geologic relationships from rocks dated by uranium-lead methods, we can determine that at least some crust was in existence about 4.0 billion years ago—a finding that compares favorably with the ages obtained from lunar crustal samples in the 4.0 to 4.1 billion-year range.

The earliest crust must have crystallized from molten material. If this was magma formed by the melting of a portion of the mantle (as is usually assumed), then we must ask how the crust differentiated into continental and oceanic portions. Theories for explaining differentiation generally follow one of two assumptions: (1) either the process of differentiation occurred very early in the earth's history in a single, catastrophic event; or (2) the process of differentiation was a continuing phenomenon over a long period of geologic time, perhaps ending at a given stage in the earth's development. However, we are not concerned here with establishing the causes of differentiation but rather with studying the results.

CONSTANT FREEBOARD OF CONTINENTS

It is possible to conclude from the rocks in the earth's geologic column whether they developed under continental or oceanic (marine) conditions. Maps of such development were examined by D. U. Wise,[1] of the University of Massachusetts, who showed that the **freeboard of continents** (the relative elevation of continents with respect to sea level) has remained fairly constant throughout much of the last 0.5 billion years of the earth's history. Wise argues that the earth arrived fairly early at essentially its present volumes of continents and oceans. Since then, a steady-state relationship has been maintained even though continental margins have been locally growing or retreating. The net effect is apparently to rework any continental-margin sedimentary material back into these continen-

[1]D. U. Wise, "Continental Margins, Freeboard and the Volumes of Continents and Oceans through Time," in C. A. Burk and C. C. Drake (eds.), *The Geology of Continental Margins*, Springer-Verlag New York Inc., New York, 1974.

tal rafts of nearly constant thickness, area, and volume. Isostasy exists between continental and oceanic crust and mantle (see Chapter 8). This equilibrium does not depend on the shape or number of continents or ocean basins, but merely on the total volume of each of them.

The only change in this picture of equilibrium seems to have happened at the time of the breakup of the earth's supercontinents, about 200 million years ago (see Section 10.4). This change in the freeboard may have been a response to the magnitude of the initial breakup of this large landmass, an occurrence unique in most of the earth's history and one not duplicated at least since Precambrian time.

AGE OF OCEAN BASINS

As we show later in this chapter, the present ocean basins are geologically young. The oldest features found in the sedimentary cover on the ocean floor are of Jurassic age (see Figure 10.1), about 150 million years old. This is not to say that there were no earlier ocean basins with sediments; but rather any older sediments that had once existed were lost as these former ocean floors were consumed along subduction zones (see Chapter 1 and Section 10.3); all evidence of earlier ocean basins was thus destroyed.

10.2 MAGNETISM AND PALEOMAGNETISM

In Chapter 1 we referred briefly to the earth's magnetic field and defined some of the terms useful in describing it. We begin this section with a more extensive discussion of magnetism and particularly of magnetism of the geological past, **paleomagnetism.** The data derived from paleomagnetic studies not only gave the subject of continental drift a new push in the 1940s and early 1950s but also provided proponents of continental drift with some of their strongest arguments. Furthermore, these studies have helped demonstrate the youthfulness of the ocean basins as compared with the continents and have been instrumental in the development of the idea of plate tectonics.

EARTH'S MAGNETISM

We are all familiar with the fact that the earth behaves as if it were a magnet and that consequently the compass needle seeks the north magnetic pole. We can picture the earth's magnetic field as a series of lines of

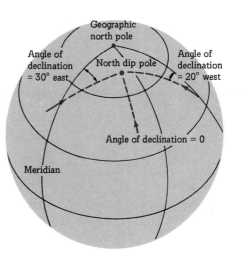

10.2 (left) The earth's magnetic field can be pictured as a series of lines of force. The arrows indicate positions that would be taken by a magnetic needle free to move in space and located at various places in the earth's field. The magnetic, or dip, poles do not coincide with the geographic poles, nor are the north and south magnetic poles directly opposite each other.

10.3 (right) Because the dip poles and geographic poles do not coincide, the compass needle does not point to true north. The angle of divergence of the compass from the geographic pole is the declination and is measured east and west of true north.

force. A magnet free to move in space will align itself parallel to one of these lines. At a point north of the Prince of Wales Island, at about 75°N and 100°W, the north-seeking end of the magnetic needle will dip vertically downward. This is the **north magnetic, or dip, pole.** Near the coast of Antarctica, at about 67°S and 143°E, the same end of our needle points directly skyward at the **south magnetic, or dip, pole.** Between these dip poles the magnetic needle assumes positions of intermediate tilt. Halfway between the dip poles the magnetic needle is horizontal and lies on the **magnetic equator.** Here the intensity of the earth's field is least, and it increases toward the dip poles, where the field is approximately twice as strong as it is at the magnetic equator (Figure 10.2).

The angle that the magnetic needle makes with the surface of the earth is called the **magnetic inclination, or dip.** The north and south dip poles do not correspond with the true north and south geographic poles as defined by the earth's rotation. Because of this, the direction of the magnetized needle will in most instances diverge from the true geographic poles. The angle of this divergence between a geographic meridian and the magnetic meridian is called the **magnetic declination,** and it is measured in degrees east and west of geographic north (see Figure 10.3).

SECULAR VARIATION OF THE MAGNETIC FIELD

As long ago as the mid-seventeenth century, it was known that the magnetic declination changed with time. Since then, we have been able to demonstrate not only slow changes in declination but also changes in inclination and intensity. These changes in magnetism have taken place over periods measured in hundreds of years. Because they are detectable only with long historical records, these changes are called **secular changes,** from the Latin *saeculum,* meaning "age" or "generation," implying a long period of time.

Figure 10.4 shows changes in the declination and inclination records at Paris during Gallo-Roman times and from 1540 to 1950. Another long-range record, from 1573 to the present, is available from London and shows a pattern very similar to that of Paris from the sixteenth century on. When records are compiled for the entire earth, it becomes apparent that the changes in the magnetic field wax and wane but move generally westward at a rate averaging approximately 0.2°/year.

THE GEOMAGNETIC POLE

The magnetic field at the earth's surface can be considered as composed of three separate components. There is, first, a small component that seems to result from activity above the earth's surface and is some-

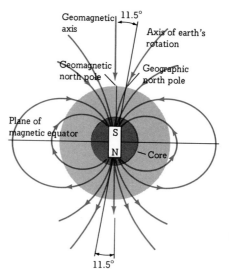

10.4 Magnetic inclination and declination vary with time. At Paris we have a continuous record of these changes since 1540. The data for inclination and declination for Gallo-Roman times (colored squares) are based on magnetic measurements of archaeological materials. [After Emile Thellier, "Recherches sur le champ magnétique terrestre," *L'Astronomie*, vol. 71, p. 182, 1957.]

10.5 The geomagnetic poles are defined by an imaginary magnetic axis passing through the earth's center and inclined 11.5° from the axis of rotation. This magnetic axis is determined by a hypothetical, earth-centered bar magnet positioned to best approximate the earth's known magnetic field.

times referred to as the <u>external field</u>. Second, there is a quantitatively more important component, the **dipole field,** best described as if it were caused by a dipole—such as a simple bar magnet—passing through the center of the earth and inclined to the earth's axis of rotation. Finally, there is what we refer to as the **nondipole field,** that portion of the earth's field remaining after the dipole field and the external field are removed.

The **dipole best approximating the earth's observed field** is one inclined 11.5° from the axis of rotation. The points at which the ends of this **imaginary magnetic axis** intersect the earth's surface are known as the **geomagnetic poles** and should not be confused with the magnetic, or dip, poles. The north geomagnetic pole is about 78.5°N, 69°W, and the south geomagnetic pole is exactly antipodal to it at 78.5°S, 111°E (see Figure 10.5).

MAGNETOSPHERE

The earth's magnetic field performs a service for terrestrial life by warding off and trapping powerful radiation from space. Radiation belts consist of electrons and very energetic ionized nuclei of atoms, mostly hydrogen. Some of the particles are believed to have been produced by cosmic-ray collisions in the atmosphere, but most are attributed to violent sprays of particles from the sun in solar flares.

The particles are diverted into a great doughnut-shaped racetrack around the earth, with the earth's axis under the doughnut's hole. They leak off and are turned back into space fairly rapidly. This region of trapped ions is called the **magnetosphere** (see Figure 10.6). It begins about 1,000 km above the earth and extends out to about 60,000 km. A similar band of charged particles has been reported encircling Jupiter.

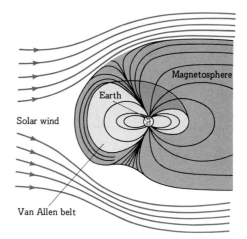

10.6 The magnetosphere: a doughnut-shaped region where the earth's magnetic field traps and wards off powerful radiation from space. The inner portion was first detected by James A. Van Allen in interpreting measurements from early United States satellites in 1958. It is sometimes referred to as the Van Allen belt. Later satellite data defined the region from 1,000 km out to about 60,000 km, now known as the *magnetosphere*.

CAUSE OF THE EARTH'S MAGNETISM

The cause of the earth's magnetism has remained one of the most vexing problems of earth study. A completely satisfactory answer to this question is still to be advanced.

As we have indicated, the earth's magnetic field is composed both of internal and external components. The external portion of the field is due largely to the activity of the sun. This activity affects the ionosphere and appears to explain magnetic storms and the northern lights. The changes and effects of the external field may be rapid and dramatic, but they have little effect on the internal field of the earth, which is of greatest concern to us.

William Gilbert, a British physicist of the sixteenth century, first showed that the earth behaves as if it were a magnet. He suggested that the earth's magnetic field results from a large mass of permanently magnetized material beneath the surface. This notion might at first seem attractive not only because large quantities of magnetic minerals have been found in the earth's crust but also because geologists think that the earth's core is made largely of iron. Since close examination reveals that the average intensity of the earth's magnetization is greater than that of the observable crustal rocks, we must therefore look deeper for the source of magnetism.

The first difficulty we face is that materials normally magnetic at the earth's surface lose their magnetism above a certain temperature. This temperature is called the **Curie temperature** and varies with each material. The Curie temperature for pure iron is about 760°C; for hematite, 680°C; for magnetite, 580°C; and for nickel, 350°C. The temperature gradient for the earth's crust is estimated to average about 30°C/km, and at this rate of increase the temperature should approximate the Curie temperature of iron about 25 km below the surface and exceed the Curie temperatures for most normally magnetic materials. Therefore, below 25 km we should not expect earth materials to be magnetic, and permanent magnetism can exist only above this level.

On the other hand, if all the earth's magnetism were concentrated in the crustal rocks, then the intensity of magnetism of these rocks would have to be some 80 times that of the earth as a whole. And yet, as we have mentioned, we know that the magnetic intensity of the surface rocks is less than that of the earth's average intensity.

From these observations we must conclude that the earth's magnetic field is not due to permanently magnetized masses either at depth or near the surface.

The theory of earth magnetism most widely entertained at present is that the earth's core acts as a self-exciting dynamo. One model of the earth's core pictures its outer portion as a fluid composed largely of iron. This core therefore not only is an excellent conductor of electrical currents but also exists in a physical state in which motions can easily occur. Electromagnetic currents are pictured as being generated and then amplified by motions within the current-conducting liquid. The energy to drive the fluid is thought to come from convection, which in turn results from temperature differences. The dynamo theory further requires that the random convective motions and their accompanying electromagnetic fields be ordered to produce a single, united magnetic field. It is thought that the rotation of the earth can impose such an order. The dynamo theory of earth magnetism still remains only a theory, but so far it has proved the most satisfactory explanation of the earth's magnetism.

PALEOMAGNETISM

Some rocks, such as iron ores of hematite or magnetite, are strongly magnetic. Most rocks, however, are only weakly so. Actually the magnetism of a rock is located in its individual minerals, and we should be more correct to speak of the magnetism of minerals rather than of the rock. By convention, however, we refer to rock magnetism. This magnetism is referred to as the rock's **natural remanent magnetism (NRM)**. This remanent magnetism may or may not agree with the present orientation of the earth's field and may have been acquired by the rock in many different ways. Identifying, measuring, and interpreting the different components of a rock's NRM form the basis of **paleomagnetism,** the study of the earth's magnetic field in the geological past.

Let us examine some of the ways an igneous rock acquires its magnetism. As a melt cools, minerals begin to crystallize. Those that are magnetically susceptible acquire a permanent magnetism as they cool below their Curie temperatures. This magnetism has the orientation of the earth's field at the time of crystallization. It is called **thermoremanent magnetism (TRM)**. This remanent magnetism remains with the minerals —and hence with the rock—unless the rock is reheated past the Curie temperatures of the minerals involved. This new heating destroys the original magnetism; and when the temperature again drops below the Curie temperatures of the magnetic minerals, the rock acquires a new TRM.

The NRM of our igneous rock may include other components. One of these is an **induced magnetism** arising from the present magnetic field of the earth. This induced magnetism is parallel to the earth's present field, but it is weak when compared with the TRM of the rock.

Sedimentary rocks acquire remanent magnetism in a different way than do igneous rocks. Magnetic particles such as hematite tend to orient themselves in the earth's magnetic field as they are deposited. Because the sedimentary particles are laid down on a nearly flat surface (such as the sea floor), they are limited in the position they can take. Therefore such particles normally reflect only the earth's magnetic declination and *not* its inclination. This orientation is retained as the soft sediments are lithified. This magnetism, known as **depositional remanent magnetism (DRM)**, records the earth's field at the time the rock particles were deposited. Of course, the sedimentary rock, like the igneous rock, may also acquire an induced magnetism reflecting the current magnetic field.

Virtual geomagnetic poles Paleomagnetic studies define the earth's magnetic field at various localities at different moments in geologic time. Instead of expressing the data on declination and inclination for a given locality, we usually express them in terms of equivalent pole positions. We refer to these poles as **virtual geomagnetic poles.** These are different from the dip poles and the geomagnetic poles we have already discussed. The pole consistent with the magnetic field as measured at any one locality is the virtual geomagnetic pole of that locality. It differs from the geomagnetic pole because it refers to the field direction of a single observational station, whereas the geomagnetic pole is the best fit of a geocentric dipole for the entire earth's field. Inasmuch as it is impossible to describe the entire earth's field at various times in the past, the virtual geomagnetic pole is commonly used in expressing paleogeomagnetic data.

Geomagnetic intensity, variations, and reversals Earlier we stated that there has been a secular variation in the position of the geomagnetic pole over the last several hundred years. In addition to this change in pole position, we also know that there has been a variation in intensity of the earth's magnetic field. Thus by 1965, the intensity had decreased nearly 6 percent from the time it was first successfully analyzed in 1835 by K. F. Gauss, a German mathematician.[2]

[2]The present magnetic field at the earth's surface is about 0.5 G (gauss), where "gauss" (or "oersted") is the unit of measure for magnetic-field intensity.

We have been able to extend our knowledge of the variation in the field's intensity by techniques first successfully used by the French physicist Emile Thellier. Samples that can be precisely dated by historical or radiocarbon techniques show that the intensity of the field has been decreasing since about the time of Christ, when it was about 1.6 times its present intensity. Previously it had risen from a low value of about half the present intensity around 3500 B.C., this having been preceded by a steady decline from an intensity of about 1.5 times that of the present around 6500 B.C. (Figure 10.7).

If the decay of magnetic intensity were to continue at its present rate, it has been estimated by Allan Cox, Stanford University geophysicist, that the intensity would reach zero about 2,000 years from now. Thereafter the direction of the magnetic field would reverse itself. There is no assurance that this will happen, and it may just as well fluctuate as it has for the last several thousand years. But if we turn farther back into earth history, we find that the poles have been reversed, not once but many times. So we assume that such a change is possible in the future.

We now have a great deal of information about past reversals of the magnetic field. That is to say, if we were able to take our ordinary magnetic compass back into time, we should find some periods in which the north-seeking end of the needle would point toward the north pole as it does today. But there would be many other times during which the needle would point to the south pole instead. The present polarity is called **normal.** A polarity at 180° to it is called **reversed.** A succession of normal and reverse fields can be pieced together for much of the recent geologic past. We call the longer intervals of dominance by a particular polarity a **polarity epoch,** and we have named these intervals after distinguished students of earth magnetism. During each polarity epoch are shorter periods in

10.7 Intensity variation of the earth's magnetic field, from Czech archaeological samples. Each circle represents a sample or samples of the same age. The vertical bars represent standard deviation from the mean ratio. [After V. Bucha, *Archaeometry*, vol. 10, p. 20, 1967.]

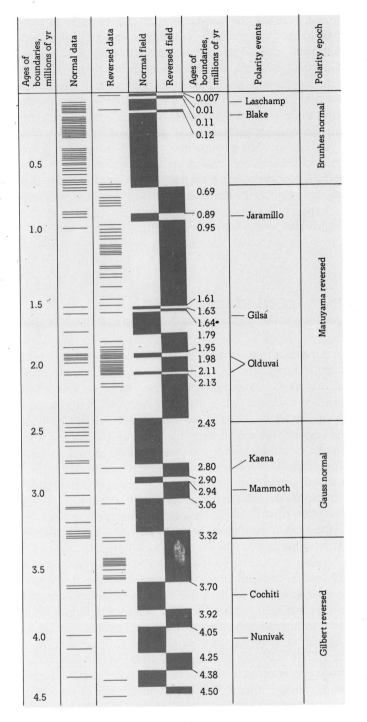

10.8 A time scale for geomagnetic reversal during the last 4.5 million years has been based on extrusive igneous rocks. Each horizontal line in the two left-hand columns represents a rock sample, whether its polarity is normal or reversed. In the normal-field column, normal-polarity intervals are shown in color; in the reversed-field column, reversed-polarity intervals are shown in color. (See also Figure 7.10.) [After Allan Cox, "Geomagnetic Reversals," *Science*, vol. 168, pp. 237–245, 1969.]

of this information when we take up the study of continental drift.

RESULTS OF PALEOMAGNETIC STUDIES

If we can measure the TRM or the DRM of a rock and relate it to the earth's present field, we can determine to what extent the orientation of the earth's magnetic field has varied at that spot through time.

Studies of ancient pole positions assume that the earth's field has been dipolar and, further, that this dipole has approximated the earth's axis of rotation. A consequence of these assumptions is that the earth's geographic poles must have coincided in the past with the earth's geomagnetic poles. Clearly this is not so at present. Why should we think that it was true in the past? A partial answer to the question lies in the dynamo theory of earth magnetism. If the dynamo theory is correct, then theoretical considerations suggest that the rotation of the earth should orient the axis of the magnetic field parallel to the axis of rotation.

Observational data support the theoretical considerations. When we plot the virtual geomagnetic poles of changing fields recorded over long periods at magnetic observatories around the world, we find a tendency for these pole positions to group around the geographic poles. More convincing are the paleomagnetic poles calculated on the basis of magnetic measurements of rocks of Pleistocene and Recent age. These materials, including lava flows and varves, reveal pole positions clustered around the present geographic pole rather than the present geomagnetic pole (see Figure 10.9).

As a result of theoretical and observational data, therefore, most authorities feel that the apparent, present-day discrepancy between magnetic and rotational poles would disappear if measurements were averaged out over a span of approximately 2,000 years. The same principle would apply for any 2,000-year period throughout geologic time. Thus, when we speak of a paleomagnetic pole, we have some confidence that it had essentially the same location as the true geographic pole of the time.

Paleomagnetic data derived from rocks of Tertiary age indicate, fairly conclusively, no significant shift of the geomagnetic poles from the Oligocene to the present. The farther backward we go in time beyond the Oligocene, however, the more convincing becomes the case for a changing magnetic pole and, on the basis of the above discussion, for a changing geo-

which the polarity is in the opposite direction, and these time intervals are referred to as **polarity events** (see Figure 10.8). We shall come back shortly to the use

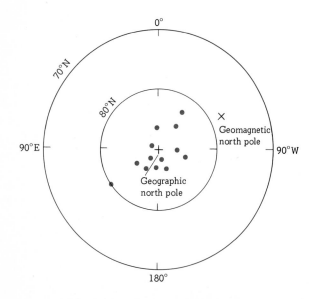

10.9 The virtual geomagnetic poles determined by magnetic measurements of earth materials of Pleistocene and Recent age cluster around the modern geographic pole rather than around the present geomagnetic pole. [After Allan Cox and R. R. Doell, "Review of Paleomagnetism," *Bull. Geol. Soc. Am.,* vol. 71, p. 734, 1960.]

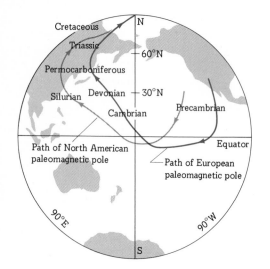

10.10 Paleomagnetic measurements of rocks from North America and Europe show the paths followed by the magnetic poles of these two continents from Precambrian times to the present. [After Allan Cox and R. R. Doell, "Review of Paleomagnetism," *Bull. Geol. Soc. Am.,* vol. 71, p. 758, 1960.]

graphic pole. The magnitude of this change is suggested in Figure 10.10, which shows that the paleomagnetic poles for Europe and North America were in the eastern Pacific during Precambrian time. Thereafter they moved southwestward and crossed the equator into the southern hemisphere before moving northwestward toward Asia and eventually to the position on the present globe.

The extensive migration of the magnetic poles (and, by extension, the geographic poles), combined with the observation that the paths of polar migration of different continents fail to coincide, raises tantalizing questions. We find ourselves, in fact, faced with the entire concept of continental drift.

10.3 CONTINENTAL DRIFT

In considering the theory of continental drift, we should bear in mind that two general types of movement may be involved: (1) movement of individual continents in relation to one another and (2) movement of the earth's poles. Actually, if the latter movement took place, most geologists feel that a slippage of the earth's crust and upper mantle is what has produced the apparent motion of the poles. In other words, the magnetic and rotational poles remain fixed within the earth, but different points on the earth's crust would at different times be located at the polar positions as the crust moved. Such motion might involve the crust as a single unit or as fragments of it.

The first coherent theory that our continents have moved as individual blocks was presented by Alfred Wegener (1880–1930), a German meteorologist and a student of the earth. Wegener, as many before and after him, was intrigued with the apparent relationship in form between the opposing coasts of South America and Africa. Was it possible that these two landmasses were once part of the same general continent but have since drifted apart? Wegener's emphatic assent and his extensive documentation of the lateral motion of continents started a spirited discussion that has continued to the present.

EVIDENCE FOR CONTINENTAL DRIFT

Wegener's theory Early in this century, Wegener postulated his theory of continental drift. He pictured the dry land of the earth as included in a single, vast continent that he named Pangaea (from the Greek for "all" and "earth"). This primeval continent, he argued, began to split asunder toward the end of the Mesozoic. The fragments began a slow drift across the earth's face, and by the Pleistocene they had taken up their positions as our modern continents.

The idea of continental drift has been argued now for more than 70 years. The early evidence for the theory was drawn almost entirely from the geologic record. In the mid-twentieth century, the development of geophysical techniques brought new data into the discussion of continental drift. Let us look at some of the evidence.

Shape of continents Anyone observing the map of Africa and South America is struck by the jigsaw-puzzle match of the two continents. If we fit the eastern nose of South America into the large western bight of Africa, the two continents have a near-perfect match. One distinguished geologist, while admitting to skepticism of continental drift, found this fit so credible that he remarked, "If the fit between South America and Africa is not genetic, surely it is a device of Satan for our frustration."[3]

If we examine the outlines of the other continents, we can perhaps be persuaded that these landmasses too can be reassembled into a single large landmass, as suggested by Wegener. If the continents are fitted at the margin of the continental slope (a level of about 900 m below sea level), there is a very impressive match indeed, as shown in Figure 10.11. (Wegener himself first suggested using the edge of the continental shelf, *not* the edge of the continent at present sea level.)

Evidence from paleomagnetism We pointedly began this section with a consideration of the earth's magnetism; for it has been the increasing body of information on rock magnetism and paleomagnetism that has kindled new interest in the old theory of continental drift. We have seen in Figure 10.10 how the paleomagnetic poles have shifted during the last 500 million years. For the two continents represented, North America and Europe, there is a divergence between the ancient poles—that is, the virtual paleomagnetic poles—of these two landmasses as we go back in time. If we fit North America and Europe back together, however, the separate polar wandering paths superpose nicely.

The divergence of the paths of polar migration among continents can be explained by the shifting of the landmasses in relation to each other. The course of polar migration suggests that the continental movement consists of a general drift of these continents

away from each other, and in some instances there is an additional continental rotation.

Magnetic measurements at sea have revealed a characteristic pattern of anomalous magnetic intensities. These magnetic anomalies are most often associated with ocean ridges and are arranged in stripes of alternating intensity parallel to the ridge. An example is provided by the magnetics of the Reykjanes Ridge, a portion of the Mid-Atlantic Ridge southwest of Iceland. Figure 10.12 shows the stripes parallel with the ridge and shows also a bilateral symmetry, the axis for which is the rift valley along the crest of the ridge.

In 1963, two British geophysicists, Fred Vine and D. H. Matthews of the University of Cambridge, suggested that the alternating stripes of high- and low-intensity magnetism are in reality zones of normal and reverse polarity in the rock of the sea floor. From this suggestion, now widely accepted, have flowed some intriguing concepts about the sea floor, the drift of continents, and the building of mountains.

Sea-floor spreading The Vine-Matthews hypothesis leads us to a consideration of sea-floor spreading. This is a mechanism that involves the active spreading of the sea floor outward, away from the crests of the main ocean ridges. As material moves from the ridge, new material replaces it along the ridge crest by welling upward from the mantle. As this mantle material cools below the Curie temperature, it will take on the magnetization of the earth's field at that time. Room for the new material is made along the ridge by the continued pulling apart of the crust and its movement laterally away from the ridge. At times of change in the earth's polarity, then, newly added material along the crest will record this change. Spreading laterally, it carries this record with it, to be followed at a later time by the record of the next polarity change. We can visualize this as a gigantic magnetic tape preserving a history of the earth's changing magnetic polarity. A diagram to suggest the mechanism is shown in Figure 10.13.

The rates at which new sea floor forms at spreading centers is seen to vary. In the North Atlantic it is about 1 cm/year, and in the South Atlantic about 2.3 cm/year. Along the Juan de Fuca Ridge off California the rate is 3 cm/year, and on the east Pacific rise it is calculated to be 4.6 cm/year, reaching a rate of 9 cm/year in some sections. (These are all "half-spreading rates": each side of the ocean floor is moving at these rates; so the continents on either side of the ocean basins would be moving apart at **twice** these rates.)

---

[3]Chester Longwell, "My Estimate of the Continental Drift Concept," in S. W. Carey (ed.), *Continental Drift, a Symposium*, p. 10, Tasmania University Press, Hobart, 1958.

10.11 A statistically determined fit of North America, South America, Africa, and Europe at the 900-m depth in the ocean. Zones of overlap are shown in the dark tone; zones of gap are shown in the lighter tone. [After Edward Bullard and others. "The Fit of the Continents around the Atlantic," *Phil. Trans Roy. Soc. London,* vol. 1088, pp. 41–51, 1965.]

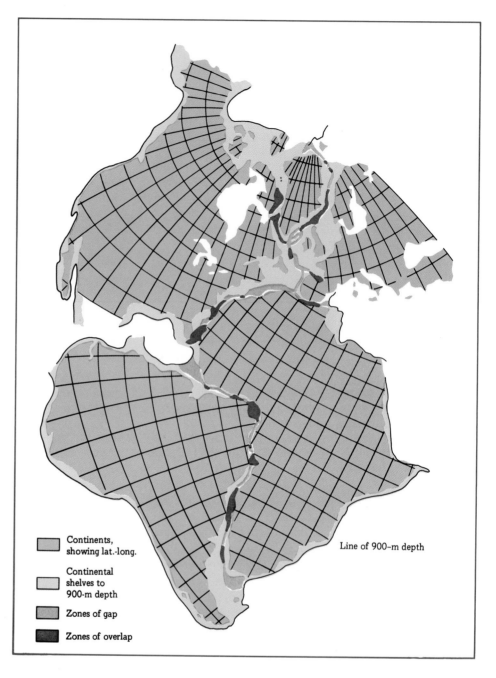

Continents, showing lat.-long.

Continental shelves to 900-m depth

Zones of gap

Zones of overlap

Line of 900-m depth

Figure 10.14 shows the effects of magnetic reversals on deep-sea sediments and associated oceanic lithosphere. If there were no sea-floor spreading, the layers of sediments would simply show alternating polarity as the earth's magnetic field changed. With sea-floor spreading, each sedimentary layer begins at the ridge with the magnetization of that particular time in the earth's history. As the lithosphere migrates away from this spreading center, the sediments are carried farther and farther away.

Sea-floor spreading indicates that the ocean floor is a dynamic system, and in terms of continental drift it suggests that expansion of ocean basins may also affect the location and position of continents. Furthermore, it raises many questions. For example: what drives the ocean floor, and where do the moving ocean floors go? The proposed answers to these questions bear on continental drift, but we shall defer their consideration until we examine the subject of plate tectonics (Section 10.4). In the meantime, let us take up additional geologic data bearing on continental drift.

10.12 Magnetic anomalies over the Reykjanes Ridge southwest of Iceland. The patterned areas display normal polarity and are separated by areas of reversed polarity. Note that the belts of normal and reversed polarity are generally symmetric about the axis of the Mid-Atlantic Ridge. Ages of the various polarity belts are based on a time scale derived from a study of magnetic anomalies on land and on the sea floor. [After F. J. Vine, "Magnetic Anomalies Associated with Ocean Ridges," in R. A. Phinney (ed.), *The History of the Earth's Crust,* Princeton University Press, Princeton, 1968. © 1968 by Princeton University Press. Reprinted by permission.]

Evidence from ancient climates Much geologic evidence cited in support of continental drift and polar wandering is based on the reconstruction of climates of the past. Their use is justified by the fact that modern climatic belts are arranged in roughly parallel zones whose boundaries are east-west and which range from the tropical equatorial climates to the polar ice climates, as suggested in Figure 10.15. Although climates at various times in earth history have been both colder and warmer than they are at present, we assume that basic climatic controls have remained the same and, therefore, that the climatic belts of the past

have always paralleled the equator and been concentric outward from the poles.

Evidence from the geologic record allows us to map the distribution of ancient climates in a very general way. The pattern of some ancient climatic zones suggests that they were related to poles and an equator with locations different from those of today and that, therefore, the landmasses and the poles have varied relative to each other since those climates existed.

Late Paleozoic glaciation On the Indian peninsula lies a sequence of rocks known as the Gondwana System, reaching in age from the late Paleozoic to the early Cretaceous. Beds of similar nature and age are recorded in South Africa, Malagasy, South America, the Falkland Islands, Australia, and Antarctica. In these other localities they are known by other names, but we can still refer to them as belonging to the Gondwana System.

Geologists who have worked on the Gondwana formations have discovered many similarities among the rocks of the various continents despite their wide geographic separation. Some of these similarities are so striking that many accept only one interpretation: that the various southern lands must once have been part of a single landmass, a great southern continent, early called **Gondwanaland** by these geologists.

The distribution of ancient glacial deposits is one of the most convincing lines of evidence for continental drift in these southern lands. During late Paleozoic time, continental ice sheets covered sections of what are now South America, Africa, the Falkland Islands, India, Antarctica, and Australia. In southwestern Africa, deposits related to these ancient glaciers are as much as 600 m thick. In many places the now-lithified deposits (tillites) rest on older rocks striated and polished by these vanished glaciers. If we plot the distribution of these deposits and the direction of ice flow on a map (see Figure 10.16), we can make two immediate observations. First, these traces of Paleozoic ice sheets occur in areas where no ice sheets (except for Antarctica) exist now or have existed during the glacial epochs of the Pleistocene Ice Age. Only an occasional towering peak may bear the scars of modern or Pleistocene ice. Continental ice sheets cannot exist in these latitudes today. Second, the direction of glacier flow is such that we can imagine the ice of Africa and South America to have been part of the same ice sheet when the two continents were one.

These observations have led most students of the earth to two conclusions. First, to account for glaciers

10.13 A diagram to suggest the development of magnetic anomalies related to ocean ridges. New oceanic crust, cooling below the Curie point, forms along the axis of the ridge and assumes the magnetic characteristics prevailing at the time of its formation. As the crust moves laterally away from the ridge crest, changes in the magnetic field of the earth are retained in strips of the oceanic crust. Serpentinite, a rock made up chiefly of the mineral serpentine, is thought by many to form the lower portion of the oceanic crust. [After F. J. Vine, "Sea-Floor Spreading—New Evidence," *J. Geol. Educ.*, vol. 17, pp. 6–16, 1969.]

Ridge model ⟵ ⟶ 1 cm/yr

■ Normal ☐ Reverse 10 km horizontally = 1 million yr

Sediments Crust

Normal magnetization

Reverse magnetization

(a) (b)

10.14 Magnetization of marine sediments: (*a*) If there were no sea-floor spreading, the layers would show alternating polarity in continuous sheets; (*b*) with sea-floor spreading, the alternate normal and reversed stripes in the crust have correspondingly magnetized sediments, beginning adjacent to that part of the crust formed at the ridge crest of the spreading center. [After P. J. Wyllie, *The Way the Earth Works*, John Wiley & Sons, Inc., New York, 1976.]

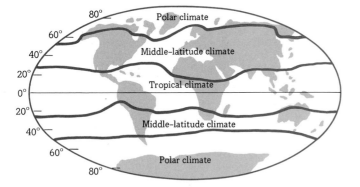

10.15 Present-day climatic boundaries are arranged concentrically around the poles and thus are approximately parallel to lines of latitude.

in present tropical and subtropical areas, they locate the south pole at this time somewhere in or near southern Africa. Second, to account for, among other things, the apparent continuity of glacier flow, these same students have postulated a single southern continent. This continent later split into several sections that drifted apart to form the modern landmasses.

Ancient evaporite deposits and coral reefs

Evaporites, sedimentary rocks composed of minerals that have been precipitated from solutions concentrated by the evaporation of the solvents, are generally accepted as evidence of an arid climate. The ancient evaporite deposits represent the great arid belts of the past. The present hot-arid belts are located in the zones of subtropical high pressure, centered at about 30° north and south of the equator. In the northern hemisphere an evaporite belt has shifted through time from a near-polar location in the Ordovician and Silurian to its present position in the modern desert belts. This again suggests a relative motion of pole and landmasses of the past as compared with present-day conditions.

Turning to another line of evidence, we find that corals mark a climate shifting geographically through time. Today true coral reefs are restricted to warm, clear marine waters between 30° north and south of the equator. If we assume that ancient reef-forming corals had similar restrictions, then plotting their distribution in the past will show the distribution of tropical waters of the past and the location of the past equator. Doing this, we find that reef-forming corals did not approximate their present distribution until halfway through the Mesozoic. Prior to that time, they lay askew to the present equator.

Plants, reptiles, and continental drift Shortly after the disappearance of the southern hemisphere's late Paleozoic ice sheets, an assemblage of primitive land plants became widespread. This group of plants, known as *Glossopteris* flora, named for the tonguelike leaves of the seed fern *Glossopteris* (Figure 10.17), has been found in South America, South Africa, Australia, India, and within 480 km of the South Pole in Antarctica. The *Glossopteris* flora is very uniform in its composition and differs markedly from the more varied contemporary flora of the southern hemisphere. Many geologists have argued that the uniformity of the *Glossopteris* flora could not have been achieved across the wide expanses of water now separating the different collecting localities. In other words, in one way or another, there must have been continuous or near-continuous land connections between now-separate continents. To many this suggests that a single continent with a single uniform flora has been split apart into smaller continents that have since migrated

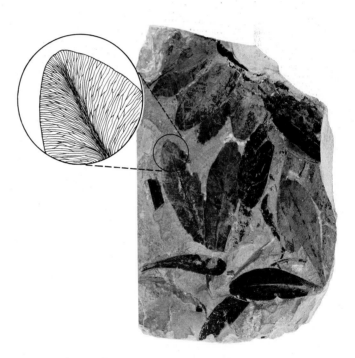

10.17 These leaves of the fossil plant *Glossopteris* come from strata of Permian age in Australia. The *Glossopteris* flora is found also in South America, Africa, India, and Antarctica. The widespread occurrence of this very uniform flora has been used as evidence both by opponents and proponents of continental drift. Diameter of detail is 2.5 cm. [Specimen from Princeton University Paleobotanical Collections, Willard Starks.]

10.3 CONTINENTAL DRIFT

229

to their present position. As we shall see, this conclusion has not gone unchallenged.

Among the vertebrate fossils of the late Paleozoic and earliest Mesozoic we find a great number of different reptilian types. Two of them provide us with arguments for continental drift. *Mesosaurus*, a toothed early reptile of the late Permian, lived in the water and thus far is known only from Brazil, South Africa, and Antarctica. Although it was aquatic, most paleontologists do not believe that it could have made the trip across the South Atlantic. If this is so, then *Mesosaurus* may offer evidence for a closer proximity of South America and Africa and thus for continental drift. A second reptile, *Lystrosaurus*, dates from the Triassic period. *Lystrosaurus*, which is about 1 m in length, was adapted to an aquatic but nonmarine environment, and its remains are reported from South Africa and from Antarctica's Alesandra Range. It is argued that *Lystrosaurus* could not have made the journey across the Antarctic Ocean separating South Africa from Antarctica and that this implies the former joining of the two continents.

Other evidence Some anciently formed mountain chains now terminate abruptly at the continental margins. Join the continents together, and some of these geological structures match up between the two landmasses. Thus the Cape Mountains of South Africa are thought to be the broken extension of the Sierra de la Ventana of Argentina in one direction and of the Great Dividing Range in eastern Australia in the other direction. The entire stretch is cited by some adherents of continental drift as a once-continuous chain of mountains now segmented and separated. Similarly, the Appalachian Mountain system ends in the sea on the northern shore of Newfoundland. Is its extension to be found in the orogenic belts of the British Isles and western Europe? Many geologists think so.

P. M. Hurley and his colleagues at the Massachusetts Institute of Technology have carried on extensive geochronologic studies of Precambrian igneous and metamorphic rocks in West Africa and the eastern bulge of South America. Radiometric-age determinations of many samples show distinct belts of roughly similar age on the two continents. If the continents are shifted back together, the several provinces of Precambrian rocks of differing ages on the rejoined landmasses match fairly well (Figure 10.18). We are tempted to believe that rocks of similar age in Africa were once continuous with rocks of the same age in South America. Sometime after they were formed, they have been separated so that we now see them on two widely separated continents.

Planetary winds probably always existed. It would be interesting to see whether ancient eolian deposits might give some indication of such major wind belts as the northeast and southeast trades of the past. Sand beds accumulated on the face of a sand dune dip in the direction that the wind blows. We can measure such dip in ancient wind-deposited sediments. If enough of these directions are available, the direction of the dune-forming wind can be determined statistically. Preliminary studies on presumed eolian deposits of late Paleozoic age in Wyoming, Utah, Arizona, and England indicate that these areas fell within the northeast-trade-wind belt when the earth's pole (as suggested by the paleomagnetic evidence) was located on the East China coast.

Reliability of evidence The reliability of evidence cited in support of continental drift has been questioned by some workers. The paleowind directions discussed above, for example, are thought by some to be not only inconclusive but actually based on deposits some of which were not truly eolian.

Some of the evidence from the fossil record is attacked as too fragmentary and open to too many interpretations to be diagnostic. The use of *Mesosaurus* as representing evidence for continental drift is one such example.

Paleomagnetic evidence is currently regarded as one of the most conclusive arguments in favor of continental drift. But some argue that the evidence may not be so strong as it first appears, and they point out that this evidence may be overextrapolated. For instance, we said earlier that discussions of paleomagnetism assume a dipolar magnetic field for the earth. Yet the more ancient fields could have had another form—four poles instead of two, perhaps. If so, the entire paleomagnetic argument in support of continental drift would have to be reevaluated.

SUMMARY OF THE PROBLEM

Prior to the 1940s, most of the supporters of the theory of continental drift lived in the southern hemisphere, particularly in South Africa. Geologists and geophysicists of the northern hemisphere, particularly in the United States and western Europe, felt that there was little or no support for the motion of continents or the wandering of the poles. In fact, it was hardly respectable to entertain the possibility of such movement.

After the 1940s, the techniques of geophysics and particularly those of paleomagnetic measurement began to bring new data into the discussion. This new evidence reopened the entire question of continental

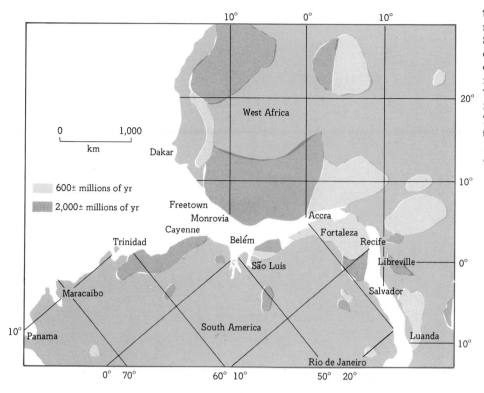

10.18 Rocks of two different ages mark the opposing shield areas of South America and Africa. If the two continents are fitted together, the zones of the same age match across the line of fit. [After P. M. Hurley and J. R. Rand, "Review of Age Data in West Africa and South America," in R. A. Phinney (ed.), *The History of the Earth's Crust,* Princeton University Press, Princeton, 1968. © 1968 by Princeton University Press. Redrawn by permission.]

drift for American and European geologists. By the 1980s, a vast amount of data bearing on continental drift had accumulated. Most workers in the field now feel that the available evidence favors the reality of continental drift. Many, having accepted that continental drift has occurred, have turned to how it might have happened—and to what implications drift might have for our understanding of earth history and earth mechanics. In the following section we take up this thread of thought and consider the concept of plate tectonics.

10.4 PLATE TECTONICS

Geologists in recent years have pointed out that the earth's outer surface can be divided into large units, or plates. We have suggested at the beginning of this chapter and elsewhere that the origin, movement, and interrelation of these plates can be coordinated with sea-floor spreading, earthquake belts, crustal movement, and continental drift.

THE WORLD SYSTEM OF PLATES

At least six major plates and a host of minor plates are recognized, as shown in Figure 10.19. The major plates include the Indian, Pacific, Antarctic, American, Afri-

can, and Eurasian plates. The boundaries of these plates are loci of present-day earthquakes and volcanic activity. Consider for example the Pacific plate. On its southern and southeastern side it is bounded by an oceanic rise that undergoes active spreading. On its northern and northwestern borders it is marked by island arcs, deep-sea trenches, and tectonic and volcanic activity. Its northeastern side along western United States is defined by the San Andreas fault, where the Pacific plate moves generally northwestward and jostles the American plate in the process.

The relationship of plates to the distribution of earthquakes is striking when one compares Figure 10.19 with Figures 8.1 and 8.2. In general, the shallow quakes mark the ocean ridges: deep earthquakes are missing in these zones. Both shallow and deep quakes are found along the arcs and trenches and along young mountains, as the Himalayas (see Table 10.1).

The plates may include both continental and oceanic areas, as for example the American plate, which embraces North and South America and Greenland as well as the western Atlantic Ocean. On the other hand, the Pacific plate is restricted to oceanic area.

Eurasian plate

African plate

Indian plate

Pacific plate

American plate

American plate

Antarctic plate

10.19 Crustal plates are outlined by zones of intensive earthquake activity. The directions of plate motion over the underlying asthenosphere are shown by arrows. Each plate moves the distance equivalent to the length of the arrow in 20 million years. [After Sheldon Judson, Kenneth S. Deffeyes, and Robert B. Hargraves, *Physical Geology*, Prentice-Hall, Inc., Englewood Cliffs, N. J., 1976.]

MAKEUP OF PLATES

The rigid lithospheric plates consist of several distinct layers. The lowest portion comprises the cooler, solid upper part of the mantle. The Moho discontinuity marks the upper contact of this layer with the crust.

Above this lowermost layer is the principal portion of the crust. As we have explained, under the oceans this crustal layer may be only a few kilometers thick, consisting of coarse gabbrolike simatic rock (rock rich in silica plus iron and magnesium). It is capped by lava flows—solidified magma extruded along the midocean ridges, rises, and other volcanic centers. Under continents this layer may be 30 km or more thick. The lower crust may include a continuation of the oceanic gabbroic layer. The upper crust is sialic (rich in silica and alumina) and includes the crystalline granitic masses of the roots of mountains covered by a relatively thin layer of sedimentary and metamorphic rocks at the continental surfaces.

Topping this layer-cake structure in the oceanic regions are the sediments of the sea floor, consisting of

fine inorganic debris plus the remains of organisms that lived in the upper portion of the ocean, died, and fell to the bottom, where they accumulated. This sedimentary layer is thin or nonexistent adjacent to the spreading centers and thickens progressively away from the ridge as a result of the outward movement of the crust away from the spreading center. The oldest crust receives sediments for periods of time longer than does the younger crust (refer to Figure 10.14 again).

CLASSIFICATION OF PLATE BOUNDARIES

Lithospheric plates originate at midocean ridges and spread laterally toward subduction zones. The motion of these plates with respect to other adjacent plates results in three principal kinds of boundary (Figures 10.20 and 10.21): Where plates move away from one another, **divergent** boundaries result; where they move toward each other, **convergent** boundaries result; and where one plate moves past another plate in a nearly parallel fashion, **transform** boundaries result.

TABLE 10.1

Plate boundaries and their interactions with other plates under tension, compression, or neither[a]

Boundary	Divergent (tensional)	Convergent (compressional)	Transform (neither tensional nor compressional)
Ocean-ocean	Ridge crest; S[b] earthquakes in narrow belt; submarine lavas	Ocean trench and volcanic island arc; S, I, and D quakes over wide belt; volcanoes common	Ridge and valley fracture zone; S quakes in narrow belt only between offset ridges; volcanoes absent
Continent-continent	Rift valley; S quakes over wide zone; volcanoes present	Young mountain ranges; S and I quakes (or just S) over wide zone; no volcanoes	Fault zone; S quakes over broad zone; no volcanoes
Ocean-continent		Ocean trench and young mountain range; S and I quakes over wide belt; D quakes occasionally; volcanoes common	

[a] Modified from P. J. Wyllie, *The Way the Earth Works*, John Wiley & Sons, Inc., New York, 1976.
[b] S, shallow-focus earthquakes; I, intermediate focus; D, deep focus.

10.20 Major features of sea-floor spreading with the three principal types of plate boundary: convergent, divergent, and transform (or parallel). [After B. Isacks, J. Oliver, and L. R. Sykes, "Seismology and the New Global Tectonics," *J. Geophys. Res.*, vol. 73, pp. 5855–5899, 1968.]

The plates may be entirely dense oceanic lithosphere or may include the lighter rafts of continental lithosphere. Movement of these plates may involve ocean-ocean, ocean-continent, or continent-continent interactions.

Because divergent boundaries represent conditions of **tension,** shallow-focus earthquakes and volcanic emanations occur along the ridge crests of ocean-ocean regions and along the rift valleys of continent-continent zones (Table 10.1).

Compression characterizes convergent boundaries and results in trenches and island-arc volcanoes, with shallow, intermediate, and deep-focus quakes along ocean-ocean contacts. Trenches and young mountain ranges occur with volcanoes and shallow-

and intermediate-focus quakes (with or without deep-focus quakes) at ocean-continent boundaries.

Where continental plates meet other continental plates, young mountain ranges result from the piling up, or doubling, of the lighter continental masses. Volcanoes are rare or nonexistent under these conditions, but shallow-focus quakes are common. Intermediate-focus quakes are rare, and deep-focus quakes entirely absent.

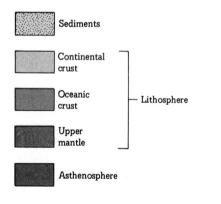

10.21 Illustrations of plate boundaries and their interactions with other plates. [From various sources.]

MOVEMENT OF PLATES

Neither tension nor compression dominates along transform boundaries. Continent-continent interactions are characterized by fault zones and shallow-focus earthquakes over a rather broad zone. Fracture zones with shallow-focus earthquakes in narrow belts mark ocean-ocean transform boundaries. Little volcanic activity is likely along any transform boundaries.

We have seen in Figure 10.19 that some of the sutures between adjacent plates are zones of active sea-floor spreading, a process described in Section 10.3. Along such zones, therefore, crustal material is considered as continuously forming, and to make room for it the sea floor is pictured as moving laterally away from the ridge crest. Transform faults, described in Chapter 8 and illustrated in Figure 8.42, cut across the ridge and produce its offset. An analysis of the first movement of the crust during earthquakes on the ridge shows two types of movement. Along the rifts that mark the axes of the oceanic ridges, motion during earthquakes indicates tensional stress, normal faulting, and dropping down of the blocks associated with the grabens along the ridges. In contrast is the movement along the fractures offsetting the ridges. Along these structures analysis of earthquake data shows that the movement is of a strike-slip nature on a plane at right angles to the ridge; and movement is away from the ridge crests, as in Figure 8.43 of the Mid-Atlantic Ridge near the equator.

If plates are actually moving away from each other through a process of sea-floor spreading, then a number of questions arise: What drives the ocean floor? Where does the moving ocean floor go?

Convection cells   Most students of the problem consider it likely that convection currents (described in Chapter 8), moving in cells, provide a mechanism for moving the ocean floor. They envision the ocean ridges as lying over a rising convection current. That seismic and volcanic activity are generally high on the ridges and that heat flow from the interior is higher at the ridge crest than it is at most other places add appeal to the suggestion. The current is thought to lie in the upper mantle, well below the Moho, and to represent ascending limbs of adjacent convection cells. At some point below the Moho the rising current is pictured as splitting into two currents flowing laterally away from the midoceanic ridge.

Plume theory   Another possible driving mechanism for the movement of plates has been suggested by Jason Morgan (of Princeton University), who earlier had pieced together much of the evidence in support of plate tectonics. He infers the existence of hot spots, or plumes, of hot rising solid mantle in zones a few hundred kilometers in diameter. This is an extension of an earlier hypothesis by J. Tuzo Wilson of Toronto, who suggested similar structures to account for volcanic island chains unrelated to plate boundaries (see Figure 3.15). (Over 120 hot spots have been active during the past 10 million years according to a study by Kevin Burke and Wilson.[4]) By this hypothesis, hot mantle rock rises and spreads out laterally in the asthenosphere in about 20 major thermal plumes around the world (Figure 10.22), creating hot spots of volcanic activity. The rest of the mantle undergoes very slow downward movement to balance the upward flow in the plumes. The movement of any plate is caused by the combination of forces exerted by the upward flow in the plume, the outward radial flow away from the plume, and the return downward flow of the mantle.

[4]Kevin C. Burke and J. Tuzo Wilson, "Hot Spots on the Earth's Surface," *Sci. Am.*, vol. 235, no. 2, pp. 46–57, 1976.

Recycling the sea floor   What happens to the old sea-floor crust as it is displaced from the ridge crest by younger and younger material? Those who explain the motion by convection see the crust and a part of the upper mantle carried laterally by convection currents toward the margins of the ocean basins. This is supported not only by the paleomagnetic measurements already discussed but also by studies of oceanic sediments. These studies indicate not only that the sedimentary deposits are thinner in the zones of the mid-oceanic ridge but also that, as they become thicker away from the ridge, the basal layers of the sedimentary accumulations become older, as we discussed earlier in this section. This all suggests that the oldest sea floor lies farthest from the oceanic ridges.

If we follow this line of reasoning, we should expect to find that belts of varying width along the active ridges mark the most recently formed oceanic crust. If we accept the reconstruction shown in Figure 10.23, then an area approaching 50 percent of the ocean floor has been formed during the Cenozoic—in approximately the last 65 million years. In this regard it is interesting to note that no rocks older than 150 million years have as yet been taken from the deep ocean basins. Moreover, present-day sedimentation rates in the oceans can account for the sedimentary accumulations there within the last 100 to 200 million years. All this leads to the suggestion that the ocean basins, as we know them, are young. It would then follow that although ocean basins have existed through most of geologic time, they have been continuously reconstituted and recycled. Furthermore, their shapes and their geographic locations have been shifting constantly (Figure 10.24).

Obduction   Some of the subducted plate may be broken off and pushed up onto the overriding plate during the process of plate collision. This subducted plate may have mafic and ultramafic igneous rocks and deep-sea sediments attached to it. In this way, formerly deep ocean rocks can be raised in mountain masses and be exposed at the earth's surface by this process called **obduction.**

DATING THE MOVEMENTS

Interpretation of earthquake data tells us lithospheric plates have been carried as distinct entities downward along subduction zones to depths of 700 km or more. Although we can only tentatively decipher the present

10.22 About 20 major mantle plumes. Thin black lines represent volcanic trails, or linear volcanic chains, tracing the hot spots. [After W. J. Morgan, "Plate Motions and Deep Mantle Convection," in R. Shagam (ed.), *Studies in Earth and Space Sciences*, pp. 7–22, Geological Society of America Memoir 132, Boulder, CO, 1972; and K. C. Condie, *Plate Tectonics and Crustal Evolution*, Pergamon Press, Inc., Elmsford, N. Y., 1976.

position of plate borders and their most recent directions and rates of movement, we know that these movements have occurred at least long enough for the generation of 700 km of new lithosphere at the spreading centers (otherwise a gap would exist in the ocean where plates have moved away from the midoceanic-ridge spreading centers).

Rates of sea-floor spreading have been determined from analyzing magnetic anomalies (Section 10.2). By relating these rates and distances of movement, we can determine the ages for various portions of the sea floor. The initial opening of several parts of the ocean basins has thus been determined for the northern hemisphere landmass (Laurasia) and for the southern hemisphere Gondwanaland. Opening of the ocean basins coincides with, and results in, the drifting of the associated continents.

Physical and biological events continuing on a given landmass might be rather abruptly terminated by the rifting apart of that landmass. Dating of these

events would permit the determination of a minimum age for the initial breaking apart of such a continent. By a combination of such methods and by the dating of the ocean-floor materials, it is possible to establish the principal sequence of events in the breakup of the supercontinents, as listed in Table 10.2.

SUMMARY SPECULATIONS

The idea of plate tectonics can be used to integrate some of the major features of the ocean basins. Thus the midoceanic ridges and rises, with their central rift valleys and a sedimentary cover that is young and thin (or lacking), reflect the upwelling of mantle material along the ascending currents of adjacent convection cells. The large fracture systems that offset the midoceanic ridges are transform faults. These faults come into being as the rising convection currents finally diverge

10.23 The shaded portion of the basins is thought to represent the area of oceanic crust formed within the last 65 million years. [After F. J. Vine, "Sea-Floor Spreading—New Evidence," *J. Geol. Educ.*, vol. 17, pp. 6–16, 1969.]

Trenches
Magnetic anomalies
Fractures
Ridge crests

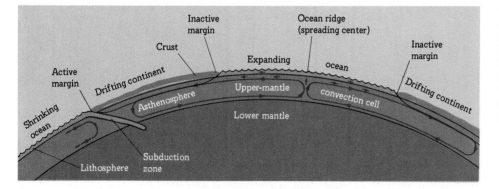

10.24 Convection cells in the upper mantle provide the energy to drive the lithospheric plates. The ocean opens up by spreading from the ocean ridge and is in turn consumed at the subduction zones. The leading edge of a drifting continent becomes an active margin, and the trailing edge an inactive margin. [From various sources.]

laterally, carrying with them the overlying crustal plates and upper mantle. Different portions of these plates are moved at different rates so that tear faults develop at their lateral margins. As the underlying currents cool and turn downward, they carry with them the oceanic crust with its overlying sediments, as well as the underlying upper mantle. To many students of the problem, the deep arcuate trenches of the oceans are the surface manifestations of the downturning of the convection currents along subduction zones. Great crustal plates, then, are seen as moving within a worldwide framework. The leading edges of the plates are consumed in the trench-island arc systems of the Pacific basin and along the Mediterranean-Himalayan belt of Eurasia. Along the trailing edges, new crust is generated in the active ridges. The drifting of continents, then, becomes a consequence of the movement of crustal plates and the appearance and disappearance of oceanic crust.

If the mantle wells up beneath a continent, we can expect that the continent will be rifted and that the two fragments will drift apart as a new ocean basin forms. Such a possibility is suggested in Figure 10.25. As the ocean basins enlarge and the continents ("tied" to the underlying but moving upper mantle) drift away from each other, their leading edges may be fronted by trenches. The continents, being lighter, override the trenches.

A modern example presents itself in the western margin of South America. Here Pacific oceanic crust appears to spread eastward toward the continent. South America's margin is marked by a deep ocean trench, a narrow continental shelf, and the towering

TABLE 10.2

Major events in continental drift and sea-floor spreading (Mesozoic and Cenozoic episodes)[a]

Era or period	Age, millions of yr	Southern hemisphere (Gondwanaland)	Northern hemisphere (Laurasia)
	—		
	—		Gulf of California opens
CENOZOIC	—	Gulf of Aden opens	Spreading direction in eastern Pacific changes
	—	Red Sea opens	Iceland begins to form
	50—	India collides with Eurasia	Arctic Basin opens
			⎯ 65 m yr
	—		Greenland begins to separate from Norway
	—		
CRETACEOUS	—	Australia begins to separate from Antarctica	
	100—		
	—		
	—	Madagascar begins to separate from Africa, and South America from Africa	North America separates from Eurasia
	—		
			⎯ 136 m yr
	150—		
JURASSIC	—		
	—		
	—	Africa begins to separate from India and Australia-Antarctica	North America begins to separate from Africa
			⎯ 190 m yr
TRIASSIC	200—		

[a]From various sources.
Note: There is growing evidence to suggest the Atlantic region opened into a proto-Atlantic Ocean (called **Iapetus Ocean**) in early Paleozoic and closed in late Paleozoic.

Andes. Earthquakes occur, becoming deeper in origin inland, the deepest being about 700 km below the surface. The earthquake foci lie along the Benioff zone that dips beneath the South American continent.[5]

[5] A seismic zone dipping beneath a continent or continental margin and having a deep-sea trench as its surface expression is called a **Benioff zone** after Hugo Benioff, an American seismologist who first described the feature. It is equivalent to our subduction zone.

The elevation of the Andean chain is interpreted in part as the result of underflow of oceanic crust. Seismic and igneous activity are seen as related to the motion between the Pacific and American plates along the Benioff zone. Farther north, the Mesozoic history of the western United States has been interpreted in terms of the underflow of 2,000 km of Pacific material beneath the American plate. The early Paleozoic mountain-building episodes in Newfoundland and Scotland have been ascribed by some workers to the interplay between two plates, one underthrusting another.

Thus mountain ranges, ocean basins, and most major features of the earth's surface owe their origin to the movement and interaction of lithospheric plates, according to this plate-tectonic theory.

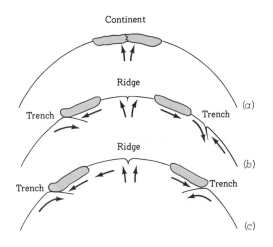

Continent

Ridge

Trench   Trench

(a)

Ridge

(b)

Trench   Trench

(c)

10.25 This sequence illustrates what might be expected if mantle material upwells beneath a continent. In (a) a continent is shown as rifting begins. In (b) the sea floor has opened between the two fragments of the original landmass. The leading edge of one is shown overriding a trench, and that of the other is still separated from an oceanic trench. Diagram (c) shows a still further stage of sea-floor spreading and continental drift. Both landmasses are shown overriding trenches. [After F. J. Vine, "Sea-Floor Spreading—New Evidence," *J. Geol. Educ.*, vol. 17, pp. 6–16, 1969.]

OUTLINE

Oldest continental crust has been dated at 3.8 billion years from highly metamorphosed rocks from the Baltic Shield, Greenland, and Minnesota.

Constant freeboard of continents is assumed from study of the relative elevation of continents with respect to sea level through at least the latter part of geologic time.

Age of ocean basins is surprisingly young because older ocean floor is destroyed at subduction zones as new floor develops at spreading centers.

The earth's magnetism results from its behaving as if it were a magnet; its changes as recorded in rocks is closely tied to the case in favor of continental drift.

Secular variation of the magnetic field has been known through long-term changes in magnetic declination since the mid-seventeenth century.

The geomagnetic pole at the earth's surface does not coincide with the geographic pole.

The magnetosphere is a region above the earth that traps powerful radiation from space and wards it off.

Cause of magnetism is both internal and external in relation to the earth. A small part is external, caused by the sun. The largest part of the earth's field originates within the earth and is best explained as caused by circulation within the earth's liquid core, which acts as if it were a giant dynamo. The **direction** and **intensity** of the magnetic field vary with time. If the intensity decreases far enough, it may eventually pass a zero point and the earth's field may become polarized in a sense the reverse of that of the present.

Paleomagnetism is the study of the earth's field in past time. **Thermal remanent magnetism** (TRM) occurs in igneous rock-forming minerals as they cool below the Curie temperature. **Depositional remanent magnetism** (DRM) forms as magnetic particles align parallel to the earth's field at time of deposition. DRM (and TRM) reflects magnetic field existing at time of rock formation.

The **virtual geomagnetic pole** is the equivalent pole position representing the magnetic field as recorded at one observation point. The discrepancy between geomagnetic poles and geographic poles is thought to average out to zero in time. The ancient poles are thought to have been, like the modern poles, dipolar; the ancient poles have migrated thousands of kilometers since the Precambrian, and the path of polar migration is different for each continent. Direction of the earth's magnetic field reversed itself many times in the geological past.

Continental drift is supported by the data provided by paleomagnetism, the shape of continents, sea-floor spreading, ancient climatic patterns, and disrupted geological structures.

Plate tectonics is a hypothesis that postulates that the earth's crust is broken into large blocks being renewed in some zones and destroyed in others.

Lithospheric plate boundaries may be **convergent**, **divergent** or **transform**, depending on whether the driving forces are compressional, tensional or parallel.

Movement of plates may be by means of convection cells or by interaction of mantle material over **hot spots**, or **plumes**.

Obduction may bring deep-sea sediments and mafic and ultramafic igneous rocks up to the earth's surface where they may be exposed in cores of eroded mountains.

Dating the movement of plates can be done by analyzing magnetic anomalies and rates of sea-floor spreading, as well as by dating structural trends and stratigraphic units that developed on continents prior to their rifting.

Plate-tectonic theory can integrate notions of continental drift, sea-floor spreading, seismic zones, continents, and ocean basins.

SUPPLEMENTARY READINGS

Clague, D. A., and R. D. Jarrard "Tertiary Pacific Plate Motion Deduced from the Hawaiian-Emperor Chain," *Bull. Geol. Soc. Am.*, vol. 84, pp. 1135–1154, 1973. *Rates of rotation of the Pacific plate relative to the hot spot presently located near Hawaii are determined from radiometric age dates obtained from the Hawaiian-Emperor and other Pacific chains of volcanoes and seamounts.*

Condie, K. C. *Plate Tectonics and Crustal Evolution*, Pergamon Press, Inc., Elmsford, N.Y., 1976. *A good basic summation of the development of oceanic and continental crust in light of plate-tectonic theory; well illustrated.*

Dewey, J. F., W. C. Pitman III, W. B. F. Ryan, and J. Bonnin "Plate Tectonics and the Evolution of the Alpine System," *Bull. Geol. Soc. Am.*, vol. 84, pp. 3137–3180, 1973. *A detailed and comprehensive summary of the history and development of the Alpine system in terms of stepwise location and motion of Africa relative to Europe and the opening of the Atlantic Ocean.*

Wyllie, P. J. *The Way the Earth Works*, John Wiley & Sons, Inc., New York, 1976. *A highly readable and well-illustrated paperback on tectonic developments, using as a central theme the new global geology with good coverage of plate tectonics and continental drift.*

QUESTIONS

1 Has the relative elevation of continents with respect to sea level changed much during geologic time? How do we know?

2 What is the approximate age of the present ocean basins? How do we know? Were there earlier oceans? What is the evidence?

3 What is meant by paleomagnetism? How is it determined?

4 What causes the earth's magnetism? Is it constant? If not, how does it vary?

5 Why is the Curie temperature of a mineral extremely critical in the use of paleomagnetic determinations?

6 What are the principal forms of remanent magnetism? How are they similar? How do they differ?

7 Have the earth's magnetic poles always been where they are in relation to present-day latitude and longitude? If not, how have they changed and how do we know that?

8 How is sea-floor spreading established as fact? How does it occur?

9 What do ancient plants and animals contribute to the discussion of continental drift?

10 What are the principal kinds of plate boundaries? Make simple sketches to show them. What stress conditions occur at each type of boundary?

11 What causes the lithospheric plates to move?

12 How fast are these plates moving? How do we measure these rates?

MASS MOVEMENT OF SLOPE MATERIAL

The earth's surface is a collection of slopes: some gentle, some steep; some long, some short; some cloaked with vegetation, some bare of plants; some veneered with soil, some with naked rock. They lead from hill crests downward to streams and rivers. Taken together, they make up our landscape (Figure 11.1). Down these slopes move rainwater and the water from melting snow to collect in rivulets, brooks, and rivers flowing to the ocean. Down them also travels earth material, which, via the rivers, finally gets to the sea, there to form sedimentary rocks. But because this material derives from the slopes, it follows that the slopes themselves must change.

We have already learned that when rocks are exposed at the earth's surface, weathering immediately sets to work to establish an equilibrium between the rock and its new environment. Among the other factors joining forces with the processes of weathering is the direct pull of gravity. It acts to move the products of weathering and even unweathered bedrock to lower and lower levels. The movement of surface material caused by gravity is known as **mass movement.** Sometimes it takes place suddenly in the form of great landslides and rock falls from precipitous cliffs, but often it

occurs almost imperceptibly as the slow creep of soil across gently sloping fields.

Water also plays a direct role in the movement of slope material. The impact of a raindrop, the flow of water in thin sheets or tiny rills, and the solution of solid material as water seeps into the ground—all contribute to the erosion of hill slopes.

11.1 FACTORS OF MASS MOVEMENT

Gravity provides the energy for the downslope movement of surface debris and bedrock, but several other factors, particularly water, augment gravity and ease its work (see Figure 11.2).

Immediately after a heavy rainstorm you may have witnessed a landslide on a steep hillside or on the bank of a river. In many unconsolidated deposits, the pore spaces between individual grains are filled partly with moisture and partly with air, and so long as this condition persists, the surface tension of the moisture gives a certain cohesion to the soil. When, however, a heavy rain forces all the air out of the pore spaces, this surface tension is completely destroyed. The cohesion of the soil is reduced, and the whole mass becomes more susceptible to downslope movement. The presence of water also adds weight to the soil on a slope, although this added weight is probably not a very important factor in promoting mass movement.

Water that soaks into the ground and completely fills the pore spaces in the slope material contributes to instability in another way: The water in the pores is under pressure, which tends to push apart individual grains or even whole rock units and to decrease the internal friction or resistance of the material to movement. Here again water assists in mass movement.

Another factor, as we shall see, is air trapped beneath rapidly moving masses of rock debris as they hurtle down steep slopes. This air, confined by the fall-

11.1 These slopes in Capitol Reef National Park, Utah, are underlain by beds of sedimentary rocks of varying resistance to erosion. Down these slopes, material moves under the influence of gravity, or is urged downward by the runoff of surface water. A massive sandstone forms the bold cliffs of the skyline. The sandstone is cut by vertical joint planes, and slabs of rock outlined by these planes of weakness fall to the slopes below. In the foreground, thin resistant beds crop out in gentler slopes underlain by shale. Note that fragments from these resistant beds break off and move downslope under the influence of gravity. The surfaces of the gentler shale slopes are scarred by rill marks, the result of surface runoff from the occasional rainstorm. [National Park Service.]

ing material, acts as a cushion to reduce the friction of the debris with the ground and to make possible high-velocity movement of rock slides.

Gravity can move material only when it is able to overcome the material's internal resistance to being set into motion. Clearly, then, any factor that reduces this resistance to the point where gravity can take over contributes to mass movement. The erosive action of a stream, an ocean, or a glacier may so steepen a slope that the earth material can no longer resist the pull of gravity and is forced to give in to mass movement. In regions of cold climate, alternate freezing and thawing of earth materials may suffice to set them in motion. The impetus needed to initiate movement may also be furnished by earthquakes, by excavating or blasting operations, by sonic booms of aircraft, or even by the

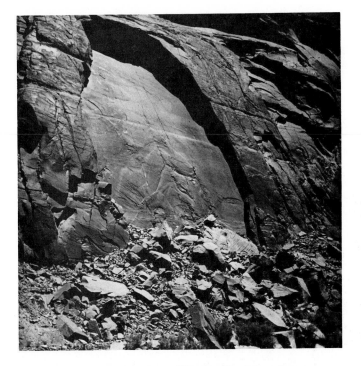

11.2 A large slab of massive Wingate Sandstone has fallen and shattered at the foot of this precipitous cliff, where the San Juan River now begins to move away the fragments. [William C. Bradley.]

gentle activities of burrowing animals and growing plants.

BEHAVIOR OF MATERIAL

In Chapter 8, in discussing the deformation of earth material, we pointed out that it may behave as an **elastic solid,** a **plastic substance,** or a **fluid.** Material moving down a slope may behave in the same manner. We could actually study any type of mass movement and classify it on the basis of these three behavior types; but we would have to assemble an excessive amount of technical data and would find the picture complicated by the fact that the material often behaves in different ways during any one movement. Therefore we shall simply classify mass movement as either **rapid** or **slow.**

11.2 RAPID MOVEMENTS

Catastrophic and destructive movements of rock and soil, the most spectacular and readily identified examples of mass movement, are popularly known as **landslides,** but the geologist recognizes slump, rock slides, debris slides, mudflows, and earthflows.

Landslides include a wide range of movements, from the slipping of a stream bank to the sudden, devastating release of a whole mountainside. Some landslides involve only the unconsolidated debris lying on bedrock; others involve movement of the bedrock itself.

Slump Sometimes called **slope failure, slump** is the downward and outward movement of rock or unconsolidated material traveling as a unit or as a series of units. Slump usually occurs where the original slope has been sharply steepened, either artificially or naturally. The material reacts to the pull of gravity as if it were an elastic solid, and large blocks of the slope move downward and outward along curved planes. The upper surface of each block is tilted backward as it moves.

Figure 11.3 shows the beginning of a slump at Gay Head, Massachusetts. The action of the sea has cut away the unconsolidated material at the base of the slope, steepening it to a point where the earth mass can no longer support itself. Now the large block has begun to move along a single curving plane, as suggested in Figure 11.4.

After a slump has been started, it is often helped along by rainwater collecting in basins between the tilted blocks and the original slope. The water drains down along the plane on which the block is sliding and promotes further movement.

A slump may involve a single block, as shown in Figures 11.3 and 11.4, but very often it consists of a series of slump blocks (Figures 11.5 and 11.6).

Rock slides The most catastrophic of all mass movements are **rock slides**—sudden, rapid slides of bedrock along planes of weakness.

One of the first major landslides to be studied in any detail occurred in 1903 and killed 70 people in the coal mining town of Frank, Alberta. Here some 36.8 million m³ of rock crashed down from the crest of Turtle Mountain, which rises over 900 m above the valley. Mining activities may have triggered this movement, but natural causes were basically responsible for it. As Figure 11.7 shows, Turtle Mountain has been sculptured from a series of limestone, sandstone, and shale beds, which have been tilted, folded, and faulted. The diagram shows that the greater part of the mountain is made of limestone and that the valley below is underlain by less-resistant beds of sandstone, siltstone,

11.3 The beginning of slump, or slope failure, along the sea cliffs at Gay Head, Massachusetts. Note that the slump block is tilted back slightly, away from the ocean. This slump block eventually moved downward and outward toward the shore along a curving plane, a portion of which is represented by the face of the low scarp in the foreground. [Harvard University, Gardner Collection.]

11.4 This diagram shows the type of movement found in a slump similar to that pictured in Figures 11.3 and 11.5. A block of earth material along the steepened cliff has begun to move downward along a plane that curves toward the ocean.

11.5 These slopes in South Dakota are underlain by a shale called the Pierre. The small stream has swung against its far bank and steepened it to the point at which the hillslope beyond has become unstable and is slumping toward the stream. [D. R. Crandall, U.S. Geological Survey.]

shale, and coal. It also shows four factors that contributed to the slide: the steepness of the mountain, the bedding planes in the limestone dipping parallel to the mountain face, the thrust plane that breaks the strata partway down the mountain, and the weak shale, siltstone, and coal beds in the valley.

The steep valley wall enhanced the effectiveness of gravity, and the bedding planes along with the fault planes served as potential planes of movement. The weak shale beds at the base of the mountain probably underwent slow plastic deformation under the weight of the overlying limestone, and as the shale was deformed, the limestone settled lower and lower. The settling action may have been helped along by the coal

11.6 Many slump movements contain several discrete units, as suggested here.

11.7 A cross section to show the conditions at Turtle Mountain that brought about the Frank, Alberta, landslide. [Modified from D. M. Cruden and J. Krahn, "A Reexamination of the Geology of the Frank Slide," *Canad. Geotech. J.*, vol. 10, no. 4, 1973.]

mining operations in the valley as well as by frost action, rain, melting snows, and earthquake tremors that had shaken the area 2 years before.

In any event, stress finally reached the point where the limestone beds gave way along their bedding planes and pushed outward toward the valley along the uppermost fault; the great mass of rock hurtled down into the valley. The rock material behaved in three different ways. (1) The shales underwent plastic deformation, producing a condition of extreme instability on the mountain slope; (2) when the strata that still held the limestone mass on the slope sheared in the manner of an elastic solid, the slide actually began; (3) once under way, the rock debris bounced, ricocheted, slid, and rolled down the mountain slope until it was literally "launched" into the air by a ledge of rock. Thereafter it arched outward and downward to the valley below. Once on the valley floor, it moved at high speed up the hills on the far side of the river. In this phase it moved like a viscous fluid with a series of waves spreading out along its front.

A more recent slide occurred in southwestern Montana on August 17, 1959. An earthquake whose focus was located just north of West Yellowstone, Montana, triggered a rock slide in the mouth of the Madison River Canyon, about 32 km to the west. A mass of rock estimated to be of over 32 million m³ fell from the south wall of the canyon. It climbed over 100 m up the opposite valley wall and dammed a lake 8 km long and over 30 m deep. More than a score of people lost their lives in the Madison River campground area below the slide (see Figures 11.8 and 11.9). Survivors reported extremely powerful winds along the margins of the slide. In fact, some persons were blown away, never again to be seen, and a 2-t automobile was observed to have been carried by the wind over 10 m before smashing against a row of trees. It seems clear that when the mass of rock fell to the valley bottom, it trapped and compressed large quantities of air. The high winds represented the extrusion of this mass of air from beneath the moving debris.

The Alaska earthquake of March 27, 1964 (see Section 9.2) had many effects on the topography and

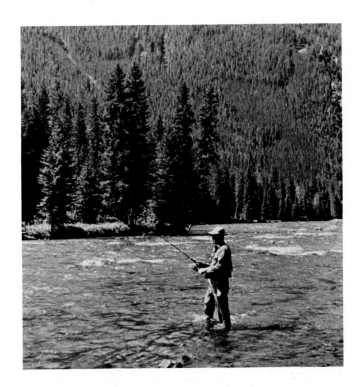

11.8 The Madison River Canyon before the landslide of August 1959. The slide began on the forested valley wall on the far side of the river. The debris clogged the river channel to form a lake. The spot where this fisherman once stood is now under 30 m of water. (Compare with Figure 11.9.) [Montana Power Co.]

TABLE 11.1

Data on some large landslides[a]

Location and date	Debris volume, millions of m³	Distance moved, km	Minimum speed, km/h	Height climbed, m
Elm, Switzerland, September 11, 1881	11.3	1.4	160	100
Sherman, Cordova, Alaska, March 27, 1964	28.3	4.3	185	145
Madison Canyon, Montana, August 17, 1959	32.6	0.8	130	110
Gros Ventre River, Wyoming, June 23, 1925	36.5	1.3	150	105
Frank, Alberta, Canada, April 29, 1903	36.8	1.7	175	120
Silver Reef, California, prehistoric	226	5.3	105	50
Blackhawk, California, prehistoric	283	8.0	120	60
Mayunmarca, Peru, April 25, 1974	~1,000	8.0	120	200
Saidmarreh, Iran, prehistoric	4,245	12.8	340	460

[a]Compiled from various sources and arranged in order of volume of material moved.

11.9 An aerial view of the Madison Canyon after the landslide. An outlet channel for the newly formed lake has been cut across the landslide dam. [William C. Bradley.]

shoreline of a large part of the southern portion of the state. Landslides were common and in inhabited places caused much damage. In the glaciated mountains of the Alaska Range, rock slides were particularly spectacular. Austin Post of the University of Washington mapped some 78 earthquake-induced slides that dumped material down glacially steepened mountain slopes onto the surfaces of the glaciers. The debris of the largest of these slides covered over 11 km², and the total debris on all glaciers amounted to more than 120 km². One large slide on the Sherman glacier near Cordova contributed 28 million m³ of rock debris to the surface of the glacier. Assuming the density of the material to be about 2.3, the mass of rock material weighed 57.5 million t (Figure 11.10).

In addition to the modern landslides described above, there are vestiges of prehistoric slides of great magnitude. One, at Saidmarreh, Iran, involved the movement of 4,245 million m³ of rock for a distance of over 12 km. Another, the Blackhawk slide of southern California, moved 283 million m³ of material over a distance of 8 km (Figure 11.11). Data on several large slides are given in Table 11.1.

Debris slides A debris slide is a small, rapid movement of largely unconsolidated material that slides or rolls downward and produces a surface of low hummocks with small intervening depressions. Movements of this sort are common on grassy slopes, particularly after heavy rains, and in unconsolidated material along the steep slopes of stream banks and those of

(a)   (b)

11.10 (a) On August 26, 1963, seven months before the Good Friday earthquake in southern Alaska, the Sherman glacier appeared like this in the camera of Austin Post of the U.S. Geological Survey. (b) The earthquake of March 24, 1964, set off a number of major rock slides, of which one occurred across the Sherman glacier. Post's photograph of August 24, 1964, shows the slide as a 4-km long tongue of dark debris against the bright background of the glacier.

shorelines. Debris slides can be quite common in sparsely vegetated country of steep slopes mantled with loose earth material.

Warnings of landslides We usually think of a landslide as breaking loose without warning, but it is more accurate to say that people in the area simply fail to detect the warning.

For example, a disastrous rock slide at Goldau, Switzerland, in 1806, wiped out a whole village, killing 457 people. The few who lived to tell the tale reported that they themselves had no warning of the coming slide but that animals and insects in the region may have been more observant or more sensitive. For several hours before the slide, horses and cattle seemed to be extremely nervous, and even the bees abandoned their hives. Some slight preliminary movement probably took place before the rock mass broke loose.

During the spring of 1935, slides took place in clay deposits along a German superhighway that was being built between Munich and Salzburg. The slides came as a complete surprise to the engineers, but for a full week the workmen had been murmuring, *Der Abhang wird lebendig* ("The slope's becoming alive").

11.11 The furrowed surface of the prehistoric Blackhawk slide, California, has been slightly eroded by desert streams. Width of slide approximately 1.8 km. [Fairchild Aerial Photography Collection at Whittier College.]

Landslides like the one on Turtle Mountain are often preceded by slowly widening fissures in the rock near the upward limit of the future movement.

Studies of debris slides in the Santa Monica Mountains and vicinity in southern California indicate that debris slides are coincident with particular patterns of rainfall. When the winter rainfall exceeds 25 cm, the stage is set for the initiation of debris flows. Then slides are most apt to occur as a result of a brief period of additional rain if the intensity is 0.5 cm/hr or greater.

MUDFLOWS

A **mudflow** is a well-mixed mass of rock, earth, and water that flows down valley slopes with the consistency of newly mixed concrete. In mountainous, desert, and semiarid areas, mudflows manage to transport great masses of material.

The typical mudflow originates in a small, steep-sided gulch or canyon where the slopes and floor are covered by unconsolidated or unstable material. A sudden flood of water, from cloudbursts in semiarid country or from spring thaws in mountainous regions, flushes the earth and rocks from the slopes and carries them to the stream channel. Here the debris blocks the channel until the growing pressure of the water becomes great enough to break through. Then the water and debris begin their down-valley course, mixing together with a rolling motion along the forward edge of the flow. The advance of the flow is intermittent. Sometimes it is slowed or halted by a narrowing of the stream channel; at other times it surges forward, pushing obstacles aside or carrying them along with it.

Eventually the mudflow spills out of the canyon mouth and spreads across the gentle slopes below. No longer confined by the valley walls or the stream channel, it splays out in a great tongue, spreading a layer of mud and boulders that ranges from a few centimeters to 1 m or more in thickness. Mudflows can move large boulders weighing 80 t or more for hundreds of meters across slopes as gentle as 5°.

EARTHFLOWS

Earthflows involve the plastic movement of unconsolidated material lying on solid bedrock and are usually helped along by excessive moisture. They move slowly but perceptibly and may involve from a few to several million cubic meters of earth material. The line at which the material pulls away from the slope is

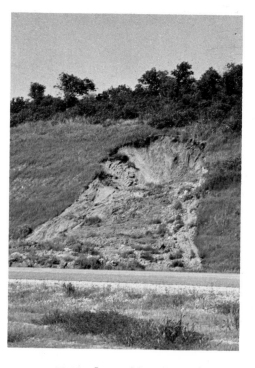

11.12 An earthflow in a road cut near Dallas, Texas, shows a sharp scar high on the slope and, farther down, the area of soil movement by flow. [C. W. Brown.]

marked by an abrupt scarp or cliff, as shown in Figure 11.12.

TALUS

Strictly speaking, a **talus** is a slope built up by an accumulation of rock fragments at the foot of a cliff or a ridge. The rock fragments are sometimes referred to as **rock waste** or **slide rock**. In practice, however, talus is widely used as a synonym for the rock debris itself.

In the development of a talus, rock fragments are loosened from the cliff and clatter downward in a series of free falls, bounces, and slides. As time passes, the rock waste finally builds up a heap or sheet of rock rubble. An individual talus resembles a half cone with its apex resting against the cliff face in a small gulch (Figure 11.13). A series of these half cones often forms a girdle around high mountains, completely obscuring their lower portions. Eventually, if the rock waste accumulates more rapidly than it can be destroyed or removed, even the upper cliffs become buried, and the growth of the talus stops. The slope angle of the talus varies with the size and shape of the rock fragments. Although angular material can maintain slopes up to 50°, rarely does a talus ever exceed angles of 40°.

A talus is subject to the normal process of chemical weathering, particularly in a moist climate. The rock waste is decomposed, especially toward its lower limit, or toe, which may grade imperceptibly into a soil.

11.3 SLOW MOVEMENTS

Slow mass movements of unconsolidated material are more difficult to recognize and less fully understood than rapid movements; yet they are extremely important in the sculpturing of the land surface. Because they operate over long periods of time, they are probably responsible for the transportation of more material than are rapid and violent movements of rock and soil.

Before the end of the nineteenth century, William Morris Davis aptly described the nature of slow movements.

The movement of land waste is generally so slow that it is not noticed. But when one has learned that many land forms result from the removal of more or less rock waste, the reality and the importance of the movement are better understood. It is then possible to picture in the imagination a slow washing and creeping of the waste down the land slopes; not bodily or hastily, but grain by grain, inch by inch; yet so patiently that in the course of ages even mountains may be laid low.[1]

CREEP

In temperate and tropical climates, a slow downward movement of surface material known as **creep** operates even on gentle slopes with a protective cover of

[1] William Morris Davis, *Physical Geography*, p. 261, Ginn & Company, Lexington, Mass., 1898.

grass and trees. It is hard to realize that this movement is actually taking place. Because the observer sees no break in the vegetative mat, no large scars or hummocks, he has no reason to suspect that the soil is in motion beneath his feet.

Yet this movement can be demonstrated by exposures in soil profile (see Figure 11.14) and by the behavior of tree roots, of large blocks of resistant rock, and of artificial objects such as fences and telephone poles (see Figure 11.15). Figure 11.16 shows a section through a hillside underlain by flat-lying beds of shale, limestone, clay, sandstone, and coal. The slope is covered with rock debris and soil. But notice that the beds near the base of the soil bend downslope and thin out rapidly. These beds are being pulled downslope by gravity and are strung out in ever-thinning bands that may extend for hundreds of meters. Eventually they approach the surface and lose their identity in the zone of active chemical weathering.

Figure 11.16 also shows other evidence that the soil is moving. Although when viewed from the surface the tree appears to be growing in a normal way, it is actually creeping slowly down the slope. Because the surface of the soil is moving more rapidly than is the soil beneath it, the roots of the tree are unable to keep up with the trunk.

Many other factors cooperate with gravity to produce creep. Probably the most important is moisture in

11.14 Alternating beds of sandstone and shale have been tilted until they stand vertically. A little over 1 m below the surface, they begin to bend downslope under the influence of gravity. The amount of downslope movement increases toward the surface. Haymond Formation in the Marathon region of Texas. [William C. Bradley.]

the soil, which works to weaken the soil's resistance to movement. In fact, any process that causes a dislocation in the soil brings about an adjustment of the soil downslope under the pull of gravity. Thus the burrows of animals tend to fill downslope, and the same is true of cavities left by the decay of organic material, such as the root system of a dead tree. The prying action of swaying trees and the tread of animals and even of human beings may also aid in the motion. The result of all these processes, aided by the influence of gravity, is to produce a slow and inevitable downslope creep of the surface cover of debris and soil.

SOLIFLUCTION

The term **solifluction** (from the Latin *solum*, "soil," and *fluere*, "to flow") refers to the downslope movement of debris under saturated conditions. Solifluction is most pronounced in high latitudes, where the soil is strongly affected by alternate freezing and thawing and where the ground freezes to great depths; but even moderately deep seasonal freezing promotes solifluction.

Solifluction takes place during periods of thaw. Because the ground thaws from the surface downward, the water that is released cannot percolate into the subsoil and adjacent bedrock, which are still frozen

and therefore impermeable to water. As a result, the surface soil becomes sodden and tends to flow down even the gentlest slopes. Solifluction is an important process in the reduction of land masses in Arctic climates, where it transports great sheets of debris from higher to lower elevations.

11.15 Telephone poles set in a talus slope south of Yellowstone National Park record movement of the slope material. Originally the poles stood vertically, but over a period of about 10 years, the talus has moved and tilted the two nearest poles 5° and the most distant pole 8°. [Sheldon Judson, 1968.]

11.16 The partially weathered edges of horizontal sedimentary rocks are dragged downslope by soil creep. The tree is also moving slowly downslope, as is evidenced by the root system spread out behind the more rapidly moving trunk. [Redrawn from C. F. S. Sharpe and E. F. Dosch, "Relation of Soil-Creep to Earthflow in the Appalachian Plateaus," *J. Geomorphol.*, vol. 5, p. 316, 1942. By permission of Columbia University Press.]

During the glacier advances of the Pleistocene, a zone of intense frost action and solifluction bordered the southward-moving ice. In some places we can still find the evidence of these more rigorous climates preserved in distorted layers of earth material just below the modern soil (see Figure 11.17).

Frost action plays strange tricks in the soils of the higher elevations and latitudes. Strange polygonal patterns made up of rings of boulders surrounding finer material, stripes of stones strewn down the face of hillsides, great tabular masses of ice within the soil, and deep ice wedges that taper downward from the surface—all are found in areas where the ground is deeply frozen (see Figure 11.18). The behavior of frozen

11.17 Beds of silt, sand, and clayey gravel have been contorted by differential freezing and thawing during the more rigorous climates of glacial times. The gravel at the base of the exposure is clayey, gravelly till deposited directly by glacier ice. Above this are the highly contorted beds of water-laid sand (dark) and silt (light), which were originally lying flat. The modern soil has developed across the contortions. The white of the silt bands is due to precipitation of calcium carbonate from soil water. The brush hook is about 0.5 m long. Exposure north of Devil's Lake, east-central North Dakota. [U.S. Geological Survey, Saul Aronow.]

11.18 Aerial photograph of polygonal patterns developed by ice wedges in the coastal plain of northern Alaska. Area shown is about 650 m wide. [U.S. Coast and Geodetic Survey.]

11.3 SLOW MOVEMENTS

11.19 Distribution of permafrost in the northern hemisphere. [Compiled by Troy L. Péwé from published and unpublished sources. For sources, see A. Lincoln Washburn, *Geocryology*, Fig. 3.3, p. 25, John Wiley & Sons, Inc., New York, 1980.]

Offshore permafrost

Continuous permafrost

Discontinuous permafrost

ground is one of the greatest barriers to the settlement of Arctic regions. The importance of these regions has increased in recent years, and studies begun by Scandinavian and Soviet investigators are now being intensively pursued by United States and Canadian scientists. The present-day distribution of permanently frozen ground is shown in Figure 11.19.

ROCK GLACIERS

Rock glaciers are tongues of rock waste that form in the valleys of certain mountainous regions. Although they consist almost entirely of rock, they bear a striking resemblance to ice glaciers. A typical rock glacier is marked by a series of rounded ridges, suggesting that the material has behaved as a viscous mass.

Observations on active rock glaciers in Alaska indicate that movement takes place through flow of interstitial ice within the mass. Favorable conditions for the development of rock glaciers include a climate cold enough to keep the ground continuously frozen, steep cliffs to supply debris, and coarse blocks that allow for large interstitial spaces.

11.4 DIRECT ACTION OF WATER ON HILL SLOPES

We have referred to the role of water in promoting mass movement. It also operates directly on slopes to move material downward to the channels of conventional streamways.

IMPACT OF RAINDROPS

Studies have been of the effectiveness of individual **raindrops in erosion.** A drop striking a wet, muddied field will splash water and suspended solid particles into the air. If there is any slope at all to the land, the material splashed uphill will not travel as far as will that splashed downhill. There will, then, be a net downslope motion of material by rain splash (see Figure 11.20).

RAIN WASH

Rain falling on a slope may soak into the ground or run off the slope. Vegetation will slow the runoff, promote infiltration, and generally reduce the erosive effectiveness of water flowing on the slopes; but if vegetation is lacking or sparse, rainwater will rapidly flow down the slope. This flow, called **rain wash,** can become an effective agent of erosion.

A thin sheet of water may accumulate on a slope and become thicker as it proceeds downslope. This is sometimes referred to as **sheet wash,** or **sheet flow.** Theoretically, a film of water would increase at a constant rate as it moved down a uniform slope, gathering more and more water. This increase occurs, but not in a uniform sheet for the simple reason that no slope is completely uniform. Initial irregularities, differing resistance to erosion, and varying lengths of flow may all contribute to the concentration of sheet wash into broad shallow troughs, or channels, which determine the location of miniature stream channels called **rills.** The rills are eroded along the axes of broad troughs; for here the water has its greatest velocity and hence its greatest erosive power. Most of us have seen the effect of such erosion on artificially created exposures along a roadway or on a slope newly prepared for planting.

SOLUTION

It is clear that as water on the surface comes into contact with earth material, it will begin to take mineral matter into solution. This will be true whether the water flows off the surface or soaks into the ground. The effect is to move material from the slopes, and we therefore must add solution as a third way in which hill slopes are reduced by the direct action of water.

11.20 A raindrop splashes onto a water-soaked surface, scattering fine particles of earth. [Wide World.]

OUTLINE

Erosion of hill slopes occurs as a direct result of gravity and by the action of water.

Mass movement of bedrock and unconsolidated material is in response to the pull of gravity. The material may behave as an elastic solid, as a plastic substance, or as a liquid. Movements are classified as either rapid or slow and may be influenced by saturation of material by water, by steepening of slopes by erosion, by earthquakes, and by the activity of animals, including human beings.

Rapid movements include landslides, mudflows, earthflows, and talus formation.

Slow movements include creep, solifluction, and rock glaciers.

Water moves material on slopes by the direct impact of raindrops, by rain wash, and by solution.

SUPPLEMENTARY READINGS

Carson, M. A., and M. J. Kirby *Hillslope Form and Process*, Cambridge University Press, New York, 1972. *An advanced text on the subject.*

Embleton, Clifford, and Cuchlaine A. M. King *Periglacial Geomorphology*, 2nd ed., John Wiley & Sons, Inc., New York, 1975. *A short but authoritative treatment of surface processes found in Arctic areas.*

Washburn, A. Lincoln *Geocryology*, John Wiley & Sons, Inc., New York, 1980. *A very fine review of periglacial processes and environments. Handsomely illustrated.*

Voight, Barry (ed.) *Rockslides and Avalanches*, Elsevier Publishing Company, Amsterdam, 1978. *This compilation of studies is in two volumes. Volume 1, Natural Phenomena, provides an excellent summary of the processes involved in rock slides and avalanches, and gives detailed descriptions of major slides, both historic and prehistoric. Volume 2 focuses on engineering problems.*

Zaruba, Quido, and Vojtech Mencl *Landslides and Their Control*, Elsevier Publishing Company, Amsterdam, 1969. *This volume gives many examples of mass movement, most of them drawn from central Europe.*

QUESTIONS

1 List some of the factors that affect the movement of earth materials down slopes.

2 Describe at least two types of rapid slope movement, using examples if possible.

3 Some movements of materials on slopes are very slow as compared with, for example, rock slides. List some different types of slow movement and describe one in some detail.

4 From your own experience and observation, can you cite examples of the downslope movements of material? If so, describe them and try to determine the causes.

12

RUNNING WATER

Through millions of years of earth history, agents of erosion have been working constantly to reduce the landmasses to the level of the seas. Of these agents, **running water** is the most important. Year after year the streams of the earth (Figure 12.1) move staggering amounts of debris and dissolved material through their valleys to the great settling basins, the oceans.

12.1 WORLD DISTRIBUTION OF WATER

The distribution of the earth's estimated water supply is given in Appendix D. The great bulk of this water, over 97 percent, is in the oceans, and less than 3 percent, most of which is ice, is on the land. Atmospheric moisture is surprisingly low, about 0.001 percent of the total. Even less is in the stream channels of the world at any one moment. We see, then, that the earth's water is in varying amounts at different places; yet it moves from one place to another. As is written in Ecclesiastes 1:7, "All the rivers run into the sea; yet the sea is not full; unto the place from whence the rivers come, thither

they return again." This is part of the circulation of water from land to ocean and back again to land that we call the **hydrologic cycle.**

12.2 HYDROLOGIC CYCLE

The streams of the world flow to the sea at a rate that would fill the ocean basins in 40,000 years. These streams, flowing as they do, must have a continuing supply of water. Furthermore, it is clear that the oceans must themselves somehow lose water in order to make room for the continuous supply that comes to them.

The identification of the source of stream water is historically recent. Well into the eighteenth century, it was the general belief that streams were replenished

12.1 Stream valleys are the single most characteristic feature of the world's landscape. Even in the desert, as here on the slopes leading down into Death Valley, streams have fashioned an intricate system of valleys. [J. P. Thayer, U.S. Geological Survey.]

by springs that drew their water by some complex system through the underground from the oceans. This belief was nourished by the assumption that rainfall was inadequate to account for the flow observed in rivers; for rivers ran continuously even though rain was intermittent. Furthermore, it was generally held that rainwater could not soak into the ground and replenish springs, a conclusion that seems to owe its origin to Seneca (3 B.C.–A.D. 65), Roman statesman and philosopher, who based his conclusions on inconclusive observations while he tended his vineyards.

In the seventeenth century, three different types of observation laid the base for the hydrologic cycle as we accept it, which is diagrammatically represented in Figure 12.2. In 1674, Pierre Perrault (1611–1680), a French lawyer and sometime hydrologist, presented the results of his measurements in the upper portion of the drainage basin of the Seine River. Over a 3-year span, he collected data on the amount of precipitation in this portion of the basin. At the same time he kept track of the amount of water discharged by the river below the portion of the river basin where he had data

on precipitation. The results showed that the flow of the river was surprisingly low when compared with the total amount of water available from precipitation. In fact, the annual precipitation was six times the total volume of river flow. There was enough rainfall, then, to account for the flow of the Seine in this part of its basin.

At about the same time, the French physicist Edmé Mariotte (1620–1684) made more exact studies of discharge in the Seine basin. His publications, appearing posthumously in 1684, verified Perrault's conclusions. Mariotte further demonstrated, by experimentation at the Paris Observatory, that seepage through the earth cover was less, but not much less, than the amount of rainfall. He also demonstrated the increase in the flow of springs during rainy weather and the decrease during time of drought. It was then evident that the earth permits penetration of moisture.

In 1693, Edmund Halley (1656–1742), of comet fame, provided data on evaporation in relation to rainfall. He roughly calculated the amount of water discharged annually by rivers into the Mediterranean Sea and added this to the amount that fell directly on

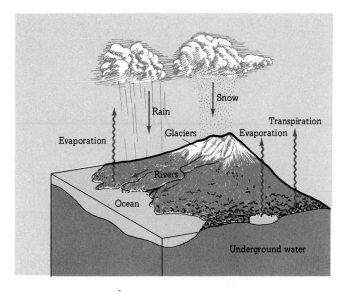

12.2 In the hydrologic cycle, water evaporated into the atmosphere reaches the land as rain or snow. Here it may be temporarily stored in glaciers, lakes, or the underground before returning by the rivers to the sea. Or some may be transpired or evaporated directly back into the atmosphere before reaching the sea.

the sea's surface. He was then able to compute the approximate amount of water that was being evaporated back into the air each year from the surface of the Mediterranean. He found that there was more than enough water being pumped into the atmosphere to feed all the rivers coming into the sea.

12.3 PRECIPITATION AND STREAM FLOW

After water has fallen on the land as precipitation, it follows one of the many paths that make up the hydrologic cycle. By far the greatest part is evaporated directly into the air or is taken up by plants and transpired ("breathed" back) to the atmosphere. A smaller amount follows the path of **runoff**, the water that flows off the land, and the smallest amount of precipitation soaks into the ground through **infiltration.**

In Chapters 13 and 14, we shall consider water at various stages of the hydrologic cycle. In Chapter 13, we look at the water that infiltrates into the ground; in Chapter 14, we study water that has been impounded on the land as glacier ice, a condition that represents a temporary halting of the water's progress through the cycle. In this chapter, we shall concentrate on the nature and effects of runoff. Bearing in mind the ways in which water that falls as precipitation proceeds

through the hydrologic cycle, we can express the amount of runoff by the following generalized formula:

Runoff = precipitation − (infiltration + evaporation and transpiration)

L. B. Leopold and W. B. Langbein of the United States Geological Survey have reported some figures on the water budget of the hydrologic cycle in the United States. There is an average of around 75 cm of precipitation per unit of area over the country each year. Of this total, about 22.5 cm are carried as runoff by the streams to the ocean. Another 52.5 cm go back into the atmosphere by evaporation and transpiration. Of course, some water filters into the ground, but all but a very small amount—about 0.25 cm—is returned to the surface, where it disappears either as runoff or as evapotranspiration. These figures indicate that across the United States there is an apparent loss of 22.5 cm annually·from the atmosphere. This is the runoff that goes to the ocean, but the debit is made up by the evaporation of moisture from the ocean and its transfer by the atmosphere to the continents.

Figures on the worldwide water budget are shown in Figure 12.3. Average precipitation for the lands of the world is variously estimated and approximates about 80 cm/year. Estimates of runoff from the continents to the oceans vary. One estimate is given in Appendix D. This does not include water that may get to the oceans in the form of ice. The figures indicate that 32.4×10^{15} l (liters) of water annually flow to the sea. This is a runoff of 24.9 cm/unit area. The difference between runoff and precipitation is the evapotranspi-

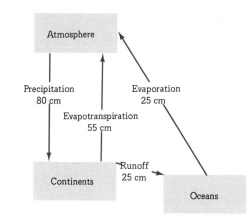

12.3 Water budget for the world. [Data from the U.S. Geological Survey.]

ration, in this instance about 55 cm. It follows that the oceans contribute about 25 cm to the yearly continental precipitation budget.

LAMINAR AND TURBULENT FLOW

Consider now the manner in which water moves. When it moves slowly along a smooth channel or through a tube with smooth walls, it follows straight-line paths that are parallel to the channel or walls. This type of movement is called **laminar flow**.

If the rate of flow increases, however, or if the confining channel becomes rough and irregular, this smooth, streamlined movement is disrupted. The water in contact with the channel is slowed down by friction, whereas the rest of the water tends to move along as before. As a result (see Figures 12.4 and 12.5), the water is deflected from its straight paths into a series of eddies and swirls. This type of movement is known as **turbulent flow.** Water in streams usually flows along in this way, its turbulent flow being highly effective both in eroding a stream's channel and in transporting materials.

When a stream reaches an exceptionally high velocity along a sharply inclined stretch or over a waterfall, the water moves in plunging, jetlike surges. This type of flow, closely related to turbulent flow, is called **jet,** or **shooting, flow.**

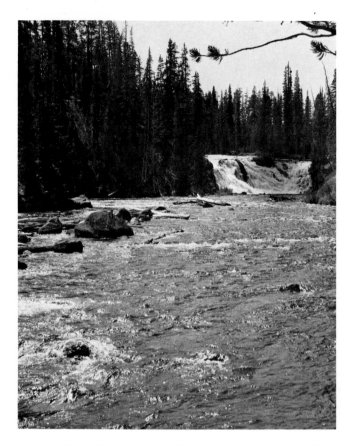

12.5 Most stream flow is turbulent; here the Lewis River in Wyoming tumbles over falls and swirls and boils through a rough, irregular channel. [Sheldon Judson.]

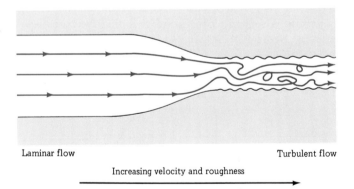

Laminar flow   Turbulent flow

Increasing velocity and roughness

12.4 Diagram showing laminar and turbulent flow of water through a section of pipe. Individual water particles follow paths depicted by the colored lines. In laminar flow, the particles follow paths parallel to the containing walls. With increasing velocity or increasing roughness of the confining walls, laminar flow gives way to turbulent flow. The water particles no longer follow straight lines but are deflected into eddies and swirls. Most water flow in streams is turbulent.

VELOCITY, GRADIENT, AND DISCHARGE

The **velocity** of a stream is measured in terms of the distance its water travels in a unit of time. A velocity of 15 cm/s is relatively low, and a velocity of about 625 to 750 cm/s is relatively high.

A stream's velocity is determined by many factors, including the amount of water passing a given point, the nature of the stream banks, and the **gradient**, or slope, of the stream bed. In general, a stream's gradient decreases from its headwaters toward its mouth; as a result, a stream's longitudinal profile is more or less concave toward the sky. We usually express the gradient of a stream as the vertical distance a stream descends during a fixed distance of horizontal flow. The Mississippi River from Cairo, Illinois, to the mouth of the Red River in Arkansas has a low gradient; for along this stretch, the drop varies between 2 and 10 cm/km. On the other hand, the Arkansas River, in its upper reaches through the Rocky Mountains in central Colorado, has a high gradient; for there the drop averages 7.5 m/km. The gradients of other rivers are even higher. The upper 20 km of the Yuba River in Califor-

12.6 Velocity variations in a stream. Both in plan view and in cross section, the velocity is slowest along the stream channel, where the water is slowed by friction. On the surface, it is most rapid at the center in straight stretches and toward the outside of a bed where the river curves. Velocity increases upward from the river bottom.

nia, for example, have an average gradient of 42 m/km; and in the upper 6.5 km of the Uncompahgre River in Colorado, the gradient averages 66 m/km.

The velocity of a stream is checked by the turbulence of its flow, by friction along the banks and bed of its channel, and to a much smaller extent by friction with the air above. Therefore, if we were to study a cross section of a stream, we should find that the velocity would vary from point to point. Along a straight stretch of a channel the greatest velocity is achieved toward the center of the stream at, or just below, the surface, as shown in Figure 12.6.

We then have two opposing forces: the **forward flow** of the water under the influence of gravity and the **friction** developed along the walls and bed of the stream. These are the two forces that create different velocities. Zones of maximum turbulence occur where the different velocities come into closest contact. These zones are very thin because the velocity of the water increases very rapidly as we move into the stream away from its walls and bed; but within these thin zones of great turbulence, a stream shows its highest potential for erosive action (see Figures 12.7 and 12.8).

There is one more term that will be helpful in our discussion of running water: **discharge,** the quantity of water that passes a given point in a unit of time. (In this book we shall measure it in cubic meters per second.) Discharge varies not only from one stream to another but also within a single stream from time to time and from place to place along its course. Discharge usually increases downstream as more and more tributaries add their water to the main channel. Spring floods may so greatly increase a stream's discharge that its normally peaceful course becomes a raging torrent.

The hydrograph The hydrograph of a river shows the variation of stream flow with time, as indicated in Figure 12.9. In the example given, we see a period of high

flow, peaking in a flood, and a generally low-water stage, which represents that portion of stream flow attributable to groundwater recharge, a flow referred to as **base flow.** In addition, the figure indicates a period of rainfall responsible for the high-water, or flood, stage, which occurs after an interval of time called the **lag time.** The shape of the hydrograph for different streams (and even for different places on the same

12.7 Zones of maximum turbulence in a stream are shown by the curly symbols in the sections through a river bed. They occur where the change between the two opposing forces—the forward flow and the friction of the stream channel—is most marked. Note that the maximum turbulence along straight stretches of the river is located where the stream banks join the stream floor. On bends, the two zones have unequal intensity; the greater turbulence is located on the outside of a curve.

12.3 PRECIPITATION AND STREAM FLOW

259

12.8 A small stream meanders across its flood plain near Steamboat Springs, Colorado. Greatest erosion takes place on the outside of the bends, and the result is a steep, cut bank at these places. [Sheldon Judson.]

stream) varies with a number of natural factors, including the infiltration rate, the relief, the geology, and the vegetative cover.

12.9 Hypothetical hydrograph, showing significant characteristics. [After Luna B. Leopold, "Hydrology for Urban Land Planning," *U.S. Geol. Surv. Circ. 554*, p. 3, 1968.]

12.4 ECONOMY OF A STREAM

Elsewhere we have seen that earth processes tend to seek a balance, to establish an equilibrium, and that there is, in the words of James Hutton, "a system of beautiful economy in the works of nature."[1] We found, for example, that weathering is a response of earth materials to the new and changing conditions they meet as they are exposed at or near the earth's surface. On a larger scale we found that the major rock groups—igneous, sedimentary, and metamorphic—reflect certain environments and that as these environments change, members of one group may be transformed into members of another group. These changes were traced in what we called the "rock cycle" (Figure 1.8). Water running off the land in streams and rivers is no exception to this universal tendency of nature to seek equilibrium.

ADJUSTMENTS OF DISCHARGE, VELOCITY, AND CHANNEL

Just a casual glance tells us that the behavior of a river during its flood stage is very different from its behavior during the low-water stage. For one thing, a river car-

[1] James Hutton, *Theory of the Earth*, vol. 2, Edinburgh, 1795.

ries more water and moves more swiftly in flood time. Furthermore, the river is generally wider during flood; its level is higher; and we should guess, even without measuring, that it is also deeper. We can relate the discharge of a river to its width, depth, and velocity:

Discharge (m³/s) = channel width (m) × channel depth (m) × water velocity (m/s)

In other words, if the discharge at a given point along a river increases, then the width, depth, velocity, or some combination of these factors must also increase. We now know that variations in width, depth, and velocity are neither random nor unpredictable. In most streams, if the discharge increases, then the width, depth, and velocity each increase at a definite rate. The stream maintains a balance between the amount of water it carries, on the one hand, and its depth, width, and velocity, on the other. Moreover, it does so in an orderly fashion, as shown in Figure 12.10.

Let us turn now from the behavior of a stream at a single locality to the changes that take place along its entire length. From our own observation we know that the discharge of a stream increases downstream as more and more tributaries contribute water to its main channel. We also know that the width and depth in-

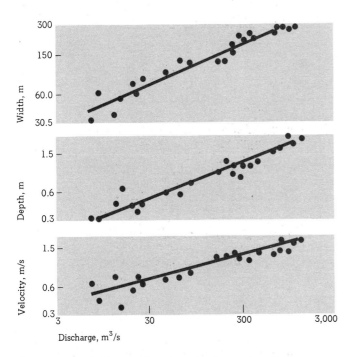

12.10 As the discharge of a stream increases at a given gauging station, so do its velocity, width, and depth. They increase in an orderly fashion, as shown by these graphs based on data from a gauging station in the Cheyenne River near Eagle Butte, South Dakota. [After Luna B. Leopold and Thomas Maddock, "The Hydraulic Geometry of Stream Channels and Some Physiographic Implications," *U.S. Geol. Surv. Prof. Paper 252*, p. 5, 1953.]

12.11 Stream velocity and depth and width of a channel increase as the discharge of a stream increases downstream. Measurements in this example were taken at mean annual discharge along a section of the Mississippi-Missouri river system. [After Luna B. Leopold and Thomas Maddock, "The Hydraulic Geometry of Stream Channels and Some Physiographic Implications," *U.S. Geol. Surv. Prof. Paper 252*, p. 13, 1953.]

crease as we travel downstream. But if we go beyond casual observation and gather accurate data on the width, depth, velocity, and discharge of a stream from its headwaters to its mouth for a particular stage of flow—for example, flood or low water—we should find again that the changes follow a definite pattern (Figure

12.4 ECONOMY OF A STREAM

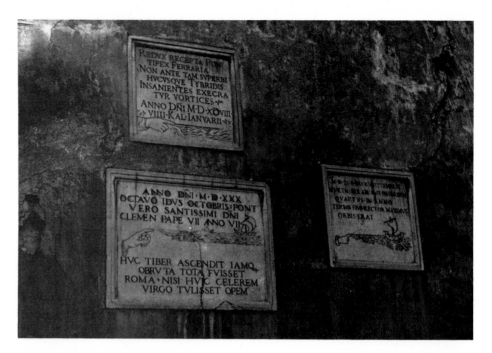

12.12 Plaques in the Piazza della Minerva near the Pantheon record the height of each of three different floods of the Tiber River at Rome before the installation of the present levees and upstream flood controls. [Sheldon Judson.]

12.13 Flood waters have made canals of the streets of this Wisconsin community along the Mississippi River. [Wisconsin Conservation Department.]

12.11) and that depth and width increase downstream as the discharge increases. And, surprisingly enough, we should also find that the stream's **velocity** increases toward its mouth. This is contrary to our expectations; for we know that the gradients are higher upstream, which suggests that the velocities in the steeper headwater areas would also be higher. But the explanation for this seeming anomaly is simple: To handle the greater discharge downstream, a stream not only must deepen and widen its channel but also must increase its velocity.

Floods When the stream discharge exceeds the channel capacity, water must rise over the channel banks and flood the adjacent low-lying land (Figures 12.12 and 12.13). Floods can be catastrophic, but disastrous as they may be, we must consider them as the expression of natural stream behavior. A study of long-term stream records shows that there is some method to the apparent madness of a stream. For instance, a stream flow that just fills its channel, that is, the **bankfull stage**, occurs every 1.5 to 2 years. Looking at

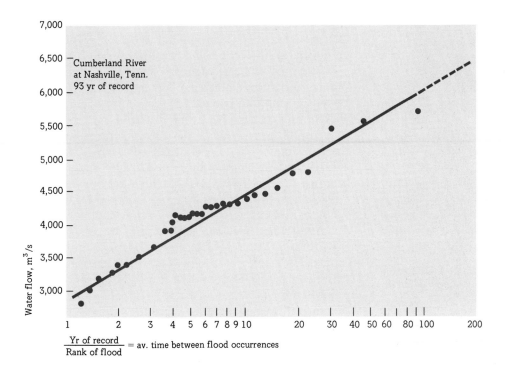

12.14 Average interval between floods of various sizes on the Cumberland River at Nashville, Tennessee, plotted from data in *U.S. Geol. Surv. Water Supply Paper 771*. The vertical scale is an ordinary linear scale, and the horizontal scale is known as a logarithmic scale.

greater volumes of flow, we find that the frequency of floods, which exceed the bank-full stage, can be statistically anticipated. The procedure is simple and involves a stream-discharge record of at least 10 years: The maximum discharge for each year is listed; then these flows are ranked according to magnitude, that is, from largest to smallest; and then the flows can be plotted, as shown in Figure 12.14.

The example in Figure 12.14 is based on floods that have been recorded at Nashville, Tennessee, over a 93-year period. The largest flow, 5,750 m³/s, is entered at 93 years because floods that large occur apparently about once in 93 years. The size of the second largest flood, 5,570 m³/s, is plotted at $^{93}\!/_2 = 46.5$ years because two floods of this size or larger have occurred in 93 years. The third largest flood is shown at $^{93}\!/_3 = 31$ years, and so on through the entire record. Because the graph shows points on a roughly straight line, we can extend the line to the upper right to estimate the size of floods to be expected once in 100 years or once in 200 years. On this basis we speak of 10-year floods, 50-year floods, and even 1,000-year floods. But remember that this is a statistical prediction. We don't know precisely when a flood of a given size will occur. The historical record merely tells us to expect a flood of a given size some time within a given span.

That floods recur and that the recurrence interval of a flood of a given size is predictable should assure us that floods are a natural, expectable phenomenon. We should expect them every so often. And indeed we should look at the river's flood plain as a part of the river's domain, useful to the river in flood times and certain to be claimed by the river from time to

time. Therefore, if we use the flood plain, we must remember that it is only on loan to us. Whatever use we put the flood plain to ought to be as compatible as possible with the use to which the river is sure to put it. Experience should teach us not only that streams will top their banks from time to time but that even the most energetic flood-control plans (including the use of dams, levees, dredging, and channel modification) sometimes fail. The problem is partly an economic one: Against what magnitude flood shall we try to protect the flood plain? The 10-year flood? The 100-year flood? The 500-year flood? Obviously the larger the flood we attempt to defend against, the greater the expense. At what point does the expense of protection outweigh the gain achieved? This is obviously a matter of public policy. An alternative to protection by engineering works is to put the flood plain to a use compatible with flooding, such as agriculture or a park system. This practice may in many instances be more appropriate than tampering with the river system.

Plans to modify the flood flow are focused on modifying the hydrograph by lowering the flood peak, increasing the lag time between the precipitation and flood crest, and spreading the flood flow over a longer period of time. A flood-control dam, then, is designed to store water during periods of high runoff and to let the excess water out slowly to downstream areas. The

12.4 ECONOMY OF A STREAM

12.15 Hypothetical hydrograph to show the effect on stream flow of a flood-control dam upstream from the hydrograph station.

result is to reduce the flood crest downstream and spread the discharge of flood waters over a longer time interval, as suggested in Figure 12.15.

Urbanization and suburbanization Building of cities and their suburbs affects stream flow and changes the hydrograph, although in the opposite direction from that of flood-control dams. Agglomerations of buildings with their associated roads, streets, sidewalks,

12.16 Hypothetical hydrograph, showing the effect of urbanization on stream flow. [After Luna B. Leopold, "Hydrology for Urban Land Planning," *U.S. Geol. Surv. Circ.* 554, p. 3, 1968.]

and paved parking areas achieve the following effects: (1) The amount of water sinking into the underground is curtailed in proportion to the amount of area sealed off by surface veneers of buildings and pavements, and even the areas of ground left open tend to be less permeable because of their compaction by extensive human activity; (2) the rate and amount of surface runoff increases; (3) the groundwater level falls because of the decreased infiltration. The effect on stream flow is shown in Figure 12.16 and can be characterized as follows: (1) The time lag between precipitation and flood peak is shortened because water is not slowed on its way to the stream channel by a vegetative cover nor does any appreciable amount sink into the ground; (2) the peak is higher because the stream must carry more water in a shorter period of time; and (3) the base flow is generally lower because of a decreased supply of groundwater on which the base flow depends. The net result of these changes is a **flashy stream,** one that has a low base flow and a high, short flood peak.

We have found that the stream channel is adjusted to the stream discharge, and this was expressed as discharge = width × depth × velocity. Observation shows that a stream rises to the limit of its bank on the average of once every 1.5 to 2 years. As noted earlier, this stage of flow is called the **bank-full stage.** The stream channel is adjusted to handle the size of flow that occurs every 1.5 to 2 years. When flow exceeds this amount, flooding begins. Studies reported by Luna Leopold of the United States Geological Survey indicate that with an increase of urbanization, the bank-full stage of flow increases. If an area is 50 percent urbanized, experience indicates that the bank-full stage of discharge increases by a factor of 2.7 over that of the same stream in an unurbanized condition. An example is shown in Table 12.1.

In the table, the stream has increased the depth and width of its channel as the discharge for bank-full stage increased. The bank-full stage increased because of increase in runoff rates caused by urbanizing one-half of the drainage area. This erosion of the channel produces sediments that are moved downstream, where they may be deposited.

TABLE 12.1

Stream flow and urbanization

	Before urbanization	After 50% urbanization
Discharge at bank-full stage	1.5 m³/s	4.0 m³/s
Velocity	0.75 m/s	0.75 m/s
Depth of channel	0.5 m	0.9 m
Width of channel	4.0 m	6.0 m
Area of drainage basin	2.5 km²	2.5 km²

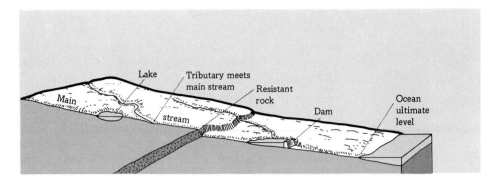

12.17 Base level for a stream may be determined by natural and artificial lakes, by a resistant rock stratum, by the point at which a tributary stream enters a main stream, and by the ocean. Of these, the ocean is considered the ultimate base level; others are temporary base levels.

Changes in vegetative cover Changes in vegetative cover can effect changes in the hydrograph. In a general way, the decrease in the vegetative cover acts in the same direction as does an increase in urbanization.

Studies of the effect of selective forest cutting on stream flow have been carried on by the United States Forest Service in the Fernow Experimental Forest near Parsons, West Virginia. The results, not entirely unexpected, are several: The removal of the trees increases the flow of streams during the growing season chiefly because the water ordinarily transpired into the atmosphere by trees makes its way to the streamways instead. The increase in stream flow is approximately proportional to the percentage of tree cover removed, the greatest increase—over 100 percent—occurring with complete cutting of the forest cover. Concurrently with the increase of stream flow, the flashiness of streams increases. As the tree cover replaces itself, the flow decreases and the flashiness of streams declines, and the return to stream-flow conditions similar to those before cutting is projected to be about 35 years.

BASE LEVEL OF A STREAM

Base level is a key concept in the study of stream activity. The **base level** is defined as the lowest point to which a stream can erode its channel. Anything that prohibits the stream from lowering its channel serves to create a base level. For example, the velocity of a stream is checked when it enters the standing, quiet waters of a lake. Consequently the stream loses its ability to erode, and it cannot cut below the level of the lake. The lake's control over the stream is actually effective along the entire course upstream because no part of the stream can erode beneath the level of the lake—at least before the lake has been destroyed. But in a geological sense, every lake is temporary, and therefore, after the lake has been destroyed (perhaps by the downcutting of its outlet), it will no longer control the stream's base level, and the stream will be free to continue its downward erosion. Because of its impermanence, the base level formed by a lake is referred to

as **temporary.** But even after a stream has been freed from one temporary base level, it will be controlled by others farther downstream; and its erosive power is always influenced by the ocean, which is the **ultimate** base level. Yet, as we shall see in Chapter 15, the ocean itself is subject to changes in level; so even the ultimate base level is not fixed.

The base level of a stream may be controlled not only by lakes but also by layers of resistant rock and the level of the main stream into which a tributary drains (see Figure 12.17).

Adjustment to changing base level We have defined base level as the lowest level to which a stream can erode its channel. If for some reason the base level is either raised or lowered, the stream will adjust the level of its channel to adapt to the new situation.

Let us see what happens when we **raise** the base level of a stream by building a dam and creating a lake across its course. The level of the lake serves as a new base level, and the gradient of the stream above the dam is now less steep than it was originally. As a result, the stream's velocity is reduced. Because the stream can no longer carry all the material supplied to it, it begins to deposit sediments at the point where it enters the lake. As time passes, a new river channel is formed with approximately the same slope as the original channel but at a higher level (Figure 12.18).

What happens when we **lower** the base level by removing the dam and hence the lake? The river will now cut down through the sediments it deposited when the lake still existed. In a short time the profile of the channel will be essentially the same as it was before we began to tamper with the stream.

In general, then, we may say that a stream adjusts itself to a rise in base level by building up its channel through sedimentation and that it adjusts to a fall in base level by eroding its channel downward.

(a)

Lake   Dam

Original stream gradient

(b)

Deposition along stream bed
upstream from lake

New stream
gradient

Deposition
in lake

Channel erosion
downstream
from dam

Old gradient

(c)

Erosion of stream
and lake deposits

Dam and lake
destroyed

Deposition in eroded
section of channel

Original stream gradient restored

12.18 A stream adjusts its channel to the changing base level. Construction of a dam across a stream raises its base level, imposes a lower velocity on the stream above the dam, and thus causes deposition in this section of the channel. Failure of the dam lowers the base level, increases the velocity, and causes erosion of the previously deposited sediments.

12.5 WORK OF RUNNING WATER

The water that flows through river channels does several jobs: (1) It **transports** debris, (2) it **erodes** the river channel deeper into the land, and (3) it **deposits** sediments at various points along the valley or delivers them to lakes or oceans. Running water may help to create a chasm like that of the Grand Canyon of the Colorado, or in flood time it may spread mud and sand across vast expanses of valley flats, or it may build deltas, as at the mouths of the Nile and Mississippi.

The nature and extent of these activities depend on the kinetic energy of the stream, and this in turn depends on the amount of water in the stream and the gradient of the stream channel. A stream expends its energy in several ways. By far the greatest part is used up in the friction of the water with the stream channel and in the friction of water with water in the turbulent eddies we discussed above. Relatively little of the stream's energy remains to erode and transport material. Deposition takes place when energy decreases and the stream can no longer move the material it has been carrying.

TRANSPORTATION

The material that a stream picks up directly from its own channel—or that is supplied to it by slope wash, tributaries, or mass movement—is moved downstream toward its eventual goal, the ocean. The amount of material that a stream carries at any one time, which is called its **load,** is usually less than its **capacity**—that is, the total amount it is capable of carrying under any given set of conditions. The maximum size of particle that a stream can move measures the **competence** of a stream.

There are three ways in which a stream can transport material: (1) solution, (2) suspension, and (3) bed load.

Solution In nature, no water is completely pure. We have already seen that when water falls and filters down into the ground, it dissolves some of the soil's compounds. Then the water may seep down through openings, pores, and crevices in the bedrock and dissolve additional matter as it moves along. Much of this water eventually finds its way to streams at lower levels. The amount of dissolved matter contained in water varies with climate, season, and geologic setting and is measured in terms of parts of dissolved matter per million parts of water. Sometimes the amount of dissolved material exceeds 1,000 ppm, but usually it is much less. By far the most common compounds found in solution in running water, particularly in arid regions, are calcium and magnesium carbonates. In addition, streams carry small amounts of chlorides, nitrates, sulfates, and silica, with perhaps a trace of potassium. It has been estimated that the total load of dissolved material delivered to the seas every year by the streams of the United States is nearly 300 million t. The rivers of the world average an estimated 115 to 120 ppm of dissolved matter, which means that annually they carry to the sea about 3.9 billion t.

Suspension Particles of solid matter that are swept along in the turbulent current of a stream are said to be in **suspension.** This process of transportation is controlled by two factors: the turbulence of the water and a characteristic known as **terminal velocity** of each individual grain. Terminal velocity is the constant rate of fall that a grain eventually attains when the acceleration caused by gravity is balanced by the resistance of the fluid through which the grain is falling. In this case, the fluid is water. If we drop a grain of sand into a quiet pond, it will settle toward the bottom at an ever-increasing rate until the friction of the water on the grain just balances this rate of increase. Thereafter it will settle at a constant rate, its terminal velocity. If we can set up a force that will equal or exceed the terminal velocity of the grain, we can succeed in keeping it in suspension. Turbulent water supplies such a force. The eddies of turbulent water move in a series of orbits, and grains caught in these eddies will be buoyed up, or held in suspension, as long as the velocity of the turbulent water is equal to, or greater than, the terminal velocity of the grains.

Terminal velocity increases with the size of the particle if its general shape and density remain the same. The bigger a particle, the more turbulent the flow needed to keep it in suspension. And because turbulence increases when the velocity of stream flow increases, it follows that the greatest amount of material is moved during flood time, when velocities and turbulence are highest. The graph in Figure 12.19 shows how the suspended load of a stream increases as the discharge increases. In just a few hours or a few days during flood time, a stream transports more material than it does during the much longer periods of low or normal flow. Observations of the area drained by Coon Creek, at Coon Valley, Wisconsin, over a period of 450 days showed that 90 percent of the stream's total suspended load was carried during an interval of 10 days, slightly over 2 percent of total time.

Silt and clay-sized particles are distributed fairly evenly through the depth of a stream, but coarser particles in the sand-size range are carried in greater amounts lower down in the current, in the zone of greatest turbulence.

Bed load Materials in movement along a stream bottom constitute the stream's **bed load,** in contrast to its suspended load and solution load. Because it is difficult to observe and measure the movement of the bed load, we have few data on the subject. Measurements on the Niobrara River near Cody, Nebraska, however, have shown that at discharges between about 6 and 30 m³/s, the bed load averaged about 50 percent of the total load. Particles in the bed load move along in three ways: saltation, rolling, or sliding.

The term **saltation** has nothing to do with salt. It is derived from the Latin *saltare,* "to jump." A particle moving by saltation jumps from one point on the

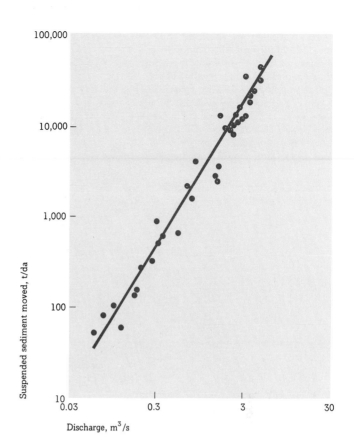

12.19 The suspended load of a stream increases very rapidly during floods, as illustrated by this graph based on measurements in the Rio Puerco near Cabezon, New Mexico. [After Luna B. Leopold and John P. Miller, "Ephemeral Streams—Hydraulic Factors and Their Relation to the Drainage Net," *U.S. Geol. Surv. Prof. Paper 282-A,* p. 11, 1956.]

stream bed to another. First it is picked up by a current or turbulent water and flung upward; then, if it is too heavy to remain in suspension, it drops to the stream floor again at some spot downstream.

Some particles are too large and too heavy to be picked up, even momentarily, by water currents, but they may be pushed along the stream bed, and depending on their shape, they move forward by either **rolling** or **sliding.**

CHANGES IN WATER QUALITY

The most obvious change that we have brought to the rivers of the world is their **pollution,** a change directly caused by the planned (or unplanned) use of the rivers

12.20 Variation in sediment yields from individual watersheds in northern Mississippi under different types of land use and changing amounts of precipitation. [After S. J. Ursic and F. E. Dendy, "Sediment Yields from Small Watersheds under Various Land Uses and Forest Covers," *Proc. Fed. Inter-Agency Sedimentation Conf. 1963, U.S. Dept. Agric. Misc. Pub. 970*, p. 51, 1965.]

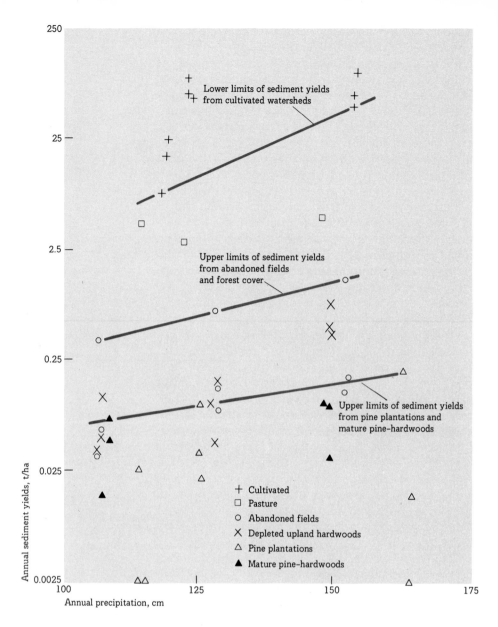

as a sewage system. Increasing technology, industrialization, and population have produced an increasing amount of refuse of all kinds. Where rivers are available, they become a handy and, at first, inexpensive way to remove our cultural refuse. This load has proved more than many rivers can handle and still maintain their original quality.

Solid load Agriculture, more than any other human activity, increases the rate at which sediments are delivered to the streams. It is true that construction activities produce a higher rate of <u>sediment production,</u> but

Amount of solids in water

these are isolated events; over the long run the continuous use of land for farming is volumetrically more important in sediment production.

There is a great deal of information showing the effect of the clearing of land and of farming practices on the production of sediments. For instance, one of the effects of forest clearance in the Fernow Experimental Forest referred to earlier in this chapter was to increase the solids delivered to the stream. With complete commercial cutting of the watershed, the stream's solid load was as much as 3,700 times that of similar streams in areas in which no lumbering occurred. In Mississippi, studies on sediment production from land under differing uses produced the startling conclusions shown in Figure 12.20. Land under cultivation produced a thousand times more sediment than did similar land under mature stands of trees.

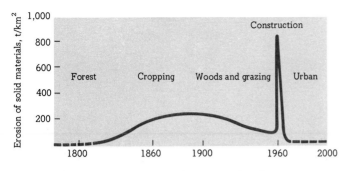

12.21 A sequence of land-use changes and sediment yield beginning prior to the advent of extensive farming and continuing through a period of construction and subsequent urban landscape. Based on experience in the Middle Atlantic region of the United States. [After M. G. Wolman, "A Cycle of Sedimentation and Erosion in Urban River Channels," *Geograf. Ann.*, vol. 49-A, p. 386, 1967.]

M. Gordon Wolman has shown the effect of changing land use on the production of sediments in the Piedmont area of Maryland from about 1800 to the present (Figure 12.21). Interestingly, urbanization has reduced sediment production to the preagricultural levels.

The increased sediments move on to the streams, which then must handle them as best they can. The results vary with the stream. In general, however, increased solid loads increase the turbidity or muddiness of a stream, decrease the depth of light penetration, and can cause silting up of valley bottoms.

To agriculture, deforestation, urban construction, and grazing we must add mining and industry as producers of excess sediments. These activities may bring other undesirable features with them in addition to increased sediment. For instance, the tailings of many coal mines contain large amounts of sulfur, which can convert to undesirable compounds like sulfuric acid.

Dissolved load We found earlier that the content of dissolved material carried by streams is 120 ppm on a worldwide average. Most of this comes from the chemical decomposition of earth materials during the normal processes of weathering. Some is recycled from the oceans through the atmosphere to the streams, and some is contributed by human activities. Figure 12.22 illustrates the increase of dissolved solids in the Passaic River at Little Falls, New Jersey, from 1948 through 1963, a period of increasing industrialization and urbanization.

It is not, however, just the increase of the dissolved load itself but the various factors involved in the increase that cause problems. That part of the **dissolved load** contributed by people covers a wide range of materials. Some are toxic to river life, some serve as nutrients promoting explosive organic proliferation, and a few are merely an affront to human sensibilities.

12.22 Increase in dissolved solids in relation to discharge in the Passaic River at Little Falls, New Jersey, from 1948 to 1963. [After Peter W. Anderson and S. D. Faust, "Quality of Water in the Passaic River at Little Falls, N.J., as Shown by Long-Range Data," *U.S. Geol. Surv. Prof. Paper 525-D*, p. 215, 1965.]

Among the most bothersome of the stream pollutants are the phosphorus and nitrogen that come from fertilizers, and sewage both treated and raw. These additions to streams spark a complicated set of reactions in the chemistry, biologic processes, and ecology of the stream, all usually undesirable (from the human point of view) and all leading to a stream of a different character than it was in its natural state. The list of chemical pollutants is as varied as the products of our industrial society. They are present in the streams because we continue to use the channels of water as a sewer system. Until we adopt other ways of cleaning our nests, the streams will continue to be polluted.

We shall see later that these same elements are critical in the quality of lake water as well.

Thermal pollution The use of river water for industrial-cooling purposes, particularly by electric power plants, has led us to recognize that the discharge of waste heat into our streams can lead to unwanted effects on the aquatic life. We denote this human activity by the term **thermal pollution**, a problem faced not only by rivers but by lakes, underground water, and shallow seawaters as well (see Figure 12.23).

Industrial cooling is a dramatic way to change the temperature of river water. But man can bring about these changes in another way. Studies of streams in urbanized areas on Long Island have been carried out

12.23 This aerial thermal-infrared image of the shore of Lake Ontario just north of Rochester, New York, shows warm water from the land in light tones, mixing with the cooler water of Lake Ontario shown in dark shades. The Genesee River, artificially contained part way into the lake, is at the right. A smaller stream, Slater Creek, discharges directly into the lake to the left. [U.S. Geological Survey.]

by E. J. Pluhowski, of the Johns Hopkins University. Comparing streams in urbanized zones with a control stream still in an undeveloped area, Pluhowski found that the streams most affected by human activities had summer temperatures from 5.5°C to 8°C above those of the control stream. Conversely, stretches of streams affected by human activities were found to be 2.2°C to 5.5°C colder in winter than were those unaffected.

Some of these variations are due to the nature of runoff between unurbanized and urbanized zones. During the summer months, runoff from heated pavements can raise the temperature of stream water appreciably, as much as 6°C as the result of a single storm, whereas an unurbanized stream shows little variation during the storm and subsequent runoff. Streams in urbanized zones are fed more by surface runoff than are streams in unurbanized zones, which depend more heavily on groundwater for their flow. In winter the surface runoff is colder than the groundwater, which tends to reflect the mean annual temperature. The stream in an urbanized area, therefore, is colder during the winter than is the stream in an unurbanized area.

EROSION

People have not always been convinced that most of the world's stream valleys are fashioned by the streams that flow in them, and by the processes that streams have encouraged. Today, however, we know enough about running water to confirm this. In the next few paragraphs we look at the various ways a stream may remove material from its channel and its banks.

But you may wish, as you continue this chapter, to keep in mind the larger question and consider what added evidence we might cite to demonstrate that a stream is responsible for its own valley.

Direct lifting In turbulent flow, as we have seen, water travels along paths that are not parallel to the bed. The water eddies and whirls, and if an eddy is powerful enough, it dislodges particles from the stream channel and lifts them into the stream. Whether or not this will happen in a given situation depends on a number of variables that are difficult to measure, but if we assume that the bed of a stream is composed of particles of uniform size, then the graph in Figure 12.24 gives us the approximate stream velocities that are needed to erode particles of various sizes, such as clay, silt, sand, granules, and pebbles. A stream bed composed of fine-sized sand grains, for example, can be eroded by a stream with a velocity of between 18 and 50 cm/s, depending on how firmly the sand grains are packed. As the fragments become larger and larger, ranging from coarse sand to granules to pebbles, increasingly higher velocities are required to move them, as we should expect.

But what we might *not* expect is that, as the particles become smaller than about 0.06 mm in diameter, they do not become more easily picked up by the stream. In fact, if the clay and silt are firmly consolidated, then increased velocities will be needed to erode the particles. The reason for this is generally that the smaller the particle, the more firmly packed the deposit is and thus the more resistant to erosion. Moreover, the individual particles may be so small that they do not project sufficiently high into the stream to be swept up by the turbulent water.

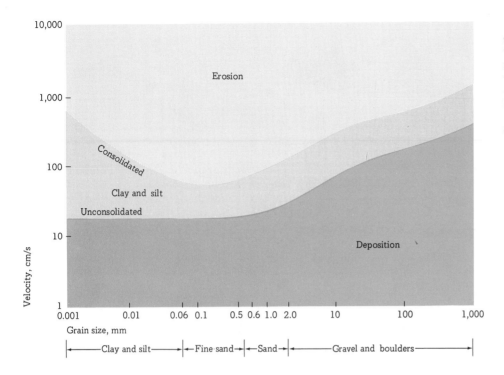

Velocity, cm/s

Grain size, mm

Clay and silt — Fine sand — Sand — Gravel and boulders

12.24 The central band in this diagram shows the velocity range at which turbulent water will lift particles of differing size off the stream bed. The width is affected by the shape, density, and consolidation of the particles. [After Ake Sundborg, "The River Klarälven, a Study in Fluvial Processes," *Geografis. Ann.*, vol. 38, p. 197, 1956.]

Abrasion, impact, and solution The solid particles carried by a stream may themselves act as erosive agents, for they are capable of abrading (wearing down) the bedrock itself or larger fragments in the bed of the stream. When the bedrock is worn by abrasion, it usually develops a series of smooth, curving surfaces, either convex or concave (Figure 12.25). Individual cobbles or pebbles on a stream bottom are sometimes moved and rolled about by the force of the current, and as they rub together, they become both rounder and smoother.

The impact of large particles against the bedrock

12.25 The solid material carried by a stream may erode the bedrock of the stream channel. Abrasion has fashioned an irregular channel in the bedrock of this small stream north of Rome, Italy. [Sheldon Judson.]

12.26 Cross-sectional sketches of typical stream valleys. The major features of valleys in cross section include divides, valley walls, river channel, and in some instances a flood plain. Divides may be flat-topped, broadly rounded, or narrow.

12.27 Big or small, wet or dry, every stream channel has its own drainage basin. Here, northwest of St. Anthony, Idaho, these upland slopes and channels form a network of drainage basins. [Austin Post, University of Washington.]

or against other particles knocks off fragments, which are added to the load of the stream.

Some erosion also results from the solution of channel debris and bedrock in the water of the stream. Most of the dissolved matter carried by a stream, however, is probably contributed by the underground water that drains into it.

DEPOSITION

As soon as the velocity of a stream falls below the point necessary to hold material in suspension, the stream begins to deposit its suspended load. Deposition is a selective process. First, the coarsest material is dropped; then, as the velocity (and hence the energy) continues to slacken, finer and finer material settles out. We shall reconsider stream deposits later.

12.6 FEATURES OF VALLEYS

CROSS-VALLEY PROFILES

Earlier in this chapter we mentioned the longitudinal profile of a stream. Now let us turn to a discussion of the cross-valley profile, that is, the profile of a cross section at right angles to the trend of the stream's valley. In Figure 12.26, notice that the channel of the river runs across a broad, relatively flat flood plain. During flood time, when the channel can no longer accommodate the increased discharge, the stream overflows its banks and floods this area. On either side of the flood plain, valley walls rise to crests called divides, separations between the central valley and the other valleys on either side. In Figure 12.26(b) no flood plain is present because the valley walls descend directly to the banks of the river. This diagram also illustrates two different shapes of divide. One is broad and flat; the other is narrow, almost knife-edged. Both are in contrast to the broadly convex divides shown in Figure 12.26(a).

DRAINAGE BASINS, NETWORKS, AND PATTERNS

A drainage basin is the entire area from which a stream and its tributaries receive their water. The Mississippi River and its tributaries drain a tremendous section of the central United States, reaching from the Rocky to the Appalachian Mountains, and each tributary of the Mississippi has its own drainage area, which forms a part of the larger basin. Every stream, even the smallest brook, has its own drainage basin (Figure 12.27). These basins are shaped differently

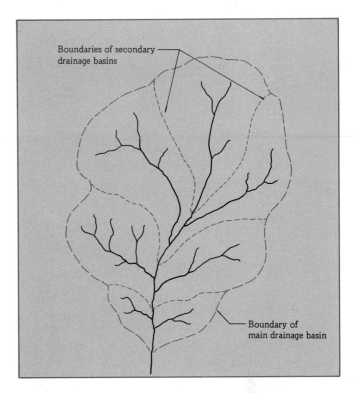

12.28 The stream's drainage basin is the area from which the stream and its tributaries receive water. This basin displays a pattern reminiscent of a tree leaf and its veins.

from stream to stream but are characteristically pear-shaped, with the main stream emerging from the narrow end (see Figure 12.28).

Individual streams and their valleys are joined together into networks. In any single network, the streams prove to have a definite geometric relationship in a way first detailed in 1945 by the United States hydraulic engineer Robert Horton. We can devise a demonstration of this relationship by first ranking the streams in a hierarchy. This ranking, shown in Figure 12.29, lists small headwater streams without tributaries as belonging to the first order in the hierarchy. Two or more first-order streams join to form a second-order stream; then two or more second-order streams join to form a third-order stream; and so on. In other words, a stream segment of any given order is formed by the junction of at least two stream segments of the next lower order. The main stream segment of the system always has the highest order number in the network, and the number of this stream is assigned to describe the stream basin. Thus the main stream in the basin shown in Figure 12.29 is a fourth-order stream, and the basin is a fourth-order basin.

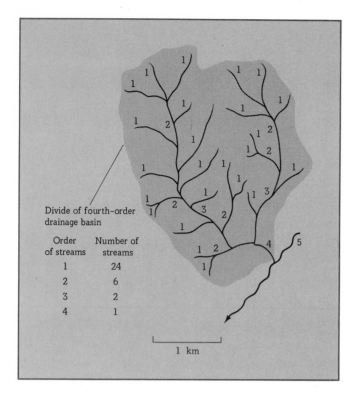

Divide of fourth-order
drainage basin

Order of streams	Number of streams
1	24
2	6
3	2
4	1

1 km

12.29 Method of designating stream orders. [After A. N. Strahler, "Quantitative Geomorphology in Erosional Landscapes," *Proc. Intern. Geol. Congr.*, sec. 13, pt. 3, p. 344, 1954.]

It is apparent from Figure 12.29 that the number of stream segments of different order decreases with increasing order. When we plot the number of streams of a given order against order, the points define a straight line on semilogarithmic paper; see Figure 12.30(a). Likewise, if we plot order against the average length of streams of a given order or plot order against areas of drainage basins of a particular order, we find a well-defined relationship; see Figure 12.30(b) and (c). Using these relationships, we can describe the river channels and basins of the United States as in Table 12.2.

We have seen that when streams are arranged in a hierarchical fashion, the arrangement bears a definite relation to the number and lengths of the stream segments in each order and to the size of the basins of the stream segments of different orders. Beyond this we find that streams develop patterns not as easily quantified as are the relationships of stream segments and stream basins based on hierarchial arrangements. We describe some of these below.

The overall pattern developed by a system of streams and tributaries depends partly on the nature of

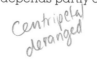

the underlying rocks and partly on the history of the streams (see Figure 12.31). Almost all streams follow a branching pattern in the sense that they receive tributaries; the tributaries, in turn, are joined by still smaller tributaries, but the manner of branching varies widely.

A stream that resembles the branching habit of a maple, oak, or similar deciduous tree is called **dendritic,** "treelike." A dendritic pattern develops when the underlying bedrock is uniform in its resistance to erosion and exercises no control over the direction of valley growth. This situation occurs when the bedrock is composed either of flat-lying sedimentary rocks or massive igneous or metamorphic rocks. The streams can cut as easily in one place as another; thus the dendritic pattern is, in a sense, the result of the random orientation of the streams.

Another type of stream pattern is **radial:** Streams radiate outward in all directions from a high central zone. Such a pattern is likely to develop on the flanks of a newly formed volcano where the streams and their valleys radiate outward and downward from various points around the cone.

A **rectangular** pattern occurs when the underlying bedrock is crisscrossed by fractures that form zones of weakness particularly vulnerable to erosion. The master stream and its tributaries then follow courses marked by nearly right-angle bends.

Some streams, particularly in a belt of the Appalachian Mountains running from New York to Alabama, follow what is known as a **trellis** pattern. This pattern, like the rectangular one, is caused by zones in the bedrock that differ in their resistance to erosion. The trellis pattern usually, though not always, indicates that the region is underlain by alternate bands of resistant and nonresistant rock.

Water gaps Particularly intriguing features of some valleys are short, narrow segments walled by steep, rocky slopes or cliffs. These are called **water gaps.** The river, in effect, flows through a narrow notch in a ridge or mountain that lies across its course, as shown in Figure 12.32. Here is one way water gaps can form: Picture an area of hills and valleys carved by differential erosion from rocks of varying resistance. Then imagine sediments covering this landscape so that the valleys and hills are buried beneath the debris. Any streams that flow over the surface of this cover will establish their own courses across the buried hills and valleys. When these streams erode downward, some of them may encounter an old ridge crest. If such a stream has sufficient erosional energy, it may cut down through the resistant rock of the hill. The course of the stream is thus superimposed across the old hill, and we call the stream a **superimposed stream.** Differential erosion may excavate the sedimentary fill from the old valleys, but the main superimposed stream

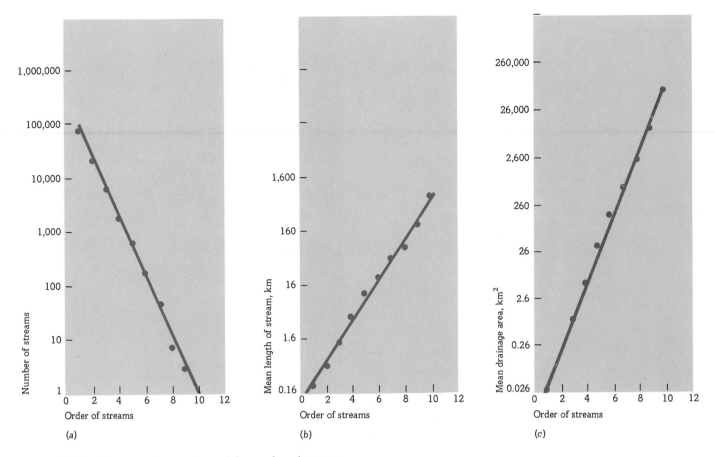

12.30 Relation between stream order and the number of streams, the mean length of streams, and the mean drainage area of streams of a location in central Pennsylvania. The system used to designate stream orders is that proposed by Robert Horton in 1945 and differs slightly from that illustrated in Figure 10.20. [After Lucien Brush, Jr., *U.S. Geol. Surv. Prof. Paper 282-F*, 1961.]

TABLE 12.2

Number and length of river channels of various sizes in the United States[a]

Order	Number	Av. length, km	Total length, km	Mean drainage area, incl. tributaries, km²	River representative of size
1[b]	1,570,000	1.6	2,512,000	2.6	
2	350,000	3.7	1,295,000	12	
3	80,000	8.5	680,000	59	
4	18,000	19	342,000	283	
5	4,200	45	189,000	1,348	Charles
6	950	102	97,000	6,400	Raritan
7	200	219	44,000	30,300	Allegheny
8	41	541	22,000	144,000	Gila
9	8	1,243	10,000	683,000	Columbia
10	1	2,880	2,880	3,238,000	Mississippi

[a]From Luna B. Leopold, "Rivers," *Am. Sci.*, vol. 50, p. 512, 1962.
[b]The size of the order 1 channel depends on the scale of the maps used; these order numbers are based on the determination of the smallest order using maps of scale 1:62,500.

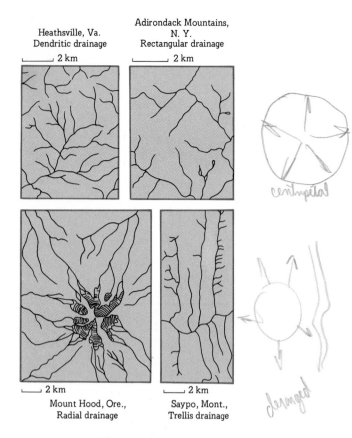

Heathsville, Va.
Dendritic drainage

⊢———⟶ 2 km

Adirondack Mountains,
N. Y.
Rectangular drainage

⊢———⟶ 2 km

centripetal

⊢———⟶ 2 km
Mount Hood, Ore.,
Radial drainage

⊢———⟶ 2 km
Saypo, Mont.,
Trellis drainage

deranged

12.31 The overall pattern developed by a stream system depends in part on the nature of the bedrock and in part on the history of the area. See the text for discussion.

12.32 An aerial photograph of a water gap in North Africa. The master stream cuts through the ridge of resistant rock. [U.S. Army Air Corps.]

flows through the hill in a new, narrow gorge, or water gap. An example of such a gap is that followed by the Big Horn River through the northern end of the Big Horn Mountains in Montana.

ENLARGEMENT OF VALLEYS

We cannot say with assurance how running water first fashioned the great valleys and drainage basins of the continents because the record has been lost in time. But we do know that certain processes are now at work in widening and deepening valleys, and it seems safe to assume that they also operated in the past.

If a stream were left to itself in its attempt to reach base level, it would erode its bed straight downward, forming a vertically walled chasm in the process. But because the stream is not the only agent at work in valley formation, the walls of most valleys slope upward and outward from the valley floor. In time, the cliffs of even the steepest gorge will be angled away from the axis of its valley.

As a stream cuts downward and lowers its chan-

nel into the land surface, weathering, slope wash, and mass movement come into play, constantly wearing away the valley walls and pushing them farther back (see Figure 12.33). Under the influence of gravity, material is carried down from the valley walls and dumped into the stream, to be moved toward the seas. The result is a valley whose walls flare outward and upward from the stream in a typical **cross-valley profile** (see Figure 12.34).

The rate at which valley walls are reduced and the angles that they assume depend on several factors. If the walls are made up of unconsolidated material that is vulnerable to erosion and mass movement, the rate will be rapid; but if the walls are composed of resistant rock, the rate of erosion will be very slow, and the walls may rise almost vertically from the valley floor (see Figure 12.35).

In addition to cutting downward into its channel, a stream also cuts from side to side, or laterally, into its banks. In the early stages of valley enlargement, when the stream is still far above its base level, downward erosion is dominant. Later, as the stream cuts its channel closer and closer to base level, downward erosion becomes progressively less important. Now a larger percentage of the stream's energy is directed toward eroding its banks. As the stream swings back and forth, it forms an ever-widening flood plain on the valley floor, and the valley itself broadens.

How fast do these processes of valley formation proceed? In some instances we can measure the ages

12.33 This gully in Wisconsin has been formed by the downward cutting of the stream combined with the slope wash and mass movement on the gully walls. [U.S. Forest Service.]

(a)

(b)

12.34 If a stream of water were the only agent in valley formation, we might expect a vertically walled valley no wider than the stream channel, as suggested by the colored lines in (a) and (b). Mass movement and slope wash, however, are constantly wearing away the valley walls, carving slopes that flare upward and away from the stream channel, as shown in the diagrams.

of gully formation in years and even months, but for large valleys we must be satisfied with rough approximations—and usually with plain guesses. For instance, one place where we can make an approximation of the rate of valley formation is the Grand Canyon of the Colorado in Arizona. Several lines of evidence suggest that it has taken something on the order of 10 million years for the Colorado to cut its channel downward about 1,800 m and expand its valley walls 6 to 20 km.

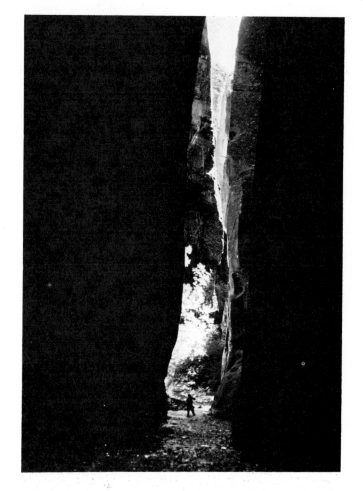

12.35 Erosion by the Virgin River has fashioned the vertically walled Virgin River Narrows in massive sandstone beds in the northwestern part of Zion National Park. Mass movement and other aspects of slope modification have yet to lower the angle of the valley walls. [National Park Service.]

12.6 FEATURES OF VALLEYS

FEATURES OF NARROW VALLEYS

Waterfalls are among the most fascinating spectacles of the landscape. Thunderous and powerful as they are, however, they are actually short-lived features in the history of a stream. They owe their existence to some sudden drop in the river's longitudinal profile—a drop that may be eliminated with the passing of time.

Waterfalls are caused by many different conditions. Niagara Falls, for instance, are held up by a relatively resistant bed of dolomite underlain by beds of nonresistant shale (see Figure 12.36). This shale is easily undermined by the swirling waters of the Niagara River as they plunge over the lip of the falls. When the undermining has progressed far enough, the dolomite collapses and tumbles to the base of the falls. The same process is repeated over and over again as time passes, and the falls slowly retreat upstream. Historical records suggest that the Horseshoe, or Canadian, Falls (by far the larger of the two falls at Niagara) have been retreating at a rate of 1.2 to 1.5 m/year, whereas the smaller American falls have been eroded away 5 to 6 cm/year. The 11 km of gorge between the foot of the falls and Lake Ontario are evidence of the headward retreat of the falls through time.

Yosemite Falls in Yosemite National Park, California, plunge 770 m over the Upper Falls, down an inter-

12.37 Rapids may represent a stage in the destruction of waterfalls, as suggested in this diagram.

mediate zone of cascades, and then over the Lower Falls. The falls leap from the mouth of a small valley high above the main valley of the Yosemite. The Upper Falls alone measure 430 m, nine times the height of Niagara. During the Ice Age, glaciers scoured the main valley much deeper than they did the unglaciated side valley. Then, when the glacier ice melted, the main valley was left far below its tributary, which now joins it after a drop of nearly a kilometer.

Rapids, like waterfalls, occur at a sudden drop in the stream channel. Although rapids do not plunge straight down as waterfalls do, the underlying cause of formation is often the same. In fact, many rapids have developed directly from preexisting waterfalls (Figure 12.37).

12.36 Niagara Falls tumble over a bed of dolomite, underlain chiefly by shale. As the less-resistant bed is eroded, the undermined ledge of dolomite breaks off, and the lip of the falls retreats.

FEATURES OF BROAD VALLEYS

If conditions permit, the various agents working toward valley enlargement ultimately produce a broad valley with a wide level floor. During periods of normal or low water, the river running through the valley is confined to its channel; but during high water, it overflows its banks and spreads out over the flood plain.

A flood plain that has been created by the lateral erosion and the gradual retreat of the valley walls is called an **erosional flood plain** and is characterized

12.38 The Laramie River in southeastern Wyoming meanders across its flood plain. Continuing migration of the meanders has left behind the scroll-like patterns marking former positions of the stream channel. [James R. Balsley, U.S. Geological Survey.]

by a thin cover of gravel, sand, and silt a few meters or a few tens of meters in thickness. On the other hand, the floors of many broad valleys are underlain by deposits of gravel, sand, and silt scores of meters thick. These thick deposits are laid down as changing conditions force the river to drop its load across the valley floor. Such a flood plain, formed by the building up of the valley floor, or aggradation, is called a **flood plain of aggradation.** Flood plains of aggradation are much more common than erosional flood plains and are found in the lower reaches of the Mississippi, Nile, Rhône, and Yellow Rivers, to name but a few.

Both erosional flood plains and flood plains of aggradation exhibit the following characteristics.

Meanders The channel of the Menderes River in Asia Minor curves back on itself in a series of broad hairpin bends, and today all such bends are called **meanders** (Figure 12.38).

Both erosion and deposition are involved in the formation of a meander. First, some obstruction swings the current of a stream against one of the banks, and the current is deflected over to the opposite bank. Next, erosion takes place on the outside of each bend, where the turbulence is greatest. The material detached from the banks is then moved downstream, there to be deposited in zones of decreased turbulence—either along the center of the channel or on the inside of the next bend. As the river swings from side to side, the meander continues to grow by erosion on the outside of the bends and by deposition on the inside.

Because a meander is eroded more on its downstream side than on its upstream side, the entire bend tends to move slowly down-valley. This movement is not uniform, however, and under certain conditions the downstream sweep of a series of meanders is distorted into cutoffs, meander scars, and oxbow lakes.

In its down-valley migration, a meander sometimes runs into a stretch of land that is relatively more resistant to erosion. But the next meander upstream continues to move right along, and gradually the neck between them is narrowed. Finally the river cuts a new, shorter channel, called a **neck cutoff,** across the neck. The abandoned meander is called an **oxbow** because of its characteristic shape. Usually both ends of the oxbow are gradually silted in, and the old meander becomes completely isolated from the new channel. If the abandoned meander fills up with water, an **oxbow lake** results. Although a cutoff will eliminate a particular meander, the stream's tendency toward meandering still exists, and before long the entire process begins to repeat itself.

We found that a meander grows and migrates by erosion on the outside of the bend and by deposition on the inside. This deposition on the inside leaves behind a series of low ridges and troughs. Swamps often form in the troughs, and during flood time the river may

12.6 FEATURES OF VALLEYS

279

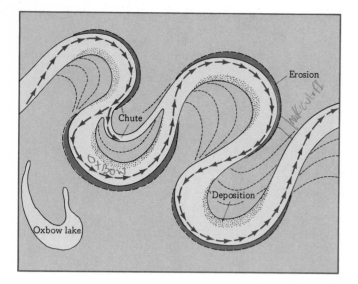

12.39 Erosion takes place on the outside of a meander bend, whereas deposition is most marked on the inside. If the neck of a meander is eroded through, an oxbow forms. A chute originates along the inside of a meander where irregular deposition creates ridges and troughs as the meander migrates.

develop an alternate channel through one of the troughs. Such a channel is called a **chute cutoff**, or simply a **chute** (see Figure 12.39).

The meandering river demonstrates a unity in ways other than the balance of erosion and deposition. The length of a meander, for example, is proportional to the width of the river, and this is true regardless of the size of the river. It holds for channels a few meters wide as well as those as large as that of the Mississippi River. As shown in Figure 12.40(*a*), this principle is also true of the Gulf Stream even though this "river" is unconfined by solid banks. A similar relationship holds between the length of the meander and the radius of curvature of the meander; see Figure 12.40(*b*). These relationships seem to be controlled largely by the discharge and the sediments carried. Increasing discharge causes an increase in the size of characteristics such as width, length, and wavelength. For a given discharge, the wavelength tends to be greater for streams carrying a high proportion of sand and gravel in their loads than for those transporting mainly fine sands.

Braided streams On some flood plains, particularly where large amounts of debris are dropped rapidly, a stream may build up a complex tangle of converging and diverging channels separated by sandbars or islands. A stream of this sort is called **braided.** The pattern generally occurs when the discharge is highly variable and the banks are easily eroded to supply a heavy load. It is characteristic of alluvial fans, glacial-outwash deposits, and certain heavily laden rivers.

In general, the gradient of a braided stream is higher than that for a meandering stream of the same discharge, as indicated in Figure 12.41. This is apparently indicative of a tendency of a stream to increase its efficiency in order to transport proportionally larger loads. Thus, if a stream with a single channel divides into two channels, the new channels will have a higher gradient and a greater total width than the original channel carrying the same amount of water; and turbulence increases. The two new channels can now carry a greater load, particularly bed load, than could the single-channeled stream.

Natural levees In many flood plains, the water surface of the stream is held above the level of the valley floor by banks of sand and silt known as **natural levees,** a name derived from the French *lever,* "to raise." These banks slope gently, almost imperceptibly, away from their crest along the river toward the valley wall. The low-lying flood plain adjoining a natural levee may contain marshy areas known as **back swamps.** Levees are built up during flood time, when the water spills over the river banks onto the flood plain. Because the muddy water rising over the stream bank is no longer confined by the channel, its velocity and turbulence drop immediately, and much of the suspended load is deposited close to the river; but some is carried farther along to be deposited across the flood plain. The deposit of one flood is a thin wedge tapering away from the river, but over many years the cumulative effect produces a natural levee that is considerably higher beside the river bank than away from it (see Figure 12.42). On the Mississippi delta, for instance, the levees stand 5 to 6 m above the back swamps. Although natural levees tend to confine a stream within its channel, each time the levees are raised slightly, the bed of the river is also raised. In time, the level of the bed is raised above the level of the surrounding flood plain. If the river manages to escape from its confining walls during a flood, it will assume a new channel across the lowest parts of the flood plain toward the back swamps.

A tributary stream entering a river valley with high levees may be unable to find its way directly into the main channel; so it will flow down the back-swamp zone and may run parallel to the main stream for many kilometers before finding an entrance. Because the Yazoo River typifies this situation by running 320 km parallel to the Mississippi, all rivers following similar courses are known as **yazoo-type** rivers.

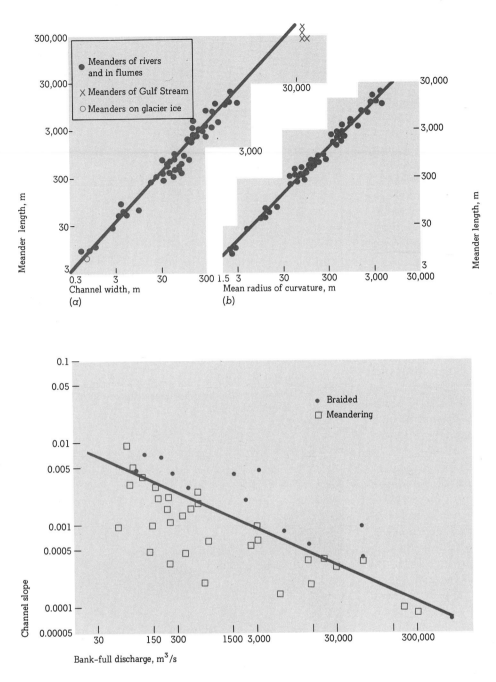

12.40 (a) Length of the meander increases with the widening, meandering stream. (b) A similar orderly relationship exists between length of the meander and the mean radius of curvature of the meander. [After Luna B. Leopold and M. Gordon Wolman, "River Meanders," *Geol. Soc. Am. Bull. 71*, p. 773, 1960.]

12.41 Meandering streams have lower gradients than do braided streams of the same discharge. In this diagram, each symbol represents measurements at a single point on a stream.

12.42 Natural levees characterize many aggrading streams. They build up during periods of flood as coarser material is deposited closest to the stream channel to form the levee and finer material is deposited in the back swamps. As the banks build up, the floor of the channel also rises.

Flood plains and their deposits Virtually all streams are bordered by flood plains. These may range from very narrow (a few meters wide) to the flood plains whose widths are measured in kilometers, as is that of the lower Mississippi River. The flood plain appears very flat, particularly when viewed in relation to the slopes of the valley walls that flank it. But the plain is not without its ups and downs. A large flood plain may

12.6 FEATURES OF VALLEYS

have differences of relief of several meters and be marked by such features as natural levees, meander scars, and oxbow lakes.

The flood plain is made up of stream-carried sediments. Two general categories of deposit are common. One of these is made up of fine silts, clays, and sands, which may be spread across the plain by a river that overflows its banks during flood. These are called **overbank deposits.** The other group of deposits is made of coarse material, gravel and sand, and is related directly to the channel of the stream. These deposits include chiefly the material that is deposited in the slack water on the inside of the bends of a winding or meandering river. These bars of sand and gravel are called **point bars.**

We found, in discussing meanders, that the loops of a channel migrate laterally so that the river swings both across the valley and down-valley. As the river migrates laterally, it leaves behind coarse deposits in point bars and in the deeper portions of the abandoned channel. While erosion takes place on one bank of the river, deposition of sediments from upstream takes place on the point bars and, during flood, by overbank deposits—the river thus builds its flood plain in some places and simultaneously destroys it in others. The flood plain becomes a temporary storage place for river sediments.

We can view the flood plain, then, as a depositional feature more or less in equilibrium. This equilibrium may be disturbed in several ways. The stream may lose some of its ability to erode and carry sediment. A net gain of deposits will occur, and the flood plain will build up. On the other hand, the flow of water or the gradient may increase or the supply of sediments may decrease, and there will be a net loss of sediments as erosion begins to destroy the previously developed flood plain.

Deltas and alluvial fans For centuries, the Nile River has been depositing sediments as it empties into the Mediterranean Sea, forming a great triangular plain with its apex upstream. This plain came to be called a **delta** because its shape is similar to the Greek letter Δ (see Figures 12.43 and 12.44).

Whenever a stream flows into a body of standing water, such as a lake or an ocean, its velocity and transporting power are quickly stemmed. If it carries enough debris and if conditions in the body of standing water are favorable, a delta will gradually be built. An ideal delta is triangular in plan, with the apex pointed upstream and with the sediments arranged according

12.43 The delta (darker area) of the Saint Clair River in Lake Saint Clair has the classic shape as well as its distributary channels.

12.44 In this picture taken from *Gemini 4,* the delta of the Nile stands out because it is well vegetated and thus contrasts with the desert country that borders it. [NASA.]

to a definite pattern. The coarse material is dumped first, forming a series of dipping beds called **foreset beds.** But the finer material is swept farther along to settle across the sea or lake floor as **bottomset beds.** As the delta extends farther and farther out into the water body, the stream must extend its channel to the edge of the delta. As it does so, it covers the delta with **topset beds,** which lie across the top of the foreset beds (see Figure 12.45).

Very few deltas, however, show either the perfect delta shape or this regular sequence of sediments. Many factors, including lake and shore currents, varying rates of deposition, the settling of delta deposits as a result of their compaction, and the downwarping of the earth's crust—all conspire to modify the typical form and sequence.

Across the top of the delta deposits, the stream spreads seaward in a complex of channels radiating from the apex. These **distributary channels** shift their position from time to time as they seek more favorable gradients.

Deltas are characteristic of many of the larger rivers of the world, including the Nile, Mississippi, Ganges, Rhine, and Rhône. On the other hand, many rivers have no deltas either because the deposited material is swept away as soon as it is dumped or because the streams do not carry sufficient detrital material to build up a delta.

An **alluvial fan** is the land counterpart of a delta. These fans are typical of arid and semiarid climates, but they may form in almost any climate if conditions are right. A fan marks a sudden decrease in the carrying power of a stream as it descends from a steep gradient to a flatter one—for example, when the stream flows down a steep mountain slope onto a plain. As the velocity is checked, the stream rapidly begins to dump its load. In the process it builds up its channel, often with small natural levees along its banks. Eventually, as the levees continue to grow, the stream may flow above the general level. Then during a time of flood, it seeks a lower level and shifts its channel to begin deposition elsewhere. As this process of shifting continues, an alluvial fan proceeds to build up (see Figure 12.46).

Stream terraces A **stream terrace** is a relatively flat surface running along a valley, with a steep bank sep-

12.46 Alluvial fans in Death Valley, California. Varying sizes of fans relate to the varying sizes of the drainage basins providing material to the fans. The larger the drainage basin, the larger the fan. The largest fan shown here has a radius of about 2.5 km. [Fairchild Aerial Photograph Collection, Whittier College.]

Bajada– series of Alluvial fans

arating it either from the flood plain or from a lower terrace. It is a remnant of the former channel of a stream that now has succeeded in cutting its way down to a lower level (see Figure 12.47).

The so-called **cut-and-fill terrace** is created when a stream first clogs a valley with sediments and then cuts its way down to a lower level (Figure 12.48). The initial aggradation may be caused by a change in climate that leads either to an increase in the stream's load or to a decrease in its discharge. Or the base level of the stream may rise, reducing the gradient and

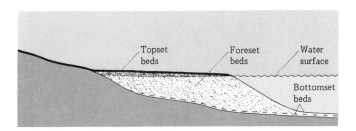

12.45 The ideal arrangement of sediments beneath a delta. Some material deposited in a lake or sea is laid on the bottom as bottomset beds. Other material is dumped in inclined foreset beds, built farther into the water and partly covering the bottomset beds. Over the foreset beds the stream lays down topset beds.

12.47 Terraces along the Ain River in the Jura Mountains of eastern France. [Sheldon Judson.]

12.48 One example of the formation of paired terraces. (a) The stream has partially filled its valley and has created a broad flood plain. (b) Some change in conditions has caused the stream to erode into its own deposits; the remnants of the old flood plain stand above the new river level as terraces of equal height. This particular example is referred to as a cut-and-fill terrace. (See also Figure 10.49.)

Flood plain

(a)

Paired terraces

(b)

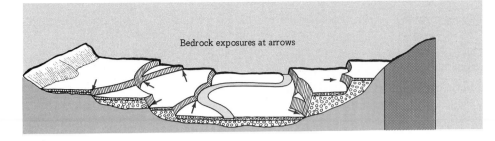

Bedrock exposures at arrows

12.49 Unpaired terraces do not match across the stream that separates them. Here is one way in which they may form. The stream has cut through unconsolidated deposits within the valley. As it eroded downward, it also swept laterally across the valley and created a sloping surface. Lateral migration was stopped locally when the stream encountered resistant bedrock (see arrows) beneath the softer valley fill. This bedrock not only deflected the stream back across the valley but also protected remnants of the valley fill from further stream erosion and allowed them to be preserved as terraces. Because they are portions of a surface sloping across the valley, however, no single remnant matches any other on the opposite side of the river.

causing deposition. In any event, the stream chokes the valley with sediment, and the flood plain gradually rises. Now, if the equilibrium is upset and the stream begins to erode, it will cut a channel down through the deposits it has already laid down. The level of flow will be lower than the old flood plain, and at this lower level the stream will begin to carve out a new flood plain. As time passes, remnants of the old flood plain, terraces, may be left standing on either side of the new one. Terraces that face each other across the stream at the same elevation are referred to as **paired terraces.** Sometimes the downward erosion by streams creates **unpaired terraces.** If the stream swings back and forth across the valley, slowly eroding as it moves, it may encounter resistant rock beneath the unconsolidated deposits; the exposed rock will then deflect the stream and prevent further erosion, and a single terrace is left behind, with no corresponding terrace on the other side of the stream (see Figure 12.49).

Terraces, either paired or unpaired, may be cut into bedrock as well. A thin layer of sand and gravel usually rests on the beveled bedrock of these terraces.

12.7 OF RIVERS AND CONTINENTS

So far we have been discussing some of the details of streams and stream action. In concluding this chapter, we shall take a more global look at some aspects of running water.

Of all the processes at work to move material from the continents to the ocean basins, the most important is running water. Like great conveyer belts, the rivers of the world move the weathered waste of the land to the oceans. They annually carry to the oceans an estimated 10 billion t of earth material, material delivered to them down the slopes by mass movement. This rate of erosion could in a few million years reduce the continents to base level, leaving the landmasses as flat as pancakes. But geologic studies tell us that sediments and sedimentary rocks have been forming for billions of years. We conclude, therefore, that continuous or repeated uplift of the earth's surface has for eons provided materials for erosion. The tectonics associated with plate motions have kept erosion from flattening the continents. Despite this, it *is* startling to observe on a world scale how close erosion comes to beveling the continents and reducing them to sea level.

A final observation about some of the major features of the world's drainage systems commands attention. If we look at a world map, we find that many of the major divides are related to plate boundaries. South America provides a spectacular example. There the Andes, formed by the collision of the Nazca and American plates, form the continental divide. Most of the continent drains to the east into the Atlantic Ocean, and only a narrow area sheds water to the west and the Pacific Ocean. The great mountain masses of the Himalaya, the Tibetan Plateau, and the mountains of Afghanistan, Iran, the Caucasus, and on to the Alps represent a complex convergence of the great Eurasian plate with several other plates along its southern border. The continental divide of North America, from southern Mexico to Alaska, is associated with plate movements, although their details are still to be worked out. And we can also note that ancient mountains, such as the Appalachians of eastern North America and the Urals of eastern Europe, mark old lines of plate convergence, and they still serve as significant drainage divides in a period that is hundreds of millions of years later.

OUTLINE

Running water has fashioned over 400 million km of stream channels, down which the bulk of material eroded from the continent moves toward the oceans.

The hydrologic cycle is the path of water circulation from oceans to atmosphere to land to oceans.

Precipitation and stream flow (runoff) are related as follows:

Runoff = precipitation − (infiltration + evaporation and transpiration)

Flow is either **turbulent** or **laminar**, but turbulent is the rule.

Water in a stream has **velocity**, flows down a **gradient**, and is measured in terms of **discharge**.

The economy of a stream is part of a dynamic equilibrium that characterizes earth processes.

It is clear that in a stream, **discharge = width × depth × velocity**. The change of one characteristic will affect one or more of the others. In adjusting to an increased discharge downstream, the gradient decreases, and width, depth, and velocity increase.

Floods are a normal, but relatively rare, stage of stream flow. Flood-recurrence intervals are statistically predictable.

Human activity can affect a stream's velocity, discharge, and channel characteristics.

Base level of a stream is the point below which it cannot erode. Lowering of the base level produces erosion, and raising of the base level produces deposition.

Work of running water includes transportation, erosion, and deposition.

Transportation by a stream involves its **load** (the amount it carries at any one time), its **capacity** (the total amount it can carry under given conditions), and its **competence** (the maximum-sized particle it can move under given conditions). Material is moved in **solution**, in **suspension**, and as **bed load**. **Human activity** can affect both the solid and suspended load of a stream.

Erosion by a stream involves **direct lifting** of material, **abrasion**, **impact**, and **solution**.

Features of valleys include those of the valley bottom, the drainage basin, and the river channel.

A **cross-valley profile** of a typical valley shows **flood plains**, **valley walls**, and **divides**.

A **drainage basin** is the area drained by a river and its tributaries. Individual streams and their valleys are joined in definite geometrical relationships. The patterns developed by stream systems include **dendritic**, **radial**, **rectangular**, and **trellis**. Some streams form **water gaps**.

Enlargement of valleys is accomplished through downward and lateral erosion by the stream and by mass movement and water erosion on the valley walls.

Features of narrow valleys include **waterfalls**, **rapids**, and **steep gradients**.

Broad valleys have **meanders**, **braided streams**, **natural levees**, **flood plains**, **deltas**, **alluvial fans**, and **stream terraces**.

Rivers are the most important agent in removing material from the continents. Movement of the earth's plates continues to supply material to be eroded. **Plate margins** provide some of the major stream divides of the continents.

SUPPLEMENTARY READINGS

Bloom, **Arthur L.** *Geomorphology*, Prentice-Hall, Inc., Englewood Cliffs, N. J., 1978. *Four chapters, 9 through 12, emphasize the development of landforms by the fluvial process.*

Gregory, **K. J.**, **and D. E. Walling** *Drainage Basin Form and Process*, Cambridge University Press, New York, 1973. *A very good advanced text devoted to the forms of drainage basins and the processes that go on in them.*

Morisawa, **Marie** *Streams: Their Dynamics and Morphology*, McGraw-Hill Book Company, New York, 1968. *A short book of intermediate level, which is a readable summary of the hydrology and work of streams.*

Ritter, Dale F. *Process Geomorphology*, William C. Brown Company, Publishers, Dubuque, Iowa, 1978. *Chapters 5, 6, and 7 deal with drainage basins, fluvial processes, and fluvial landforms in more detail than is possible in this volume.*

Stanley A. Schumm *The Fluvial System*, John Wiley & Sons, Inc., New York, 1977. *An excellent, although somewhat specialized, treatment.*

1 With the aid of a diagram, describe the hydrologic cycle. You can make your answer quantitative by adding data from the text or Appendix Table D.2.

2 You will get a better grasp of the hydrologic cycle by putting some numbers on it. Try this. If the precipitation in a drainage basin equals 10^9 m³/yr and the sum of the runoff and infiltration is 4×10^8 m³/yr, how much water is returned annually to the atmosphere by evaporation and transpiration? What is this in terms of percentage of the total precipitation?

(We got our answer for total evaporation and transpiration as 6×10^8 m³/yr by substituting in the formula):

Evaporation + transpiration = precipitation − (runoff + infiltration)

3 What is the difference between laminar and turbulent flow? Which is most important in streams?

4 Draw and label a stream hydrograph. Explain how human intervention in a stream's natural regime may modify its hydrograph.

5 Referring to Figure 12.14, what is a reasonable suggestion for the recurrence interval of a flood of magnitude of flow of about 6,250 m³/sec?
(We got about 200 years.)

6 Base level is an important concept in understanding the behavior of a stream. What is the base level of a stream? What is the ultimate base level? Is it constant or changing, and why do you answer as you do? Give two examples of temporary base level. (Here is another place where a simple sketch can sharpen your answer.)

7 What happens to the activity of a stream when its base level rises? When it falls?

8 Distinguish between load, capacity, and competence in a stream.

9 What are the ways by which material is carried by a stream?

10 How does land use affect the availability of sediments to a stream?

11 Stream order is one way of describing the organization of streams, stream networks, and their drainage basins. Referring to Figure 12.30(b) and (c), what would be the approximate length of an 11th-order stream? What would be the approximate size of the drainage basin of this stream?
(We assumed that we could extend the graph in both instances. We got an 11th-order stream of about 1,600 km in length, with a drainage basin of about 1.1×10^6 km². Remember that the vertical axes on the graphs are marked in logarithmic scales.)

12 Nature provides a number of different stream patterns when seen on a map or from the air. List two different patterns and discuss what each means in terms of the geology of the country on which they have developed.

13 Many different features distinguish streams and valleys. You should be familiar enough with them to define and discuss the following: waterfalls, rapids, flood plain, meander, oxbow, cutoff, braided stream, natural levee, point bar, delta, alluvial fan, stream terrace.

13

UNDERGROUND WATER

Tremendous quantities of water lie exposed in oceans, lakes, rivers, and—in the solid state—glaciers on the earth's surface. But beneath the surface, hidden from our sight, is another great store of water. In fact, for the earth as a whole, the water beneath the surface exceeds by more than 66 times the amount of water present in streams and freshwater lakes (see Appendix D, Table D.1). In the United States, streams each year discharge an estimated 30,000 km³ of water into the oceans. In contrast, there lie beneath the surface an estimated 7,575,000 km³ of water. It is certainly true that underground reservoirs in the United States contain far more usable water than do all our surface reservoirs and lakes combined, and the United States depends on this underground supply for about one-fifth of total water needs.

13.1 BASIC DISTRIBUTION

Underground water, subsurface water, and **subterranean water** are all general terms used to refer to water in the pore spaces, cracks, tubes, and crevices of the consolidated and unconsolidated material beneath our feet. The study of underground water is largely an investigation of these openings and of what happens to the water that finds its way into them. In Chapter 12, we found that most of the water beneath the surface comes from rain and snow that fall on the face of the earth. Part of this water soaks directly into the ground, and part of it drains away into lakes and streams and thence into the underground.

ZONES OF SATURATION AND AERATION

Some of the water that moves down from the surface is caught by rock and earth materials and is checked in its downward progress. The zone in which this water is

288

held is known as the **zone of aeration,** and the water itself is called **suspended water.** The spaces between particles in this zone are filled partly with water and partly with air. Two forces operate to prevent suspended water from moving deeper into the earth: (1) the molecular attraction exerted on the water by the rock and earth materials and (2) the attraction exerted by the water particles on one another (see Figures 13.1 and 13.2).

The zone of aeration can be subdivided into three belts: **belt of soil moisture, intermediate belt,** and **capillary fringe.** Some of the water that enters the belt of soil moisture from the surface is used by plants, and some is evaporated back into the atmosphere. But

13.1 A drop of water held between two fingers illustrates the molecular attraction that prevents downward movement. Water is similarly suspended within the pore spaces of the zone of aeration shown in Figure 13.2.

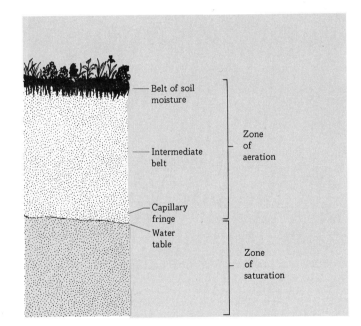

13.2 Two major zones of underground water: zone of aeration and zone of saturation. The water table marks the upper surface of the zone of saturation. Within the zone of aeration is a belt of soil moisture, the source of moisture for many plants. Also from here some moisture is evaporated back to the atmosphere. In many instances, this belt lies above an intermediate belt, where water is held by molecular attraction and little movement occurs except during periods of rain or melting snow. In the capillary fringe, just above the water table, water rises a few centimeters to a meter or more from the zone of saturation, depending on the size of the interstices.

some water also passes down to the intermediate belt, where it may be held by molecular attraction (as suspended water). Little movement occurs in the intermediate belt except when rain or melting snow sends a new wave of moisture down from above. In some areas the intermediate belt is missing, and the belt of soil moisture lies directly above the third belt, the capillary fringe. Water rises into the capillary fringe from below, to a height ranging from a few centimeters to 2 or 3 m.

Beneath the zone of aeration lies the **zone of saturation.** Here the openings in the rock and earth materials are completely filled with **groundwater,** and the surface between the zone of saturation and the zone of aeration is called the **groundwater table,** or simply the **water table.** The level of the water table fluctuates with variations in the supply of water coming down from the zone of aeration, with variations in the rate of discharge in the area, and with variations in the amount of groundwater that is drawn off by plants and human beings.

It is the water below the water table, within the zone of saturation, that we shall focus on in this chapter.

THE WATER TABLE

The water table is an irregular surface of contact between the zone of saturation and the zone of aeration. Below the water table lies the groundwater; above it lies the suspended water. The thickness of the zone of aeration differs from one place to another, and the level of the water table fluctuates accordingly. In general, the water table tends to follow the irregularities of the ground surface, reaching its highest elevation beneath hills and its lowest elevation beneath valleys. Although the water table reflects variations in the ground surface, the irregularities in the water table are less pronounced.

In looking at the topography of the water table, let us consider an ideal situation. Figure 13.3 shows a hill underlain by completely homogeneous material. Assume that, initially, this material contains no water at all. Then a heavy rainfall comes along, and the water soaks slowly downward, filling the interstices at depth. In other words, a zone of saturation begins to develop. As more and more water seeps down, the upper limit of this zone continues to rise. The water table remains horizontal until it just reaches the level of the two valley bottoms on either side of the hill. Then, as additional water seeps down to the water table, some of it seeks an outlet into the valleys. But this added water is "supported" by the material through which it flows, and the water table is prevented from maintaining its flat surface. The water is slowed by the friction of its movement through the interstices and even, to some degree, by its own internal friction. Consequently, more and more water is piled up beneath the hill, and the water table begins to reflect the shape of the hill. The water flows away most rapidly along the steeper slope of the water table near the valleys and most slowly on its gentler slope beneath the hill crest.

We can modify the shape of the groundwater surface by providing an artificial outlet for the water. For example, we can drill a well on the hill crest and extend it down into the saturated zone. Then, if we pumped out the groundwater that flowed into the well, we should create a dimple in the water table. The more we pumped, the more pronounced the depression—a **cone of depression**—would become.

Returning to our ideal situation, we find that, if the supply of water from the surface were to be completely stopped, the water table under the hill would slowly flatten out as water discharged into the valleys. Eventually it would almost reach the level of the water table under the valley bottoms; then the flow would stop. This condition is common in desert areas where the rainfall is sparse.

13.2 MOVEMENT OF UNDERGROUND WATER

The preceding chapter stated that the flow of water in surface streams could be measured in terms of so many meters per second. But in dealing with the flow of underground water, we must change our scale of measurement; for here, although the water moves, it usually does so very, very slowly. Therefore we find that centimeters per day and, in some places, even centimeters or meters per year provide a better scale of measurement. The main reason for this slow rate of flow is that the water must travel through small, confined passages if it is to move at all. It will then be worthwhile for us to consider the porosity and permeability of earth materials.

(a)

(b)

(c)

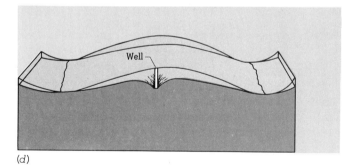

Well

(d)

13.3 Ideally, the water table is a subdued reflection of the surface of the ground. In (a) and (b) the water table rises as a horizontal plane until it reaches the level of the valley bottoms on either side of the hill. Thereafter, as more water soaks into the ground, it seeks an outlet toward the valleys. Were the movement of the water not slowed down by the material making up the hill, it would remain essentially horizontal. The friction caused by the water passing through the material (and even to some extent, the internal friction of the water itself) results in a piling up of water beneath the hill; the bulge is highest beneath the crest and lowest toward the valleys (c). The shape of the water table may be altered by pumping water from a well (d). The water flows to this new outlet and forms a cone of depression.

POROSITY

The **porosity** of a rock is measured by the percentage of its total volume that is occupied by voids or interstices. The more porous a rock is, the greater the amount of open space it contains. Through these open spaces underground water must find its way.

Porosity differs from one material to another. What is the porosity of a rock made up of particles and grains derived from preexisting rocks? Here porosity is determined largely by the shape, size, and assortment of these rock-building units. A sand deposit composed of rounded quartz grains with fairly uniform size has a high porosity. But if mineral matter enters the deposit and cements the grains into a sandstone, the porosity is reduced by an amount equal to the volume of the cementing agent. A deposit of sand poorly sorted with finer particles of silt and clay mixed in has low porosity because the smaller particles fill up much of the space between the larger particles.

Even a dense massive rock, such as granite, may become porous as a result of fracturing. And a soluble massive rock, such as limestone, may have its original planes of weakness enlarged by solution.

Clearly, then, the range of porosity in earth materials is extremely great. Recently deposited muds (called **slurries**) may hold up to 90 percent by volume of water, whereas unweathered igneous rocks such as granite, gabbro, or obsidian may hold only a fraction of 1 percent. Unconsolidated deposits of clay, silt, sand, and gravel have porosities ranging from about 20 to as much as 50 percent. But when these deposits have been consolidated into sedimentary rocks by cementation or compaction, their porosity is sharply reduced. Average porosity values for individual rock types have little meaning because of the extreme variations within each type. In general, however, a porosity of less than 5 percent is considered low; from 5 to 15 percent represents medium porosity; and over 15 percent is considered high.

PERMEABILITY

Whether or not we find a supply of fresh groundwater in a given area depends on the ability of the earth materials to transmit water as well as on their ability to contain it. The ability to transmit underground water is termed **permeability.**

The rate at which a rock transmits water depends not only on its total porosity but also on the size of the interconnections between its openings. For example, although a clay may have a higher porosity than a sand, the particles that make up the clay are minute flakes and the interstices between them are very small. Therefore water passes more readily through the sand than through the more porous clay simply because the molecular attraction on the water is much stronger in the tiny openings of the clay. The water moves more freely through the sand because the passageways between particles are relatively large and the molecular attraction on the water is relatively low. Of course, no matter how large the interstices of a material are, there must be connections between them if water is to pass through. If they are not interconnected, the material is impermeable.

A permeable material that actually carries underground water is called an **aquifer,** from the Latin for "water" and "to bear." Perhaps the most effective aquifers are unconsolidated sand and gravel, sandstone, and some limestones. The permeability of limestone is usually due to solution that has enlarged the fractures and bedding planes into open passageways. The fractured zones of some of the denser rocks such as granite, basalt, and gabbro also act as aquifers, although the permeability of such zones decreases rapidly with depth. Clay, shale, and most metamorphic and crystalline igneous rocks are generally poor aquifers.

Because the flow of underground water is usually very slow, it is largely laminar; in contrast the flow of surface water is largely turbulent. (There is one exception, however: the turbulent flow of water in large underground passageways formed in such rocks as cavernous limestone.) In laminar flow, the water near the walls of interstices is presumably held motionless by the molecular attraction of the walls. Water particles farther away from the walls move more rapidly, in smooth, threadlike patterns because the resistance to motion decreases toward the center of an opening. The most rapid flow is reached at the very center.

The energy that causes underground water to flow is derived from gravity. Gravity draws water downward to the water table; from there it flows through the ground to a point of discharge in a stream, lake, or spring. Just as surface water needs a slope to flow on, so must there be a slope for the flow of groundwater. This is the slope of the water table, the **hydraulic gradient.** It is measured by dividing the length of flow (from the point of intake to the point of discharge) into the vertical distance between these two points, a distance called **head.** Therefore, hydraulic gradient is expressed as h/l, where h is head and l is length of flow. Thus, if h is 10 m and l is 100 m, the hydraulic gradient is 0.1 or 10 percent.

An equation to express the rate of water movement through a rock was proposed by the French engineer Henri Darcy in 1856. What is now known as **Darcy's law** is essentially the same as his original equation. The law may be expressed as follows:

$$V = K\frac{h}{l}$$

where V is velocity, h the head, l the length of flow, and K a coefficient that depends on the permeability of the material, the acceleration of gravity, and the viscosity of water. But because h/l is simply a way of expressing the hydraulic gradient, we may say that in a rock of constant permeability, the velocity of water will increase as the hydraulic gradient increases. Remembering that the hydraulic gradient and the slope of the groundwater table are the same thing, we may also say that the velocity of groundwater varies with the slope of the water table. Other things being equal, the steeper the slope of the water table, the more rapid the flow. In ordinary aquifers, the rate of water flow has been estimated as not faster than 1.5 m/day and not slower than 1.5 m/year, although rates of over 120 m/day and as low as a few centimeters per year have occasionally been recorded.

The movement of underground water down the slope of the water table is only a part of the picture because the water is also in motion at depth. Water moves downward from the water table in broad looping curves toward some effective discharge agency, such as a stream, as suggested in Figure 13.4. The water feeds into the stream from all possible directions, including straight up through the bottom of the channel. We can explain this curving path as a compromise between the force of gravity and the tendency of water to flow laterally in the direction of the slope of the water table. This tendency toward lateral flow is actually the

result of the movement of water toward an area of lower pressure, the stream channel in Figure 13.4. The resulting movement is neither directly downward nor directly toward the channel but is, rather, along curving paths to the stream.

13.3 GROUNDWATER IN NATURE

So far we have assumed that groundwater is free to move on indefinitely through a uniformly permeable material of unlimited extent. Subsurface conditions actually fall far short of this ideal situation. Some layers of rock material are more permeable than others are, and the water tends to move rapidly through these beds in a direction more or less parallel to bedding planes. Even in a rock that is essentially homogeneous, the groundwater tends to move in some preferred direction.

SIMPLE SPRINGS AND WELLS

Underground water moves freely downward from the surface until it reaches an impermeable layer of rock or until it arrives at the water table. Then it begins to move laterally. Sooner or later it may flow out again at the surface of the ground in an opening called a **spring.**

Springs have attracted human attention throughout history. In early days they were regarded with superstitious awe and were sometimes selected as sites for temples and oracles. To this day many persons feel that spring water possesses special medicinal and therapeutic values. Water from "mineral springs" contains salts in solution that were picked up by the water as it percolated through the ground. The same water pumped up out of a well would be regarded as merely hard and not desirable for general purposes.

Springs range from intermittent flows that disappear when the water table recedes during a dry season, through tiny trickles, to an effluence of 3.8 billion l daily, the abundant discharge of springs along a 16-km stretch of Fall River, California.

This wide variety of spring types is the result of underground conditions that vary greatly from one place to another. As a general rule, however, a spring results wherever the flow of groundwater is diverted to a discharge zone at the surface (see Figure 13.5). For example, a hill made up largely of permeable rock may contain a zone of impermeable material, as shown in Figure 13.6. Some of the water percolating downward will be blocked by this impermeable rock,

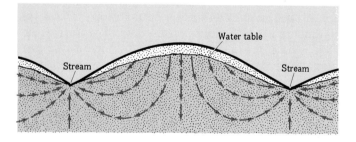

13.4 The flow of groundwater through uniformly permeable material is suggested here. Movement is not primarily along the groundwater table; rather, particles of water define broadly looping paths that converge toward the outlet and may approach it from below.

13.10 Mammoth Hot Springs, Yellowstone National Park, are thermal springs that have built these terraces by depositing travertine. [U.S. Geological Survey.]

cooling. In the eastern group, however, the circulation of the groundwater carries it so deep that it is warmed by the normal increase in earth heat (see "Thermal gradient" in the Glossary).

The well-known spring at Warm Springs, Georgia, is heated in just this way. Long before the Civil War, this spring was used as a health and bathing resort by the people of the region. Then, with the establishment of the Georgia Warm Springs Foundation for the treatment of victims of infantile paralysis, the facilities were greatly improved. Rain falling on Pine Mountain, about 3 km south of Warm Springs, enters a rock formation known as the Hollis. At the start of its journey downward, the average temperature of the water is about 16.5°C. It percolates through the Hollis Formation northward under Warm Springs at a depth of around 100 m and then follows the rock as it plunges into the earth to a depth of 1,140 m, 1.6 km farther north. Normal rock temperatures in the region increase about 1.8°C/100 m of depth, and the water is warmed as it descends along the bottom of the Hollis bed. At 1,140 m, the bed has been broken and shoved against an impervious layer that turns the water back. This water is now hotter than the water coming down from above, and it moves upward along the top of the Hollis Formation, cooling somewhat as it goes. Finally it emerges in a spring at a temperature of 36.5°C.

Less than 1 km away is Cold Spring, whose water comes from the same rainfall on Pine Mountain. A freak of circulation, however, causes the water at Cold Spring to emerge before it can be conducted to the depths and be warmed. Its temperature is only about 16.5°C.

GEYSERS

A **geyser** is a special type of thermal spring that ejects water intermittently with considerable force. The word comes from the Icelandic name of a spring of this type, *geysir*, probably based on the verb *gjosa*, "to rush forth."

Although the details of geyser action are still not understood, we know that, in general, a geyser's behavior is caused by the arrangement of its plumbing and the proximity of a good supply of heat. Here is probably what happens: Groundwater moving downward from the surface fills a natural pipe, or conduit, that opens upward to the surface. Hot igneous rocks, or the gases given off by such rocks, gradually heat the column of water in the pipe and raise its temperature toward the boiling point. Now the higher the pressure on water, the higher its boiling point; and because water toward the bottom of the pipe is under the greatest pressure, it must be heated to a higher temperature than the water above before it will come to a boil. Eventually the column of water becomes so hot that either a slight increase in temperature or a slight decrease in pressure will cause it to boil. At this critical point, the water near the base of the pipe is heated to the boiling point. The water then passes to steam and, as it does so, expands, pushing the water above it to-

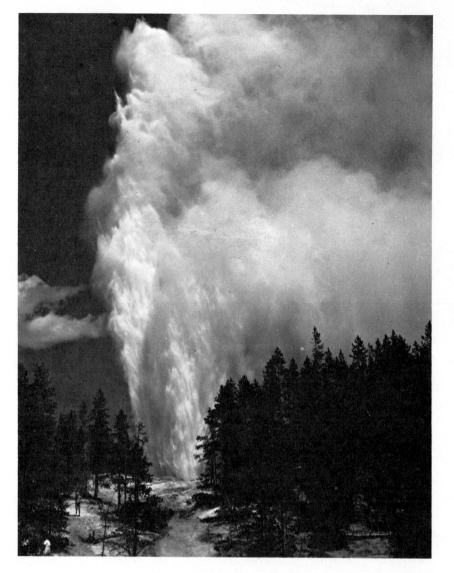

13.11 Steamboat Geyser, Yellowstone National Park, Wyoming. The periodic eruption of a geyser is due to the particular pattern of its plumbing and its proximity to a liberal supply of heat and groundwater. [National Park Service.]

ward the surface. But this raising of the heated column of water reduces the pressure acting on it, and it too begins to boil. The energy thus developed throws the water and steam high into the air, producing the spectacular action characteristic of many geysers (see Figure 13.11). After the eruption has spent itself, the pipe is again filled with water and the whole process begins anew.

We can compare this theoretical cycle with that of Old Faithful Geyser in Yellowstone National Park. The first indication of a coming eruption at Old Faithful is the quiet flow of water in a fountain some 1 to 2 m high. This preliminary activity lasts for a few seconds and then subsides. It represents the first upward push of the column of water described in our theoretical case. This

push reduces the pressure and thereby lowers the boiling point of the water in the pipe. Consequently, the water passes to steam, and less than 1 min after the preliminary fountain, the first of the violent eruptions takes place. Steam and boiling water are thrown 45 to 50 m into the air. The entire display lasts about 4 min. Emptied by the eruption, the tube then gradually refills with groundwater, the water is heated, and in approximately 1 h the same cycle is repeated. The actual time between eruptions at Old Faithful averages about 65 min but may be from 30 to 90 min.

13.4 RECHARGE OF GROUNDWATER

As we have seen, the ultimate source of underground water is precipitation that finds its way below the surface of the land.

Some of the water from precipitation seeps into the ground, reaches the zone of saturation, and raises the water table. Continuous measurements over long periods of time at many places throughout the United States show an intimate connection between the level of the water table and rainfall (see Figure 13.12). Because water moves relatively slowly in the zone of aeration and the zone of saturation, fluctuations in the water table usually lag a little behind fluctuations in rainfall.

Several factors control the amount of water that actually reaches the zone of saturation. For example, rain that falls during the growing season must first replenish moisture used up by plants or passed off through evaporation. If these demands are great, very little water will find its way down to **recharge** the zone of saturation. Then too, during a very rapid, heavy rainfall, most of the water may run off directly into the streams instead of soaking down into the ground. A slow, steady rain is much more effective than a heavy, violent rain in replenishing the supply of groundwater. High slopes, lack of vegetation, or the presence of impermeable rock near the surface may promote runoff and reduce the amount of water that reaches the zone of saturation. It is true, however, that some streams are themselves sources for the recharge of underground water. Water from the streams leaks into the zone of saturation, sometimes through a zone of aeration (see Figure 13.13).

In many localities, the natural recharge of the underground supplies cannot keep pace with human demands for groundwater. Consequently, attempts are sometimes made to recharge these supplies artificially. On Long Island, New York, for example, water that has been pumped out for air-conditioning purposes is returned to the ground through special recharging wells or, in winter, through idle wells that are

13.13 The groundwater may be recharged by water from a surface stream leaking into the underground.

used in summer for air conditioning. In the San Fernando Valley, California, surplus water from the Owens Valley aqueduct is fed into the underground in an attempt to keep the local water table at a high level.

13.5 CAVES AND RELATED FEATURES

Caves are probably the most spectacular examples of the handiwork of underground water. In dissolving great quantities of solid rock in its downward course, the water fashions large rooms, galleries, and underground stream systems as the years pass. In many caves, the water deposits calcium carbonate as it drips off the ceilings and walls, building up fantastic shapes of **dripstone.**

Caves of all sizes tend to develop in highly soluble rocks such as limestone, $CaCO_3$, and small ones occur in the sedimentary rock dolomite, $CaMg(CO_3)_2$. Rock salt, $NaCl$, gypsum, $CaSO_4 \cdot 2H_2O$, and similar rock types are the victims of such rapid solution that underground caverns usually collapse under the weight of overlying rocks before erosion exposes them.

Calcite, the main component of limestone, is very insoluble in pure water. But when the mineral is attacked by water containing small amounts of carbonic acid, it undergoes rapid chemical weathering; most natural water contains carbonic acid, H_2CO_3, the combination of water, H_2O, with carbon dioxide, CO_2. The carbonic acid reacts with the calcite to form calcium bicarbonate, $Ca(HCO_3)_2$, a soluble substance that is then removed in solution. If not redeposited, it eventually reaches the ocean.

Let us look more closely at underground water as it brings about the decay of calcite. Calcite contains the complex carbonate ion, $(CO_3)^{2-}$ (built by packing three oxygen atoms around a carbon atom), and the

13.12 Relationship between the water level in an observation well near Antigo, Wisconsin, and precipitation, as shown by records from 1945 to 1952. The water table reflects the changes in precipitation. The graphs represent 3-year running monthly averages. For example, 5.8 cm of precipitation for May means that precipitation averaged 5.8 cm/month from May 1947 to May 1950, inclusive. [After A. H. Harder and William J. Drescher, "Ground Water Conditions in Southwestern Langlade County, Wisconsin," *U.S. Geol. Surv. Water Supply Paper 1294*, 1954.]

calcium ion, Ca^{2+}. These two ions are combined in much the same way as sodium and chlorine in forming salt, or halite. The weathering or solution of calcite takes place when a hydrogen ion, H^+, approaches $(CO_3)^{2-}$. Because the attraction of hydrogen for oxygen is stronger than the attraction of carbon for oxygen, the hydrogen ion pulls away one of the oxygen atoms of the carbonate ion and, with another hydrogen ion, forms water. The two other oxygen atoms remain with the carbon atom as carbon dioxide gas, CO_2. The calcium ion, Ca^{2+}, now joins with two negative bicarbonate ions, $(HCO_3)^-$, to form $Ca(HCO_3)_2$ in solution. We can express these activities as follows:

Two parts water *plus* two parts carbon dioxide *yield* two parts carbonic acid
$$2H_2O + 2CO_2 \rightleftharpoons 2H_2CO_3$$

Two parts carbonic acid *yield* two hydrogen ions *plus* two bicarbonate ions
$$2H_2CO_3 \rightleftharpoons 2H^+ + 2(HCO_3)^-$$

Two hydrogen ions *plus* two bicarbonate ions *plus* calcite *yield*
$$2H^+ + 2(HCO_3)^- + CaCO_3 \rightleftharpoons$$

water *plus* carbon dioxide *plus* calcium bicarbonate in solution
$$H_2O + CO_2 + Ca^{2+} + 2(HCO_3)^-$$

The principal chemical reaction in the formation of caves is given in the above equations. However, even though the reaction can be demonstrated in the laboratory and occurs in nature, its extension to the formation of caves is not as direct as it might first appear.

There is a strong difference of opinion as to whether caves form above, at (or just below), or deep beneath the water table. Those who believe that caves form below the water table call upon some mechanism that would lower the water table and drain the cave, thus filling it with air. Drilling records, in fact, show that there are large openings in limestone that lie flooded with water beneath the water table. Presumably, if land keeps rising and erosion proceeds until valleys are deep enough to drain the caves in the surrounding countryside, the caves will become filled with air and will be ready for discovery and exploration.

We have seen that water and carbon dioxide will combine to provide a solution that will corrode lime-

stone. Such a solution, seeping into the ground, will begin to dissolve limestone as soon as it meets it just below the soil. Here the waters are rapidly neutralized and, seeping deeper, can no longer dissolve the rock. In this regard, we can note that the waters seeping down from the surface into caves today are depositing rather than dissolving calcite. Yet it is also clear that the chambers we now call "caves" were formed at an appreciable depth below the surface. How, then, could this happen? The details of the answer are yet to be worked out, but studies by John Thrailkill of the University of Tennessee point a probable way to the answer. He suggests mechanisms by which the flow of groundwater through limestone would allow concentrations of large amounts of fresh water undersaturated with respect to calcite.

One such mechanism is shown in Figure 13.14. In this diagram, the water table is shown as being almost horizontal, a condition often encountered in limestone rock and in contrast to the sloping water table shown as an idealized case in Figure 13.4. The flow lines lead to the stream, and the water table is one of these flow lines. Water reaches the underground chiefly via an enlarged opening at the surface. This opening could be formed by differential solution from the surface downward along particularly well-developed or favorably placed fractures in the limestone rock. In any event, the opening allows the rapid injection of large amounts of surface water into the underground. Because the surface water is injected rapidly and in large amounts, it may pass through the zone above the water table without coming into effective contact with the limestone and will remain undersaturated with respect to calcite as it joins the groundwater. If this happens, there is a zone beneath the inlet and below the

13.14 Diagram through a small limestone aquifer, showing approximate flow lines and the location of region of solution. See the text for discussion. Vertical exaggeration, 4×. [After John Thrailkill, "Chemical and Hydrologic Factors in the Excavation of Limestone Caves." *Bull. Geol. Soc. Am.*, vol. 79, p. 39, 1968.]

13.15 Stalactites grow downward from the cave ceiling and stalagmites grow upward from the cave floor. Temple of the Sun in the Big Room of Carlsbad Caverns, New Mexico. [National Park Service.]

water table where groundwater may still carry enough carbonic acid to effect solution in the underground.

Regardless of where and how caves are originally formed, we know that the weird stone formations characteristic of most of them must have developed above the water table when the caves were filled with air. These bizarre shapes are composed of calcite deposited by underground water that has seeped down through the zone of aeration. They develop either as **stalactites,** looking like stony icicles hanging from the cave roof (Greek *stalaktos,* "oozing out in drops"), or as **stalagmites,** heavy posts growing up from the floor (Greek *stalagmos,* "dripping"). When a stalactite and stalagmite meet, a **column** is formed (see Figure 13.15). A stalactite forms as water charged with calcium bicarbonate in solution seeps through the cave roof.

Drop after drop forms on the ceiling and then falls to the floor. But during the few moments that each drop clings to the ceiling, a small amount of evaporation takes place; some carbon dioxide is lost, and a small amount of calcium carbonate is deposited. Over the centuries a large stalactite may gradually develop. Part of the water that falls to the cave floor runs off, and part is evaporated. This evaporation again causes the deposition of calcite, and a stalagmite begins to grow upward to meet the stalactite.

On the ground surface above soluble rock material, depressions sometimes develop that reflect areas where the underlying rock has been carried away in solution. These depressions, called **sinkholes** or merely **sinks,** may form in one of two different ways. In

13.16 Underground solution of limestone not only produces caves with stalactites and stalagmites, as suggested here, but in addition may develop karst topography at the surface with sinkholes as well as valleys that drain into the subsurface.

13.17 The "December Giant," a sinkhole near Montevallo in central Alabama, is believed to have formed December 2, 1972. It is about 45 m deep and measures about 100 m by 130 m at the surface. [U.S. Geological Survey.]

one case, the limestone immediately below the soil may be dissolved by the seepage of waters downward. The process of solution may be focused by local factors, such as a more abundant water supply or greater solubility of the limestone. Eventually a depression—a sink—evolves. Probably more commonly, sinks form when the surface collapses into a large cavity below. In either event, surface water may drain through the sinkholes to the underground; or if the sinks' subterranean outlets become clogged, they may fill with water to form lakes. An area with numerous sinkholes is said to display **karst topography,** from the name of a plateau in Yugoslavia and northeastern Italy where this type of landscape is well developed (see Figures 13.16 and 13.17).

POLLUTION

Water underground is subject to contamination of the same type as is surface water. We have noted in Section 13.3 that water in limestone is more susceptible to pollution than is that flowing more slowly in, for example, a sandstone aquifer. Underground water may be subject to thermal pollution as well as microbial or chemical pollution. The very process of recycling water may give it the added undesirable quality of increased temperature; water used for air conditioning is withdrawn from the ground at one temperature, used in cooling, and returned to the ground at a higher temperature.

SALTWATER INVASION

A particular type of pollution to which underground freshwater supplies are subject is the **invasion of salt water**. This may be salt water of an adjacent ocean environment, or it may be salt water trapped in rocks lying beneath the freshwater aquifers.

Fresh water has a lower specific gravity than does salt water: Normal seawater has specific gravity of about 1.025, compared with 1.000 for fresh water. Therefore fresh water will float on top of salt water. If there is little or no subsurface flow, this can lead to an equilibrium in which a body of fresh water is buoyed on salt water. In a groundwater situation in which fresh water and salt water are juxtaposed, we might have a situation such as that shown in Figure 13.18. Here an island (or it could as well be a peninsula) is shown underlain by a homogeneous aquifer that extends under the adjacent ocean. Fresh water falling on the land has built up a prism of fresh water in hydrostatic equilibrium with the salt water that surrounds it. The height of a column of fresh water is balanced by an equal mass of salt water. In the example given, the column of fresh water (H) is equal to h_1, the height of the water table above sea level, plus h_2, the thickness of the groundwater below the sea level. A column of salt water equivalent in length to h_2 will balance a column of fresh water of length H, or $h_1 + h_2$. Working this out, we find that $h_2 = 40h_1$. This means that if the water table is 10 m above sea level, then the freshwater-saltwater contact will be 400 m below sea level. It also means that as we lower the water table by whatever means, we raise the elevation of the saltwater-freshwater interface at the rate of 40 m of rise for each

meter of lowering of the water table. It is very possible, then, that a well that originally bottomed in fresh water could, by a lowering of the water table, turn into a saltwater well, as suggested in Figure 13.18.

LAND SUBSIDENCE

Excess pumping may deplete groundwater supplies and in certain places bring about saltwater invasion of wells. But another bothersome problem connected with heavy pumping of groundwater is **land subsidence**. It expresses itself at the surface either as broad, gentle depressions of the land or as catastrophic collapse and the formation of sinkholes.

Zones of broad subsidence have been created in California over large areas of the San Joaquin Valley and secondarily in the Santa Clara Valley south of Oakland and San Francisco (Figure 13.19). The phenomenon is related to the pumping of groundwater and the resulting compaction of sediments. The Tulare-Wasco area of the San Joaquin Valley is a good example of what happens. In 1905, intensive pumping for irrigation began in the area. By 1962, more than 20,000 km² of irrigable land had undergone subsidence of more than 0.3 m, and as much as 4 m of subsidence had occurred in some places. Subsidence was slow and hardly noticeable to residents. But the casings of many wells have been damaged, the problems of surveying and construction have been complicated, and the operation of irrigation districts has been endangered.

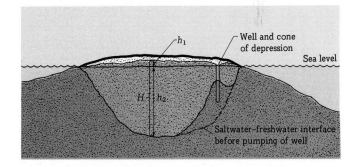

13.18 Cross section through an oceanic island (or a peninsula) underlain by homogeneous, permeable material. The fresh water forms a prism floating in hydrostatic balance with the neighboring salt water. Reduction of the groundwater table on the island will cause a change in the position of the saltwater-freshwater interface. See the text for discussion.

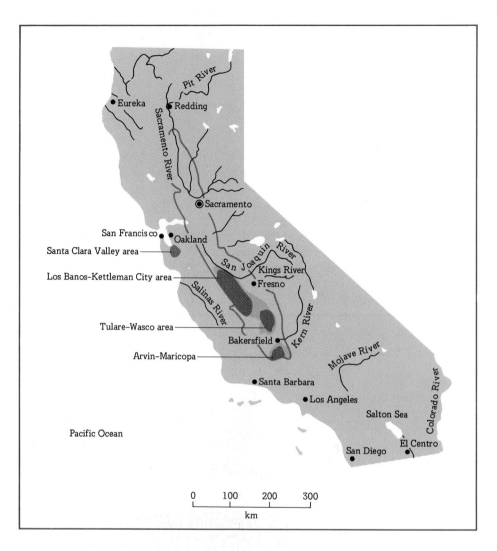

13.19 Principal areas in California where withdrawal of groundwater has caused subsidence of the land surface. [After B. E. Lofgren and R. L. Klausing. "Land Subsidence Due to Groundwater Withdrawal, Tulare-Wasco Area, California," *U.S. Geol. Surv. Prof. Paper* 437-B, p. 2, 1969.]

OUTLINE

Underground water is over 66 times as abundant as the water in streams and lakes.

Distribution is in the **zone of aeration** and in an underlying **zone of saturation.** The irregular surface that separates these zones is called the **groundwater table.**

Movement of underground water is usually by **laminar flow.**

> **Porosity** is the total percentage of void space to a given volume of earth material.
>
> **Permeability** is the ability of earth material to transmit water. Flow is driven by gravity according to **Darcy's law: $V = K\,(h/l)$.**

Groundwater in nature moves in aquifers. It comes to the surface as springs and wells.

> A **spring** occurs when the ground surface intersects the water table. **Simple wells** draw water from the zone of saturation.
>
> **Artesian wells** are those in which gravity drives the water level above the top of the aquifer.
>
> **Thermal springs** derive their heat from the cooling of igneous rocks or by a normal increase of the earth's heat with depth.
>
> **Geysers** are thermal springs marked by periodic, violent eruptions and are controlled by a particular arrangement of underground passages.

Recharge of groundwater is either natural or artificial.

Caves are usually created by the solution of limestone. The chemical reaction involves carbonic acid (formed by the combination of water and carbon dioxide) with limestone.

13   UNDERGROUND WATER

Formation of caves apparently takes place in two stages: (1) the creation of chambers, galleries, and tunnels, and (2) the decoration of these large openings by forms growing from the ceilings (**stalactites**) and forms growing from the floor (**stalagmites**). Solution of limestone develops an irregular "pitted" topography of depressions with no exterior surface drainage. This is called **karst topography** and is marked by numerous **sinkholes**.

Misuse of groundwater can cause numerous problems including **pollution, saltwater invasion,** and **land subsidence.**

Fetter, C. W., Jr. *Applied Hydrogeology,* Charles E. Merrill Publishing Company, Columbus, Ohio, 1980. *A textbook designed for an undergraduate course on underground water.*

Freeze, R. Allen, and **John A. Cherry** *Groundwater,* Prentice-Hall, Inc., Englewood Cliffs, N. J., 1979. *A fairly advanced text on the subject.*

Sweeting, Marjorie M. *Karst Landforms,* Columbia University Press, New York, 1973. *An excellent volume on caves, karst, and associated landscape.*

Todd, D. K. *Ground Water Hydrology,* John Wiley & Sons, Inc., New York, 1959. *An old but still good basic text on the subject.*

SUPPLEMENTARY READINGS

QUESTIONS

1 The streams of the conterminous United States discharge approximatley 30,000 km³ of water into the oceans each year. How long would it take them to carry a volume of water equal to that stored as underground water in the 48 states? You will find the second number you need for the answer in the first paragraph of this chapter.
   (We got 252.5 years.)

2 Define water table. In a landscape underlain by a uniformly permeable material, what is the shape of the water table? Draw a labeled diagram through two hills separated by a stream valley to show this relationship.

3 Distinguish between porosity and permeability in earth materials.

4 What is Darcy's law?

5 How is the slope of the groundwater table defined?

6 Draw a labeled diagram to show the nature of flow of groundwater through a uniformly permeable material. The diagram you used in Question 2 would be a good place to start.

7 What is a perched water table? Diagram and label an example.

8 Myth has it that a well must be very deep to qualify as an artesian well. Although no particular depth defines an artesian well, there are several conditions that must be met before we can say we have an artesian well. What are they? Here again a neat, simple, concisely labeled diagram will help to give precision to your answer.

9 You should understand the chemistry involved in the solution of limestone caves and in the deposition of stalactites and stalagmites that decorate them. Do you?

10 What is karst topography?

11 You find that the water table in your island well is 5 m above sea level. How far down will fresh water persist before salt water is encountered? Your well is 50 m deep, and drawdown due to pumping produces a cone of depression of 4-m depth. What problems do you face?
   (We got a depth of 200 m of fresh water beneath sea level—5m × 40 m. With a 4-m cone of depression, salt water will rise 160 m and thus will be 5 m above the bottom of the well, and that is a problem!)

14

GLACIATION

The seas and rivers of moving ice known as **glaciers** have attracted the attention of the inquisitive deep in the Arctic, Antarctic, and mountainous regions of the world. There they have discovered that glaciers are active agents of erosion, transportation, and deposition and that these impressive masses of ice were far more widespread in the past than they are today. Geologists have learned too that the ice of the last great glacial period has modified and molded great stretches of landscape in what are now the temperate zones.

14.1 FORMATION OF GLACIER ICE

A **glacier** is a mass of ice that has been formed by the recrystallization of snow and that flows forward—or has flowed at some time in the past—under the influ-

ence of gravity. This definition eliminates the pack ice formed from seawater in polar latitudes and, by convention, the icebergs, even though they are large fragments broken from the seaward end of glaciers.

Like surface streams and underground reservoirs, glaciers depend on the oceans for their nourishment. Some of the water drawn up from the oceans by evaporation falls on the land in the form of snow. If the climate is right, part of the snow may last through the summer without melting. Gradually, as the years pass, the accumulation may grow deeper and deeper until at last a glacier is born. In areas where the winter snowfall exceeds the amount of snow that melts away during the summer, stretches of perennial snow known as **snowfields** cover the landscape. At the lower limit of a snowfield lies the **snow line**. Above the snow line, glacier ice may collect in the more sheltered areas of the snowfields. The exact position of the snow line varies from one climatic region to another. In polar regions, for example, it reaches down to sea level, but near the equator it recedes to the mountain tops. In the high mountains of East Africa it ranges from elevations of 4,500 to 5,400 m. The highest snow lines in the world are in the dry regions known as the "horse latitudes," between 20° and 30° north and south of the equator. Here the snow line reaches higher than 6,000 m.

Fresh snow falls as a feathery aggregate of complex and beautiful crystals with a great variety of patterns. All the crystals are basically hexagonal, however, and all reflect their internal arrangement of hydrogen and oxygen atoms (see Figure 14.1). Snow is not frozen rain; rather, it forms from the condensation of water vapor that is at temperatures below the freezing point.

After snow has been lying on the ground for some time, it changes from a light fluffy mass to a heavier granular material called **firn** or **névé.** *Firn* derives from a German adjective meaning "of last year," and *névé* is a French word from the Latin for "snow." Solid remnants of large snowbanks, those tiresome vestiges of winter, are largely firn. Several processes are at work in the transformation of snow into firn. The first is **sublimation,** a general term for the process by which a solid material passes into the gaseous state without first becoming liquid. In sublimation, molecules of water vapor escape from the snow, particularly from the edges of the flakes. Some of the molecules attach themselves to the center of the flakes, where they adapt themselves to the structure of the snow crystals. Then, as time passes, one snowfall follows another, and the granules that have already begun to grow as a result of sublimation are packed tighter and tighter together under the pressure of the overlying snow.

Water has the rare property of increasing in volume when it freezes; conversely, it decreases in volume as the ice melts. But the cause and effect may be interchanged: If added pressure on the ice squeezes the molecules closer together and reduces its volume, the ice may melt. In fact, if the individual granules are in contact, they begin to melt with only a slight increase in pressure. The resulting meltwater trickles down and refreezes on still lower granules at points where they are not yet in contact. And all through this process the basic hexagonal structure of the original snow crystals is maintained.

Gradually, then, a layer of firn granules, ranging from a fraction of a millimeter to approximately 3 or 4 mm in diameter, is built up. The thickness of this layer varies, but the average seems to be 30 m on many mountain glaciers.

The firn itself undergoes further change as continued pressure forces out most of the air between the granules, reduces the space between them, and finally transforms the firn into **glacier ice,** a true solid composed of interlocking crystals. In large blocks it is usually opaque and takes on a blue-gray color from the air and the fine dirt it contains.

The ice crystals that make up glacier ice are minerals; the mass of glacier ice, made up of many interlocking crystals, is a metamorphic rock because it has been transformed from snow into firn and eventually into glacier ice. Later we shall see that glacier ice itself undergoes further metamorphism.

CLASSIFICATION OF GLACIERS

The glaciers of the world fall into four principal classifications: valley glaciers, piedmont glaciers, ice sheets, and continental glaciers.

Valley glaciers are streams of ice that flow down the valleys of mountainous areas (see Figure 14.2). Like streams of running water, they vary in width, depth, and length. A branch of the Hubbard Glacier in Alaska is 120 km long, whereas some of the valley glaciers that dot the higher reaches of our western mountains are only a few hundred meters in length. Valley glaciers that are nourished on the flanks of high mountains and that flow down the mountain sides are sometimes called **mountain glaciers,** or **Alpine glaciers.** Very small mountain glaciers are referred to as **cliff glaciers, hanging glaciers,** or **glacierets.** A particular type of valley glacier sometimes grows up in areas where large masses of ice are dammed by a mountain barrier along the coast. Some of the ice escapes

14.1 Snowflakes exhibit a wide variety of patterns, all hexagonal and all reflecting the internal arrangement of hydrogen and oxygen. It is from snowflakes that glacier ice eventually forms.

14.1 FORMATION OF GLACIER ICE

14.2 Aerial view of the Hoh Glacier, a valley glacier on Mt. Olympus, Olympic National Park, Washington. [National Park Service.]

through valleys in the mountain barrier to form an **outlet glacier,** as it has done along the coasts of Greenland and Antarctica.

Piedmont glaciers form when two or more glaciers emerge from their valleys and coalesce to form an apron of moving ice on the plains below.

Ice sheets are broad, moundlike masses of glacier ice that tend to spread radially under their own weight. The Vatna Glacier of Iceland is a small ice sheet measuring about 120 by 160 km and 225 m in thickness. A localized sheet of this sort is sometimes called an **ice cap** (see Figure 14.3). The term **continental glacier** is usually reserved for great ice sheets that obscure the mountains and plains of large sections of a continent, such as those of Greenland and Antarctica. On Greenland, ice exceeds 3,000 m in thickness near the center of the ice caps. Ice in Antarctica averages about 2,300 m in thickness, and in some places this ice is over 4,000 m thick.

14.3 A small ice cap on Axel Heiberg Island, Northwest Territories, Canada. [Austin Post, University of Washington.]

DISTRIBUTION OF MODERN GLACIERS

Modern glaciers cover approximately 10 percent of the land area of the world. They are found in widely scattered locations in North and South America, Europe, Asia, Africa, Antarctica, and Greenland, and on many of the north polar islands and the Pacific islands of New Guinea and New Zealand. A few valley glaciers are located almost on the equator. Mount Kenya in East Africa, for instance, only 0.5° from the equator, rises over 5,100 m into the tropical skies and supports at least ten valley glaciers.

The total land area covered by existing glaciers is estimated as 17.9 million km², of which the Greenland and Antarctic ice sheets account for about 96 percent. The Antarctic ice sheet covers approximately 15.3 million km², and the Greenland sheet covers about 1.8 million km². Small ice caps and numerous mountain glaciers scattered around the world account for the remaining 4 percent.

NOURISHMENT AND WASTAGE OF GLACIERS

When the weight of a mass of snow, firn, and ice above the snow line becomes great enough, movement begins and a glacier is created. The moving stream flows downward across the snow line until it reaches an area where the loss through evaporation and melting is so large that the forward edge of the glacier can push no farther. A glacier, then, can be divided into two zones: a **zone of accumulation** and a **zone of wastage**. Both are illustrated in Figure 14.4.

14.4 A glacier is marked by a zone of accumulation and a zone of wastage. Within a glacier, ice may lie either in the zone of fracture or deeper in the zone of flow. A valley glacier originates in a basin, the *cirque*, and it is separated from the headwall of the cirque by a large crevasse known as the *bergschrund*.

The position of the front of a glacier depends on the relationship between the glacier's rate of nourishment and its rate of wastage. When nourishment just balances wastage, the front becomes stationary, and the glacier is said to be in equilibrium. This balance seldom lasts for long, however, for a slight change in either nourishment or wastage will cause the front to advance or recede.

Today most of the glaciers of the world are receding. With only a few exceptions this process has been going on since the latter part of the nineteenth century, although at varying rates. A striking feature of modern glaciers is that they follow the same general pattern of growth and wastage the world over and serve as indicators of widespread climatic changes.

Valley glaciers are nourished not only in the zone of accumulation but also by great masses of snow that avalanche down the steep slopes along their course. In fact, according to one interpretation, avalanches caused by earthquakes have enabled certain glaciers to advance in a single month as far as they would have if fed by the normal snowfall of several years.

Below the snow line, wastage takes place through a double process of evaporation and melting known as **ablation**. If a glacier terminates in a body of water, great blocks of ice break off and float away in a process called **calving**. This is the action that produces the **icebergs** of the polar seas.

GLACIER MOVEMENT

Except in rare cases, glaciers move only a few centimeters or at most a few meters per day. That they actually move, however, can be demonstrated in several ways. The most conclusive test is to measure the movement directly, by emplacing a row of stakes across a valley glacier. As time passes, the stakes move down-valley with the advancing ice, the center stakes more rapidly than those near the valley walls. A second source of evidence is provided by the distribution of rock material on the surface of a glacier. When we examine the boulders and cobbles lying along a valley glacier, we find that many of them could not have come from the walls immediately above and that the only possible source lies up-valley. We can infer, then, that the boulders must have been carried to their present position on the back of the glacier. Another indication of glacier movement is that when a glacier melts, it often exposes a rock floor that has been polished, scratched, and grooved. It is simplest to explain this

14.5 The flow patterns in the Malaspina Glacier, Alaska, are emphasized by the dark bands of ice-carried debris. [Austin Post, University of Washington.]

surface by assuming that the glacier actually moved across the floor, using embedded debris to polish, scratch, and groove it.

Clearly, then, a glacier moves (see Figure 14.5). In fact, different parts of it move at different rates. But although we know a good bit about how a glacier flows forward, certain phases are not yet clearly understood. In any event, we can distinguish two zones of movement: (1) an upper zone, between 30 and 60 m thick, which reacts like a brittle substance—that is, it breaks sharply rather than undergoing gradual, permanent distortion; and (2) a lower zone, which because of the pressure of the overlying ice behaves like a plastic substance. The first is the **zone of fracture;** the second is the **zone of flow** (see Figure 14.4).

As plastic deformation takes place in the zone of flow, the brittle ice above is carried along. But the zone of flow moves forward at different rates—faster in some parts, more slowly in others—and the rigid ice in the zone of fracture is unable to adjust itself to this irregular advance. Consequently, the upper part of the glacier cracks and shatters, giving rise to a series of deep, treacherous **crevasses** (see Figures 14.4 and 14.6).

A glacier attains its greatest velocity somewhere above the valley floor, in midstream; for the sides and bottom are retarded by friction against the valley walls and beds. In this respect, the movement of an ice stream resembles that of a stream of water. The mechanics of ice flow, however, are still a matter of study—a study made difficult by the fact that we cannot actually observe the zone of flow because it lies concealed within the glacier. Yet the ice from the zone

14.6 Crevassed surface of the Columbia glacier, Chugach Mountains, Alaska. [Austin Post, University of Washington.]

of flow eventually emerges at the snout of the glacier, and there it can be studied. We find that by the time it has emerged, it is brittle; but it retains the imprint of movement by flow. The individual ice crystals are now several centimeters in size; in contrast, crystals in ice newly formed from firn measure but a fraction of a centimeter. We can conclude that the ice crystals have grown by recrystallization as they passed through the zone of flow. The ice at the snout is also marked by bands that represent shearing and differential movement within the glacier. Recrystallization has taken place along many of the old shear planes, and along others the debris carried forward by the ice has been concentrated. These observations suggest that some movement in the zone of flow has taken place as a result of shearing.

Measurements of the direction of flow of ice in modern glaciers show that in the zone of wastage, the flow of ice is upward toward the ice surface at a low angle. In the zone of accumulation, the direction of movement is downward from the ice surface at a low angle (see Figure 14.4).

No matter how the glacier ice moves in the interior of a glacier, there is also a movement involving the entire ice tongue or sheet. The glacier literally slips along its base, moving across the underlying ground surface. This **basal slip** is added to the internal flow of the glacier to give the total movement of the ice body (see Figure 14.7).

14.8 Rapid surges along the Yanert glacier in the Alaska Range have displaced morainal loops down-valley. The photograph was taken in 1964, and the latest surge at that time had been in 1942. [Austin Post, University of Washington.]

Although most glacier motion is slow, some glaciers at times display a very rapid motion called a **surge**. Surges, or catastrophic advances, are now known from certain mountain and valley glaciers around the world (Figure 14.8). The largest surge so far reported seems to be that of the Bråsellbreen, a portion of a roughly circular ice cap on Spitzbergen. Here the ice advanced up to 20 km along a 21-km front some time between 1935 and 1938, as revealed by aerial photographs taken in those two separate years. The most rapid advance authenticated by actual observation is 110 m/day on the Kutiah Glacier, which is located in the mountains of northern India.

Glaciers surge because of an instability that allows gravity to tear them loose from the ground surface on which they lie. The cause of this instability is not yet completely understood, but it involves the accumulation of an excess of water at the glacier's base. One

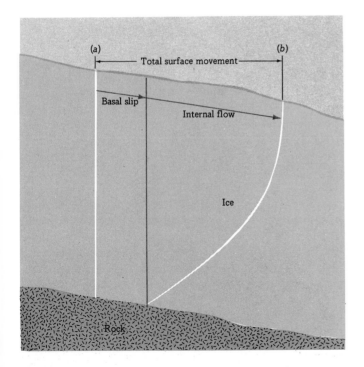

14.7 Forward motion of a glacier consists in part of the internal flow of the ice and in part of the slippage of the entire glacier over the base. [After Robert P. Sharp, *Glaciers*, Oregon System of Higher Education, Eugene, 1960.]

way this water may accumulate is through a thickening of the ice and thus an increase of pressure in the glacier. At the base this increase may be sufficient to allow some melting there.

Surges are as yet unrecorded from the Antarctic ice cap. Nevertheless, it has been estimated that should a surge occur there, it might well involve the movement into the Antarctic Ocean of enough ice to raise the world's sea level between approximately 10 and 30 m during a period of a few years or a few tens of years. The results would make some science fiction tame reading by comparison.

TEMPERATURES OF GLACIERS

All glaciers are cold, but some are colder than others. Therefore students of glaciers speak of "warm" and "cold" glaciers. A **cold glacier** is one in which no surface melting occurs during the summer months; its temperature is always below freezing. A **warm glacier** is one that reaches the melting temperature throughout its thickness during the summer season.

Obviously no glacier can exist above the melting point of ice; yet there are glaciers that hover at this temperature. All glaciers must form at a subfreezing temperature, and therefore we must ask how a glacier warms from its formation temperature to the melting point. The sun's heat cannot penetrate more than a few meters into the poorly conducting ice. Some other mechanism must operate to warm a glacier. One such mechanism involves the downward movement of water from the surface of the glacier. There the heat of the sun melts surface and near-surface ice, and the meltwater percolates downward into the glacier. When eventually it freezes, it gives up heat at the rate of 80 cal/g (gram) of water. (Because each gram of ice requires 0.5 cal to rise 1°C, these 80 cal increase the temperature of 160 g by 1°C.) In addition, there is the normal heat flow from the earth to the surface and the heat generated by the friction of the moving ice.

14.2 RESULTS OF GLACIATION

MOVEMENT OF MATERIAL

Glaciers have special ways of eroding, transporting, and depositing earth materials. A valley glacier, for example, acquires debris by means of **frost action**,

landsliding, and **avalanching.** Fragments pried loose by frost action clatter down from neighboring peaks and come to rest on the back of the glacier. And great snowbanks, unable to maintain themselves on the steep slopes of the mountainsides, avalanche downward to the glacier, carrying along quantities of rock debris and rubble. This material is either buried beneath fresh snow or avalanches, or tumbles into gaping crevasses in the zone of fracture and is carried along by the glacier.

When a glacier flows across a fractured or jointed stretch of bedrock, it may lift up large blocks of stone and move them off. This process is known as **plucking,** or **quarrying.** The force of the ice flow itself may suffice to pick up the blocks, and the action may be helped along by the great pressures that operate at the bottom of a glacier. Suppose the moving ice encounters a projection or rock jutting up from the valley floor. As the glacier ice forces itself over and around the projection, the pressure on the ice is increased, and some of the ice around the rock may melt. This meltwater trickles toward a place of lower pressure, perhaps into a crack in the rock itself. There it refreezes, forming a strong bond between the glacier and the rock. Continued movement by the glacier may then tear the block out of the valley floor.

At the heads of valley glaciers, plucking and frost action sometimes work together to pry rock material loose. Along the back walls of the collection basins of mountain glaciers, great hollows called **cirques,** or **amphitheaters,** develop in the mountainside (see Figure 14.9). As the glacier begins its movement downslope, it pulls slightly away from the back wall, forming a crevasse known as a **bergschrund.** One wall of the bergschrund is formed by the glacier ice; the other is formed by the nearly vertical cliff or bedrock. During the day, meltwater pours into the bergschrund and fills the openings in the rock. At night the water freezes, producing pressures great enough to loosen blocks of rock from the cliff. Eventually these blocks are incorporated into the glacier and are moved away from the headwall of the cirque.

The streams that drain from the front of a melting glacier are charged with **rock flour,** very fine particles of pulverized rock. So great is the volume of this material that it gives the water a characteristically grayish blue color similar to that of skim milk. Here then is further evidence of the grinding power of the glacier mill.

Glaciers also pick up rock material by means of abrasion. As the moving ice drags rocks, boulders, pebbles, sand, and silt across the glacier floor, the bedrock is cut away as though by a great rasp or file. And the cutting tools themselves are abraded. It is this mutual abrasion that produces rock flour and gives a high **polish** to many of the rock surfaces across which a glacier has ridden. But abrasion sometimes produces scratches, or **striations,** on both the bedrock floor and on the grinding tools carried by the ice (see Figure 14.10). More extensive abrasion creates deep gouges, or **grooves,** in the bedrock. The striations and grooves along a bedrock surface show the direction of the glacier's movement. At Kelleys Island, in Lake Erie north of Sandusky, Ohio, the bedrock is marked by grooves 0.3 to 0.6 m deep and 0.6 to 1 m wide. In the Mackenzie Valley, west of Great Bear Lake in Canada, grooves as wide as 45 m have been described, with an average depth of 15 m and lengths ranging from several hundred meters to over 1 km.

EROSIONAL EFFECTS

The erosional effects of glaciers are not limited to the fine polish and striations mentioned above, however; glaciers also operate on a much grander scale, producing spectacularly sculptured peaks and valleys in the mountainous areas of the world (Figure 14.11).

Cirques As we have seen, a **cirque** is the basin from which a mountain glacier flows, the focal point for the glacier's nourishment. After the glacier has disappeared and all its ice has melted away, the cirque is revealed as a great amphitheater, or bowl, with one side partially cut away. The back wall rises a few scores of meters to over 900 m above the floor, often as an almost vertical cliff. The floor of a cirque lies below the level of the low ridge separating it from the valley of the glacier's descent. The lake that forms in the bedrock basin of the cirque floor is called a **tarn.**

A cirque begins with an irregularity in the mountainside formed either by preglacial erosion or by a process called **nivation,** a term that refers to erosion beneath and around the edges of a snowbank. Nivation works in the following way: When seasonal thaws melt some of the snow, the meltwater seeps down to

the bedrock and trickles along the margin of the snowbank. Some of the water works its way into cracks in the bedrock, where it freezes again, producing pressures that loosen and pry out fragments of the rock. These fragments are moved off by solifluction, by rill wash, and perhaps by mass wasting, forming a shallow basin. As this basin gradually grows deeper, a cirque eventually develops. Continued accumulation of snow leads to the formation of firn; if the basin becomes deep enough, the firn is transformed into ice.

14.10 Striations on bedrock about 1.5 km west of Fair Haven, Vermont. The pen points about N, 10°W and marks the general direction of the younger of two sets of striations. The older set is oriented about N, 10°E. [Paul MacClintock.]

14.11 The Grand Teton Mountains of Wyoming illustrate a landscape typical of mountain glaciation. [National Park Service.]

Finally the ice begins to flow out of the cirque into the valley below, and a small glacier is born.

The actual mechanism by which a cirque is enlarged is still a matter of dispute. Some observers claim that frost action and plucking on the cirque wall within the bergschrund are sufficient to produce precipitous walls hundreds of meters in height. Others, however, point out that the bergschrund, like all glacier crevasses, remains open only in the zone of fracture, 60 m at most. Below that depth, pressures cause the ice to deform plastically, closing the bergschrund.

This debate has led to the development of the so-called **meltwater hypothesis** to explain erosion along the headwalls below the base of the bergschrund. The proponents of this theory explain that meltwater periodically descends the headwalls of cirques, melts its way down behind the ice and into crevices in the rock, and there freezes at night and during cold spells. The material thus broken loose is then removed by the glacier, and cirque erosion proceeds mainly by this form of headwall recession.

Horns, arêtes, and cols A **horn** is a spire of rock formed by the headward erosion of a ring of cirques around a single high mountain. When the glaciers originating in these cirques finally disappear, they leave a steep, pyramidal mountain outlined by the headwalls of the cirques. The classic example of a horn is the famous Matterhorn of Switzerland (Figure 14.12).

An **arête** (French for "fishbone," "ridge," or "sharp edge") is formed when a number of cirques gnaw into a ridge from opposite sides. The ridge becomes knife-edged, jagged, and serrated (see Figure 14.12).

A **col** (from the Latin *collum*, "neck"), or pass, is fashioned when two cirques erode headward into a ridge from opposite sides. When their headwalls meet, they cut a sharp-edged gap in the ridge (Figure 14.12).

Glaciated valleys Instead of fashioning their own valleys, glaciers probably follow the course of preexisting valleys, modifying them in a variety of ways; usually the resulting valleys have a **broad U-shaped cross profile**, whereas mountain valleys created exclusively by streams have **narrow V-shaped cross profiles**. Because the tongue of an advancing glacier is rela-

(a)

(b)

Horn — Arête

Col   Cirque

(c)

(a)

(b)

(c)

14.12 (left) The progressive development of cirques, horns, arêtes, and cols. (a) Valley glaciers have produced cirques; but since erosion has been moderate, much of the original mountain surface has been unaffected by the ice. The result of more extensive glacial erosion is shown in (b). In (c) glacial erosion has affected the entire mass and has produced not only cirques but also a horn, jagged, knife-edged arêtes, and cols. [After William Morris Davis, "The Colorado Front Range," *Ann. Assoc. Am. Geo.*, vol. 1, p. 57, 1911.]

14.13 (right) A mountainous area before, during, and after glaciation. [Redrawn from William Morris Davis, "The Sculpture of Mountains by Glaciers," *Scot. Geog. Mag.*, vol. 22, pp. 80–83, 1906.]

tively broad, it tends to broaden and deepen the V-shaped stream valleys, transforming them into broad U-shaped troughs. And because the moving body of ice has difficulty manipulating the curves of a stream valley, it tends to straighten and simplify its course. In this process of straightening, the ice snubs off any spurs of land that extend into it from either side. The cliffs thus formed are shaped like large triangles, or flatirons, with their apex upward, and are called **truncated spurs** (see Figure 14.13).

A glacier also gives a mountain valley a characteristic longitudinal profile from the cirque downward. The course of a glaciated valley is marked by a series

of **rock basins,** probably formed by plucking in areas where the bedrock was shattered or closely jointed. Between the basins are relatively flat stretches of rock that was more resistant to plucking. As time passes, the rock basins may fill up with water, producing a string of lakes that are sometimes referred to as **paternoster lakes** because they resemble a string of beads.

14.2   RESULTS OF GLACIATION

14.14 The coastline of British Columbia and the panhandle of Alaska are intricately laced by fiords, glaciated valleys flooded by the sea. This section of British Columbia shoreline is just north of Vancouver Island.

Queen Charlotte Sound

0   25   50

km

Hanging valleys are another characteristic of mountainous areas that have undergone glaciation. The mouth of a hanging valley is left stranded high above the main valley through which a glacier has passed. As a result, streams from hanging valleys plummet into the main valley in a series of falls and plunges. Hanging valleys may be formed by processes other than glaciation, but they are almost always present in mountainous areas that formerly supported glaciers and are thus very characteristic of past valley glaciation. What has happened to leave these valleys stranded high above the main valley floor? During the time when glaciers still moved down the mountains, the greatest accumulation of ice would tend to travel along the central valley. Consequently, the erosive action there would be greater than in the tributary valleys with their relatively small glaciers, and the main valley floor would be cut correspondingly deeper. This action would be even more pronounced where the main valley was underlain by rock that was more susceptible to erosion than was the rock under the tributary valleys. Finally, some hanging valleys were probably created by the straightening and widening action of a glacier on the main valley. In any event, the difference in level between the tributary valleys and the main valley does not become apparent until the glacier has melted away.

Cutting deep into the coasts of Alaska, Norway, western Canada, Alaska, Greenland, Labrador, and New Zealand are deep, narrow arms of the sea—**fiords** (Figure 14.14). Actually these inlets are stream valleys that were modified by glacier erosion and then were partially filled by the sea. The deepest known fiord, Vanderford in Vincennes Bay, Antarctica, has a maximum depth of 2,287 m.

Some valleys have been modified by continental glaciers rather than by the valley glaciers that we have been discussing so far. The valleys occupied by the Finger Lakes of central New York State are good examples. These long, narrow lakes lie in basins that were carved out by the ice of a continental glacier. As the great sheet of ice moved down from the north, its progress seems to have been checked by the northern scarp of the Appalachian Plateau. But some of the ice moved up into the valleys that had previously drained the plateau. The energy concentrated in the valleys was so great that the ice was able to scoop out the basins that are now filled by the Finger Lakes.

Asymmetric rock knobs and hills In many places, glacier erosion of bedrock produces small, rounded, asymmetric hills with gentle, striated, and polished slopes on one side and steeper slopes lacking polish and striations on the opposite side. The now-gentle slope faced the advancing glacier and was eroded by abrasion. The opposite slope has been steepened by the plucking action of the ice as it rode over the knob (Figure 14.15).

Large individual hills have the same asymmetric profiles as the smaller hills have. Here, too, the gentle slope faced the moving ice.

TYPES OF GLACIAL DEPOSITS

The debris carried along by a glacier is eventually deposited either because the ice that holds it melts or, less commonly, because the ice smears the debris across the land surface.

The general term **drift** is applied to all deposits that are laid down directly by glaciers or that, as a result of glacial activity, are laid down in lakes, oceans, or streams. The term dates from the days when geologists thought that the unconsolidated cover of sand and gravel blanketing much of Europe and America had been made to drift into its present position either by the sea or by icebergs. Drift can be divided into two general categories: **unstratified** and **stratified**.

Deposits of unstratified drift Unstratified drift laid down directly by glacier ice is called **till**. It is composed of rock fragments of all sizes mixed together in random fashion, ranging from boulders weighing several tonnes to tiny clay and colloid particles (see Figures 14.16 and 14.17). Many of the large pieces are striated, polished, and faceted as a result of the wear they underwent while transported by the glaciers (see Figures 14.18 and 14.19). Some of the material picked up

14.15 The smooth, gentle slopes of these rock knobs have been produced by glacial abrasion, and the steep slopes have been produced by plucking. Ice movement was from right to left. [Geological Survey of Canada.]

along the way was smeared across the landscape during the glacier's progress, but most of it was dumped when the rate of wastage began to exceed the rate of nourishment and the glacier gradually melted away.

The type of till varies from one glacier to another. Some tills, for instance, are known as **clay tills** because clay-sized particles predominate, with only a scattering of larger units. Many of the most recent tills in northeastern and eastern Wisconsin are of this type, but in many parts of New England, the tills are composed for the most part of large rock fragments and boulders. Deposits of this sort are known as **boulder,** or **stony, tills**.

Some till deposits seem to have been worked over by meltwater. The materials have begun to be sorted out according to size, and some of the finer particles may even have been washed away. This is the sort of winnowing action we should expect to find near the nose of a melting glacier, where floods of meltwater wash down through the deposits.

Till is deposited by glaciers in a great variety of topographic forms, including moraines, drumlins, erratics, and boulder trains.

Moraines Moraine is a general term used to describe many of the landforms that are composed largely of till.

A **terminal moraine,** or **end moraine,** is a ridge of till that marks the utmost limit of a glacier's advance. These ridges vary in size from ramparts scores of meters high to very low, interrupted walls of debris. A terminal moraine forms when a glacier reaches the critical point of equilibrium—the point at which it wastes away at exactly the same rate as it is nourished. Although the front of the glacier is now stable, ice continues to push forward; and so it will deliver a continuous supply of rock debris. As the ice melts in the zone of wastage, the debris is dumped, and the terminal moraine grows. At the same time, water from the melting ice pours down over the till and sweeps part of it out in a broad flat fan that butts against the forward edge of the moraine like a giant ramp (Figure 14.18).

The terminal moraine of a mountain glacier is crescent-shaped, with the convex side extending down-valley. The terminal moraine of a continental ice sheet is a broad loop or series of loops traceable for many miles across the countryside.

Behind the terminal moraine and at varying distances from it, a series of smaller ridges known as **recessional moraines** may build up. These ridges mark the position where the glacier front was stabilized temporarily during the retreat of the glacier.

Not all the rock debris carried by a glacier finds its way to the terminal and recessional moraines, however. A great deal of till is laid down as the main body of the glacier melts to form gently rolling plains across the valley floor. Till in this form, called a **ground moraine,** may be a thin veneer lying on the bedrock, or it may form a deposit scores of meters thick, partially or completely clogging preglacial valleys.

Finally, valley glaciers produce two special types of moraine. While a valley glacier is still active, large amounts of rubble keep tumbling down from the valley walls, collecting along the side of the glacier. When the ice melts, this debris is stranded as a ridge along each side of the valley, forming a **lateral moraine.** At its down-valley end, the lateral moraine may grade into a terminal moraine. The other special type of deposit produced by valley glaciers is a **medial moraine,** created when two valley glaciers join to form a single

14.16 The range in size of the particles composing till is very large. In this photograph of an exposure of till near Guilford, Connecticut, note boulders and cobbles mixed with smaller particles ranging all the way down to colloid size. Spade gives scale. [Sheldon Judson.]

14.17 This moraine composed of till lies along the East Rosebud River and records the last major advance of glaciers from the Beartooth Mountains, Montana. The largest boulders measure 3 m in maximum dimension. [Sheldon Judson.]

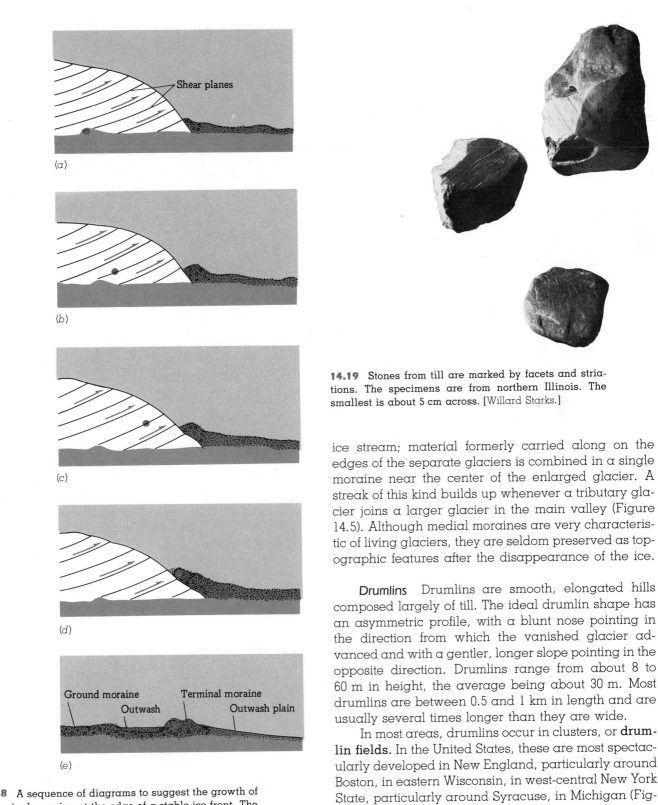

14.18 A sequence of diagrams to suggest the growth of a terminal moraine at the edge of a stable ice front. The progressive movement of a single particle is shown. In (a) it is moved by the ice from the bedrock floor. Forward motion of ice along a shear plane carries it ever closer to the stabilized ice margin, where it is finally deposited as part of the moraine in (d). Diagram (e) represents the relation of the terminal moraine, ground moraine, and outwash after the final melting of the glacier.

14.19 Stones from till are marked by facets and striations. The specimens are from northern Illinois. The smallest is about 5 cm across. [Willard Starks.]

ice stream; material formerly carried along on the edges of the separate glaciers is combined in a single moraine near the center of the enlarged glacier. A streak of this kind builds up whenever a tributary glacier joins a larger glacier in the main valley (Figure 14.5). Although medial moraines are very characteristic of living glaciers, they are seldom preserved as topographic features after the disappearance of the ice.

Drumlins Drumlins are smooth, elongated hills composed largely of till. The ideal drumlin shape has an asymmetric profile, with a blunt nose pointing in the direction from which the vanished glacier advanced and with a gentler, longer slope pointing in the opposite direction. Drumlins range from about 8 to 60 m in height, the average being about 30 m. Most drumlins are between 0.5 and 1 km in length and are usually several times longer than they are wide.

In most areas, drumlins occur in clusters, or **drumlin fields.** In the United States, these are most spectacularly developed in New England, particularly around Boston, in eastern Wisconsin, in west-central New York State, particularly around Syracuse, in Michigan (Figure 14.20), and in certain sections of Minnesota. In Canada, extensive drumlin fields are located in west-

14.20 Map of part of the drumlin-field area south and east of Charlevoix, Michigan. Ice moved toward the south-southeast. [After Frank Leverett and F. B. Taylor, "The Pleistocene of Indiana and Michigan and the History of the Great Lakes," *U.S. Geol. Surv. Monog. 53*, p. 311, 1915.]

14.21 This aerial photograph shows the drumloidal pattern of hills and intervening grooves formed parallel to the flow of glacier ice on the Nechako Plateau, British Columbia. Flow of ice was from right to left. [Geological Survey of Canada.]

ern Nova Scotia and in northern Manitoba and Saskatchewan; Figure 14.21 shows a drumlin field in British Columbia.

Just how drumlins were formed is still not clear. Because their shape is a nearly perfect example of streamlining, it seems probable that they were formed deep within active glaciers in the zone of plastic flow.

Erratics and boulder trains A stone or a boulder that has been carried from its place of origin by a glacier and left stranded on bedrock of different composition is called an **erratic**. The term is used whether the stone is embedded in a till deposit or rests directly on the bedrock. Some erratics weigh several tonnes, and a few are even larger. Near Conway, New Hampshire, there is a granite erratic 27 m in maximum dimension, weighing close to 9,000 t. Although most erratics have traveled only a limited distance, many have been car-

ried along by the glacier for hundreds of kilometers. Chunks of native copper torn from the Upper Peninsula of Michigan, for example, have been transported as far as southeastern Iowa, a distance of nearly 800 km, and to southern Illinois, over 960 km.

Boulder trains consist of a series of erratics that have come from the same source, usually with some characteristic that makes it easy to recognize their common origin. The trains appear either as a line of erratics stretching down-valley from their source or in a fan-shaped pattern with the apex near the place of origin. By mapping boulder trains that have been left behind by continental ice sheets, we can obtain an excellent indication of the direction of the ice flow (see Figure 14.22).

Deposits of stratified drift Stratified drift is ice-transported sediment that has been washed and sorted by glacial meltwaters according to particle size. Because water is a much more selective sorting agent than is ice, deposits of stratified drift are laid down in recognizable layers, unlike the random arrangements of particles typical of till. Stratified drift occurs in outwash and kettle plains, eskers, kames, and varves—all discussed below.

Outwash sand and gravel The sand and gravel that are carried outward by meltwater from the front of a glacier are referred to as **outwash** (see Figure 14.23). As a glacier melts, streams of water heavily loaded with reworked till or with material washed directly from the ice weave a complex, braided pattern of channels across the land in front of the glacier. These

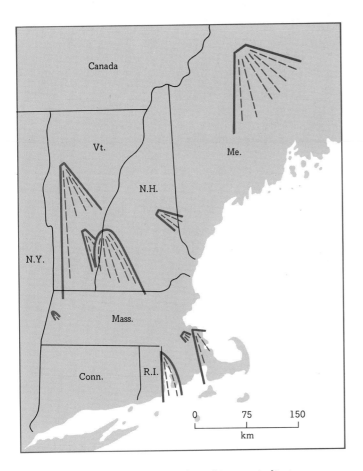

14.22 The boulder trains plotted on this map indicate the general direction of ice movement across New England. The apex of the fan indicates the area from which the boulders were derived; the fan itself covers the area across which they were deposited. [After J. W. Goldthwait in R. F. Flint, "Glacial Map of North America," *Geol. Soc. Am. Spec. Paper 60*, 1945.]

14.23 Outwash forms at the snout of Brady Glacier, Fairweather Range, Alaska. [Austin Post, University of Washington.]

streams, choked with clay, silt, sand, and gravel, rapidly lose their velocity and dump their load of debris as they flow away from the ice sheet. In time, a vast apron of bedded sand and gravel is built up, which may extend for kilometers beyond the ice front. If the zone of wastage happens to be located in a valley, the outwash deposits are confined to the lower valley and compose a **valley train**. But along the front of a continental ice sheet, the outwash deposits stretch out for kilometers, forming what is called an **outwash plain** (see Figure 14.24).

14.24 These sand and gravel deposits represent the outwash from a now-vanished continental glacier. North of Otis Lake, Wisconsin. [W. C. Alden, U.S. Geological Survey.]

Kettles Sometimes a block of stagnant ice becomes isolated from the receding glacier during wastage and is partially or completely buried in till or outwash before it finally melts. When it disappears, it leaves a **kettle**, a pit, or depression, in the drift (see Figure 14.25). These depressions range from a few meters to several kilometers in diameter and from a few meters to over 30 m in depth. Many outwash plains are pockmarked with kettles and are referred to as **pitted outwash plains**. As time passes, water sometimes fills the kettles to form lakes or swamps, features found through much of Canada and the northern United States.

(a)

(b)

14.25 Kettle formation: (a) Stagnant ice is almost buried by outwash. (b) Eventual melting forms a depression.

Eskers and crevasse fillings   Winding, steep-walled ridges of stratified gravel and sand, sometimes branching and often discontinuous, are called **eskers** (see Figure 14.26). They usually vary in height from about 3 to 15 m, although a few are over 30 m high. Eskers range from a fraction of a kilometer to over 160 km in length, but they are only a few meters wide. Most investigators believe that eskers were formed by the deposits of streams running through tunnels beneath stagnant ice. Then, when the body of the glacier finally disappeared, the old stream deposits were left standing as a ridge.

Crevasse fillings are similar to eskers in height, width, and cross profile, but unlike the sinuous and branching pattern of eskers, they run in straight ridges. As their name suggests, they were probably formed by the filling of a crevasse in stagnant ice.

Kames and kame terraces   In many areas, stratified drift has built up low, relatively steep-sided hills called **kames,** either as isolated mounds or in clusters. Unlike drumlins, kames are of random shape and the deposits that compose them are stratified. They were formed by the material that collected in openings in stagnant ice. In this sense they are similar to crevasse fillings but without the linear pattern (see Figure 14.27).

A **kame terrace** is a deposit of stratified sand and

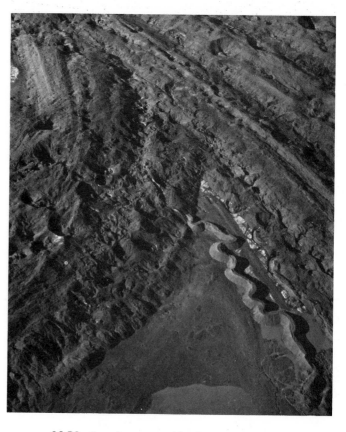

14.26   An esker exposed by the retreat of the Malaspina Glacier, Alaska. Note depressions in the areas of rough topography. They have formed by melting of ice blocks. [Austin Post, University of Washington.]

14.27   This isolated kame in eastern Wisconsin was formed by the partial filling of an opening in stagnant glacier ice. The melting of the ice has left this steep-sided hill of stratified material. [Raymond C. Murray.]

(a)

(b)

(c)

14.28 The sequence in the development of a kame terrace. (a) Ice wasting from an irregular topography lingers longest in the valleys. (b) While the ice still partially fills one of these valleys, outwash may be deposited between it and the valley walls. (c) The final disappearance of the ice leaves the outwash in the form of terraces along the sides of the valley.

gravel that has been laid down between a wasting glacier and an adjacent valley wall. When the glacier disappears, the deposit stands as a terrace along the side of the valley (see Figure 14.28).

Varves A **varve** is a pair of thin sedimentary beds, one coarse and one fine (see Figure 7.7). This couplet of beds is usually interpreted as representing the deposits of a single year and is thought to form in the following way: During the period of summer thaw, waters from a melting glacier carry large amounts of clay, fine sand, and silt out into lakes along the ice margin. The coarser particles sink fairly rapidly and blanket the lake floor with a thin layer of silt and silty sand. But as long as the lake is unfrozen, the wind creates currents strong enough to keep the finer clay particles in suspension. When the lake freezes over in the winter, these wind-generated currents cease, and the fine particles sink through the quiet water to the bottom, covering the somewhat coarser summer layer. A varve is usually a few millimeters to 1 cm thick, although thicknesses of 5 to 8 cm are not uncommon. In rare instances, thicknesses of 30 cm or more are known.

COMPARISON OF VALLEY- AND CONTINENTAL-GLACIATION FEATURES

Some of the glacial features that we have been discussing are more common in areas that have undergone valley glaciation; others usually occur only in regions that have been overridden by ice sheets; many other features, however, are found in both types of area. Table 14.1 lists and compares the features that are characteristic of the two types.

INDIRECT EFFECTS OF GLACIATION

The glaciers that diverted rivers, carved mountains, and covered half a continent with debris also gave rise to a variety of indirect effects that were felt far beyond the glaciers' immediate margins. Sea level fell with the

TABLE 14.1

Valley- and continental-glaciation
features compared

Features	Valley glacier	Continental ice sheet
Striations, polish, etc.	Common	Common
Cirques	Common	Absent
Horns, arêtes, cols	Common	Absent
U-shaped valley, truncated spurs, hanging valleys	Common	Rare
Fiords	Common	Absent
Till	Common	Common
Terminal moraine	Common	Common
Recessional moraine	Common	Common
Ground moraine	Common	Common
Lateral moraine	Common	Absent
Medial moraine	Common, easily destroyed	Absent
Drumlins	Rare or absent	Locally common
Erratics	Common	Common
Stratified drift	Common	Common
Kettles	Common	Common
Eskers, crevasse fillings	Rare	Common
Kames	Common	Common
Kame terraces	Common	Present in hilly country

accumulation of glacier ice and rose as glaciers melted. Desert and near-desert areas were better watered during maximum glaciation. In now-temperate latitudes beyond the ice margin, the climate also changed, the wind laid down vast blankets of dust (called **loess**), and the slope processes were speeded by solifluction. We defer the discussion of these subjects to Chapter 16.

14.3 DEVELOPMENT OF THE GLACIAL THEORY

Geologists have made extensive studies of the behavior of modern glaciers and have carefully interpreted the traces left by glaciers that disappeared thousands of years ago. On the basis of their studies, they have developed the **glacial theory:** In the past, great ice sheets covered large sections of the earth where no ice now exists, and many existing glaciers once extended far beyond their present limits.

THE BEGINNINGS

The glacial theory took many years to evolve, years of trying to explain the occurrence of erratics and the vast expanses of drift strewn across northern Europe, the British Isles, Switzerland, and adjoining areas. The exact time when inquisitive minds first began to seek an explanation of these deposits is shrouded in the past, but by the beginning of the eighteenth century, explanations of what we now know to be glacial deposits and features were finding their way into print. According to the most popular early hypothesis, a great inundation had swept these deposits across the face of the land with cataclysmic suddenness or else had drifted them in by means of floating icebergs. Then, when the flood receded, the material was left where it is now.

By the turn of the nineteenth century, a new theory was in the air—the theory of ice transport. We do not know who first stated the idea or when it was first proposed, but it seems quite clear that it was not hailed immediately as a great truth. As the years passed, however, more and more observers became intrigued with the idea. The greatest impetus came from Switzerland, where the activity of living glaciers could be studied on every hand.

In 1821, J. Venetz, a Swiss engineer delivering a paper before the Helvetic Society, presented the argument that Swiss glaciers had once expanded on a great scale. It has since been established that from about 1600 to the middle of the eighteenth century, there was actually a time of moderate but persistent glacier expansion in many localities. Abundant evidence in the Alps, Scandinavia, and Iceland indicates that the climate was milder during the Middle Ages than it is at present, that communities existed and farming was carried on in places later invaded by advancing glaciers or devastated by glacier-fed streams. We know, for example, that a silver mine in the valley of Chamonix was being worked during the Middle Ages and that it was subsequently buried by an advancing glacier, where it lies to this day. And the village of St. Jean de Perthuis has been buried under the Brenva Glacier since about 1600.

Although Venetz's idea did not take hold immediately, by 1834 Jean de Charpentier was arguing in its support before the same Helvetic Society. Yet the theory continued to have more opponents than defenders. It was one of the skeptics, Jean Louis Rodolphe Agassiz (1807–1873), who did more than anyone else to develop the glacial theory and bring about its acceptance.

AGASSIZ

As a young zoologist, Louis Agassiz had listened to Charpentier's explanation; afterward, he undertook to demonstrate to his friend and colleague the error of his ways. During the summer of 1836 the two men made a trip together into the upper Rhône valley to the Getrotz Glacier. Before the summer was over, it was Agassiz who was convinced of his error. In 1837, he spoke before the Helvetic Society, championing the glacial theory and suggesting that during a "great Ice Age" not only the Alps but much of northern Europe and the British Isles were overrun by a sea of ice.

Agassiz's statement of the glacial theory was not accepted immediately, but in 1840 he visited England and won the support of leading British geologists. In 1846, he arrived in the United States, where in the following year he became professor of zoology at Harvard College and later founded the Museum of Comparative Zoology. In this country he convinced geologists of the validity of the glacial theory; by the third quarter of the nineteenth century, the theory was firmly entrenched. The last opposition died with the turn of the twentieth century.

PROOF OF THE GLACIAL THEORY

What proof is there that the glacial theory is valid? The most important evidence is that certain features produced by glacier ice are produced by no other known process. Thus Agassiz and his colleagues found isolated stones and boulders quite alien to their present surroundings. They noticed too that boulders were actually being transported from their original location by modern ice. Some of the boulders they observed were so large that rivers could not possibly have moved them, and others were perched on high places that a river could have reached only by flowing uphill. They also noticed that when modern ice melted, it revealed a polished and striated pavement unlike the surface fashioned by any other known process. To explain the occurrence of these features in areas where no modern glaciers exist, they postulated that the ice once extended far beyond its present limits.

Notice that the development of this theory sprang from the concept of uniformitarianism that we mentioned in Section 1.1: "The present is the key to the past." The proof of glaciation lies not in the authority of the textbook nor in the lecture. It lies in observing modern glacial activity directly and in comparing the results of this activity with features and deposits found beyond the present extent of the ice.

THEORY OF MULTIPLE GLACIATION

Even before universal acceptance of the glacial theory, which spoke of a single great Ice Age, some investigators were coming to the conclusion that the ice had advanced and retreated not just once but **several times in the recent geological past.** By the early twentieth century, a broad fourfold division of the Ice Age, or Pleistocene, had been demonstrated in this country and in Europe. According to this theory, each major

14.29 Analysis of deep-sea sediments shows that the average surface temperature of the ocean fluctuated many times from warm to cold in the recent geologic past. Here is the record for the surface waters of the central Caribbean during the last 700,000 years; the most recent cooling is labeled to represent the last major glaciation in North America. [After Cesare Emeliani and Nicholas J. Shackleton, "The Brunhes Epoch: Isotopic Paleotemperatures and Geochronology," *Science*, vol. 183, p. 513, 1974.]

glacial advance was followed by a retreat and a return to climates that were sometimes even warmer than that of the present. Early in this century, four stages of ice advance had been discovered in the United States and were called **Nebraskan, Kansan, Illinoian,** and **Wisconsin** for the midwestern states where deposits of that particular period were first studied or where they are well exposed. Evidence from Alaska, the Sierra Nevada, and western Europe indicates that one or two major advances preceded the Nebraskan. In central Europe, successive sheets of loess suggest at least 17 periods of Pleistocene glaciation before the last major advance. Finally, analysis of sediment cores taken from the deep oceans reveals repeated cooling and warming of the oceans during the last 2 million years (Figure 14.29). About 20 of these cycles have been identified, and we believe that they represent the climatic changes accompanying repeated glaciation and deglaciation on land.

14.4 Extent of Pleistocene Glaciation

During the maximum advance of the glaciers of the Wisconsin age (the last of the four great ice advances during the Pleistocene), 39 million km² of the earth's surface—about 27 percent of the present land areas—were probably buried by ice. Approximately 15 million km² of North America were covered. Greenland was also under a great mass of ice, as it is now. In Europe, an ice sheet spread southward from Scandinavia across the Baltic Sea and into Germany and Poland, and the Alps and the British Isles supported their own ice caps. Eastward in Asia, the northern plains of Russia were covered by glaciers, as were large sections of Siberia and of the Kamchatka Peninsula and the high plateaus of Central Asia (see Figure 14.30).

In eastern North America, ice moved southward out of Canada to New Jersey, and in the Midwest it reached as far south as Saint Louis. The western mountains were heavily glaciated by small ice caps and valley glaciers. The southernmost glaciation in the United States was in the Sierra Blanca of south-central New Mexico. The maximum extent of the Wisconsin glaciation in the United States and the maximum limit of glaciation during the Pleistocene are shown in Figure 14.31.

14.30 Extent of Pleistocene glaciation (darker areas) in the northern hemisphere. [After Ernst Antevs, "Maps of the Pleistocene Glaciations," *Bull. Geol. Soc. Am.*, 40, p. 636, 1929.]

14.31 Extent of Pleistocene glaciation in the United States. Darker zones indicate area covered at various times during the Wisconsin glaciation. Diagonal lines represent area glaciated during pre-Wisconsin stages but not covered by the later Wisconsin ice. Tint area unglaciated. [After R. F. Flint, "Glacial Map of North America," *Geol. Soc. Am. Spec. Paper 60*, 1945.]

14   GLACIATION

14.32 In Permian time, toward the end of the Paleozoic Era, glaciation affected the ancient continent of Gondwanaland. The supercontinent has since fragmented, and traces of the Permian glaciation are found in South America, South Africa, India, Antarctica, and here in southern Australia. Glacier ice striated, polished, and gouged this slightly metamorphosed siltstone. The knife blade points in the direction of ice movement. [William C. Bradley.]

14.5 Pre-pleistocene Glaciations

So far we have discussed only the glaciers that exist today and those that moved within the last few million years, the Pleistocene. Geologists have found evidence that glaciers appeared and disappeared in other periods as well. The record is fragmentary, as we should expect; for time tends to conceal, jumble, and destroy the various manifested effects of glaciation.

Recent evidence indicates that glaciation and deglaciation, so characteristic of the Pleistocene, reached back into the late Miocene in Alaska. Glacial deposits in Antarctica have been dated by the potassium-argon method as 10 million years in age. Farther back in the geologic-time scale, it seems certain that there were other extensive glaciations. Toward the end of the Paleozoic Era (230 to 290 million years ago), glaciation was widespread in what is now South Africa, South America, India, Australia (Figure 14.32), and Antarctica. There is evidence of glaciation during the Silurian and Devonian in South America; and earlier, in the Ordovician, ice spread across what is now the Sahara Desert. In the late Precambrian, a little over 600 million years ago, glaciation affected various landmasses now in the northern hemisphere. Still earlier, perhaps 850 million years ago, large parts of the ancient continent of Gondwanaland (see Chapter 10) were ice-covered. Some Russian geologists believe that sections of northwestern Siberia were glaciated some 1.2 billion years ago. Extensive glaciation, dating from 2.2 billion years ago, is recorded in what is now south-central Ontario, Canada. At about the same time, glaciation occurred in South Africa.

Much of this glaciation took place in localities where no ice can exist today. The geologic evidence tells us that this is because the landmasses have since drifted from latitudes hospitable to glaciation to new locations (see Chapter 10).

14.6 Causes of Glaciation

As Agassiz did over 100 years ago, we can travel about the world today and observe modern glaciers at work, and we can reason convincingly that glaciers were more extensive in the past than they are at present. We can even make out a good case for the belief that glacier ice advanced and receded many times in the immediate and more remote geological past. But can we explain why glaciation takes place? The answer to that question is a resoundingly equivocal "maybe."

The geologic record has contributed some basic data that any theory of the causes of glaciation must take into account. Among these are the following:

1 Glaciation is related to the elevation and arrangement of continents. During much of geologic time, the continents have been lower than they are today, and shallow seas have flooded across their margins. Such conditions were unfavorable for widespread glaciation. But for several million years the continents have been increasing in elevation until now, in the Pleistocene, they stand on the average an estimated 450 m higher than they did in the mid-Cenozoic. We have

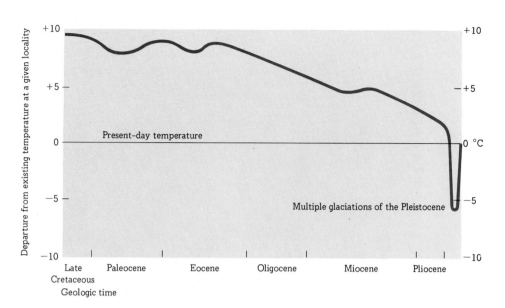

14.33 Temperatures trended downward from the Late Cretaceous to the Pliocene. This general decline was followed by the several glacial and interglacial episodes, the details of which cannot be shown at the scale of this diagram. The values in degrees Celsius represent approximate departures from today's temperatures of localities in the middle latitudes. [Data in part from Erling Dorf.]

Vertical axis: Departure from existing temperature at a given locality

Present-day temperature

Multiple glaciations of the Pleistocene

Late Cretaceous — Paleocene — Eocene — Oligocene — Miocene — Pliocene

Geologic time

already found that the last great Ice Age came with the Pleistocene. We believe too that the other great glaciations coincided with high continents. In addition to having adequate elevations, to allow glaciation to begin, continents must be located in high latitudes, either over the pole of rotation or clustered about it.

2 Glaciation is not due to a slow, long-term cooling off of the earth since its creation. We have already found that extensive glaciation occurred several times during the geological past. But these glacial periods are unusual; for during most of geologic history the climate has been nonglacial.

3 There has been a cooling of the earth's climate beginning during the Tertiary and climaxing with the glacial fluctuations of the Pleistocene. Although the earth has been generally warm during most of its history, evidence now shows that its mean temperature had dropped an estimated 8 to 10°C from the Eocene to the end of the Pliocene and that the glacial and interglacial epochs of the Pleistocene are short-term fluctuations at the end of the long-term cooling (see Figure 14.33).

4 The advance and retreat of glaciers have probably been broadly simultaneous throughout the world. For instance, dating by means of radioactive carbon has demonstrated that geologically recent fluctuations of the continental glaciers in North America occurred at approximately the same time as similar fluctuations in Europe. Furthermore, observations indicate that the general retreat of mountain glaciers now recorded in North America and Europe is duplicated in South America.

One way to approach the problem is to consider that we are dealing with two great classes of causes: one is long-term and has caused the general decline of temperatures since the Eocene; the other is shorter range and accounts for the comings and goings of the ice sheets during the Pleistocene. One operates on a

scale of approximately 50 million years, and the other on the scale of perhaps 100,000 years. In considering these two different types of cause, we can envisage the long-term cause as cooling the climate to the point where one or more other causes of shorter wavelength and lower amplitude can trigger glaciation and deglaciation.

LONG-TERM CHANGES IN CLIMATE

Geologic evidence shows that the earth's "normal" climate has been more equable than that of the present. Through most of the earth's history, the climate has been warmer than the present, and landmasses presently in the poleward zones enjoyed average annual temperatures about 10°C higher than today. At some time during the early Tertiary, probably in the Eocene, average temperatures began to drop.

Causes of long-term climatic fluctuation What could have caused the decline of temperatures during the last 50 or 60 million years? We can make a number of suggestions.

Increasing continentality Today the average height of continents is computed to be about 840 m, as we have stated, about 450 m higher than it was in the beginning of the Eocene. Rising landmasses would result in a general drop in temperature. It is estimated that about one-third of the 10°C decline in average temperatures might be accounted for by the increase in the height of continents since the Eocene.

The rise of continents not only would lower global temperatures but also could interfere with the patterns of worldwide atmospheric circulation. It is conceivable that the transfer of heat from equatorial latitudes toward the poles would be hindered and that the poleward latitudes would become colder. A corollary to this would be an interruption and change in the oceanic circulation and a lessened efficiency in the transfer of heat poleward.

Continental drift It has been suggested by some geologists that continental drift has positioned the landmasses in such a manner as to favor glaciation (see Chapter 10). Thus Antarctica centers on the southern pole and is in an ideal position for the growth of ice sheets. In the northern hemisphere, Europe, Asia, and North America are arranged around the Arctic Ocean, which has restricted communication with the Pacific and the Atlantic. It is argued by some that this arrangement is conducive to the glaciation of the continents. We know, however, that the present pole-continent arrangements have not changed appreciably since the early Tertiary. Therefore, if continental drift is a factor in glaciation, its effect seems to lie in bringing the continents into their present relationship to the poles and to each other, thus facilitating the slow decline in temperature that eventually ended in the glaciations of the Ice Age.

Other causes Some have suggested that an increase in the number of particles—cosmic dust—in space has decreased the amount of energy received from the sun by the earth. The suggestion is based on the fact that we know that there are great amounts of interstellar material forming vast clouds—dust clouds—in space. It is argued that perhaps the passage of the planetary system through such a cloud has caused glaciation. It is an argument that lacks proof or any forseeable way to test it.

Instead of dust in space, it has been suggested that dust from volcanic eruptions has reduced the amount of solar energy that could penetrate the atmosphere and thus caused climatic cooling. However, the evidence indicates that there has been no appreciable variation in volcanism through a large part of geologic time for the earth as a whole. We cannot, therefore, with any confidence, assign climatic changes of large scale to volcanism.

There is, however, the possibility that the amount of heat received by the earth has varied appreciably through time. This heat, known as the **solar constant,** is measured as if it were falling on a plane at the outer edge of the atmosphere and at right angles to the sun's rays. At present it is about 2 cal/cm²-min. Measurements carried on since 1918 suggest that this figure has decreased approximately 3 percent in 50 years. Assuming that the solar radiation changes, some argue that a long-range decrease might account for a cooling of the earth's climate. All we can say is that there appears to have been a measurable change in the solar radiation during historic time. Thus larger changes

may have happened in the past and may have affected the climate.

We have seen that the Ice Age was marked by the coming and going of glaciers that covered up to one-third of the earth's surface. The duration of these glacial and interglacial intervals has been much shorter than the long-term cooling of climate that we have discussed. The ratio of a glacial-interglacial cycle to the longer period of cooling that preceded the fluctuations of the Ice Age is roughly 1/500.

We are now confident that the climate had long been cooling before the first glaciation of the continents of the northern hemisphere took place. It is quite possible, however, that Antarctica supported an ice cap before this, even as it does today in a period that we can call **interglacial.** Whatever the cause or causes of the several glaciations of the Pleistocene, it seems reasonable that they were superimposed on the long-term cooling, which lasted some 50 million years.

Causes of short-term climatic fluctuation An impressive array of suggestions to explain the successive glaciations of the Pleistocene has emerged. These include variation in the radiant energy produced by the sun, fluctuations in the amount of volcanic ash thrown into the atmosphere, variation in the amount of cosmic dust intervening between the sun and the earth, surging of the Antarctic ice sheet, fluctuation of the carbon dioxide content of the atmosphere, variation in the patterns of oceanic circulation, and variation of the earth's position in relation to the sun, that is, variation in the geometry of the earth's orbit around the sun. Each of these theories has some support in fact, but only the last one can be extrapolated backward in time to predict the distribution of heat received at the earth's surface and thus to predict the frequency of Pleistocene glaciations. Furthermore, we can now begin to test these predicted frequencies from geologic evidence.

The changing geometry of the earth's orbit produces a variation in the geographical and seasonal distribution of solar energy on the earth. We refer to the hypothesis using this changing geometry as the **orbital hypothesis,** sometimes called the **Milankovitch hypothesis** after the Yugoslavian astronomer instrumental in its development. It holds that variations of the earth's position in relation to the sun produced climatic changes that brought on the great Ice Age.

Three factors enter into the changing position of the earth in relation to the sun, each of which can be measured and its rate of change determined. One is the **eccentricity** of the earth's orbit around the sun. The earth is somewhat closer to the sun at some epochs than others. This motion has a period of about 93,000 years. A second factor is the **obliquity of the ecliptic,** which produces seasons and is determined by the angle that the earth's axis makes with the plane in which the earth circles the sun. This angle, or obliquity, changes through about 3° every 41,000 years. Finally, there is the **precession of the equinoxes,** which merely means that the axis of the earth wobbles because of the gravitational effect of the sun, the moon, and the planets. Like a giant top, the earth completes one wobble every 21,000 years. All these factors affect the relation of the earth to the sun—hence the amount of heat received at different places on the earth at any one time. The variation can be calculated backward in time to any desired date for any latitude. These calculations have been made, and a curve showing maxima and minima of heat received at the earth during the Pleistocene is the result. The periods of low heat receipt are said to coincide with, and be the cause of, the various ice advances.

The orbital hypothesis has been around in ever-increasing sophistication for over 100 years. In the last 25 years, more and more bits of geologic history have accumulated and now suggest a tantalizing fit between the astronomical prediction and the actual earth record. Many workers have been involved, but here we cite only three: James D. Hays (Columbia University), John Imbrie (Brown University), and N. J. Shackleton (Cambridge University) have recently marshalled the climatic evidence found in deep-sea cores extending back for nearly 0.5 million years, and from these they have constructed a curve not unlike that of Figure 14.29. The correlation of this curve with the climatic fluctuations predicted from the orbital motions of the earth is striking. And their conclusion that "changes in the earth's orbital geometry are the fundamental cause of" Pleistocene ice ages is extremely persuasive.

SUMMARY

We have touched on only a few of the theories advanced to explain glaciation, and there are many more that we have not even mentioned. To date, we do not know which, if any, of these hold the right answer.

It seems reasonable, however, to expect that the cause of Pleistocene glaciation is multiple. Clearly the

world climate has cooled slowly during the 50 million years or more from the early Tertiary to the Pleistocene. Whatever the cause or causes of this cooling, we can theorize that the climate reached the point at which some other cause or causes took over and produced the glaciations and deglaciations of the Pleistocene.

14.7 IMPLICATIONS FOR THE FUTURE

If geology teaches us nothing else, it demonstrates that our globe is in constant change—that the face of the earth is mobile. Mountains rise only to be laid low by erosion; seas lap over the continents; entire regions progress through various stages; and glaciers come and go. We still live in the Pleistocene Ice Age, a pinpoint in time that has been preceded, and will be followed, by extensive climatic changes.

Of course, we cannot predict with any surety anything but change. On the other hand, some feel that it is reasonable to extrapolate the earth's orbital variations into the future and to assume that these will control major climatic fluctuations, as they appear to have controlled them during the Pleistocene. If we accept these premises, then the long-term outlook is toward widespread glaciation in the northern hemisphere.

Figure 14.34 is one such prediction. It shows the climatic fluctuations in temperature for the last 150,000 years and projects present temperatures along a cooling course that reaches a full-fledged glacial stage in

14.34 The mean global temperature for the last 150,000 years is shown. If the orbital variations of the earth control the climate of the future as many believe they have controlled the past, then we will enter a full-fledged glacial period in about 23,000 years. See text for discussion of "superinterglacial." [After J. Murray Mitchell, Jr., "Carbon Dioxide and Future Climate," *Environmental Data Service,* p. 8, March 1977.]

23,000 years. The author of this particular prediction also suggests that we may inadvertently extend the present interglacial by continued burning of coal, oil, and gas. This would increase the carbon dioxide in the atmosphere, create a global "hothouse" effect, and perhaps produce a **superinterglacial** period warmer than today and warmer than previous interglacials. Among other effects, this might extend the present interglacial for some 2,000 years, before the decline in temperatures leads to the next glaciation.

OUTLINE

Glacier formation is a low-temperature, metamorphic process that converts snow to firn to ice.

> **Classification of glaciers** includes **valley glaciers, piedmont glaciers, ice sheets,** and **continental glaciers.**

> **Nourishment of glaciers** occurs in the **zone of accumulation.** The **zone of wastage** lies below the snow line, where melting and evaporation exceed snowfall.

> **Glacier movement** is usually a few centimeters per day, but some glaciers move rapidly in **surges.** Below a 60-m **brittle zone,** glacier ice is in a zone of flow. The glacier moves by slipping along its base, by recrystallization of individual crystals, and by shearing.

> **Temperatures of glaciers** vary. A warm glacier reaches the pressure melting point throughout the summer. In a cold glacier, no melting occurs even in the summer.

Results of glaciation are both erosional and depositional.

> **Erosion** of the ground surface takes place by **plucking** or by **abrasion.** Features formed include **cirques, horns, arêtes, cols, U-shaped valleys, hanging valleys, fiords, striations, grooves,** and **polish.**

> **Glacial deposits** include unstratified material called **till,** which is found in **moraines** and **drumlins. Boulder trains** are special depositional features.

> **Stratified deposits** include **outwash** of sand and gravel, which is usually found in **outwash plains, eskers, crevasse fillings, kames,** and **kame terraces. Varves** usually consist of clay and silt, and form in glacial lakes.

The glacial theory was born in Switzerland and was first clearly stated by Louis Agassiz. **Proof of the glacial theory** rests on the principle of uniformitarianism.

Multiple glaciations mark the Pleistocene, and four glaciations—**Nebraskan, Kansan, Illinoian,** and **Wisconsin**—are recognized in the central United States.

Pre-Pleistocene glaciations are known to have occurred several times in the Paleozoic and the Precambrian.

Causes of glaciation are not yet known, but for the Pleistocene, they seem to involve a long, slow cooling of climate that began in the early Tertiary and a later set of short-term fluctuations that produced the glacial and interglacial epochs.

Orbital variations of the earth's orbit form the basis for the currently favored explanation of the short-term fluctuations.

Implications for the future are several, and the forecast is that climatic change is certain.

SUPPLEMENTARY READINGS

Agassiz, Louis *Studies on Glaciers,* Neuchâtel, 1840; trans. and ed. A. V. Carozzi, Harper & Row, Publishers, Inc., New York, 1967. *No better exposition of the proof of the former extent of glaciers can be found.*

Embleton, Clifford, and Cuchlaine A. M. King *Glacial Geomorphology,* 2nd ed., John Wiley & Sons, Inc., New York, 1975. *A very good volume stressing glacier ice, erosion, and deposition.*

Flint, Richard Foster *Glacial and Quaternary Geology,* John Wiley & Sons, Inc., New York, 1971. *A standard text in the field and a very good one.*

Imbrie, John, and Katherine P. Imbrie *Ice Ages: Solving the Mystery,* Enslow Publishers, Short Hills, N.J., 1979. *A very readable account of the ice ages and, as the preface states, "what they were like, why they occurred and when the next one is due." It tells the story of the search for a solution to the "ice-age mystery" that has engaged so many people over the last 150 years.*

Paterson, W. S. B. *The Physics of Glaciers*, Pergamon Press Ltd., London, 1968. *If you are interested in physics, this is the best summary available on its application to glacier ice.*

Ritter, Dale F. *Process Geomorphology*, William C. Brown Company, Publishers, Dubuque, Iowa, 1978. *Chapter 9 treats glacier and glacier mechanics; Chapter 10, glacial erosion, deposition, and landforms.*

Sugden, David E., and Brian S. John *Glaciers and Landscape*, John Wiley & Sons, Inc., New York, 1976. *Discusses (in intelligible style) glacier movement and distribution and the effects of glacial erosion and deposition. Well illustrated; good bibliography.*

QUESTIONS

1 Define a glacier and describe how snow is converted to glacier ice.

2 With reference to the zones of accumulation and wastage, explain why a glacier's snout may move forward, retreat, or stand still.

3 Glacier movement may take place in several ways. What are they?

4 Erosion by glaciers produces a variety of features and you should be familiar with them. Briefly and succinctly, what are the following: cirque, tarn, striation, horn, arête, col, U-shaped valley, hanging valley, fiord, asymmetric knob or hill?

5 Glacial deposits are both stratified and unstratified. What is the term applied to unstratified material? Why is it unstratified?

6 Unstratified glacial material accumulates in certain characteristic forms. What are some of them and how do they form?

7 What are each of the following: drumlin, erratic, boulder train?

8 Why are stratified glacial deposits stratified? Describe the nature and origin of the following, all of which are made of stratified material: outwash plain, valley train, esker, crevasse filling, kame, kame terrace, varve.

9 In what way is the development of the glacial theory related to the doctrine of uniformitarianism?

10 What are some of the proofs of multiple glaciation?

11 Why glaciation has happened not once but many times in earth history has long been a puzzle. In extended form, outline an essay on the causes of glaciation.

OCEANS

More then 70 percent of the earth lies beneath the oceans. The understanding of this portion of the earth is of utmost importance in geology. It was in the oceans of the past that most of the sedimentary rocks were formed—rocks that today cover three-quarters of the continental land masses—and the great ocean basins are clearly linked to the history and origin of the continents.

In the decades since World War II, our knowledge of the oceans has expanded enormously. In this chapter, we will examine briefly ocean water and its movement, the topography of the ocean floor and its sedimentary cover, and the processes that shape the shorelines.

15.1 OCEAN WATER

We refer to the northern hemisphere as the **land hemisphere** because north of the equator, oceans and seas cover only about 60 percent of the earth's surface, in contrast with the southern hemisphere, 80 percent of which is flooded by marine waters. Between 45° and 70°N, the ocean occupies only 38 percent of the surface, whereas 98 percent of the earth's surface is covered by oceans between 35° and 65°S (see Figure 15.1).

The greatest ocean depth so far recorded is in the Pacific Ocean between Guam and Yap. Here the ocean bottom lies more than 11 km below the surface, considerably greater than the height of Mt. Everest, which is not quite 9 km above sea level. The average ocean depth is about 3,800 m, and the mean elevation of the continents is only 840 m (see Figure 15.2).

ORIGIN OF OCEAN WATER

Some minerals contain water as part of their crystal, structure, hornblende being an example. It has been suggested that, over millions of years, weathering of rock containing hydrous minerals may have been the origin of the ocean waters. But a closer examination of this suggestion shows this process could not have ac-

Land hemisphere

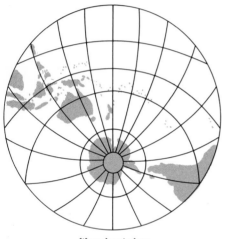

Water hemisphere

15.1 On the land hemisphere map, centered on western Europe, land and sea are about evenly divided. But an indisputable predominance of the seas is seen on the water hemisphere map, centered on New Zealand.

counted for the volume of oceanic water, nor for the chlorine in the dissolved salt (NaCl) we now see. Two other possibilities seem more probable.

One of these possibilities is that volcanic eruptions have been continuously adding water and other elements from great depths below the surface, depths that far exceed the depths of weathering. This process, called **continuous degassing** of the earth, has the drawback that apparently most of the material now coming to the surface through volcanic eruptions is recycled material and is not being added to the hydrologic cycle for the first time.

The other possibility assigns the accumulation of most oceanic water to an early stage of earth history, a period for which we have little tangible evidence. In this suggestion, the excess hydrogen combined with oxygen to form water and bubbled to the surface during the first few hundred million years of earth time when temperatures were high and the present gross stratification of the earth (see Section 9.12) was still incomplete. The excess chlorine came to the surface at the same time.

At the present, we do not have sufficient evidence to choose between these two latter possibilities. In fact, both processes may have contributed to the creation of oceanic waters.

THE NATURE OF SEAWATER

Most of the naturally occurring elements have been identified in the ocean. On the average, 1,000 g of water contain 35 g of dissolved inorganic salts, but this varies in the open ocean between 33 and 38 parts per thousand. The salinity of the ocean is thus frequently expressed as 35 ‰. Chlorine and sodium together account for over 86 percent (or about 30 parts per thousand) of the dissolved material in seawater (see Table 15.1).

TABLE 15.1

The major constituents dissolved in seawater[a]

Ion		Percentage of all dissolved material
Chlorine	Cl^-	55.04
Sodium	Na^+	30.61
Sulfate	SO_4^{2-}	7.68
Magnesium	Mg^{2+}	3.69
Calcium	Ca^{2+}	1.16
Potassium	K^+	1.10
Bicarbonate	HCO_3^-	.41
		99.69

[a] From H. U. Sverdrup, Martin W. Johnson, and Richard H. Fleming, *The Oceans*, p. 166, Prentice-Hall, Inc., Englewood Cliffs, N. J., 1942.

The salinity of the ocean varies from place to place, although the proportion of the major components of the salinity remains constant. High rainfall in equatorial climates reduces salinity. In the subtropical belts north and south of the equator, low rainfall and high evaporation tend to increase salinity (see Figure 15.3). In the Arctic and Antarctic areas, on the other hand, the melting of glacier ice reduces the saltiness of the seas.

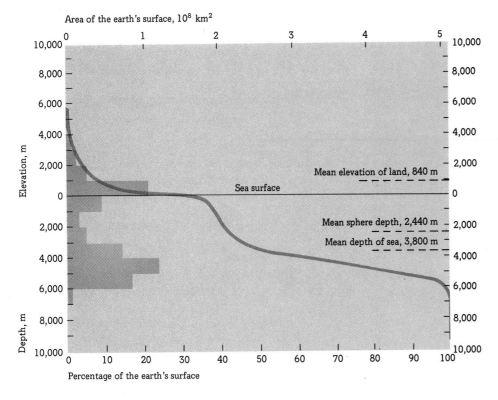

Area of the earth's surface, 10^8 km²

15.2 The relative distribution of land and sea. Note that the mean ocean depth is 3,800 m, whereas the mean elevation of the land is only 840 m. [After H. U. Sverdrup, Martin W. Johnson, and Richard H. Fleming, *The Oceans*, p. 19, Prentice-Hall, Inc., Englewood Cliffs, N. J., 1942.]

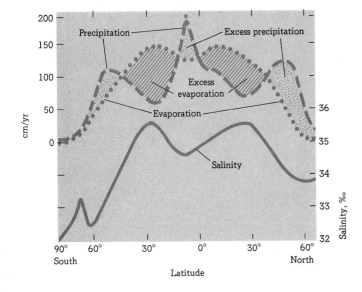

15.3 In general, the total salinity of seawater varies inversely with precipitation and directly with evaporation. [After M. Grant Gross, *Oceanography*, 2nd ed., Fig. 6.12, p. 154, Prentice-Hall, Inc., Englewood Cliffs, N. J., 1977.]

it has been estimated that there is a complete mixing of all water of the oceans about once every 2,000 years. The currents of the ocean waters are caused by tides, by the differences in density of seawater from place to place, and by the wind systems of the earth.

TIDAL CURRENTS

The attractive forces that operate between the sun, the moon, and the earth set the waters of the ocean in motion to produce tidal currents (see Figure 15.4). The speed of these currents may reach several kilometers per hour if local conditions are favorable. Velocities of 20 km/h develop during extreme tides in Seymour Narrows, between Vancouver Island and the mainland of British Columbia. Tidal currents of half this velocity are not uncommon.

Tidal currents close to shore, often confined by the topographic configuration of the shore and the near-shore topography, are stronger than those in the open ocean. Constricted as they are, the near-shore currents are **reversing,** flowing landward with the rising tide and seaward with the falling tide. In the open waters of the ocean, the currents are much weaker, gen-

15.2 CURRENTS

Seawater is in constant movement, in some places horizontally, in others downward, and in still others upward. Its rate of movement varies from spot to spot, but

15.4 Twice a day, tides cause the withdrawal of ocean water and the exposure of this muddy bottom along a Maine inlet. [Sheldon Judson.]

erally not more than 1 km/h, and have a rotary motion, clockwise in the northern hemisphere.

DENSITY CURRENTS

The density of seawater varies from place to place with changes in salinity, temperature, and the amount of material held in suspension. These changes in density cause movement of seawater. Thus water of high salinity is heavier than water of low salinity and sinks beneath it; cold, heavy water sinks below warmer and lighter water; and heavy, muddy water sinks beneath light, clear water. An example of each of the three types of density currents is given in the following paragraphs.

Water flowing through the Straits of Gibraltar illustrates density currents caused by differences in salinity. A pair of currents, one above the other, flow in opposite directions between the Mediterranean Sea and the Atlantic Ocean. They originate as follows. The Mediterranean, lying in a warm, dry climatic belt, loses about 1.5 m of water every year through evaporation. Consequently, the saltier, heavier water of the Mediterranean moves outward along the bottom of the straits and sinks downward into the less salty, lighter water of the Atlantic. At the same time, the lighter surface water of the Atlantic moves into the Mediterranean basin. The water flowing from the Mediterranean settles to a depth of about 1,000 m in the Atlantic and then spreads slowly outward beyond the equator on the south, the Azores on the west, and Ireland on the north. It has been estimated that as a result of this activity the water of the Mediterranean basin is changed once every 75 years (see Figure 15.5).

Differences in temperature are responsible for

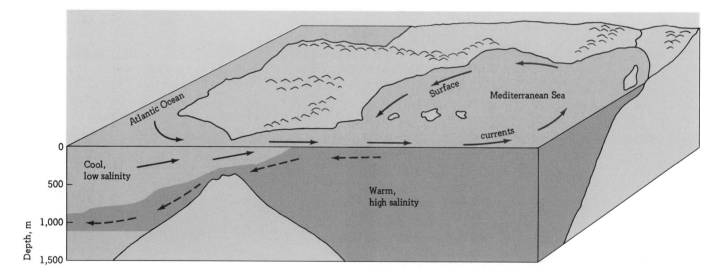

15.5 A density current flows from the Mediterranean Sea through the Straits of Gibraltar and spreads out into the Atlantic Ocean. High evaporation and low rainfall in the Mediterranean area produce a more saline and hence heavier water than the water of the neighboring Atlantic Ocean. As a result, Mediterranean water flows out through the lower portion of the straits, and lighter Atlantic water moves above it and in the opposite direction to replace it. The higher temperature of the Mediterranean water is more than counteracted by its greater salinity, and the water sinks to a level in the Atlantic somewhat lower than 1,000 m below the surface.

15 OCEANS

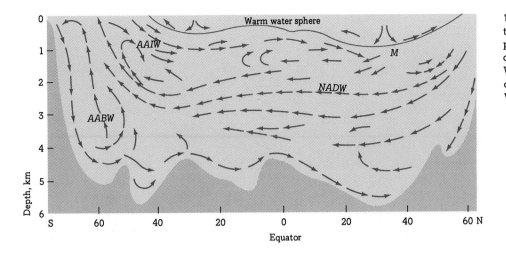

15.6 Circulation of water in the Atlantic Ocean shown along a north–south profile as determined by temperature and salinity. *NADW*, North Atlantic Deep Water; *AAIW*, Antarctic Intermediate Water; *AABW*, Antarctic Bottom Water; and *M*, Mediterranean water.

major circulation of water through much of the deep ocean. For example, as a result of such variations, water from the cold Arctic and Antarctic regions creeps slowly toward the warmer environment near the equator. The cold, relatively dense water from the Arctic sinks near Greenland in the North Atlantic and can be traced to the equator and beyond as far as 60°S. This water is called **North Atlantic Deep Water.** Colder and denser water (**Antarctic Bottom Water**) moves downward to the sea floor off Antarctica and creeps northward, pushing beneath the North Atlantic Deep Water. In fact, the Antarctic Bottom Water reaches well north of the equator before it loses its identity. Lighter than these two types of water is the **Antarctic Intermediate Water,** which sinks in the area of 50°S and flows northward above the North Atlantic Deep Water until it loses its identify about 30°N in the area where the Mediterranean water flows from east to west. These relationships are shown in Figure 15.6.

A third type of density current, known as a **turbidity current,** is caused by the fact that turbid or muddy water has a greater density than clear water and therefore sinks beneath it. We can demonstrate how these currents operate on a small scale in the laboratory, and they have actually been observed in freshwater lakes and reservoirs. Turbidity currents explain several different phenomena in the present-day oceans as well as in the oceans of the geological past.

Turbidity currents offer the most plausible explanation of certain deposits that have been studied from the deep ocean basins. Samples of sediments from these basins contain thin layers of sand that could not have been carried so far from shore by the slow drift of water along the ocean floor, but rapidly moving turbidity currents would be quite capable of moving the sand down the slopes of the ocean basins to the deep floors.

Analysis of other samples from the ocean floor reveals that particles become progressively finer from bottom to top of the deposit. The best explanation of this **graded bedding** is that the deposit was laid down by a turbidity current carrying particles of many different sizes and that the larger particles were the first to be dropped on the ocean floor. Such deposits are called **turbidites.**

Turbidity currents may be set in motion by the slumping and sliding of material along the slopes of the ocean basin under the influence of gravity, or aided by the jarring of an earthquake. Or the currents may be created by the churning up of bottom sediments under the influence of violent storms.

MAJOR SURFACE CURRENTS

The major movements of water near the ocean's surface occur in such currents as the Gulf Stream, the Japanese current, and the equatorial currents. These great currents are caused by a variety of factors, including the prevailing winds, the rotation of the earth, variations in the density of seawater, and the shape of ocean basins. Let us examine, by way of illustration, the surface currents of the Atlantic Ocean in both the northern and southern hemispheres, as shown in Figure 15.7.

The equatorial currents lie on each side of the equator, and they move almost due west. They derive their energy largely from the trade winds that blow constantly toward the equator, from the northeast in the northern hemisphere and from the southeast in the

15.7 Major surface currents of the world.

southern hemisphere. The westerly direction of the currents is explained by the **Coriolis effect.** The rotation of the earth causes moving objects to veer to the right in the northern hemisphere and to the left in the southern hemisphere. As the water driven by the trade winds moves toward the equator, it is deflected to the right (toward the west) in the northern hemisphere and to the left (also toward the west) in the southern hemisphere. As a result, the north and south equatorial currents are formed. As these currents approach South America, one is deflected north and the other south. This deflection is caused largely by the shape of the ocean basins, but it is aided by the Coriolis effect and by the slightly higher level of the oceans along the equator where rainfall is heavier than elsewhere.

The north equatorial current moves into the Caribbean waters and then northeastward, first as the Florida current and then as the Gulf Stream. The Gulf Stream, in turn, is deflected to the east (to the right) by the Coriolis effect. This easterly movement is strengthened by prevailing westerly winds between 35° and

45°N, where it becomes the North Atlantic current. As it approaches Europe, the North Atlantic current splits. Part of it moves northward as a warm current past the British Isles and parallel to the Norwegian coast. The other part is deflected southward as the cool Canary current and eventually is caught up *again* by the northeast trade winds, which drive it into the north equatorial current.

In the South Atlantic, the picture is very much the same—a kind of mirror image of the currents in the North Atlantic. After the south equatorial current is deflected southward, it travels parallel to the eastern coast of South America as the Brazil current. Then it is bent back to the east (toward the left) by the Coriolis effect and is driven by prevailing westerly winds toward Africa. This easterly moving current veers more and more to the left until finally, off Africa, it is moving northward. There it is known as the Buenguela current, which in turn is caught up by the trade winds and is turned back into the south equatorial current.

The cold surface water from the Antarctic regions moves along a fairly simple course, uncomplicated by large landmasses. It is driven in an easterly direction by the prevailing winds from the west. In the northern hemisphere, however, the picture is complicated by continental masses. Arctic water emerges from the polar seas through the straits on either side of Greenland to form the Labrador current on the west and the Greenland current on the east. Both currents subsequently join the North Atlantic current and are deflected easterly and northeasterly.

We need not examine in detail the surface currents of the Pacific and Indian Oceans. We can note, however, that the surface currents of the Pacific follow the same general patterns as those of the Atlantic. Furthermore, the surface currents of the Indian Ocean differ only in detail from those of the South Atlantic (see Figure 15.7).

15.3 THE OCEAN BASINS

Most of the seawater surrounding the continents is held in one great basin which girdles the southern hemisphere and branches northward into the Atlantic, Pacific, and Indian Oceans. The Atlantic and Pacific Oceans, in turn, are connected with the Arctic Ocean through narrow straits.

As we saw in Section 1.3, the floors of the ocean basins are underlain by a layer of basaltic material approximately 5 km thick. This is covered by a layer of sedimentary deposits, which varies in thickness from 2 km to a feather edge near the spreading centers of the midoceanic ridges.

The ocean basins have not always been as we see them today. In Sections 1.3 and 10.1, we found that they have changed in shape and size through time. These changes have been the result of sea-floor spreading as new oceanic crust formed along the active oceanic ridges and old crust was consumed along the margins of colliding plates. The geologic record allows us to reconstruct the locations of continents and ocean basins at various times in the past (Figure 15.8) and even to predict where they might be at some future time. Figure 15.8 thus shows the changing size and shape of the Atlantic Ocean and its bordering land masses through 200 million years of time.

15.4 TOPOGRAPHY OF THE OCEAN FLOOR

The more we learn about the ocean floor, the more spectacular we find its topography to be, even more so than that of the continents. To be sure, the landmasses present us with some impressive landscapes, but most of the dry land is less than a kilometer above sea level. Continuous erosion has worn down most of the continental masses close to sea level. In contrast, volcanism and tectonic forces have created large, bold features on the ocean floor, features that have been unaffected by the vigorous erosion almost everywhere active above sea level. Only where sedimentation has blanketed the oceanic topography can the sea floor be said to be nearly featureless.

The three major topographic divisions of the oceanic realm are the **continental margins**, the **floors of the ocean basins**, and the **oceanic ridges**, which we describe below. Two additional striking forms bear mention: **seamounts** and **deep-sea trenches**.

4/30 THE CONTINENTAL MARGINS

We have already noted that ocean waters spill up over the edges of the continents. In this zone of transition between the continents and the deep oceans, the subsea topograpy is commonly a sequence of shelf-slope-rise from land to ocean deep. The **continental shelf** is the drowned edge of the continent and slopes seaward on a gentle gradient of about 1 m/1,000 m. Where they are present, shelves are broad features. Off Florida the shelf reaches out 240 km and off South America opposite the Rio de la Plata the shelf is 560 km wide.

At a depth that averages about 200 m, the outer edge of the shelves give way to the steeper gradients, the **continental slopes**. These slopes lead downward at rates of 1 m/40 m or greater to the **continental rises**, which approximate the gradients of the continental shelves.

This shelf-slope-rise sequence characterizes the continental margin of much of the Atlantic basin. Here the sequence forms the **trailing edges** of continents moving away from the axis of spreading in the mid-Atlantic Ocean.

The situation is different where an oceanic plate collides with a continental plate, as along the western coast of South America; the continental shelf is missing. Here subduction of oceanic floor beneath the continent creates a steep slope from the coastline into a deep submarine trench. A still different continental margin forms where continent and ocean slip by each other in opposite directions. Here the continental margin is broken into a series of linear, tectonic blocks more or less parallel with the shoreline, as off the coast of California from San Diego to north of San Francisco.

15.8 The size and shape of the Atlantic Ocean basin at various moments in time are shown from 200 million years ago to the present. Lines of ancient latitude (paleolatitudes) are shown in the first five frames. Modern latitude and longitude are shown in the map of the present. [After J. D. Phillips and D. Forsyth, "Plate Tectonics, Paleomagnetism, and the Opening of the Atlantic," *Bull. Geol. Soc. Amer.*, vol. 83, p. 1579, Fig. 1, 1972.]

Drowned valleys occur on the continental shelves. Many of these seem to be extensions of large valleys on the adjacent land. One of the best-known of these valleys is the submerged extension of the Hudson River valley off the eastern coast of North America. This valley is cut some 60 m into the gently sloping shelf and widens down-valley from about 5 km to 24 km at its seaward end. It is generally agreed that this and similar valleys were cut during periods of Pleistocene glaciation when the sea level was 100 m or more lower than it is at present.

More difficult to explain than these valleys on the continental shelves are the deep submarine canyons along the continental slopes. Some canyons, such as those on the slopes off the Hudson and Congo Rivers, appear to be extensions of valleys on land or on the continental shelves. Others appear to have no association with such valleys. In any event, these canyons have V-shaped cross profiles and gradients that decrease along the lower sections of the continental slopes. Their slightly winding courses may extend 235 km out to sea, as does the Congo canyon. Some cut down more than 1 km below the surface of the continental slopes. Turbidity currents moving down the continental slope are most generally cited as the erosional agents responsible for these canyons.

FLOORS OF THE OCEAN BASINS

Seaward of the continental margins lie the floors of the ocean basins with a mean depth of 3,800 m. Near the foot of the continental rise lie the **abyssal plains,** which except for the deep-sea trenches are the deepest parts of the ocean. Here settle the sediments carried from the land across the continental margins. Because of these sediments, the topography is subdued, mantled by clay and silt that obscures at least in part the forms created by volcanism and crustal fracturing.

OCEANIC RIDGES

A network of broad ridges segment the deep oceanic basins. There are two general types of ridges. One we refer to as a **midoceanic ridge** after its best-known example, the Mid-Atlantic Ridge. The second type is called an **aseismic ridge,** because in contrast to the seismically active midoceanic ridges, this type exhibits little or no earthquake activity.

Midoceanic ridges The Mid-Atlantic Ridge bisects the Atlantic Ocean, generally paralleling the continental borders to the east and west. It rises toward the ocean surface and in some places, as for example in Iceland, emerges above sea level. In width, it is between 1,500 and 2,000 km wide, and it slopes gently downward and away from its central axis. Shallow-focus earthquakes and volcanic eruptions of basalt mark the crest as does a trough or rift valley 25 to 50 km wide and 1 to 2 km deep. Transform faults offset and segment the ridge and its axial valley. From the summit of this long, restless ridge, the plates of the Atlantic basin diverge west and east. The ridge is really a heat bulge along which thermal energy and new crustal material are transferred upward from the mantle. As new crust is transferred laterally, it cools; and the more it is cooled, the deeper it subsides. This process accounts for the broad flanks leading from the ridge crest toward the abyssal plains on either side.

The Mid-Atlantic Ridge is but one segment of a network of spreading ridges in the world's oceans.

Aseismic ridges The second type of ridge, the **aseismic ridge,** is also a thermal feature. Unlike the midoceanic ridge, however, where heat is applied all along the ridge, volcanic activity takes place only at the extreme tip of the aseismic ridge, which also is the highest point along the ridge.

The Hawaiian Islands can serve as an example (see Section 3.1). This chain of islands, built on top of an aseismic ridge, begins with the active volcanic island of Hawaii. We can follow the ridge northwestward past dead volcanic islands such as Molokai and Oahu and, finally, 2,000 km from Hawaii, to the coral reef islands of Midway and Kure. The ridge becomes lower and continues another 500 km beyond Kure, at which point it turns almost due north. The ridge then all but disappears, but its trend is marked by a series of submarine peaks that can be traced into the north Pacific where the most northerly peaks are being consumed, along with adjacent sea floor, in the Kurile trench off southern Kamchatka.

The ridge, its associated islands, and submarine peaks trace the direction of plate movement across a stationary plume or hot spot. In this case, the active volcanoes of Hawaii are the surface expressions of a plume. The hot, rising mantle of the plume fuels the volcanic activity and bulges up the adjacent sea floor in a domal feature 1,000 km in diameter. As the plate moves over the plume, it takes the earlier-formed heat bulge and volcanoes farther and farther from the heat source. With continued cooling, the ridge and its volcanoes subside, and eventually the volcanoes disappear below sea level. The Hawaiian Ridge, its islands, and submerged peaks mark, then, the direction of movement of the Pacific plate over tens of millions of years. The hot spot trace tells us that the plate has been moving northwestward and that at an earlier time it moved northward (see Figure 15.9).

Other aseismic ridges characterize the ocean basins and we think their origin was similar to that of the Hawaiian Ridge.

SEAMOUNTS AND DEEP-SEA TRENCHES

Seamounts are found on the oceanic ridges and across the ocean floors. **Deep-sea trenches** occur on the ocean floor and along the margins of some continents.

Seamounts We have referred to seamounts, thousands of which occur on the sea floor. Some have the characteristic conical form we associate with volcanoes, as indeed they are. Others are flat-topped cones.

15.4 TOPOGRAPHY OF THE OCEAN FLOOR

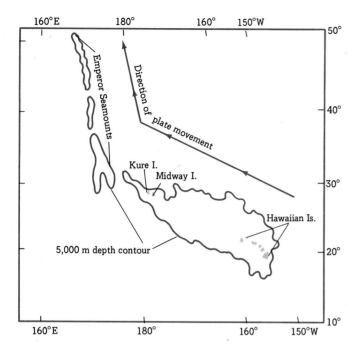

Deep-sea trenches The greatest ocean depths are found in the **deep-sea trenches.** Most of these lie around the margins of the Pacific Ocean (see Figure 15.11). Usually arcuate in plan, they are as much as 200 km wide and up to 2,400 km in length. They reach 2 to 3 km below the adjacent ocean floor. These deeps, as we have seen, mark the convergence of plates where the oceanic crust is pushed beneath continental masses or beneath other oceanic crust.

15.5 SEDIMENTS OF THE OCEANS

In earlier chapters, we have spoken from time to time of the processes by which earth materials are weathered, eroded, transported, and finally deposited to be transformed into sedimentary rocks. The great ocean basins of the world constitute the ultimate collection area for the sediments and dissolved material that are carried from the land, and the great bulk of the sedimentary rocks found on our modern landmasses were once deposits on the ocean floors of the past. In this section we shall speak briefly of the sediments being laid down in the modern oceans—the sediments destined to become the sedimentary rocks of the future.

DEPOSITS ON THE CONTINENTAL MARGINS

Theoretically, when particles of solid material are carried out and deposited in a body of water, the largest particles should fall out nearest the shore and the finest particles farthest away from the shore, in a neatly graduated pattern. But there are a great many exceptions to this generalization. Many deposits on the continental shelves show little tendency to grade from

15.9 The Island of Hawaii, at the southeastern end of the aseismic Hawaiian ridge, marks the position of a mantle plume. To the northwest, the other islands of the Hawaiian chain, together with the drowned volcanic peaks on a narrowing Hawaiian ridge and the Emperor seamounts, trace the track of the Pacific plate across the Hawaiian plume over the course of tens of millions of years.

These latter, called **guyots,** after Arnold Guyot, a nineteenth-century Swiss-American geologist-geographer, were once volcanic islands truncated by wave action before they sank beneath the ocean surface (see Figure 15.10). Their subsidence was caused by the cooling related to either a midoceanic ridge or an aseismic ridge.

15.10 Soundings in the Pacific Ocean during World War II revealed the presence of many flat-topped submarine mountains called "guyots." The guyot in this profile is located at 8.8°N., 163.1°E. [After H. H. Hess, "Drowned Ancient Islands of the Pacific Basin," *Am. J. Sci.*, vol. 244, p. 777, 1946.]

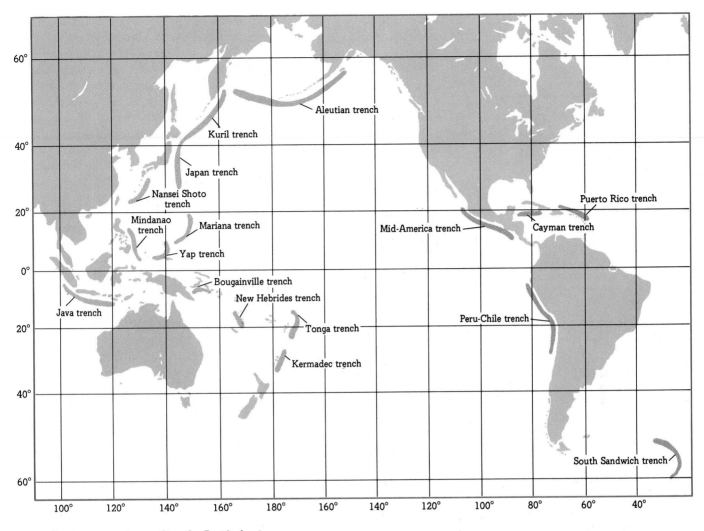

15.11 Deep-sea trenches outline the Pacific basin.

coarse to fine away from the shoreline. We would expect to find sand only close to the shore—actually it shows up from place to place all along the typical continental shelf right up to the lip of the continental slope. It is particularly common in areas of low relief on the shelf. In fact, on glaciated continental shelves, sand mixed with gravel and cobbles comprises a large part of the total amount of deposited material.

Also common on the continental shelves are deposits of mud, especially off the mouths of large rivers and along the course of ocean currents that sweep across the river-laid deposits. Mud also tends to collect in shallow depressions across the surface of the shelves, in lagoons, sheltered bays, and gulfs.

Where neither sand nor mud collects on the shelves, the surface is often covered with fragments of rock and gravel. This is commonly the case on open stretches of a shelf where strong ocean currents can winnow out the finer material, and off rocky points and exposed stretches of rocky shoreline. In narrow straits running between islands or giving access to bays, the

energy of tidal and current movements is often so effectively concentrated that the bottom is scoured clean and the underlying solid rock is exposed.

Sediments not trapped on the shelves continue on out beyond the shelf break. Most of these are deposited on the slopes and rises. Here clay, silt, sand, and even gravel accumulate. In general, however, the sediments are finer than on the shelf. Nevertheless, fully as much if not more sedimentation occurs on slopes and rises as on the shelves. The continental margins, considered as a unit, receive over 90 percent of the land-derived sediments. The rest of it goes on to the deep ocean basins.

DEPOSITS ON THE DEEP-SEA FLOORS

The deposits that spread across the floors of the deep sea are generally much finer than those on the slopes, rises, and shelves, although occasional beds of sand have been found even in the deeps. One would expect this. One would also expect that the rates of sedimentation would be slower than on the continental margins, and indeed they are. The rates on the continental margins are from 20 to 200 times that of the deep sea.

Deep-sea deposits of material derived from the continents are referred to as **terrigenous** ("produced on the earth") **deposits.** Those formed of material derived from the ocean itself are **pelagic** ("pertaining to the ocean") **deposits.**

Among the terrigenous deposits are beds of windborne volcanic ash that has fallen over the ocean and has settled through thousands of meters of water to the sea bottom. In the polar regions, silts and sands from glacier ice make up much of the bottom deposit, and around the margins of the continents we often find deep-sea mud deposits of silt and clay washed down from the landmasses. Much of the ocean bottom, especially in the Pacific, is covered by an extremely fine-grained deposit known as **brown clay,** which may have originated on the continents and then drifted out into the open ocean.

In the Antarctic and Arctic Oceans, some of the deep ocean sediments are believed to have been ice-rafted from the land to the oceans by icebergs. These **glacial marine deposits** are chiefly silt containing coarse fragments. Near the landmasses are extensive layers of sand and coarse silt deposited by turbidity currents, the turbidites, discussed previously.

Most of the pelagic deep-sea deposits consist of very fine-grained muds, often called **oozes,** sediments of which at least 30 percent by volume is made up of the hard parts of very small, sometimes microscopic organisms. These hard parts are constructed of mineral matter extracted from the seawater by tiny plants and animals. With the death of the organism, the remains sink slowly to the sea floor. In composition, the oozes are either **calcareous** or **siliceous.**

The calcareous muds are often referred to as **foraminiferal muds** because a significant percentage of their volume is made up of the tests of small one-celled animals called *foraminifera.*

Siliceous muds include **radiolarian muds,** made up largely of the delicate and complex hard parts of minute marine protozoa called *radiolaria,* and by **diatomaceous mud,** made up of the siliceous cell walls of one-celled marine algae known as *diatoms.* Radiolarian mud predominates in a long east–west belt in the Pacific Ocean just north of the equator, and the greatest concentration of diatomaceous mud occurs in the North Pacific and in the Antarctic Ocean. Figure 15.12 shows the distribution of these various deposits.

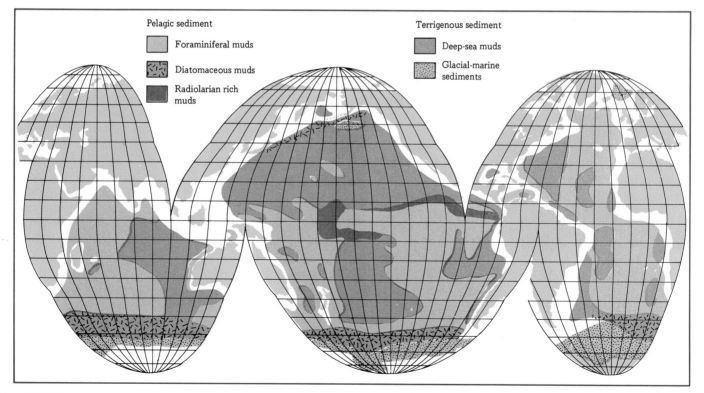

15.12 Type and distribution of deep-sea sediments. [After M. Grant Gross, *Oceanography,* 2nd ed., Fig. 4.12, Prentice-Hall, Inc., Englewood Cliffs, N. J., 1977.]

15.13 (above) This photograph of the deep-sea floor was taken in the Atlantic Ocean at a depth of nearly 5,400 m. The objects are manganese nodules. The largest is about 15 cm in maximum dimension. [Lamont Geological Observatory.]

15.14 (right) This section through a manganese nodule shows that the concentric layers deposited as the nodule grew. The nodule, 6 cm across, was dredged from a depth of 5,400 m off the southeast coast of South Africa. [Raymond C. Gutschick.]

More spectacular deposits are the **manganese nodules** (see Figures 15.13 and 15.14). These average 25 cm in diameter and when cut open exhibit the patterns of concentric growth. Composed largely of oxides of manganese and iron, they contain in addition the oxides of many of the rarer elements. It is estimated that they are abundant across some 20 percent of the floor of the Pacific. Their composition and abundance represent a commercial resource not only as an ore of manganese but also of copper and of such rarer elements as cobalt, titanium, and zirconium.

15.6 SHORELINES 4/30

Not all of us have the occasion to study firsthand the currents of the oceans, or the forms and processes of the ocean floor. But most of us have the opportunity from time to time to observe the activity of water along the shorelines of oceans or lakes (see Figure 15.15).

THE PROCESSES

Wind-formed waves Most water waves are produced by the friction of air as it moves across a water surface. The harder the wind blows, the higher the water is piled up into long **wave crests** with intervening troughs, both crests and troughs at right angles to the

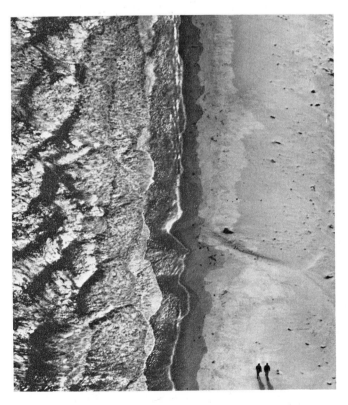

15.15 Two persons walk this sandy strip between high and low tide. [Standard Oil Co. of N. J.]

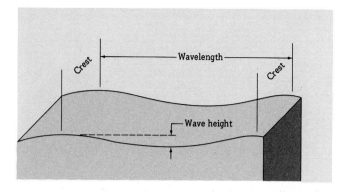

15.16 Diagrammatic explanation of terms used to describe water waves.

wind. The distance between two successive wave crests is the **wavelength**, and the vertical distance between the wave crest and the bottom of an adjacent trough is the **wave height** (see Figure 15.16). When the wind is blowing, the waves it generates are called a **sea.** But wind-formed waves persist even after the wind that formed them dies. Such waves, or **swells,** may travel for hundreds or even thousands of kilometers from their zone of origin.

We are concerned with both the movement of the waveform and the motion of water particles in the path of the wave. Obviously the waveform itself moves forward at a measurable rate. But in deep water, the water particles in the path of the wave describe a circular orbit: Any given particle moves forward on the crest of the wave, sinks as the following trough approaches, moves backward under the trough, and rises as the next crest advances. Such a motion can best be visualized by imagining a cork bobbing up and down on the water surface as successive wave crests and troughs pass by. The cork itself makes only very slight forward progress under the influence of the wind. Wave motion extends downward until at a depth equal to about one-half the wavelength it is virtually negligible. But between this level and the surface, water particles move forward under the crest and backward under the trough of each wave, in orbits that decrease in diameter with depth (Figure 15.17).

As the wave approaches a shoreline and the water becomes more shallow, definite changes take place in the motion of the particles and in the form of the wave itself. When the depth of water is about half the wavelength, the bottom begins to interfere with the motion of water particles in the path of the wave, and their orbits become increasingly elliptical. As a result, the length and velocity of the wave decrease, and its

front becomes steeper. When the water becomes shallow enough and the front of the wave steep enough, the wave crest falls forward as a breaker, producing what we call **surf** (see Figure 15.18). At this moment, the water particles within the wave are thrown forward against the shoreline. The energy thus developed is then available to erode the shoreline or to set up currents along the shore that are able to transport the sediments produced by erosion.

Wave refraction and coastal currents Although most waves advance obliquely toward a shoreline, the influence of the sea floor tends to bend or **refract** them until they approach the shore nearly head-on.

Let us assume that we have a relatively straight stretch of shoreline with waves approaching it obliquely over an even bottom that grows shallow at a constant rate. As a wave crest nears the shore, the section closest to land feels the effect of the shelving bottom first and is retarded, while the seaward part continues along at its original speed. The effect is to swing the wave around and to change the direction of its approach to the shore, as shown in Figure 15.19.

As a wave breaks, not all its energy is expended on the erosion of the shoreline. Some of the water thrown forward is deflected and moves laterally, parallel to the shore. The energy of this water movement is partly used up by friction along the bottom and partly by the transportation of material.

Refraction also helps explain why, on an irregular shoreline, the greatest energy is usually concentrated on the headland and the least along the bays. Figure 15.20 shows a bay separating two promontories and a series of wave crests sweeping in to the shore across a bottom that is shallow off the headland and deep off the mouth of the bay. Where the depth of the water is greater than one-half the wavelength, the

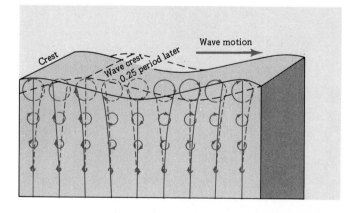

15.17 The motion of water particles relative to wave motion in deep water. Water particles move forward under the crest and backward under the trough, in orbits that decrease in diameter with depth. Such motion extends downward to a distance of about one-half the wavelength. [After *U.S. Hydrog. Off. Pub. 604*, 1951.]

15.18 Storm waves break on the shore at Madison, Connecticut. [Sheldon Judson.]

15.19 Wave crests that advance at an angle on a straight shoreline and across a bottom that shallows at a uniform rate are bent shoreward, as suggested in this diagram. Refraction is caused by the increasing interference of the bottom with the orbits of water-particle motion within the wave.

crest of the advancing wave is relatively straight. Closer to shore, off the headlands, however, the depth becomes less than half the wavelength, and the wave begins to slow down. In the deeper water of the bay, the wave continues to move rapidly shoreward until there too the water grows shallow and the wave crest slows. This differential bending of the wave tends to

make it conform in a general way to the shoreline. In so doing, the wave energy is concentrated on the headland and dispersed around the bay, as suggested in Figure 15.20.

SHORELINE FEATURES

Erosion and deposition work hand in hand to produce most of the features of the shoreline. An exception to this generalization is an offshore island that is merely the top of a hill or a ridge that was completely surrounded by water as the sea rose in relation to the land. But even islands formed in this way are modified by erosion and deposition.

A composite profile of a shoreline from a point above high tide seaward to some point below low tide reveals features that change constantly as they are influenced by the nature of waves and currents along the shore. Not all features are present on all shorelines, but several are present in most shore profiles. The **offshore** section extends seaward from low tide. The **shore**, or **beach**, section reaches from low tide to the foot of the **sea cliff** and is divided into two segments: In front of the sea cliff is the **backshore**, characterized by one or more **berms**, resembling small terraces with low ridges on their seaward edges built up by storm waves; seaward from the berms to low tide is the **foreshore**. Inland from the shore lies the **coast**. Deposits of the shore may veneer a surface cut by the waves on bedrock and known as the **wave-cut terrace**. In the offshore section too, there may be an accumula-

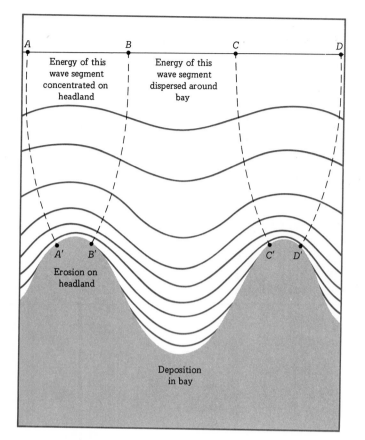

15.20 Refraction of waves on an irregular shoreline. It is assumed that the water is deeper off the bay than off the headlands. Consider that the original wave is divided into three equal segments, *AB*, *BC*, and *CD*. Each segment has the same potential energy. But observe that by the time the wave reaches the shore, the energy of *AB* and *CD* has been concentrated along the short shoreline of headlands *A'B'* and *C'D'*, whereas the energy of *BC* has been dispersed over a greater front (*B'C'*) around the bay. Energy for erosion per unit of shoreline is therefore greater on the headlands than it is along the bay.

In the figure: Energy of this wave segment concentrated on headland; Energy of this wave segment dispersed around bay; Erosion on headland; Deposition in bay.

tion of unconsolidated deposits comprising a **wave-built terrace** (Figure 15.21).

The shoreline profile is ever-changing. During great storms the surf may pound in directly against the sea cliff, eroding it back and at the same time scouring down through the beach deposits to abrade the wave-cut terrace. As the storm (and hence the available energy) subsides, new beach deposits build up out in front of the sea cliff. The profile of a shoreline at any one time, then, is an expression of the available energy: It changes as the energy varies. This relation between profile and available energy is similar to the changing of a stream's gradient and channel as the

15.21 Some of the features along a shoreline and the nomenclature used in referring to them. [After F. P. Shepard, *Submarine Geology*, 2nd ed., p. 168, Harper & Row, Publishers, Inc., New York, 1963.]

In the figure: Offshore; Shore, or beach; Coast; Foreshore; Backshore; Low water; High water; Berms; Sea cliff; Wave-built terrace; Wave-cut terrace; Bar; Trough.

discharge (and therefore the energy) of the stream varies (see Chapter 12).

Features caused by erosion **Wave-cut cliffs** are common erosional features along a shore, particularly where it slopes steeply down beneath the sea (see Figure 15.22). Here waves can break directly on the shoreline and thus can expend the greatest part of their energy in eroding the land. Wave erosion pushes the wave-cut cliff steadily back, producing a wave-cut terrace or platform at its foot. Because the surging water of the breaking waves must cross this terrace before reaching the cliff, it loses a certain amount of energy through turbulence and friction. Therefore the farther the cliff retreats and the wider the terrace becomes, the less effective are the waves in eroding the cliff. If sea level remains constant, the retreat of the cliffs becomes slower and slower.

Waves pounding against a wave-cut cliff produce various features as a result of the differential erosion of the weaker sections of the rock. Wave action may hollow out cavities, or **sea caves,** in the cliff, and if this erosion should cut through a headland, a **sea arch** is formed. The collapse of the roof of a sea arch leaves a mass of rock, a **stack,** isolated in front of the cliff (Figure 15.23).

Features caused by deposition Features of deposition along a shore are built of material eroded by the waves from the headlands and of material brought down by the rivers that carry the products of weathering and erosion from the landmasses. For example, part of the material eroded from a headland may be drifted by currents into the protection of a neighboring bay, where it is deposited to form a sandy beach.

The coastline of northeastern New Jersey (Figure 15.24) illustrates some of the features caused by deposition. Notice that the Asbury Park–Long Branch section of the coastline is a zone of erosion that has

15.22 Wave-cut cliff along the Oregon coast. [Kenneth S. Deffeyes.]

15.23 The sea has cut arches through this promontory. To the far right is a stack, a rock mass that erosion has cut off from the mainland. Arches State Park, California. [Sheldon Judson.]

been formed by the destruction of a broad headland area. Erosion still goes on along this part of the coast, where the soft sedimentary rocks are easily cut by the waves of the Atlantic. The material eroded from this section is moved both north and south along the coastline. Much additional material is derived from just offshore. Storm waves move material shoreward until it is added to the sand supply moving either northward or southward. Sand swept northward is deposited in Raritan Bay and forms a long sand beach projecting northward, a **spit** known as Sandy Hook.

Just south of Sandy Hook, the flooded valleys of the Navesink River and of the Shrewsbury River are bays that have been almost completely cut off from the open ocean by sandy beaches built up across their mouths. These beaches are called **bay barriers.**

Sand moved southward from the zone of erosion has built up another sand spit. Behind it lies a shallow lagoon, Barnegat Bay, that receives water from the sea

through a **tidal inlet,** Barnegat Inlet. This passage through the spit was probably first opened by a violent storm, presumably of hurricane force. Just inside the inlet, a delta has been formed of material deposited partly by the original breakthrough of the bar and partly by continued tidal currents entering the lagoon.

Long stretches of the shoreline from Long Island to Florida and from Florida westward around the Gulf Coast are marked by shallow, often marshy lagoons separated from the open sea by narrow sandy beaches. Many of these beaches are similar to those that enclose Barnegat Bay, apparently elongated spits attached to broad headlands. Others, such as those that enclose Pamlico Sound at Cape Hatteras, North Carolina, have no connection with the mainland.

15.6 SHORELINES

347

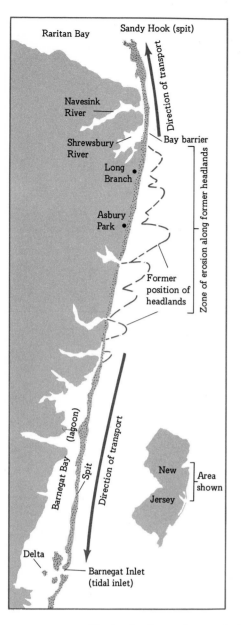

15.24 Erosion by the sea has pushed back the New Jersey coastline as indicated on this map. Some of the material eroded from the headlands has been moved northward along the coast to form Sandy Hook, a spit. To the south, a similar but longer feature enclosed Barnegat Bay, a lagoon with access to the open ocean through a tidal inlet. [After an unpublished map by Paul MacClintock.]

These sandy beaches are best termed **barrier islands**.

Human activity can affect shoreline processes. Figure 15.25(*a*) and (*b*) illustrates deposition caused by the Santa Monica pier and breakwater, Los Angeles.

15.7 THE CHANGING SEA LEVEL

Sea level is not constant. We find the evidence of marine animals in the rocks of our highest mountains, and deep drilling has recorded shallow marine sediments hundreds and even thousands of meters below the modern sea level. In Section 9.2, we found that large earthquakes can change the relative position of land and sea in local areas. In the following paragraphs we look at some geologically recent changes of sea level felt over wide areas.

Isostasy (see Section 1.3) accompanying continental glaciation and deglaciation has led to some shoreline changes. The weight of a continental glacier is so great as to cause a downward warping of the land it covers. When the glacier melts and the load of ice is removed, the land slowly recovers and achieves its original balance (see Figure 15.26).

We have good evidence of the recoil of land following glacial retreat along the shores of the Gulf of Bothnia and the Baltic Sea. Accurate measurements show that from 1800 to 1918, the land rose at rates ranging from 0.0 cm/year at the southern end to 1.1 cm/year at the northern end. Studies indicate further that the land has been rising at a comparable rate for 5,000 years. Areas that were obviously sea beaches are now elevated from a few meters to 240 m above sea level. A comparison of precise measurements along railroads and highways in Finland from 1892 to 1908, with measurements from 1935 to 1950, shows that uplift there has proceeded at the rate of from 0.3 to 0.9 m/century. The greatest change has been on the Gulf of Bothnia, at about 64°N. Most of this movement has been caused by the recoil of the land following the retreat of the Scandinavian ice cap.

Similar histories of warping and recoil have been established in other regions, including the Great Lakes area in the United States. At one stage in the final retreat of the North American ice sheet, the present Lakes Superior, Michigan, and Huron had higher water levels than they do now and were joined together to form glacial Lake Algonquin. Beaches around the borders of Lake Algonquin were horizontal at the time of their formation; today they are still horizontal south of Green Bay, Wisconsin, and Manistee, Michigan. But 290 km north, at Sault Sainte Marie, Michigan, the oldest Algonquin beach is 108 m higher than it is at Manistee.

The water that is now locked up in the ice of glaciers originally came from the oceans. It was transferred landward by evaporation and winds, precipitated as snow, and finally converted to firn and ice. If all this ice were suddenly to melt, it would find its way

15.25 (a) Santa Monica pier, Los Angeles, California, 1931. Construction of the pier has caused increased deposition of sand on either side of its base. (b) Santa Monica pier with breakwater, 1949. Construction of the breakwater has caused growth of the beach seaward on the north side of the pier. [Fairchild Aerial Photography Collection, Whittier College.]

(a)

(b)

back to the ocean basins and would raise the sea level an estimated 60 m. A rise of this magnitude would transform the outline of the earth's landmasses and would submerge towns and cities along the coasts. For the last several thousand years, melting glaciers have been raising the sea level (see Figure 15.27), and modern records of sea level indicate that the sea is still rising at about 1 mm/year.

During most of the glacial periods of the past, water impounded on the land in the form of ice was more extensive than it is at present. Consequently, the sea level must have been lower than it is now. It has in fact been estimated that during the maximum extent of Pleistocene glaciation, the sea level was from 105 to 120 m lower. Most geologists accept this estimate, although some feel that it is too conservative. During the height of the Wisconsin glaciation, the last of the major ice advances, the sea level is usually estimated as having been between 70 and 100 m lower than at present. On the other hand, during the interglacial periods, when glaciers were somewhat less widespread than they are today, the sea level must have been higher.

15.7 THE CHANGING SEA LEVEL

349

15.26 Change between land and sea in northern Ungava, Quebec, Canada. Land has risen over 125 m since the formation of the last glacial beach in the area more than 7,000 years ago. This change was very rapid until about 6,000 years ago, when it slowed abruptly and continued at a much lower rate to the present. The change in the relationship between land and sea has been due to the isostatic recovery of the land after the disappearance of the ice sheet that had depressed it. [After B. Matthews, "Late Quaternary Land Emergence in Northern Ungava, Quebec," *Arctic*, vol. 20, no. 3, p. 186, 1967.]

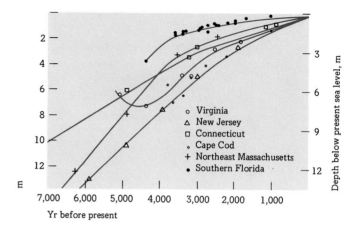

15.27 Sea level has been rising for the last several thousand years. Here radiocarbon dates on samples of wood, shell, and peat originally deposited at or close to sea level have been combined to produce these curves of sea-level rise at various points along the eastern coast of the United States. [After David Scholl and Minze Stuirer, "Recent Submergence of Southern Florida," *Bull. Geol. Soc. Am.*, vol. 78, p. 448, 1967.]

OUTLINE

Oceans cover more than 70 percent of the earth's surface.

Origin of seawater was probably by "degassing" of the earth's interior. This may have been fairly continuous by volcanic eruptions throughout earth history, or may have been confined to an early thermal stage.

Dissolved material gives oceans their saltiness. The material is moved through the rock cycle from lands to oceans and back to lands.

Currents of the ocean mix seawater once about every 2,000 years.

Tidal currents, wind-driven (surface) **currents,** and **density currents** are the three major types of currents.

One worldwide **ocean basin** contains the world's oceans and branches northward as the Pacific, Atlantic, and Indian Oceans.

Topography of the ocean floor is divisible into three major units.

The **continental margins** include the continental shelves, slopes, and rises.

The floors of the ocean basins lie on the average 3,800 m below the surface. Aside from the trenches, the **abyssal plains** are the deepest parts of the basins.

Oceanic ridges segment the ocean basins. **Midoceanic ridges** mark the divergence of two plates. **Aseismic ridges** mark the direction of plate movement as it crosses a plume or hot spot fixed in the mantle.

Seamounts are drowned volcanic peaks. **Guyots** are volcanoes that have been truncated by wave action and have then subsided below sea level. **Deep-sea trenches** mark convergence of plates where one is subducted beneath the other.

Sediments in the oceans are most rapidly accumulated on the continental margins. Sediments can be divided into **pelagic** and **terrigenous,** the latter dominating the continental margins.

Shorelines acquire characteristics generated by the energy of wind-driven waves.

Shoreline features include those formed by **erosion** and those formed by **deposition.** Erosional features include cliffs, caves, arches, stacks, and tidal inlets. Depositional features include beaches, spits, bay barriers, and barrier islands.

Sea level is not constant. Isostasy attendant upon glaciation and deglaciation changes sea level. The sea level also rises and falls as glaciers melt or expand.

Bird, E. C. F. *Coasts*, The M.I.T. Press, Cambridge, Mass., 1968. *A good, brief, introductory volume on the subject.*

Gross, M. Grant *Oceanography*, 2nd ed., Prentice-Hall, Inc., Englewood Cliffs, N. J., 1977. *A very good general reference that covers ocean circulation, geology, marine life, and shorelines.*

Ross, David A. *Introduction to Oceanography*, 2nd. ed., Prentice-Hall, Inc., Englewood Cliffs, N. J., 1977. *A sound text written to be understandable to the nonscientist.*

Shepard, Francis P., and Harold R. Wanless *Our Changing Shorelines*, McGraw-Hill Book Company, New York, 1971. *A wonderfully illustrated and extensive study of the coastlines of the United States. It should convince even the casual browser that the shoreline has been, is, and will continue to be, in constant change.*

Turekian, Karl K. *Oceans*, 2nd ed., Prentice-Hall, Inc., Englewood Cliffs, N. J., 1976. *Brief, authoritative, and a good place to begin further reading.*

1   What is degassing of the earth and what bearing does it have on the origin of ocean water?

2   Most of the known elements have been identified in dissolved form in seawater. Which are the volumetrically most important?

3   What are the three types of density currents found in the ocean? Give an example of each.

4   How is water moved in the major surface currents of the ocean?

5   Referring to the bar graph in Figure 15.2, what is a reasonable explanation for the doubly-peaked distribution of the earth's surface below and above sea level.

6   By now you should be familiar with the major topographic subdivisions of the ocean floor, so write a brief description of each.

7   Explain why the Hawaiian Islands and their drowned continuation to the northwest are cited as an indication of the direction of movement of the Pacific plate?

8   Here's a question the answer to which none of us will live long enough to prove right or wrong. Sediments eroded from the continents are being added to the ocean floor at a rate of about 10^{10} t/yr. How long will it be before erosion wears the continents down to nearly level surfaces approximating sea level?

Here are some added data: From Figure 15.2, the average elevation of the continents is 840 m and from Appendix Table D.8, the land area of the world is 149×10^6 km^2. We can use 2.65 for the average density of sediments.

(We got 32 million years, as follows):

$$\text{Years to flatten continents} = \frac{\text{land above sea level (tonnes)}}{\text{tonnes eroded per year}}$$

$$840 \text{ m} \times 149 \times 10^6 \text{ km}^2 \times 10^6 \text{ m}^2 \times 2.65 \text{ t/m}^2 = 33.2 \times 10^{16} \text{ t}$$

$$\frac{33.2 \times 10^{16} \text{ t}}{10^{10} \text{ t}} = 32 \times 10^6 \text{ y}$$

9   How do sediments of the deep ocean differ from those of the continental margins? Why?

10   With a diagram, show how the refraction of waves advancing on an irregular shoreline concentrates more energy per unit length of shoreline on the headlands than in the bays.

11   How many meters would sea level rise if the ice on Antarctica were to melt? From Appendix Table D.8, the area of the oceans is 361×10^6 km^2. The volume of ice on Antarctica is variously estimated; we use 22.5×10^6 km^3.

(Our answer was 62 m of sea level rise. Because an answer in meters was asked for, we first converted km to m. We then divided ice volume by ocean area to get sea-level rise):

$$\frac{22.5 \times 10^6 \text{ km}^3 \times 10^9 \text{ m}^3}{361 \times 10^6 \text{ km}^2 \times 10^6 \text{ m}^2} = 62 \text{ m}$$

16

WIND AND OTHER DESERT PROCESSES

All living things, including human beings, need moisture to survive, so it is not strange that the dry regions of the world are sparsely populated and that our knowledge of these regions is scanty at best. Still, deserts and near-deserts cover nearly one-third of the land surface of the earth, and we cannot measure their importance solely in terms of population density.

16.1 DISTRIBUTION OF DESERTS

Although there is no generally accepted definition of a **desert,** we can at least say that a desert is characterized by a lack of moisture, leading to—among other things—a restriction on the number of living things that can exist there. There may be too little initial moisture, or the moisture that occurs may be evaporated by extremely high temperatures or locked up in ice by ex-

treme cold. Because we are not concerned here with polar deserts, we shall consider only those in the hotter climates. The distribution of middle- and low-latitude deserts is shown in Figure 16.1; they fall into two general groups.

The first are the so-called **topographic deserts,** deficient in rainfall either because they are located toward the center of continents, far from the oceans, or more commonly because they are cut off from rain-bearing winds by high mountains. Takla Makan, north of Tibet and Kashmir in extreme western China, is an example of a desert located deep inside a continental landmass. The desert climate of large sections of Nevada, Utah, Arizona, and Colorado, on the other hand, is caused by the Sierra Nevada range of California, which cuts off the rain-bearing winds blowing in from the Pacific. A similar, though smaller, desert area in western Argentina has been created by the Andes.

Much more extensive than the topographic deserts are the **subtropical deserts** lying in zones that range 5° to 30° north and south of the equator. Their origin is best explained by the general circulation of the earth's atmosphere. Air heated at the equator rises and is diverted north and south. At about 30° on either side of

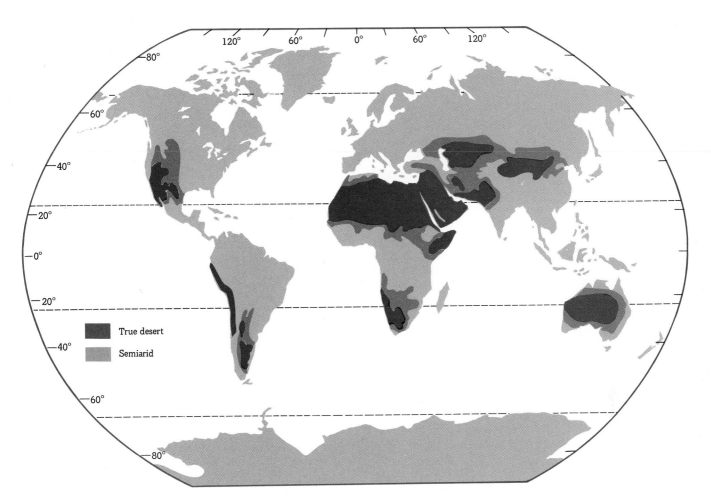

16.1 Deserts and near-deserts cover nearly one-third of the land surface of the earth. Middle- and low-latitude deserts, but not polar deserts, are shown here.

True desert
Semiarid

the equator, this air begins to pile up and form permanent atmospheric high-pressure zones. In these zones, known as the **subtropical high-pressure zones,** the air settles.

As the air settles, it warms. The warmer the air becomes, the more moisture it can hold and the less likely it is to release moisture as precipitation. Furthermore, this descending and warming air is generally clear and without clouds. Therefore the sun is all the more effective in warming the earth and the air that comes in contact with it.

At the earth's surface, the descending air moves laterally away from the high-pressure zones. Poleward, the winds become the **prevailing westerlies.** Toward the equator, the winds are the **northeast trades** (in the northern hemisphere) and the **southeast trades** (in the southern hemisphere). These trade winds blow toward the equator where they become even more heated. They rise and cool, thus giving up moisture. Eventually in the upper atmosphere, they are diverted north and south toward the subtropical high-pressure zones, to begin the path all over again.

It is this continuing circulation of the air that creates the great deserts on either side of the equator. These include the Sahara Desert of North Africa; the Arabian Desert of the Middle East; the Victoria Desert of Australia; the Kalahari Desert of South West Africa and Botswana; the Sonora Desert of northwestern Mexico, southern Arizona, and California; the Atacama Desert of Peru and Chile; and the deserts of Afghanistan, Baluchistan, and northwestern India.

Some of the smaller deserts along the tropical coastlines have been created by the influence of oceanic currents bordering the continents. Winds blow in across the cool water of the ocean and suddenly strike the hot tropical landmass. There the air is heated, and its ability to hold water is increased. The resulting lack of precipitation gives rise to desert conditions, as along the coast of southern Peru and northern Chile, where the cool Humboldt current flows north toward the equator.

16.2 Work of the Wind

In humid lands, a virtually continuous vegetative cover blankets the earth's surface; but in deserts, the lack of moisture ensures that vegetation is absent or sparse, and the wind can then directly affect the land surface. Therefore we find good examples of erosion, transportation, and deposition by wind in deserts.

MOVEMENT OF MATERIAL

Wind velocities increase rapidly with height above the ground surface, just as the velocity of running water increases at levels above the channel floor. Furthermore, like running water, most air moves in turbulent flow. But wind velocities increase at a greater rate than water velocities, and the maximum velocities attained are much higher.

The general movement of wind is forward, across the surface of the land; but within this general movement, the air is moving upward, downward, and from side to side. In the zone about 1 m above ground surface, the average velocity of upward motion in an air eddy is approximately one-fifth the average forward velocity of the wind. This upward movement greatly affects the wind's ability to transport small particles of earth material, as we shall see.

Right along the surface of the ground there is a thin but definite zone where the air moves very little or not at all. Field and laboratory studies have shown that the depth of this zone depends on the size of the particles that cover the surface. On the average, the depth of this **zone of no movement** is about one-thirtieth the average diameter of the surface grains (see Figure 16.2). Thus, over a surface of evenly distributed pebbles with an average diameter of 30 mm, the zone of no movement would be about 1 mm deep. This fact too has a bearing on the wind's ability to transport material.

Dust storms and sandstorms Material blown by the wind usually falls into two size groups: **sand** and **dust**. The diameter of wind-driven sand grains averages between 0.15 and 0.30 mm, with a few grains as fine as 0.06 mm. All particles smaller than 0.06 mm are classified as dust.

In a true **dust storm** (see Figure 16.3), the wind picks up fine particles and sweeps them upward hundreds or even thousands of meters into the air, forming

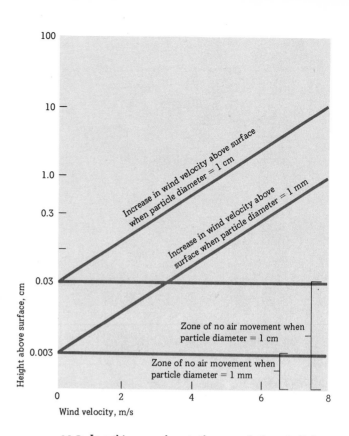

16.2 In a thin zone close to the ground, there is little or no air movement, regardless of the wind velocity immediately above. This zone is approximately one-thirtieth the average diameter of surface particles. Two zones are shown in the graph: one for surfaces on which the particles average 1 mm in diameter and one for surfaces with 1-cm particles. Diagonal lines represent the increase in velocity of a wind of given intensity blowing over surfaces covered with particles of 1-mm and 1-cm average diameter. [By permission, after R. A. Bagnold, *The Physics of Blown Sand and Desert Dunes*, p. 54, Methuen & Co., Ltd., London, 1941.]

a great cloud that may blot out the sun and darken the sky. In contrast, a **sandstorm** is a low moving blanket of wind-driven sand with an upper surface 1 m or less above the ground. The greatest concentration of moving sand is usually just a few centimeters above the ground surface, and individual grains seldom rise even as high as 2 m. Above the blanket of moving sand, the air is quite clear, and a person on the ground appears to be partially submerged, as though standing in a shallow pond.

Often, of course, the dust and sand are mixed together in a wind-driven storm (Figure 16.4); but the wind soon sweeps the finer particles off, and eventually the air above the blanket of moving sand becomes clear. Apparently, then, the wind handles particles of different sizes in different ways. A dust-sized grain is swept high into the air, and a sand-sized grain is driven along closer to the ground. The difference arises from the strength of the wind and the terminal velocity of the grain.

16.3 Turbulent air sweeps silt-sized particles high into the air in this dust storm on the Great Plains. [National Oceanographic and Atmospheric Administration.]

16.4 Sand drifts across this railroad in Texas, while finer particles are winnowed out and are carried upward, partially obscuring the sky. [U.S. Department of Agriculture.]

We defined the **terminal velocity** of a grain as the constant rate of fall attained by the grain when the acceleration due to gravity is balanced by the resistance of the fluid—in this case, the air—through which the grain falls (see Section 12.5). Terminal velocity varies only with the size of a particle when shape and density are constant. As the particle size increases, both the pull of gravity and the air resistance increase too. But the pull of gravity increases at a faster rate than the air resistance: A particle with a diameter of 0.01 mm has a terminal velocity in air of about 0.01 m/s; a particle with a 0.2 mm diameter has a terminal velocity of about 1 m/s; and a particle with a diameter of 1 mm has a terminal velocity of about 8 m/s.

To be carried upward by an eddy of turbulent air, a particle must have a terminal velocity that is less than the upward velocity of the eddy. Close to the ground surface, where the upward currents are particularly strong, dust particles are swept up into the air and are carried in suspension. Sand grains, however, have terminal velocities greater than the velocity of the upward-moving air; they rise for a moment and then fall back to the ground. But how does a sand grain get lifted into the air at all if the eddies of turbulent air are unable to support it?

Movement of sand grains Careful observations, both in the laboratory and on open deserts, show sand grains moving forward in a series of jumps, in a process known as **saltation.** We used the same term to describe the motion of particles along a stream bed, but there is a difference: An eddy of water can actually lift individual particles into the main current, whereas wind by itself cannot pick up sand particles from the ground.

Sand particles are thrown into the air only under the impact of other particles. When the wind reaches a critical velocity, grains of sand begin to roll forward along the surface. Suddenly one rolling grain collides with another; the impact may lift either the second particle or the first into the air.

Once in the air, the sand grain is subjected to two forces. First, gravity tends to pull it down to earth again, and eventually it succeeds. But even as the grain falls, the horizontal velocity of the wind drives it forward. The resulting course of the sand grain is parabolic from the point where it was first thrown into the air to the point where it finally hits the ground. The angle of impact varies between 10° and 16° (see Figure 16.5).

When the grain strikes the surface, it may either bounce off a large particle and be driven forward once again by the wind, or bury itself in the loose sand, perhaps throwing other grains into the air by its impact.

In any event, it is through the general process of saltation that a sand cloud is kept in motion. Countless grains are thrown into the air by impact and are driven along by the wind until they fall back to the ground. Then they either bounce back into the air again or else pop other grains upward by impact. The initial energy that lifts each grain into the air comes from the impact of another grain, and the wind contributes additional energy to keep it moving. When the wind dies, all the individual particles in the sand cloud settle to earth.

Some sand grains, particularly the large ones, never rise into the air at all, even under the impact of other grains. They simply roll forward along the ground, very much like the rolling and sliding of particles along the bed of a stream of water. It has been estimated that between one-fifth and one-quarter of the material carried along in a sandstorm travels by rolling and the rest by means of saltation.

Notice that after the wind has started the sand grains moving along the surface, initiating saltation, the wind no longer acts to keep them rolling. The cloud of saltating grains obstructs the wind and shields the ground surface from its force; thus, as soon as saltation begins, the velocity of near-surface winds drops rapidly. Saltation continues only because the impact of the grains continues. The stronger the winds blow during saltation, the heavier will be the blanket of sand and the less the possibility that surface grains will be rolled by the wind.

Movement of dust particles As we have seen, dust particles are small enough and have low enough terminal velocities to be lifted aloft by currents of turbulent air and to be carried along in suspension. But just how does the wind lift these tiny particles in the first place?

Laboratory experiments show that under ordinary conditions, particles smaller than 0.03 mm in diameter cannot be swept up by the wind after they have settled to the ground. In dry country, for example, dust may lie undisturbed on the ground even though a brisk wind is blowing, but if a flock of sheep passes by and kicks loose some of the dust, a dust plume will rise into the air and move along with the wind.

The explanation for this seeming reluctance of dust particles to be disturbed lies in the nature of air movement. The small dust grains lie within the thin zone of negligible air movement at the surface. They are so small that they do not create local eddies and disturbances in the air, and the wind passes them by—or the particles may be shielded by larger particles against the action of the wind.

Some agent other than the wind must set dust particles in motion and lift them into a zone of turbulent air—perhaps the impact of larger particles or sudden downdrafts in the air movement. Irregularities in a plowed field or in a recently exposed stream bed may help the wind begin its work by creating local turbulence at the surface. Also, vertical downdrafts of

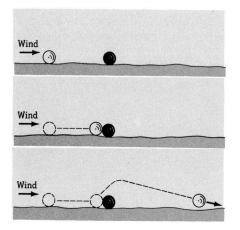

16.5 A sand grain is too heavy to be picked up by the wind but may be put into the air by saltation. Here a single grain is rolled forward by the wind until it bounces off a second grain. Once in the air, it is driven forward by the wind and is then pulled to the ground by gravity. It follows a parabolic path, hitting the ground at an angle between 10° and 16°.

chilled air during a thunderstorm sometimes strike the ground with velocities of 40 to 80 km/h and churn up great swaths of dust.

EROSION

Erosion by the wind is accomplished through two processes: **abrasion** and **deflation.**

Abrasion   Like the particles carried by a stream of running water, saltating grains of sand driven by the wind are highly effective abrasive agents in eroding rock surfaces. As we have seen, wind-driven sand seldom rises more than 1 m above the surface of the earth, and measurements show that most of the grains are concentrated in the 0.5 m closest to the ground. In this layer their abrasive power is greatest.

Although evidence of abrasion by sand grains is rather meager, there is enough to indicate that this erosive process takes place. For example, we sometimes find fence posts and telephone poles abraded at ground level and bedrock cliffs with a small notch along their base. In desert areas the evidence is more impressive; for here the wind-driven sand has in some places cut troughs or furrows in the softer rocks. The cross profile of one of these troughs is not unlike that of a glaciated mountain valley in miniature, the troughs ranging from a few centimeters to perhaps 8 m in depth. They run in the usual direction of the wind, and their deepening by sand abrasion has actually been observed during sandstorms.

The most common products of abrasion are certain pebbles, cobbles, and even boulders that have been

16.6   A ventifact with three well-developed facets, on the floor of the Mojave Desert, California. The pocket knife to the right of the ventifact gives the scale. [Sheldon Judson.]

eroded in a particular way. These pieces of rock are called **ventifacts,** from the Latin for "wind" and "made." They are found not only on deserts but also along modern beaches—in fact, wherever the wind blows sand grains against rock surfaces (see Figures 16.6 and 16.7). Their surface is characterized by a rela-

8 cm

(a)

Approximate height of densest zone in the layer of wind-driven sand

Ground surface

(b)

(c)

10-16°

(d)

16.7   A facet on a ventifact is cut by the impact of grains of wind-driven sand. [After Robert P. Sharp, "Pleistocene Ventifacts East of the Big Horn Mountains, Wyoming," *J. Geol.,* vol. 57, p. 182, 1949.]

tively high gloss, or sheen, and by facets, pits, gouges, and ridges.

The face of an individual ventifact may display one facet, or 20 facets, or more—sometimes flat but more commonly curved. Where two facets meet, they often form a well-defined ridge, and the intersection of three or more facets gives the ventifact the appearance of a small pyramid. Apparently the surface becomes pitted when it lies across the direction of wind movement at an angle of 55° or more; but it becomes grooved when it lies at angles of less than 55°.

Deflation Deflation (from the Latin for "to blow away") is the erosive process of the wind carrying off unconsolidated material. The process creates several recognizable features in the landscape. For example, it often scoops out basins in soft, unconsolidated deposits ranging from a few meters to several kilometers in diameter. These basins are known as **blowouts**, for obvious reasons. Even in relatively consolidated material, the wind will excavate sizable basins if some other agency is at work loosening the material. We find such depressions in the almost featureless High Plains of eastern New Mexico and western Texas, where the bedrock is loosely cemented by calcium carbonate. Several times during the Pleistocene, the climate in this area shifted back and forth between moist and dry. During the moist periods, water dissolved some of the calcium carbonate and left the sandstone particles lying on the surface. Then during the dry periods, the wind came along and removed the loosened material. Today we find the larger particles piled up in sand hills on the leeward side of the basin excavated by the wind. The smaller dust particles were carried farther along and spread in a blanket across the plains to the east (see Figure 16.8).

Deflation removes only the sand and dust particles from a deposit and leaves behind the larger particles of pebble or cobble size. As time passes, these stones form a surface cover, a **desert pavement**, that cuts off further deflation (see Figure 16.9).

Whenever the wind loses its velocity and hence its ability to transport the sand and dust particles it has picked up from the surface, it drops them back to the ground. The landscape features formed by wind-deposited materials are of various types, depending on the size of particles, the presence or absence of vegetation, the constancy of wind direction, and the amount of material available for movement by the wind. We still have a great deal to learn about this sort of deposit, but certain generalizations seem valid.

Calcium carbonate cement removed by solution

Sandstone loosely cemented by calcium carbonate

Moist climate

Deflation of basin

Coarse material deposited as sand dune

Fine material blown from area

Dry climate

16.8 The High Plains of eastern New Mexico and western Texas are pockmarked with broad, shallow depressions fashioned in loosely consolidated sandstone. In this instance, wind deflation has created blowouts, but only after the calcite cement of the sandstone was destroyed by downward-percolating waters. Destruction of the cement took place during moist periods in the Pleistocene, and deflation occurred in intervening dry periods. [After Sheldon Judson, "Geology of the San Jon Site, Eastern New Mexico," *Smithsonian Misc. Coll.*, vol. 121, no. 1, p. 13, 1953.]

Loess Loess is a buff-colored, unstratified deposit composed of small, angular mineral fragments. Loess deposits range in thickness from a few centimeters to 10 m or more in the central United States to over 100 m in parts of China. A large part of the surface deposits across some 0.5 million km^2 of the Mississippi River basin is made up of loess, and this material has produced the modern fertile soils of several midwestern states, particularly Iowa, Illinois, and Missouri (see Figure 16.10).

Most geologists, though not all, believe loess to be material originally deposited by the wind. They base their conclusion on several facts. The individual particles in a loess deposit are very small, strikingly like the particles of dust carried about by the wind today. Moreover, loess deposits stretch over hill slopes, valleys, and plains alike, an indication that the material has settled from the air. And the shells of air-breathing snails present in loess strongly impugn the possibility

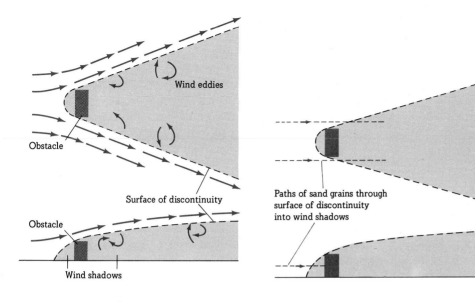

Obstacle

Wind eddies

Surface of discontinuity

Obstacle

Wind shadows

Paths of sand grains through surface of discontinuity into wind shadows

16.13 (left) The shaded area indicates the wind shadow created by an obstacle. The wind is diverted over and around the obstacle. Within the wind shadow, wind velocity is low and air movement is marked by eddies. A surface of discontinuity separates the air within the wind shadow from the air outside. [By permission, after R. A. Bagnold, *The Physics of Blown Sand and Desert Dunes*, p. 190, Methuen & Co., Ltd., London, 1941.]

16.14 (right) Because of its momentum, the sand in the more rapidly moving air outside the wind shadow either passes through the surface of discontinuity to settle in the wind shadow behind the obstacle or strikes the obstacle and falls in the wind shadow in front of the obstacle. [By permission, after R. A. Bagnold, *The Physics of Blown Sand and Desert Dunes*, p. 190, Methuen & Co., Ltd., London, 1941.]

(a)

(b)

(c)

16.15 Sand falling in the wind shadow tends to be gathered by wind eddies within the shadow to form a shadow dune, as shown in this sequence of diagrams. [By permission, after R. A. Bagnold, *The Physics of Blown Sand and Desert Dunes*, p. 190, Methuen & Co., Ltd., London, 1941.]

builds up—a process aided by eddying air that tends to sweep the sand in toward the center of the wind shadow (see Figures 16.14 and 16.15). This process is familiar in dry sandy regions, where sand piles up before and in the lee of obstacles, but it is also seen in

snowy climates. The principle is illustrated by snow fences (devices that have also been used effectively against windblown sand).

Wind shadow of a dune   A sand dune itself acts as a barrier to the wind; and by disrupting the flow of air, it may cause the continued deposition of sand. A profile through a dune in the direction toward which the wind blows shows a gentle slope facing the wind and a steep slope to the leeward. A wind shadow exists in front of the leeward slope, and it is here that deposition is active. The wind drives the sand grains up the gentle windward slope to the dune crest and then drops them into the wind shadow. The leeward slope is called the **slip face** of the dune because of the small sand slides that take place there.

The slip face is necessary for the existence of a true wind shadow. Here is how the slip face is formed: A mound of sand affects the flow of air across it, as shown in the topmost diagram of Figure 16.16. Notice that the wind flows over the mound in streamlined patterns. These lines of flow tend to converge toward the top of the mound and diverge to the leeward. In the zone of diverging air flow, velocities are less than in the zone of converging flow. Consequently, sand tends to be deposited on the leeward slope just over the top of the mound where the velocity begins to slacken. This slope steepens because of deposition, and eventually

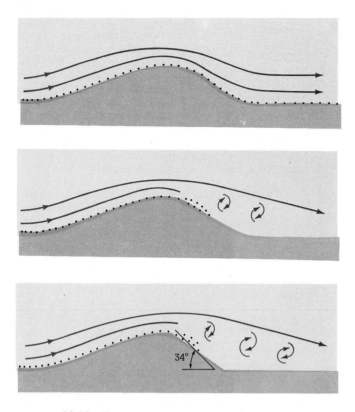

16.16 The development of a slip face on a dune. Wind converges on the windward side of the dune and over its crest and diverges to the lee of the dune. The eventual result is the creation of a wind shadow in the lee. In this wind shadow, sand falls until a critical angle of slope (about 34°) is reached. Then a small landslide occurs, and the slip face is formed. [By permission, after R. A. Bagnold, *The Physics of Blown Sand and Desert Dunes*, p. 202, Methuen & Co., Ltd., London, 1941.]

the sand slumps under the influence of gravity. The slump usually takes place at at angle of about 34° from the horizontal. A slip face is thus produced, steep enough to create a wind shadow in its lee (Figure 16.17). Within this shadow, sand grains fall like snow through quiet air. Continued deposition and periodic slumping along the slip face account for the slow growth or movement of the dune in the direction toward which the wind blows.

Shoreline dunes Not all dunes are found in deserts. Along the shores of the ocean and of large lakes, ridges of windblown sand called **fore dunes** are built up even in humid climates. They are well developed along the southern and eastern shores of Lake Michi-

gan, along the Atlantic coast from Massachusetts southward, along the southern coast of California, and at various points along the coasts of Oregon and Washington (Figure 16.18).

These fore dunes are fashioned by the influence of strong onshore winds acting on the sand particles of the beach. On most coasts, the vegetation is dense enough to check the inland movement of the dunes (Figure 16.19), and they are concentrated in a narrow belt that parallels the shoreline. These dunes usually have an irregular surface, sometimes pockmarked by blowouts (see the subsection on Deflation, which is on page 357).

Sometimes, however, in areas where vegetation is scanty, the sand moves inland in a series of ridges at right angles to the wind. These **transverse dunes** exhibit the gentle windward slope and the steep leeward slope characteristic of other dunes. Transverse dunes are also common in arid and semiarid country where sand is abundant and vegetation scarce. In the Sahara Desert, some transverse forms measure over 100 km in length and 1 to 3 km in width. They may carry on their crests smaller forms of varying shapes (Figure 16.20).

Barchans Barchans are sand dunes shaped like a crescent, with their horns pointing downwind. They move slowly with the wind, the smaller ones at a rate of about 15 m/year, the larger ones about 7.5 m/year. The maximum height obtained by barchans is about 30 m, and their maximum spread from horn to horn is about 300 m (see Figure 16.21).

Just what leads to the formation of a barchan is still a matter of dispute. However, certain conditions seem essential, including: a wind that blows from a fixed direction, a relatively flat surface of hard ground, a limited supply of sand, and a lack of vegetation.

Parabolic dunes Long, scoop-shaped, **parabolic dunes** look rather like barchans in reverse; that is, their horns point upwind rather than downwind. They are usually covered with sparse vegetation, which permits limited movement of the sand. Parabolic dunes are quite common in coastal areas and in various places throughout the southwestern states. Ancient parabolic dunes, no longer active, exist in the upper Mississippi valley and in central Europe.

Longitudinal dunes Longitudinal dunes are long ridges of sand running in the general direction of wind movement. The smaller types are about 3 m high and about 60 m long. In the Libyan and Arabian Deserts, however, they commonly reach a height of 100 m and may extend for 100 km across the country. There they are known as **seif dunes,** from the Arabic word for "sword." As with large transverse dunes, the large longitudinal forms may be decorated by smaller dunes along their crests (see Figure 16.22).

16.17 The slip face of a dune as illustrated in Figure 16.16 is shown here in dunes of the Great Sand Dunes National Monument near Alamosa, Colorado. The steep slip faces are in shadow and the gentler upwind slopes are well lighted. Some asymmetric ripples of sand move up the windward slopes toward the slip faces. [Macpherson Raymond, Jr.]

Stratification in dunes The layers of sand within a dune are usually inclined. The layers along the slip face have an angle of about 34°, whereas the layers along the windward slope are gentler.

Because the steeper beds along the slip face are analogous to the **foreset beds** in a delta (see Figure 12.45), we can use the same term in referring to them in dunes. These beds develop if there is a continuous deposition of sand on the leeward side of the dune, as in barchans and actively moving transverse dunes.

Backset beds develop on the gentler slope to the windward. These beds constitute a large part of the total volume of a dune, especially if there is enough vegetation to trap most of the sand before it can cross over to the slip face. **Topset beds** are nearly horizontal beds laid down on top of the inclined foreset or backset beds.

16.18 These shoreline dunes form complex patterns behind the beach at Coos Bay, Oregon. The beach serves as a source of sand, and this source is continuously renewed by longshore currents of ocean water. Onshore winds (from the left) drive the beach sand inland. The photograph shows about 450 m of shoreline.

16.2 WORK OF THE WIND

16.19 Dunes along the shore of Tusket Island near Yarmouth, Nova Scotia, are partially anchored by grass and bushes. [Information Canada Photothèque.]

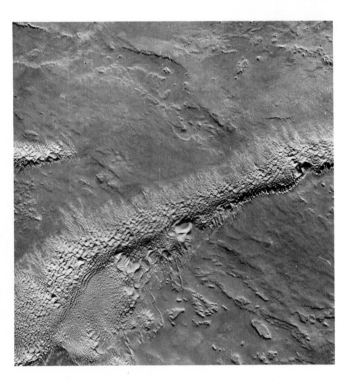

16.20 This transverse dune in the Sahara Desert south of the Atlas Mountains is nearly 1 km wide. Its surface exhibits the complex pattern of many smaller dunes. [U.S. Army Air Corps.]

16.21 These barchans are moving across the Pampa de Islay, Peru, in the direction in which their horns point. [Aerial Explorations, Inc.]

16.22 These seif dunes are located in the Rub ál Khali district of southern Saudi Arabia and are oriented parallel to the direction of the prevailing winds. [Arabian-American Oil Company.]

16.3 OTHER DESERT PROCESSES

WEATHERING AND SOILS

Because of the lack of moisture in the desert, the rate of weathering, both chemical and mechanical, is extremely slow. As most of the weathered material consists of unaltered rock and mineral fragments, we can conclude that mechanical weathering predominates.

Some **mechanical weathering** is simply the result of gravity, such as the shattering of rock material when it falls from a cliff. Wind-driven sand can bring about some degree of mechanical weathering. Sudden flooding of a desert by a cloudburst also moves material to lower elevations, reducing the size of the rock fragments and scouring the bedrock surface in the process. And in almost every desert in the world, temperatures fall low enough to permit frost action. But here again the deficiency of moisture slows the process. Finally, the wide temperature variations characteristic of deserts cause rock materials to expand and contract and may produce some mechanical weathering.

This low rate of weathering is reflected in the soils of the desert. Seldom do we find extensive areas of residual soil; for the lack of protective vegetation permits the winds and occasional floods to strip away the soil-producing minerals before they can develop into true soils. Even so, soils sometimes develop in local areas, but they lack the humus of the soils in moister climates, and they contain concentrations of such soluble substances as calcite, gypsum, and even halite because there is insufficient water to carry these minerals away in solution. In the deserts of Australia, rock-like concentrations of calcite, iron oxide, and even silica sometimes form a crust to be found on the surface.

One minor weathering feature of the desert is of special interest because it seems to have its counterpart on the surface of Mars. A thin, shiny, reddish-brown to blackish coating called **desert varnish** occurs on some desert stones. The coating is a thin film—a fraction of a millimeter thick—made up largely of iron and manganese oxides. Interestingly enough, there has long been disagreement as to how this varnish forms on earth. Some suggest that it is the result of entirely inorganic processes. Other workers believe that some organic activity is also involved as well.

WATER

Although rainfall is extemely sparse in desert areas, there is still enough water present to act as an important agent of erosion, transportation, and deposition. In fact, water is probably more effective than even the driving winds in molding desert landscape.

Very few streams flowing through desert regions ever find their way to the sea, and the few that do, such as the Colorado River in the United States and the Nile of Egypt, originate in areas from which they receive sufficient water to sustain them through their long courses across the desert. Exploitation of such rivers for agricultural, domestic, and industrial use can drastically affect their flow. In the case of the Colorado River, for instance, diversion of water to human use in California, Arizona, Nevada, and Colorado reduces that river to but a trickle by the time it reaches the Gulf of California. Also the flow of the Nile has been modified by the construction of Lake Nasser just above the first cataract and by the diversion of water to the Fayum depression west of Cairo.

Most desert stream beds, however, are dry over long periods of time and flow only when an occasional flood comes along. Even then, the flow is short-lived; for the water either evaporates rapidly or vanishes into

(a)

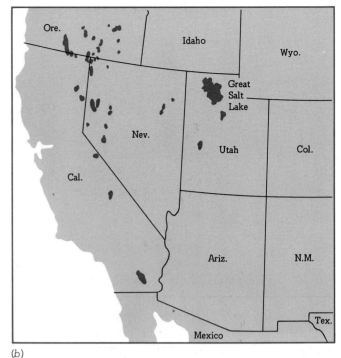

(b)

16.23 During glaciation, many basins of the western United States held lakes where only intermittent water bodies exist today; modern saltwater lakes are remnants of basins formerly filled with lakes. Modern lakes (b) are compared with those of the glaciation of higher mountains and the western United States (a). [After O. E. Meinzer, "Map of the Pleistocene Lakes of the Basin-and-Range Province and Its Significance," *Bull. Geol. Soc. Am.*, vol. 33, pp. 543, 545, 1922.]

the highly permeable rubble and debris of the desert. In some places, however, such as the western United States, broad desert plains slope down from the mountain ranges toward central basins, called **playas,** where surface runoff collects from time to time. But the **playa lakes** formed in these basins usually dry up in a short time or at best exist as shallow, salty lakes, of which Great Salt Lake is the best-known example.

Desert floods are unlike floods in humid areas. The typical desert flood, like the rain that produces it, is local in extent and of short duration. In moist regions, most floods arise from a general rain falling over a relatively long period of time; consequently these floods affect large areas. Because of the widespread vegetative cover, the floods tend to rise and fall slowly. But on the bare ground of the desert, the runoff moves swiftly and floods rise and fall with great rapidity. These **flash floods** give little warning, and the experienced desert traveler has a healthy respect for them. He will never pitch camp on a dry stream floor even though the stream banks offer protection from the wind. He knows that at any moment a surging wall of debris-laden water may sweep down the stream bed, destroying everything in its path.

The amount of moisture in desert areas does

change. Thus during the Ice Age, when glaciers lay across Canada and the northern United States and draped the flanks of most of the higher mountains of the West, the climate of the arid and semiarid areas of the western United States was quite different from what it is today. Glaciations produced **pluvial periods** (from the Latin *pluvia*, "rain"), when the climate was undoubtedly not so moist as that of the eastern United States today but certainly more hospitable than we now find in the arid regions. During any single pluvial period, rainfall was greater, evaporation less, and vegetation more extensive. At this time, large sections of the southwestern states were dotted with lakes known as **pluvial lakes** (Figure 16.23).

What was responsible for this pluvial climate? The presence of continental glaciers in the north is thought to have modified the general wind circulation of the globe. The belt of rain-bearing winds was moved to the south, and the temperatures were lowered. Consequently, the rates of evaporation decreased, and at the same time the amount of precipitation increased. When the ice receded, the climate again became very much what it is today.

Deserts cover one-third of the earth's surface. A **topographic** desert owes its aridity to the distance from a source of moisture or to protection from rain-bearing winds by a mountain mass. A **tropical** desert lies between 5° and 30° north or south of the equator in one of two zones of subtropical high pressure developed by planetary atmospheric circulation.

Work of the wind is more effective in the desert than anywhere else on the lands. Even here, however, it is subordinate to the work of running water in shaping the landscape.

   Movement of material by wind depends in part on the size of the particle: Dust-sized particles move differently than do sand-sized particles. Dust particles are carried high into the atmosphere in **suspension**. Sand moves along in continuous contact with the ground or moves a few centimeters above the surface in **saltation**.

   Erosion by wind consists of **abrasion**, of which **ventifacts** are the most common example, and **deflation**, of which **blowouts** are examples.

   Deposition consists of dust deposits (**loess**) and sand deposits (usually **dunes**). Sand deposition in dunes begins in the **wind shadow** in the lee of an obstacle. Once established, the dune provides its own wind shadow. In the dune, material is moved up the windward slope and is deposited on the steep **slip face** in the dune's wind shadow. Dune types include **fore dunes, transverse dunes, barchans, parabolic dunes,** and **longitudinal dunes.**

Other desert processes are also important.

   Weathering, both mechanical and chemical, is slow because of lack of water.

   Soils are rubbly and sometimes marked by crustlike accumulations of calcite, iron oxide, and silica.

   Water, despite its scarcity, is abundant enough to create streamways in even the driest desert. Flow of streams is short-lived but often catastrophic.

Bagnold, R. A. *The Physics of Blown Sand and Desert Dunes,* Methuen & Co., Ltd., London, 1941. *Although this book appeared four decades ago, it is yet to be replaced by a better one on the subject. The study of the movement of sand and dust by wind and of sand dunes begins here.*

Bloom, Arthur L. *Geomorphology,* Prentice-Hall, Inc., Englewood Cliffs, N. J., 1978, *Chapter 13 is a discussion of arid, semiarid, and savanna landscapes. Chapter 14 deals with the work of the wind and the forms that result from it.*

Mabbutt, J. A. *Desert Landforms,* The M.I.T. Press, Cambridge, Mass., 1977. *A short but comprehensive volume of intermediate level, which covers the various desert processes and landforms including, of course, wind action and its results.*

Petrov, M. P. *Deserts of the World,* John Wiley & Sons, Inc., New York, 1976. *Translated from the Russian, this volume surveys the deserts of the world and considers, as well, the physical and biological processes of the desert.*

1   What are some of the different types of deserts and their causes?

2   Distinguish between dust storms and sandstorms.

3   How does wind move sand-sized grains?

4   What is a ventifact and how is it formed?

5   Silt- and clay-sized particles dominate loess deposits. What are two different origins for loess deposits?

6   What is a wind-shadow and what function does it play in the deposition of sand?

7   On a dune, what is a slip face and what is its function in the dune-building process?

8   There are several different dune forms. Be able to write a brief, informative paragraph about each of the following: fore dune, barchan, parabolic dune, longitudinal dune, and transverse dune.

9   What is a pluvial climate? What significance would it have in western United States were one to occur there?

17

ENERGY

By about 1750, the coal deposits in England began to be fully exploited, to displace water, wind, and wood, and to fuel the beginnings of the Industrial Revolution; for industrial society is dependent on the availability of very large supplies of **energy.** Industrialism spread rapidly throughout the Western world and then to some eastern nations and today moves on to as yet unindustrialized countries. This change in social organizations has been driven by energy, at first in the form of coal and later as petroleum and natural gas (Figure 17.1). Now, more than three-quarters of the way through the twentieth century, industrial society finds itself in a crisis, seeking ways to maintain reliable and economically feasible supplies of energy. We make no attempt here to offer solutions to this crisis, but as a basis for dealing with solutions, we look at the various forms of energy that are available on earth. Our energy, both actual and potential, has three major sources: the sun, the nuclei of some elements, and gravity.

The **fossil fuels** (coal, petroleum, and natural gas) represent the remains of plants or animals that gathered their energy from the sun millions of years ago. Energy from the sun drives the hydrologic cycle and combined with gravity provides us with **water power.** The sun heats the atmosphere differentially which in turn causes its movement and produces **wind power.** Present-day plant life creates **wood,** and we can use modern organic matter to produce the gaseous fuel, **methane,** or the liquid fuel, **alcohol.** And recently we have begun to capture the sun's energy with artificial devices to produce what is popularly known as **solar power.** The nuclei of some elements give off energy by splitting, or **fission,** as does the radioactive form of uranium. The combination, or **fusion,** of the nuclei of some forms of hydrogen also produces energy. Much energy from the earth (**geothermal energy**) is derived from the heat created by the long-continued decay of radioactive elements in the earth's crust. Gravity as represented by the rising and falling of the ocean surface produces **tidal power.**

By far the most important sources of energy are the fossil fuels. Thus in the United States, they account for about 95 percent of the energy consumed. The other 5

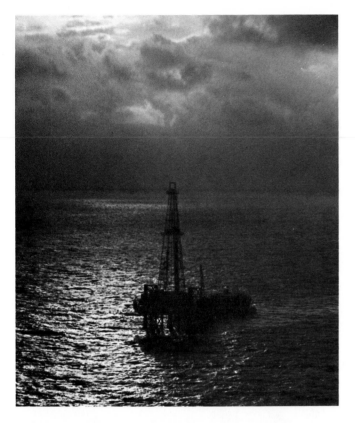

17.1 Petroleum and natural gas are the major sources of the world's energy supply today. In recent years, explorations for oil and gas have expanded from dry land to the submerged borders of the continents. Offshore reserves are increasingly important and are tapped by wells drilled from platforms set at sea, as is this one in the Arabian Gulf. [Arabian-American Oil Company.]

percent is predominantly in hydroelectric and nuclear power (Figure 17.2).

17.1 NATURAL RESOURCES

Before proceeding to a consideration of energy sources, we should pause briefly to make a few general observations about natural resources. The world's natural resources are concentrated in relatively small areas and are unevenly distributed throughout the crust; deposits that have been located may soon be exhausted unless the natural reserves are huge. These natural resources have been concentrated by the geologic processes we have been studying, such as weathering, running water, igneous activity, and metamorphism; but only a few geologic environments have favored their formation. And after they are used, they are gone forever. Unlike our practice with plant resources, we cannot grow another crop.

To put this another way, most of the natural resources that we exploit for energy and minerals are **nonrenewable**. Once they have been used up, they cannot be replenished—at least in our lifetime. Coal and oil, once burned, are vanished into water and carbon dioxide. A copper vein, once mined out, cannot be redeposited.

Estimating the amount of a nonrenewable resource available to us is at best a difficult and often

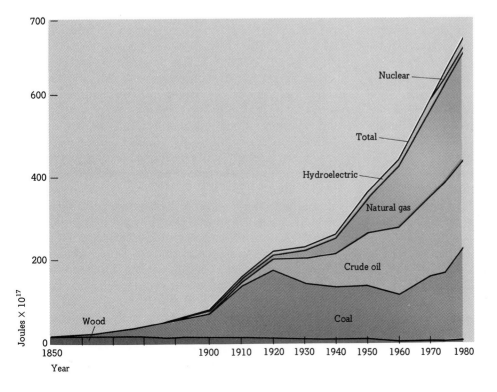

17.2 Energy consumption in the United States has expanded dramatically from 1850 to 1980. During this time the major source of energy has shifted from wood to coal to oil and gas. [Data from various sources.]

imprecise exercise. Most of our earth resources lie hidden beneath the surface. Direct evidence of their abundance comes from surface occurrences, existing mines, and exploratory drilling. Most estimates, however, are heavily dependent on indirect geological, geophysical, and geochemical data. Today's estimates may be modified by new discoveries. New technology may make possible extraction of resources that are today impractical or impossible to bring to the surface. Or new uses for an earth material may raise a once useless deposit to the rank of a valuable resource. Despite these considerations, many of our natural resources are finite. Changing estimates can only change the useful life of the resource.

In contrast, some resources are replenishable. Water can be pumped from an underground supply as long as the pumpage does not exceed the infiltration of water into the underground; and we can grow a new forest to replace one that we harvest. **The sun is our ultimate source of renewable energy;** on our human time scale, the atomic fires that send us light and heat across nearly 150 million km are limitless.

17.2 COAL

Coal, as we found in Section 5.8, is a sedimentary and metamorphic rock. It is formed from plants that grew in ancient swamps. Conditions favored preservation, which means that the remains of the plants accumulated in a nonoxidizing environment and were eventually buried by other sediments, usually sand or mud, which are now the sandstone and shale typically associated with coal beds (Figure 17.3).

The coal deposits start off as organic materials made chiefly of carbon, oxygen, and hydrogen. With rising temperature and pressure, due to the burial of the deposits, the hydrogen and oxygen are gradually lost. The higher the percentage of carbon in the remaining deposit, the higher the rank of the coal. We find, therefore, that the lowest grade of coal, **lignite,** has about 70 percent carbon; that **bituminous,** or **soft,** coal has about 80 percent carbon; and that the highest rank of coal, **anthracite,** contains from 90 to 95 percent carbon (Figure 17.4).

It takes about 2 m of peat, the organic deposits of swamps, to produce 1 m of soft coal or 0.5 m of anthracite. Some coal beds, or **seams,** reach 100 m in thick-

17.3 A coal seam exposed in a road cut near the Monongahela River in western Pennsylvania. The seam is about 2 m thick. [U.S. Bureau of Mines.]

ness. The thickest coal seam known is a bituminous deposit in Australia, which is nearly 200 m thick. On the other hand, the average coal seam is less than a meter in thickness, and seams greater than 3 m are rare.

In addition to carbon, oxygen, and hydrogen, coal contains many other elements, even though in small amounts. One, sulfur, has made some coal a dangerous pollutant of both air and water. It enters the air as sulfur dioxide gas, SO_2, which can be harmful to life if it reaches certain concentrations. Furthermore, SO_2, when combined with the moisture of the atmosphere, forms sulfuric acid, H_2SO_4. This in turn, when washed out of the atmosphere, produces a rain of higher than normal acidity. **Acid rain** has recently become a problem for certain areas downwind from regions burning large amounts of both coal and oil. Thus many lakes in New England and Scandinavia have become so acid that fish life has become adversely affected. In addition, drainage from coal mines may be so high in H_2SO_4 that streams receiving the mine drainage have become unsuitable for many fish species.

The geology of coal is well understood, and the search for coal is governed by the relatively simple

17.4 Increasing calorific value of coal with increasing rank. [After Brian J. Skinner, *Earth Resources*, p. 26, Prentice-Hall, Inc., Englewood Cliffs, N. J., 1976.]

characteristics of flat-lying to moderately folded sedimentary rocks. Coal occurs widely but unevenly around the world, with the United States being particularly well endowed (Figures 17.5 and 17.6). Most of the major coal-bearing basins have been identified, and we have a pretty good idea of both the known reserves as well as potential resources yet to be proved. Estimates of the total recoverable coal in the world vary, but something like 7.1×10^{12} t seems to be a reasonable figure. For the United States, total recoverable coal is about 1.5×10^{12} t. These are impressive figures and represent a resource that can last for a few hundred years, a much longer time than we can project for oil and natural gas.

We usually think of coal as a solid fuel, but for well over a century, methane gas (CH_4) has been produced from coal. Today, natural gas has generally replaced coal gas because natural gas has been more economical. But gasification of coal remains an alternative source of gas. Indeed, the increasing costs of natural gas are making the gasification of coal more and more attractive.

Not only can gas be produced from coal but gasification can serve as an intermediate step to the production of oil. Direct liquefaction of coal is also possible, and commercial installations producing oil directly from coal may not be far in the future.

Coal will have an increasingly important place in our total energy supply for many decades. But we still rely most heavily on oil and gas in the underground, a subject to which we now turn.

17.3 OIL AND NATURAL GAS

Oil and natural gas have proved to be economical, efficient, and relatively clean fuels. As a result, by 1950, they had overtaken coal as the primary source of energy in the United States, and today some four times as much energy is produced in this country by burning oil and natural gas as is produced from coal.

ORIGIN

Almost without exception, **petroleum and natural gas are associated with sedimentary rocks of marine origin.** Petroleum and natural gas are mixtures of hydrocarbon compounds (composed largely of hydrogen and carbon) with minor amounts of sulfur, nitrogen, and oxygen. Virtually all geologists are now convinced that they were derived originally from organic remains—both plants and animals.

The hydrocarbons found in our oil and gas fields, however, differ somewhat from those we know in living things. So some changes take place between the organic remains and the end product. The first step is the accumulation of an oceanic sediment rich in remains of plant and animal life. This is not so simple as it sounds; for most sea environments provide oxygen to destroy the organic remains before they can be buried by subsequent sediments. We, however, have some examples of basins in which the circulation of water is so slow that the bottom waters become depleted in oxygen, and organic-rich sediments can accumulate. Even this oxygen-poor environment may not suffice to protect the organic sediments because some bacteria can utilize the sulfates of seawater to oxidize organic matter. Therefore we need to have a marine basin that accumulates organic sediments in an oxygen- *and* sulfate-deficient environment.

The sediments destined to provide a source of petroleum start off as muds and silts, which eventually end up as shales, siltstones, and fine-grained limestones. Their organic content varies between 0.5 and 5

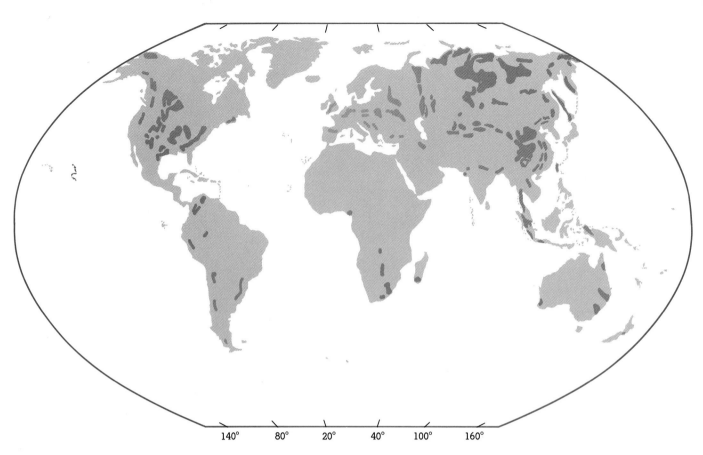

140° 80° 20° 40° 100° 160°

17.5 **Known and assumed occurrences of coal in the world.** [After Günter B. Fettweis, *World Coal Resources*, Fig. J, p. 415, Elsevier Scientific Publishing Co., New York, 1979.]

percent, and averages about 1.5 percent. A small amount of this may be in the form of hydrocarbons from the beginning, but the bulk of hydrocarbons, over 90 percent, form after the initial deposition of sediments.

Hydrocarbon production takes place in two stages. First, biological, chemical, and physical processes begin to break down the organic matter into what is called **kerogen,** a precursor to oil and gas. The second stage is marked by the thermal alteration of kerogen to hydrocarbons as the deposit is buried deeper and deeper by younger, overlying sediments. The production of hydrocarbons begins at a temperature of about 50° to 60°C and a depth of 2 to 2.5 km. Hydrocarbon formation continues to depths of 6 to 7 km and temperatures of 200° to 250°C. Formation of oil dominates in the lower range of temperature and burial, and gas in the higher range, as suggested for the oil and gas field diagrammed in Figure 17.7. Whether the hydrocarbons, once formed, will gather into an oil and/or gas field depends on a number of additional factors discussed in the following paragraphs.

ACCUMULATION

We have found that a sedimentary unit rich in organic material and buried to depths between 2 and 7 km is needed for the creation of oil and gas. This rock becomes the **source rock** for petroleum, defined as a fine-grained sedimentary rock that has naturally generated enough hydrocarbons to form a commercial accumulation of oil and gas. The hydrocarbons in the source rock then move to more permeable rocks, as follows: In the first place, there is the weight of the overlying rocks that tends to squeeze the oil and gas through pores and cracks in the rock; in the second place, there is usually water in the source bed as well as in adjacent rocks, and if the pores are large enough, the low density oil and gas tend to move upward through the water-saturated layers. Eventually the oil and gas reach the surface and are lost there, unless something intervenes to collect and hold them in the underground.

An impermeable rock unit will block or divert oil and gas in their migration. If the impermeable unit is

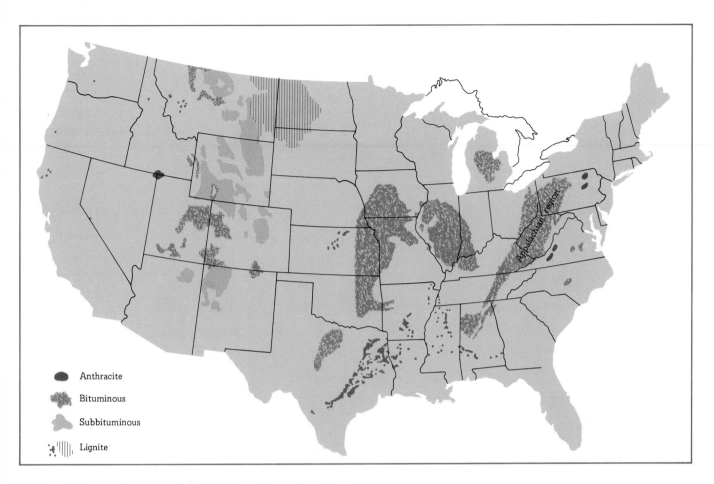

17.6 The coal fields of the United States are large. Extensions of deposits in the western United States are found beneath the plains of Canada, and coals typical of the eastern United States are mined in Nova Scotia. Deposits of lignitic through bituminous rank occur widely in Alaska. ["Strippable Reserves of Bituminous Coal and Lignite in the United States," *U.S. Bur. Mines Info. Circ. 8531*, 1971.]

Legend:
- Anthracite
- Bituminous
- Subbituminous
- Lignite

Increasing generation of hydrocarbons

17.7 Data from an oil and gas field in the Aquitaine basin in southwestern France show that the maximum generation of oil occurs at a shallower depth and a lower temperature than does that of gas. In this example, oil generation peaks at a depth of 3.5 km and a temperature of 100°C. This is followed by the intense generation of gas at greater depths and temperatures. Note that the rate of temperature rise increases with depth. [Generalized from K. Le Tran, "Geochemical Study of Hydrogen Sulfide Sorbed on Sediments," in *Advances in Organic Geochemistry*, Fig. 2, p. 722, eds. H. R. Gaertner and H. Werner, Pergamon Press, Inc., Elmsford, N. Y., 1971.]

17.3 OIL AND NATURAL GAS

17.8 The common types of oil traps, drawn here in cross section, include (a) anticlines, (b) faults, (c) unconformities, (d) reefs, (e) sand lenses, and (f) salt domes. Oil accumulation is shown in color.

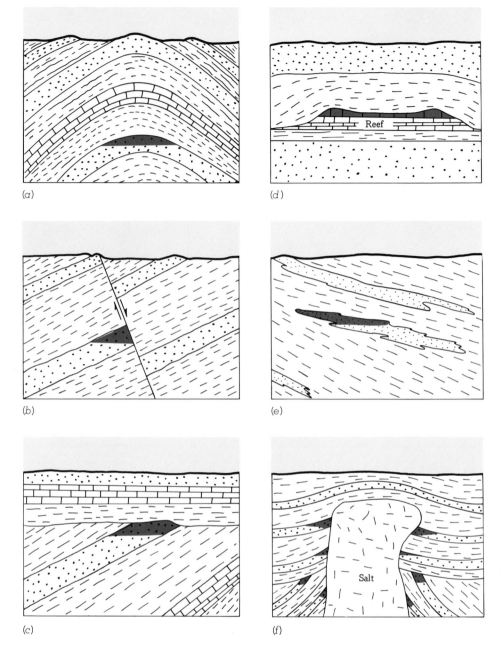

(a)

(d) Reef

(b)

(e)

(c)

(f) Salt

properly positioned, it will halt the hydrocarbons on their way toward the surface and keep them underground. Associated with this impermeable **capping bed** is a permeable rock unit into which the hydrocarbons have moved. We call this the **reservoir rock** because it contains the accumulated oil and gas. The capping bed and related reservoir rock together form a **trap** and serve as a natural tank. The traps can be of several types, as illustrated in Figure 17.8.

Once the seal provided by the capping bed is broken by the drill, the oil and natural gas move from the pores of the reservoir rock to the drill hole and then can be brought to the surface for processing and distribution. Of course, the trap may be broken by natural means as well. For instance, earth movements after the formation of the trap may so disturb it that fractures develop, and the hydrocarbons will escape either to another trap or to the surface. Or, again, surface erosion may eventually reach and destroy the trap. Therefore one would expect that the older the rock, the less chance there is that it still contains oil, and in a general way this is true. The greatest amount of production comes from rocks of Cenozoic age, followed by fields of

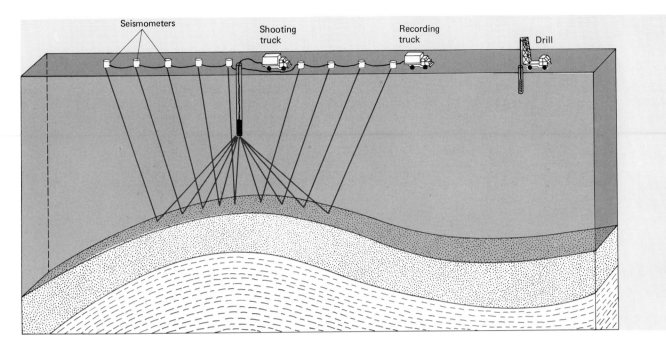

17.9 The operation of a seismic-exploration effort on land is shown in this diagram. First a drill prepares a hole about 50 m deep. A shooting truck then loads the bottom of the hole with explosives, plugs the hole, and explodes the charge electrically on a command from the recording truck. Sound waves from the explosion follow paths shown by the colored lines from the explosion to a reflecting layer at depth and back up to miniature seismometers connected by cable to the recording truck. Computer processing of the seismic records produces a diagram that resembles a geologic cross section. [After Sheldon Judson, Kenneth S. Deffeyes, and Robert B. Hargraves, *Physical Geology*, p. 443, Prentice-Hall, Inc., Englewood Cliffs, N. J., 1976.]

oil and gas in Mesozoic and then Paleozoic rocks. There is essentially no production from Precambrian rocks.

THE SEARCH FOR PETROLEUM

The search for petroleum is carried on largely by indirect methods. Most of the geologic structures that contained oil and gave surficial indication of their presence have long since been drilled. So now we must resort to means other than surface observation to discover potential traps of oil and gas buried beneath kilometers of sedimentary rocks and contained in the beds submerged beneath the oceans along the continental margins. One exploration technique is shown in Figure 17.9.

Figure 17.10 shows the areas in the world favorable to the exploration for oil and natural gas, both onshore and offshore. It also shows the producing areas. The conterminous United States has been very effectively explored, as have the Canadian provinces of Alberta and Saskatchewan. Probably half the oil and

gas in these areas has been found and used, and all the giant fields have been discovered. The new petroleum frontiers for the North American continent lie in the arctic lands of Canada and Alaska and along the offshore margins of the continent.

How long will the supply of oil and natural gas last? That is a question we cannot answer with precision. We know that there is a finite amount of these liquid and gaseous hydrocarbons in the ground, and we also know that they are not renewable resources like trees or rivers. We have several estimates of the supply duration, and each has roughly the same conclusion: namely, that early in the twenty-first century, we shall run out of oil and natural gas. Two estimates for crude-oil production in the United States are given in Figure 17.11, one made in 1956 and the other in 1980. The 1956 estimate suggested that the United States would reach its peak production of crude oil between 1960 and 1970. By 1980, experience told us that the United States production of crude oil peaked in 1968. Furthermore, it became apparent that by the year 2000, we shall have produced 80 percent of the crude oil available to us in the United States, exclusive of Alaska.

On a worldwide scale, for both oil and gas, the picture is similar to that in the United States, although over the world there is a larger percentage of the original hydrocarbons yet to be found than is true for the

17.3 OIL AND NATURAL GAS

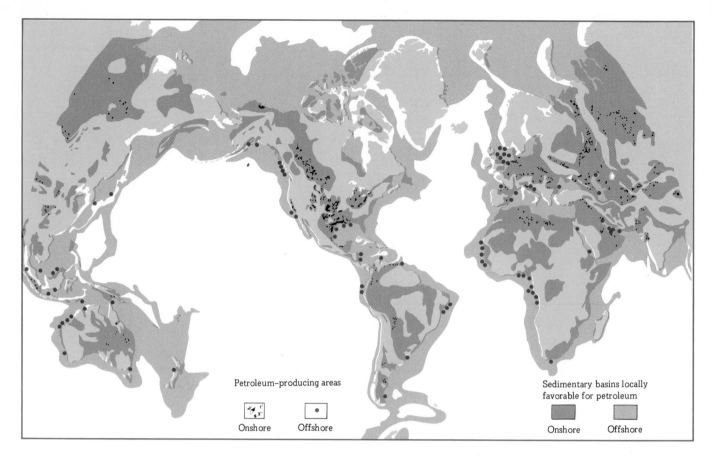

Petroleum–producing areas

Onshore Offshore

Sedimentary basins locally favorable for petroleum

Onshore Offshore

17.10 Areas favorable to the accumulation of petroleum both on the continents and subsea. Only small portions of the areas shown as favorable actually contain producible accumulations. [After V. E. McKelvey and others, "World Subsea Mineral Resources," *U.S. Geol. Surv. Misc. Geol. Invest. Map I-632,* sheet 3, 1969.]

extensively explored United States. Nevertheless, when we consider the "big three" of our present sources of energy—coal, oil, and natural gas—we find that the world supply will be consumed over a very short time span in terms of human history (see Figure 17.12). So the fossil fuels as we currently know them will become exhausted. What will replace them?

OIL SHALE AND TAR SANDS

Oil shale and tar sands are potential sources of additional fossil fuels. In fact, the potential resources outstrip the known and postulated supplies of natural gas and oil.

Oil shale Most fine-grained sedimentary rocks contain some organic matter. Given the proper conditions of burial, these rocks may produce commercial amounts of oil and gas. But some organic-rich rocks are never deeply enough buried to allow the conversion of the kerogen they contain into hydrocarbons. One can,

however, produce oil and gas from these deposits by heating them. We thus utilize an "immature" source rock to make an artificial oil field. Thus far the production of petroleum from shale is at best on a small scale, with some production in China, Estonia, and other parts of the Soviet Union.

The richest known deposits of high-grade oil shale are in the United States. During the Eocene Epoch, large freshwater lakes spread across portions of what are now Colorado, Utah, and Wyoming (Figure 17.13). In them accumulated great masses of organic-rich sediments, today collectively known as the Green River oil shales. Some shale is rich enough to produce nearly 240 l/t (liters/ton), and the potential reserves are large. Just how large is not presently clear, but they certainly are equal to all proved and predicted reserves of oil and natural gas still left in the United States. For many years, pilot plants have been experimenting with the technology of processing the Green River shales. So far, large-scale exploitation of the resource has stalled in part because the oil from processed shale is more expensive than the traditional oil-field petroleum and in part because of the environmental problems associ-

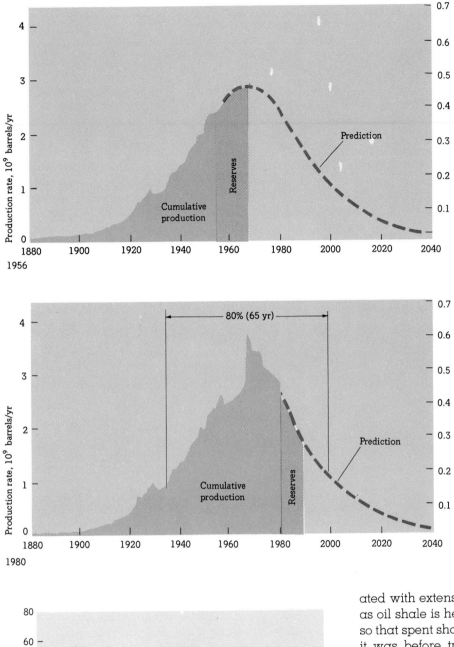

17.11 Crude-oil production histories and predictions for the United States and adjacent continental shelves (exclusive of Alaska) made in 1956 and 1980. Past production is shown in tint color, reserves in gray, and oil predicted to be discovered in white. [Earlier diagram after M. King Hubbert, "Nuclear Energy and the Fossil Fuels," Publication 95, p. 23, Shell Development Company, Houston, 1956; 1980 diagram updated from M. King Hubbert, "Energy Resources," in *Resources and Man*, W. H. Freeman & Company, Publishers, San Francisco, 1969. Copyright National Academy of Sciences.]

17.12 The period during which most of the fossil fuels will be consumed is short compared to human history and is an instant in terms of geologic time. This graph, showing the use of fossil fuels as a pimplelike bump on the scale of earth's history, has become known as "Hubbert's pimple" after M. King Hubbert, who originally published this graph in 1962. [After M. K. Hubbert, "Energy Resources: A Report to the Committee on Natural Resources," *Natl. Res. Council Pub. 1000-D*, National Academy of Sciences, Washington, 1962.]

ated with extensive shale oil operations. For instance, as oil shale is heated to extract oil, the shale expands, so that spent shale is 20 percent greater in volume than it was before treatment. Furthermore, immense volumes of shale must be used in any practical exploitation of the oil shale. Therefore the problem remains of what to do with the leftover shale once the oil is produced. Beyond this, the extraction process itself demands large amounts of water, already in short supply in the western states.

Besides the Green River shales, the United States possesses vast amounts of additional organic-rich shales, although their concentration of organic matter by no means compares with that in the Green River

17.3 OIL AND NATURAL GAS

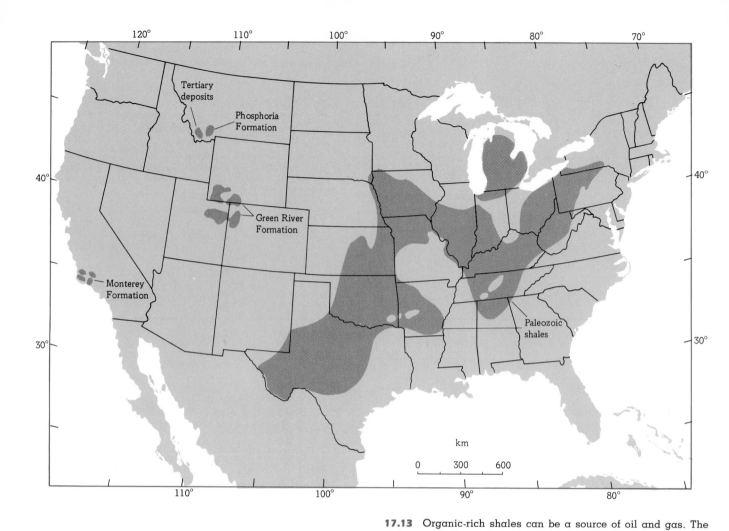

shales. Those shales close enough to the surface for mining are shown in Figure 17.13. However, present technology is far from making these deposits useful to us as an energy source, even though they may be eventually exploited.

Tar sands   Earlier in this section we pointed out that hydrocarbons formed in the underground work their way upward and, if not contained by a trap, will eventually reach the surface and there be lost. But in some situations, certain hydrocarbons may be brought to, or close to, the surface and survive for a considerable time. These hydrocarbons are the dark, heavy, low-volatility, high-viscosity ones. We know them as **tar** or **asphalt.** And in some localities, they are preserved as the dark, sticky cement of a sandstone. It is as if we were dealing with a reservoir bed from which all the flowing oil had been removed and in which only the heavy tar remained.

17.13   Organic-rich shales can be a source of oil and gas. The richest in the United States are found in the Green River Formation of Colorado, Utah, and Wyoming. Additional shales, accessible to mining and capable of yielding 40 l/t or more, are widespread in the United States. [Data from U.S. Geological Survey.]

Three very large deposits of tar sands are known, of which the Canadian Athabascan tar sands are probably the largest (Figure 17.14). Estimates of recoverable oil from the Canadian deposits range between 250 and 500 billion barrels. Presently there is a small amount of production from the Athabascan tar sands at Fort MacMurray, where about 50,000 barrels of oil are extracted per day. Two other deposits, comparable in size to the Canadian sands, are now known. One is in Venezuela along the Orinoco River; the other, in the Soviet Union.

When these large deposits plus several small deposits are considered together, they appear to have the potential of yielding as much energy in the form of crude oil as the total energy represented by oil and natural gas originally available to us in the traditional petroleum fields of the world.

Some conclusions   Before leaving this discussion of oil and gas, we may find it worthwhile to draw some conclusions about the future of fossil fuels in general. As we have stated, we know that all fossil fuels are nonrenewable, and once they are used up, they cannot be replaced. We also know that the traditional sources of oil and natural gas are very limited and will be essentially gone in a few decades. Coal, being more abundant, will extend the life of fossil fuels as an energy source, and so may oil shales and tar sands. But whatever we do, the prediction in Figure 17.12 will not be modified much if we continue to use fossil fuels at the ever-increasing rates that we do now. Somewhere 100 to 200 years down the line, we shall have used up half the fossil fuels that were originally available to us.

What then are the alternatives to our dependence on the fossil fuels? We look at some of them in the following sections.

17.4 WATER POWER

Water power has been used as a source of energy for at least 2,000 years and perhaps longer. Water has the advantage of being a renewable resource since it is

17.14 The areas of Alberta known to be underlain by tar sands. Some production now comes from deposits near Fort McMurray.

the basis of the hydrologic cycle: The sun pumps it by evaporation from the sea, and the atmosphere delivers it over the landmasses, where it falls as precipitation and flows off in rivers and streams back to the sea. Water power is clean and relatively cheap. It has the drawback that the dams generally necessary for the storage of water have limited lifetimes and usefulness because sediments silt up the lakes formed behind them. There is also the objection that damming floods out a great deal of prized scenery by drowning valleys and their associated rivers.

About 25 percent of the water-power capacity of the United States has already been developed. Worldwide, about 6 percent of the capacity has been developed. And it is estimated that if the world's water power were fully developed, it would produce energy equal to about 30 percent of the total energy now consumed. Africa and South America are the continents with the greatest water power potential; thus far very little, about 1 percent, has been developed there.

17.5 NUCLEAR POWER

The nucleus of an atom is, as we discussed in Chapter 2, made up of protons or protons and neutrons. It is the binding energy, holding these particles together in the nucleus, that forms the basis for nuclear power. This energy can be released either by breaking the nucleus of an atom into nuclei of lighter-weight elements, in a process known as **fission,** or by joining the nuclei of two elements together to form the nucleus of a heavier element, a process called **fusion.** Both of these processes have been achieved at the explosive level in the atomic bomb (fission) and the hydrogen bomb (fusion). We have been able to develop technology that controls fission so that the slow release of the energy can be converted to the generation of electric power. We cannot yet control the fusion process.

FISSION

People have been able to tap only a small fraction of the atom's potential energy. At present, we are using only that available from the heavy element uranium. It is not the common isotope of **uranium,** with an atomic weight of **238,** that fissions under neutron bombardment, but rather the lighter isotope, **uranium 235.** This is found in nature intermixed with ^{238}U in the proportion of about 1 part in 140. Because the two isotopes are chemically identical, the ^{235}U can be separated only by a physical process capitalizing on the difference in mass. By making uranium into a gaseous compound and then pumping it against plates pierced with billions of tiny holes less than 0.000001 cm in diameter, the process causes the lighter ^{235}U to move through the holes to concentrate a product more than 90 percent ^{235}U.

In nuclear fission, a slow-moving neutron smashes into the nucleus of the ^{235}U atom. It splits the nucleus into two new atomic nuclei of less mass, plus several neutrons. Because the mass of the fission products is less than that of the original atom, some of the mass has been converted into energy. For nuclear fission to take place spontaneously, the piece of material must exceed a certain critical mass. That is, there must be a sufficient number of nuclei so that neutrons bombarding the material cannot fail to hit a nucleus and split it, producing more active neutrons. After the process is started, it thus proceeds by a chain reaction. This reaction, so explosive in the atomic bomb, is controllable and is used in generating electrical energy. Scores of nuclear reactors are now operating throughout the world on a commercial basis.

The fuel used in the nuclear plant is the ^{235}U extracted from the more abundant ^{238}U. It is a nonrenewable resource, and so the question arises as to how much ^{238}U (and hence ^{235}U) is available and how inexpensively it can be obtained. The evidence points to a limited supply of fissionable uranium. If this is true, then we are facing a difficulty in respect to nuclear power similar to that we face with the fossil fuels.

There is, however, a particular situation that may help to solve the declining reserves of uranium. We can, in effect, replace ^{235}U with artificially produced fissionable atoms. The fairly abundant ^{238}U and ^{232}Th (thorium) not naturally fissionable, can be converted into fissionable elements. Thus ^{238}U may absorb a neutron in its nucleus and convert into ^{239}Pu (plutonium), which is fissionable and can sustain the chain reaction necessary to our purposes. In the same way, neutrons will convert ^{232}Th into fissionable ^{236}U. We then need to use the neutrons of decaying ^{235}U to create radioactive plutonium or ^{236}U at a rate equal to, or faster than, the fissioning of ^{235}U. This creation of new radioactive isotopes is called **breeding,** and the reactors in which they are used are called **breeder reactors.** There are still technical problems to be solved before the process can become commercial, but solutions are hoped for.

In the United States, our richest uranium deposits occur in sedimentary rocks in Utah, New Mexico, Wyoming, and Arizona. Lower-grade deposits are found with phosphate deposits, lignite beds, or shales.

In addition to the technological problems that continue to surround fission as a source of energy, the political and health-related factors concern many for this type of power generation.

FUSION

The fusion process relies on the fusion of the nuclei of hydrogen to form helium, the same process that produces the energy in our sun and the other stars.

Shortly after the first hydrogen bomb was exploded in 1952, research began on the possibility of controlling the fusion process for peaceful uses. Most experiments now rely on **deuterium,** a heavy isotope of hydrogen (Section 2.1 and Figure 2.5), as a fuel (Figure 17.15). If controlled hydrogen fusion can be achieved (and there is little assurance that it can be), then fuel for the process presents no problem because there is an essentially limitless source of heavy hydrogen in the ocean. Compared with fission, the fusion process is relatively clean. But if controlled fusion becomes feasible, there will still be questions to be answered. For instance, a product of fusion is another isotope of hydrogen, **tritium,** which is radioactive. Its short half-life of 12.26 years makes it much less of a problem than the radioactive products of fission. However, the walls of any fusion reactor vessels will become radioactive as they are bombarded by neutrons in the fusion process. This bombardment will be so intense that few alloys will be able to sustain it for long. So disposing of "used" radioactive reactor walls will become a problem, although not comparable to that of disposing of spent fuel rods from a fission reactor.

Assuming that all the very real problems of the control and commercialization of fusion power can be solved, we cannot expect fusion to be a usable source of energy before the first decade of the twenty-first century.

17.6 GEOTHERMAL POWER

The volcanoes that plague sections of our earth are expressions of geothermal energy, as are the geysers and hot springs in such areas as Yellowstone National Park, Iceland, and New Zealand. It would be useful if this energy could be extensively harnessed. Our use of

Ohmic heating coils

Shielding

Variable curvature coils

Neutral beam nozzle

Neutral beam

Toroidal field coils

Equilibrium field coils

Vacuum vessel

17.15 This cutaway picture shows an experimental device called a "tokamak," designed to test the feasibility of controlled nuclear fusion. The fuel used is deuterium, a heavy isotope of hydrogen. This is contained as a plasma, or ionized gas, in the toroidally shaped vacuum vessel. Strong magnetic fields keep the plasma away from the walls of the vessel as a strong current heats the plasma toward the temperature of fusion. This particular device has been designed for a "break even" experiment in which as much energy is obtained by fusion as is expended in producing the fusion. [Princeton Plasma Physics Laboratory.]

geothermal energy is already a reality on a small scale. For instance, most of the space heat for Reykjavik, the capital of Iceland, comes from hot springs on the island; the geyser basin north of San Francisco is generating electricity on a commercial scale; and the geothermal field of Larderello in central Italy has been producing power since before World War II.

The decay of radioactive elements within the earth continues to produce heat and to support a slow heat flow toward the surface (Section 3.4). In trying to harness this energy, we run into a number of problems, the first of which is that the heat flow in the earth is so diffuse that general exploitation is impractical. We look, therefore, for those areas where heat flow is unusually high. These areas are of two kinds: the plate boundaries and the hot spots away from the plate boundaries. Within such areas we can consider two possibilities in the use of geothermal power to generate electricity.

The first possibility is already being exploited, where the source of energy is pools of geothermally produced hot water and steam. The energy comes from hot, igneous rocks in still active or quiescent igneous areas. The geothermal pool is not unlike an oil trap in some of its characteristics, but instead of hydrocarbons, we are dealing with heated water. The permeable-rock units containing the water are sealed off, at least in part, from the surface by impermeable rocks. The water in the aquifer is heated by the igneous rocks below. Some of the heated water may leak to the surface in the form of hot springs or geysers (Figure 17.16). That which remains trapped may be tapped by drill holes and brought to the surface to produce electrical power. Commercial installations are of this type, and the bulk of them rely on geothermal steam as the source of energy.

A second possibility can be considered, but as yet it is only in the experimental stage. It may be that there are hot igneous rocks at shallow depth, but the overlying rocks lack water usually because they are too impermeable. In this situation, it may be possible to drill

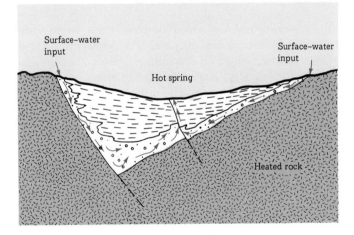

Surface-water input

Hot spring

Surface-water input

Heated rock

17.16 A geothermal system involves the heating of water by hot, igneous rocks at some depth below the surface. The system shown here involves a down-faulted basin in which sediments have accumulated. Surface water percolates downward as groundwater in the coarse sediments at the margins of the basin and is held in by finer, impermeable sediments in the center of the basin. Some heated water is shown as escaping back to the surface as a hot spring. Wells drilled into the basin would be able to tap the heated water at depth and bring it to the surface as an energy source.

17.6 GEOTHERMAL POWER

to some depth where the rocks are hot enough and then create an artificial permeability by subsurface explosions. The next step is to introduce water through the drill hole into the fractured zone where the water is to be heated. A recovery hole is drilled, and the heated water is brought to the surface to power electric generators. In such a system, the water, after being used for power generation, is recycled into the ground, heated again, and once more used to produce electrical power.

The drawbacks to geothermal power include the fact that its use will be restricted to those areas where geothermal flow is particularly high. A major technological problem involves the mineral-forming elements carried in the hot water or steam. These can precipitate out in the generating installation, thereby rapidly reducing effectiveness and increasing maintenance costs. Also, a considerable amount of noise is generated in the process of using geothermal steam. And finally, geothermal pools of steam are not a renewable resource, or at least we have so far been unable to develop a renewal process.

17.7 SOLAR ENERGY

We have already discussed the fossil fuels, which are in effect stored sunshine, and we have mentioned water power, which is the indirect effect of solar energy. In the United States, we have seen a renewed interest in devices for directly collecting solar energy. So we can now buy solar hot-water heaters and construct houses that can be heated, at least in part, by sunlight. The big question is whether we can harness the sun's energy for large-scale power production.

Two techniques can be used for the generation of electricity. The first is the use of solar-thermal systems that employ mirrors or lenses to focus sunlight onto receivers. These in turn heat a fluid, which in its turn is used to drive a turbine and generate electricity. The second approach is the use of photosensitive material to convert some of the sunlight directly into electrical energy. The problems of both these approaches lie in the enormous area needed for the installation of the collectors and the cost not only of installation but also—perhaps more importantly—of maintenance of the collector units. Authorities are in disagreement as to whether the direct use of sunshine as an energy source will satisfy a major part of our needs in the future.

17.8 OTHER SOURCES OF ENERGY

We should not close this chapter without mentioning some other possibilities.

Wind continues to have a small place in the energy-supply picture. Windmills on land can supply small energy needs of individual homes or farms, and it is conceivable that we can occasionally return to sail for supplementary transportation over the oceans.

The **tides** of the world represent the gravitational pull of moon and sun, the moon exerting the greatest influence. Using the rise and fall of tides to generate electricity has often seemed attractive. The Rance River estuary in northern France is producing tidally generated electricity, as is a plant in the northern Soviet Union along the coast of the White Sea. The Passamaquoddy area on the Maine-Canadian border has long been cited as a possible place to install a tidal plant, although thus far it has not proved economically or politically feasible.

Decay of organic matter will produce methane gas. Small methane-generating plants utilizing organic wastes are now in operation. It is suggested by some that this process on a large scale can produce methane, which could in large part replace natural gas. **Alcohol**, another fuel, is produced from the fermentation of plant material and is used in some places as an energy source for automobiles.

Finally, the **geothermal gradient** in the ocean has been suggested by some as a large-scale source of energy. A difference of about $20°C$, for instance, is not uncommon between the cold bottom waters and the warm surface waters. Theoretically this difference in temperature should be convertible into power; but the development of this source of energy is still in the future.

OUTLINE

Fossil fuels (oil, gas, and coal) are presently the most important sources of energy in the industrialized world.

Coal is a sedimentary rock that represents ancient swamps and associated peat deposits. Coal increases in carbon content from **lignite** through **bituminous coal** to **anthracite**.

Oil and natural gas are today the primary sources of energy.

The **origin** of oil and gas involves organic-rich, fine-grained sediments. First the organic matter is transformed to **kerogen**. This turns to oil and gas as the sediment is buried to depths between 2–2.5 km and 6–7 km and reaches temperatures between 50°–60°C and 200°–250°C.

Accumulation of oil and natural gas in the underground demands a **source bed**, a permeable **reservoir rock**, an impermeable **capping bed**, and a **trap** to contain the hydrocarbons.

The **search** for petroleum relies largely on indirect methods of geophysics and geology. In the conterminous United States, we appear to have found and used over half of the hydrocarbons originally available. Worldwide, the supplies of oil and natural gas will be close to exhaustion in the early part of the twenty-first century.

Oil shale and **tar sands** are potential sources of large amounts of hydrocarbons and can extend somewhat the supply from conventional petroleum fields.

Water power is renewable, clean, and relatively inexpensive. One drawback is that the dams needed to impound water have a limited useful life because of siltation.

Nuclear power involves either the splitting of an atomic nucleus (**fission**) or the joining of two nuclei (**fusion**). Controlled fission is in commercial use. We do not yet know if controlled fusion is feasible. Artificial breeding of fissionable material (plutonium) could alleviate the supply problem for fission but does not answer questions of health, environment, or politics.

Geothermal energy can locally supply some energy needs by the use of steam and hot water generated at shallow depths by still-hot igneous rocks.

Solar energy: In addition to the stored solar energy in the form of fossil fuels and water power, sunlight may be used to heat water, which in turn may drive electrical generators. Another method uses photosensitive material to create electricity directly from sunlight.

Other sources of energy include the **wind** (another indirect effect of the sun), **tides**, the **thermal gradient in the oceans**, **methane** from organic wastes; and **alcohol** from plants.

Armstead, H. Christopher *Geothermal Energy*, John Wiley & Sons, Inc., New York, 1978. *A comprehensive view of geothermal energy discussing sources of energy, exploration methods, and environmental and engineering aspects.*

Fettweis, Günter B. *World Coal Resources*, Elsevier Scientific Publishing Co., New York, 1979. *This is a translation from the original German edition of 1976. There is a wealth of information in this book, although sometimes the lack of an index makes it a bit hard to find.*

Hunt, John M. *Petroleum Geochemistry and Geology*, W. H. Freeman & Company Publishers, San Francisco, 1979. *This is an excellent text that integrates the geochemistry of oil and gas with geology. The serious reader can learn a great deal from this book.*

Skinner, Brian J. *Earth Resources*, 2nd ed., Prentice-Hall, Inc., Englewood Cliffs, N. J., 1976. *Two of the chapters of this excellent and concise book focus on energy resources.*

Tissot, B. P., and D. H. Welte *Petroleum Formation and Occurrence*, Springer-Verlag, New York, 1978. *As one advocate of the book writes: "A clear, practical, readable account by two of the world's leading researchers in the field, of the formation, migration, and accumulation of petroleum, primarily from the geochemical viewpoint, but adequately seasoned with geology."*

QUESTIONS

1 Distinguish between renewable and nonrenewable natural resources.

2 Explain why the various grades of coal might be called members in a metamorphic series.

3 Review the steps involved in the formation of oil and gas in nature.

4 You are involved in deep drilling for hydrocarbons and you strike them at a depth of 6,500 m. Is it more likely that you have hit oil or gas? Why?

5 Even though nature has produced much oil and gas, there must be some place to accumulate them if they are going to be of use. Draw well-labeled diagrams to illustrate three different types of geologic structures in which oil and gas are known to have accumulated.

6 If liquid petroleum has a limited future among our supplies of energy, then what role might be played by oil shales and tar sands? Discuss them as sources of energy.

7 What are the advantages and disadvantages of hydroelectric power?

8 Comment on fusion as a source of useful energy.

9 How does a nuclear fission plant produce energy?

10 Discuss geothermal power.

18

USEFUL MATERIALS

In the previous chapter we focused on sources of energy used in our modern technology. Here we look at the variety of minerals and rocks that we extract from the earth and out of which we fabricate the vast number of things so characteristic of twentieth-century civilization. Roads and buildings, cameras and furnaces, automobiles and planes, belt buckles and pins—an almost unending list of things depends on our ability to find and extract raw materials from the ground.

The processes by which nature has produced minerals useful to us are not unique; they are the same processes involved in all other geologic phenomena. For example, the sand shoveled from a sand deposit to mix with cement to form concrete was laid down by some agent such as a stream or the seawater sloshing on a beach. The gold of a brooch may have concentrated in veins of quartz during the final stages of igneous activity. The ore that provided the aluminum for your soft-drink can may have been mined from a deposit formed by long-continued weathering that produced a soil-related deposit high in bauxite, an oxide of aluminum. What is unique about these deposits is that we find them of value.

18.1 MINERAL DEPOSITS

We can consider **mineral deposits** as either **metallic** or **nonmetallic.** The metallic deposits include those mined for such common elements as iron and aluminum as well as for the rarer elements such as chromium, tin, uranium, silver, mercury, and gold, to name but a few. The numerous nonmetallic products we take from the earth include salt, building stone, sand and gravel, phosphorous, and sulfur.

MINING

Mining is the removal of economically useful materials from the earth. The operation may go on either above or below ground. For instance, the familiar surface process of **quarrying** is used to extract building stone and rock aggregate. **Open-pit** or **strip mining,** is also carried on from the surface. The procedure often involves the removal of some amount of valueless material before the sought-for material is reached. Much of the coal mining in this country is this type of mining. **Placer mining** usually involves the separation of valuable metals or nonmetals from unconsolidated, near-surface, or surface deposits of sand and gravel. Some diamond mining is of this type and some gold, tin, and titanium come from placers (Figure 18.1).

When we think of mining, however, we usually think of **underground mining** (Figure 18.2), a process that has gone on for two thousand years and more. Underground mines may be a few tens of meters into a hillside or may extend to two or more kilometers below the surface.

Whether a mineral deposit is mined or not will depend on economics. Will it be worthwhile to mine? If it is, then the product is called an **ore,** a mineral deposit that can be worked at a profit.

GEOLOGIC PROCESSES INVOLVED

A convenient way of thinking about mineral deposits is to consider the **geologic processes** that have been responsible for them, as we outline below.

Igneous activity You will remember that, in Chapter 3, we discussed the formation of igneous rocks from a mixture of elements in a solution called a **magma.** Some magmas, however, also contain elements that, because of the size of their ions, do not readily combine with the common rock-forming minerals. Sometimes these elements crystallize early in the cooling of the magma and settle out of the solution. Sometimes they form late and are trapped in the crystallized magma.

Hydrothermal activity Igneous activity often gives rise to the circulation of heated fluids that migrate beyond the immediate area of igneous activity. These solutions, some of which may carry metallic ions, may move through surrounding country rock. There, as they cool in a new chemical and physical environment, mineral deposition may result.

Weathering The process of weathering has been responsible for the creation of some of our most important mineral deposits. The process operates in one of two ways: It may leach out undesirable elements from the weathering earth material and leave the desirable elements concentrated in the upper soil zones; or it may move the desirable elements in low concentration near the surface to deeper zones, where they are redeposited in an enriched zone.

Placer deposits Some minerals are heavy enough and resistant enough to abrasion that they can be concentrated in selected locales in stream-bed deposits. Being heavier than the usual sand and gravel of a stream, the minerals may collect in pockets, many of which prove rich enough to mine. Such deposits are called **placers.**

Rock formation In many instances, the mineral deposits comprise the whole rock and not just a small enriched fraction of it. The processes that create, for example, limestone or marble or basalt are creating an often-useful material, the rock itself.

18.2 METALS

One way of approaching the discussion of mineable metals is to divide them into those that are geochemically abundant in the crust, and those that are geochemically scarce, as suggested by Brian J. Skinner of Yale University.[1] He defines the **abundant metals** as those with crustal abundances greater than 0.1 percent. The **scarce metals** have crustal abundances less than 0.1 percent.

GEOCHEMICALLY ABUNDANT METALS

The technologically important abundant metals are iron (5.00 percent), aluminum (8.13 percent), manganese (0.01 percent), magnesium (2.09 percent), and titanium (0.44 percent). Not surprisingly, we consume larger amounts of these metals than we do the scarce metals. Furthermore, the future supplies of these metals seem fairly secure in contrast to those of the scarce metals.

Iron Iron has been, and continues to be, the workhorse metal of the industrial world. It is present in appreciable amounts in all rocks, but iron ore deposits represent accumulations in which iron has been concentrated at five to ten times its crustal abundance. Nature has accomplished this enrichment in three ways: namely, **igneous activity, weathering,** and **sedimentation.**

[1]See Supplementary Readings at the end of this chapter.

18.1 Placer mining for diamonds along the coast of South West Africa. In this case, rivers have transported diamonds to the shoreline, where wave action has distributed them in gravels along the beach. Later a cover of sand buried the diamond-bearing deposits. Here the covering sands are being removed from the gravels. [De Beers Consolidated Mines.]

18.2 More than 750 m beneath the surface of the Grace mine in eastern Pennsylvania, a miner operates a load-haul-dump vehicle to transport iron ore. Predominantly magnetite, the ore is the result of contact metamorphism caused by a gabbroic intrusion into limestone. [Bethlehem Steel Corporation.]

One of the most famous iron mining districts, the Kiruna in northern Sweden, owes its existence to igneous activity, specifically fractionation of the magma during cooling as discussed in Section 3.6. This process is also known as **magmatic segregation**. In the case of the Kiruna, one of the early formed minerals in the cooling magma was magnetite (Fe_3O_4). Being heavier than the remaining melt, the magnetite settled to the floor of the magma chamber. The result was an accumulation of a high-grade ore (over 60 percent iron on average), which was segregated out from the main igneous mass.

The red and yellow hues of soil are iron oxides. If weathering is long-continued and intense, as it is in some tropical climates, the iron may be so concentrated as to form a lateritic soil with up to 25 to 30 percent iron. In some places, laterites are mined today, but their chief economic value is as a future resource.

In some situations, where weathering has operated on parent material of very high iron content, a much richer residuum has formed. This was the case in the Lake Superior district. The first ores to be mined —the so-called **soft** ores—were nothing more than the weathering residuum of a sedimentary deposit called the **banded iron formation**, which we will refer to again in the following paragraph. These old weathering zones were very rich, up to 70 percent iron in some instances, and were easy to mine. But they had been all but mined out by World War II.

Today, the United States and most other iron-producing countries mine most of their iron ore from sedimentary deposits. One type of deposit is the banded iron formation already mentioned. It is also known as **taconite**. The rock, the result of chemical sedimentation, is characteristically thinly bedded or laminated with iron-rich zones alternating with silica—usually chert—zones. The iron content is from 15 to 35 percent.

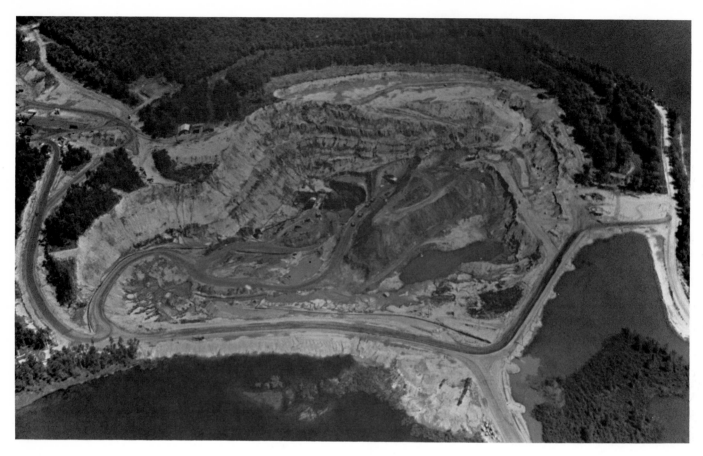

The origin of the banded iron formation is not well understood. It does represent some type of chemical sedimentation in Precambrian oceans. It is a widespread formation and is mined not only in the Lake Superior district (Minnesota, Wisconsin, and Michigan) but also in eastern Canada, Venezuela, Brazil, western Australia, the Soviet Union, India, and the west African countries of Gabon, Liberia, and Mauretania (Figure 18.3).

A second type of sedimentary iron ore has long been important in England, Germany, France, and Belgium and more recently in Canada, the United States, and Egypt. These deposits are younger than Precambrian in age, and although sedimentary in origin, are different from the banded iron formations. They are, generally, oolitic deposits (see Section 5.3), associated with marine sandstone and shale, often calcareous. The Clinton ores of Silurian age have provided much of the ore for the iron industry of Birmingham, Alabama. The iron ores of this type in western Europe are of Jurassic age, those of Newfoundland, of Ordovician age; and those of Egypt, of Cretaceous age.

The world's resources of iron are immense and should suffice us for centuries to come.

Aluminum Aluminum, after oxygen and silicon, is the most abundant element in the earth's crust. Because of its light weight, its ease of fabrication, and its electrical conductivity, aluminum is forging for itself an increasingly important role as a general-purpose metal (Figure 18.4).

Despite its crustal abundance and its presence as a major constituent of the feldspars and other aluminum silicates, only the hydroxides of aluminum serve as an aluminum ore at present. The ore is called **bauxite**, which we found in Section 4.5 to be the residuum of long-continued weathering of aluminum silicate minerals in tropical climates. As such, most bauxite deposits are restricted to present-day tropical regions. The few that are not, were formed in areas that once were tropical in climate.

One of the main problems associated with the refining of aluminum is the large amount of electrical energy that is needed. It is therefore desirable to ship the ore to places where inexpensive hydroelectric power is available. In general, most bauxite is mined

18.4 A continuous sheet of aluminum moves through this rolling mill. Bauxite, a product of tropical weathering, was the raw material from which this aluminum was refined. [Reynolds Metals Company.]

in the developing or undeveloped countries of the tropics, and most refining of the metal in areas with extensive water power as, for example, British Columbia, and New Zealand.

One further fact of interest characterizes the aluminum industry: Aluminum is one of the metals in which recycling is economically very important. For instance, in the United States, approximately 25 percent of the aluminum produced represents recycled metal.

The resources of bauxite are large, perhaps 20×10^9 t, and we can expect that additional amounts will be discovered. Still, given the fact that the present rate of production of new aluminum is approaching 20×10^6 t per year, the supply is not unlimited. Recognizing this, various nations, including the United States, Japan, the Soviet Union, Germany, and the Scandinavian countries, have already begun to consider sources other than bauxite for aluminum. Some of these already successful, although expensive, alternate sources of aluminum are available. These sources include clay minerals, feldspars, and certain other less well-known minerals. However, until bauxite ores are extensively depleted, or until political considerations turn the industrial nations to less conventional sources of aluminum, we will continue to rely on the bauxitic soils of such countries as Jamaica, Guyana, Surinam, and the Republic of Guinea.

Manganese Manganese is critical in the manufacture of steel, where it is used as a deoxidizing and desulfurizing agent. No substitute has yet been found. It is also used widely in making certain metallic alloys, including manganese bronze and manganese brass.

The major sources of manganese are the minerals pyrolusite (MnO_2) and psilomene ($Mn_2O_3 \cdot 2H_2O$). The world's richest deposits of manganese are in the Soviet Union. The mineral mined there is pyrolusite, which occurs in sedimentary deposits. Both pyrolusite and psilomene are also formed as the result of chemical weathering, and are mined in South Africa, Australia, India, Brazil, Cuba, China, and Gabon.

The known resources of manganese now being worked are not large, and in time we will need to turn to new sources for the metal. One such source has already been located in the manganese nodules on the deep-sea floor (Section 15.5).

Titanium Titanium metal and its alloys are light weight and exhibit very high tensile strength at high temperatures. These characteristics have established titanium as an important metal in the aircraft and spacecraft industry. Most titanium, however, is used in the oxide form as a white pigment in paint, replacing lead, which is toxic.

Two minerals of titanium are economically important: ilmenite ($FeTiO_3$) and rutile (TiO_2). Economic deposits of ilmenite have accumulated through the process of magmatic segregation. The largest known deposits are being mined at Sanford Lake in the Adirondack Mountains of New York State, in the Allard Lake district, Quebec, and in Norway. Both ilmenite and rutile may also be concentrated in placer deposits. Currently these are being mined in New Jersey, Australia, India, and Sri Lanka.

Future supplies of titanium are adequate for any foreseeable needs.

Magnesium Magnesium is a light-weight metal whose alloys are resistant to corrosion. It is produced from seawater, each cubic meter yielding approximately 1.2 kg of magnesium (Figure 18.5). The common rock dolomite ($CaMg(CO_3)_2$) also serves as a source. Stable economical supplies of magnesium are, therefore, essentially limitless.

GEOCHEMICALLY SCARCE METALS

The geochemically scarce metals, that is, those with less than 0.01 percent crustal abundance, are numerous. We discuss here only a few: namely, copper, lead, zinc, nickel, molybdenum, silver, gold, platinum, chromium, tin, and tungsten.

18.2 METALS

18.5 Seawater is the major source of magnesium. Magnesium hydroxide is precipitated from seawater and then treated with hydrochloric acid to give a solution of magnesium chloride. These evaporators boil off the water in the solution, leaving a moist but solid product of magnesium chloride.

Copper Copper is an excellent conductor of electricity. As such, its chief use is in electrical appliances and in electrical transmission lines. Alloyed with tin, it forms bronze; and with zinc, it forms brass.

The economically useful copper minerals are sulfide minerals, and they occur in a variety of situations.

The United States is the world's leading producer of copper, and virtually all of this comes from bulk mining of low-grade copper deposits. These are known as **porphyry copper ores.** Copper, usually in the form of the mineral chalcopyrite ($CuFeS_2$), is very finely disseminated through granite porphyry and the adjoining intruded rock by hydrothermal solutions. Rock with copper content as low as 0.4 percent is mined in open-pit operations (Figure 18.6). The porphyry copper ores of the United States are part of a great belt that stretches along the western parts of the North and South American continents from Canada to Chile (Figure 18.7). The position of this belt and its geologic age suggest that the deposits may be related to converging plate boundaries between the Pacific basins and the American landmasses.

The other major economic deposits of copper are associated with sedimentary rocks. The origin of the mineralization is still debated. Some students of these accumulations believe them to be chemical precipi-

tates related to the original deposition of the sediments. Others see strong evidence that the mineralization is due to postdepositional hydrothermal solutions. The richest of these sedimentary deposits are found in Africa in a zone on either side of the Zambian-Zaire border. Copper-bearing minerals occur in a sequence of slightly metamorphosed sedimentary rocks laid down along ancient shorelines. Here the ore minerals are chalcopyrite, bornite (Cu_5FeS_4), and chalcocite (Cu_2S). Deposits may run several percent copper. Similar and more extensively studied deposits occur in a black shale that spreads from England through the Netherlands and Germany into Poland. Known as the Kupferschiefer, the deposit is never more than a meter thick and usually much less. These ore-bearing sediments are thought to have been laid down in the stagnant bottom waters of a Permian seaway.

Presently the mines of the world produce about 10×10^6 t of copper annually, and consumption has been doubling about every 15 years. So even though known copper reserves are large and more discoveries can be expected, it will not be long before we will be looking for new sources of copper. The first source that comes to mind lies in the ferromanganese nodules of the deep sea. In some places the nodules contain up to 2 percent copper. It is impossible to estimate with precision the total resources represented by deep-sea nodules, but they are thought by some to hold a greater potential than anything known on land.

Lead and zinc The ore minerals of lead and zinc commonly occur together. Lead, no longer used as a paint pigment, is used chiefly in storage batteries. Zinc, alloyed with copper, forms brass; but its greatest single use is in galvanizing iron and steel to give them a protective coating against corrosion.

The chief mineral of lead is galena (PbS), and of zinc, sphalerite (ZnS). One of the richest areas for both minerals is in the Mississippi valley drainage of the central United States. These minerals occur over a wide area from Oklahoma to southern Wisconsin and are found in limestones, into which they have been introduced by hydrothermal solutions. Similar deposits are known in Canada, the Soviet Union, and Australia. Other deposits of commercial grade are associated with zones of contact metamorphism. Finally, the Kupferschiefer deposits also contain, in addition to copper ores, ores of lead and zinc.

The major producers of lead and zinc are the United States, Canada, the Soviet Union, Australia, and Peru, more or less in that order. We can expect to find additional occurrences of relatively high-grade deposits. But so far, no low-grade deposits of potential commercial grade are known.

Nickel Nickel is used to make stainless steel and high-temperature alloys of steel. It is concentrated in

18.6 An open-pit copper mine at Butte, Montana. [The Anaconda Company.]

nature by magmatic segregation of the mineral pentlandite $((FeNi)_9S_8)$. The world's richest known deposits occur in the Sudbury district, Ontario, Canada. Weathering also concentrates nickel in lateritic soils, some of which are being worked today in New Caledonia, Cuba, and Oregon.

The known occurrences of nickel seem to be sufficient to meet industrial needs for many generations.

Molybdenum Molybdenum is used as an alloying metal to make steel, for which extreme toughness and heat resistance are demanded.

Climax, Colorado, has provided over half of the world's needs for molybdenum for over 50 years. The mineral mined is molybdenite (MoS_2), and it occurs in a granitic porphyry. Molybdenum is also a byproduct of mining porphyry copper ores.

The known supplies of molybdenum, although not limitless, should last us far into the future.

Silver Silver's chief use today is in photography and in the electrical industries, both of which place a continually increasing demand on the supplies of the metal. It is also used in jewelry and in the manufacture of silver service.

Silver-bearing minerals are found in hydrothermal deposits. Sometimes silver occurs uncombined with other elements, that is, as native silver; but usually it is in the form of a chemical compound. Silver minerals are most commonly associated with minerals of copper, lead, and zinc; silver is a byproduct obtained during the refining of these primary ores. Seldom is silver mined for its own sake. This means that the supply of silver is tied to the markets for other metals. This in turn has meant that silver supplies do not expand appreciably with increasing demand for it, so the possibility for increased production of silver is not bright.

Gold For centuries, gold has been a tangible symbol of human wealth and as such has been the basis for monetary systems at various times. It is used in jewelry, and is in constant demand for dentistry and electronics.

It occurs most often in the native state, but it is also found as an alloy with other metals, such as silver and platinum, and is sometimes combined with tellurium to form the rare telluride minerals. It occurs in hydrothermal deposits, where it is often associated with the sulfides of other metals. Placer deposits are another source of gold.

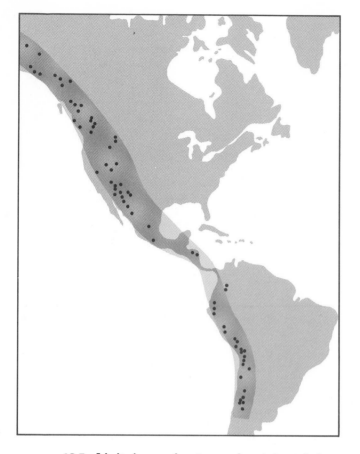

18.7 A belt of copper-bearing porphyry is located along the western margins of the North American and South American continents. A similar though less extensive zone occurs in the western Pacific, extending from northern Australia through New Guinea, Borneo, and into the Phillipines and Taiwan. [After Brian J. Skinner, *Earth Resources*, 2nd ed., Fig. 6.7, p. 84, Prentice-Hall, Inc., Englewood Cliffs, N. J., 1976.]

The best-known gold deposits of hydrothermal origin in the United States are worked in the Homestake Mine of South Dakota. And placer deposits in stream gravels derived from the granitic rocks of the Sierra Nevada set off the famous California gold rush.

By far the world's richest deposits of gold are found in the Witwatersrand of South Africa, where it occurs as native gold in Precambrian conglomerates of very wide extent. Although the association between the gold and the old gravels is inescapable, it is not yet agreed whether the gold represents a Precambrian placer deposit or whether it was introduced into the conglomerates by hydrothermal action at a later time. Regardless of origin, South Africa supplies nearly 60 percent of the world's gold. Canada and the Soviet

Union place a distant second and third, respectively, as gold suppliers.

It is not clear that the continuing rise in the demand for gold will be met by increased production. Certainly, however, the price of gold makes it possible to work extremely low grades of ore. In fact, in some deposits of commercial grade, gold is so finely disseminated that it cannot be seen even with a microscope, a fact that does not make prospecting any easier.

Platinum Alloys of platinum are used in surgical instruments, laboratory apparatus, and jewelry. Platinum is used in catalytic converters in automobile exhaust systems. The International prototype kilogram kept at Sèvres, France, is a cylinder of platinum-iridium alloy.

Platinum is found in deposits formed by magmatic segregation and in placers. In magmatic rocks rich in iron and magnesium, platinum may be swept out of the cooling melt when early formed minerals of nickel, iron, and chromium settle to the chamber floor. This is the case in the Sudbury district of Ontario, which we have already mentioned in the discussion of nickel. Here platinum is recovered as a byproduct. Similar concentrations in the Bushveld igneous complex of South Africa are rich enough to allow mining of platinum for its own sake. Placer deposits are worked in the Ural Mountains of the Soviet Union.

Low-grade deposits of platinum are large, but development may be economically difficult because the production of platinum will most likely be tied to the demand for other minerals in the deposits. In this case, the future for platinum is similar to that for silver, and increased demand may not produce commensurately larger supplies.

Chromium Chromium is used primarily as an alloying metal in the manufacture of stainless steel and high-speed tools.

The mineral chromite $((MgFe_2)CrO_4)$ is the ore mineral for chromium. It is associated with mantlelike rocks, where it accumulates by magmatic segregation. The richest deposits occur in South Africa's Bushveld complex and in a similar complex called the Great Dike in Zimbabwe (formerly Southern Rhodesia). Low-grade deposits are known in many places around the world, but the rich mines of Africa can be expected to dominate chromium production for many years to come.

Tin Tin when alloyed with copper gives bronze. It is also used in solders and in protective coatings for iron and steel.

It occurs as the mineral cassiterite (SnO_2) in contact metamorphic zones and hydrothermal veins. Being heavy and abrasive-resistant, it also collects in placers. Once the mines of Cornwall in southwestern England were the world's chief producers. Today, over 60 percent of the world's supply comes from a mineral belt that includes Malaya, Thailand, and Indonesia. The largest producer in the western hemisphere has been Bolivia, with about 15 percent of the world total. The United States has virtually no production and can expect none. The outlook worldwide is only slightly more encouraging.

Tungsten Tungsten is used as an alloying metal to produce very tough steel. Tungsten is also familiar as the filament in incandescent light bulbs.

The minerals of tungsten are most commonly wolframite ($FeWO_4$) and scheelite ($CaWO_4$). They often occur with the tin mineral cassiterite in hydrothermal and contact metamorphic zones. They may also occur in placers. The largest suppliers of tungsten are Korea, China, and Malaya. Like tin, other extensive sources are unknown.

18.3 AGRICULTURAL AND CHEMICAL MINERALS

Minerals used in the agricultural and chemical industries seem in many ways less exciting and more mundane than the metals we have just considered. In general, they are more abundant than the metals and do not pose the same problems of supply. Despite this, they are critical to some very fundamental aspects of our society. We can say some of the same things about building materials, which will be considered in a following section. Fertilizers, chemicals for industry, and building materials are sometimes grouped under the general heading of **nonmetallic** deposits.

FERTILIZERS

Oxygen and carbon are obviously necessary for vegetative growth because they make up the bulk of the plant. But many other elements are needed for healthy, vigorous growth. If these are not available to the plant from the soil, then the plant's growth slows, unless these elements are added artificially. The increased productivity of agricultural land during the last 40 years has been due in large part to the enormous increase in the use of chemical fertilizers. The most important elements for the fertilizer industry are nitrogen, potassium, and phosphorous.

Nitrogen Nitrogen is the most abundant element in the atmosphere, but plants are unable to use it directly. It must first be **fixed** from the atmosphere (that is, combined with other elements) before it can be used in the life process. The production of nitrogenous fertilizers involves the artificial fixation of atmospheric nitrogen in compounds of ammonium or as nitrates of potassium or calcium. This process demands large amounts of energy, and this means that the price of nitrogenous fertilizers is closely tied to the cost of energy.

Naturally occurring nitrates of commercial grade are few, and most that have formed have been mined out. Such deposits as do exist are found as nitrates in soils of some desert regions, particularly of Chile, and guano deposits, the dried excrement of birds and bats, which also may be preserved in very arid climates.

Potassium Potassium is widely distributed in the silicate minerals of the earth's crust, but it is unusable in organic processes until turned into a soluble form. We obtain the potassium compounds for fertilizer, then, not directly from such minerals as orthoclase or potassic plagioclase but from a group of minerals known as **marine evaporites.**

Seawater, as we found in Section 15.1, contains dissolved salts in the proportion of about 3.5 parts per thousand. When the seawater evaporates, this dissolved material precipitates out as mineral salts. With increasing evaporation, the salts become more and more concentrated in the remaining water. Eventually, the salts begin to crystallize in a particular sequence, a sequence inverse to their solubility. Calcium carbonate ($CaCO_3$), being the most insoluble, precipitates first. This is followed by gypsum ($CaSO_4 \cdot 2H_2O$), halite ($NaCl$), and finally when the original solution reaches about 4 percent of its original volume, by a complex mixture of potassium and magnesium salts (Figure 18.8).

The minerals providing the bulk of our resources of potassium are sylvite (KCl) and carnalite ($KCl \cdot MgCl_2 \cdot 6H_2O$), two of the end members of the marine evaporite series. These minerals, along with the earlier-formed marine evaporites, have been deposited in ancient, restricted shallow-water marine basins at various times in the geologic past. Large deposits of Permian age near Carlsbad, New Mexico, are the richest deposits in the United States. Additional reserves are found in the Paradox basin in southeastern Utah and southwestern Colorado, and in the Williston Basin in North Dakota and Montana. These latter deposits extend northward into Saskatchewan. The largest known deposits, however, are in the Soviet Union, and the

18.8 The sequence of solids precipitated with the continuing evaporation of seawater. [After Brian J. Skinner, *Earth Resources*, 2nd ed., Fig. 7.2, p. 107, Prentice-Hall, Inc., Englewood Cliffs, N. J., 1976.]

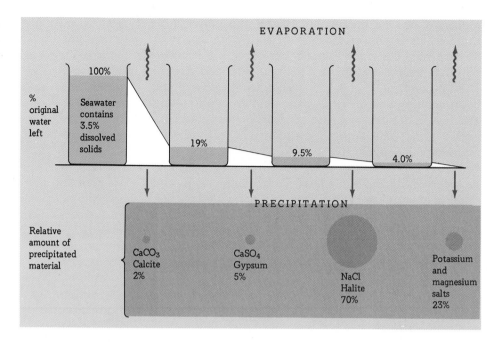

two Germanies have very extensive deposits as well. These geologic deposits should last for a long time to come, and the potassium of seawater assures us of an unlimited supply of the element.

Phosphorous Phosphorous, another vital element for plant growth, is found in the mineral apatite ($Ca_5(PO_4)_3$ OH), widespread in minor amounts in virtually all rocks. Most apatite, however, is concentrated in marine sedimentary rocks, and these are our major source of supply of phosphorous.

Although the chemistry of deposition of apatite is not well understood, we know that precipitation is slow. Where apatite does occur in abundance, there have been very few clastic sediments introduced into the basin to dilute the accumulation of apatite. In the United States, the Phosphoria Formation of Permian age contains mineable deposits in Montana, Idaho, and Wyoming. Most phosphate production in this country, however, comes from Miocene deposits in Florida. Even more extensive deposits in North Carolina may make that state our chief producer before long.

In general, phosphate resources are extensive, and more intensive exploration can be expected to expand these to the point where future supplies should be very adequate.

Sulfur Sulfur is heavily used in the fertilizer industry to produce **superphosphate** and **ammonium sulfate**. In fact, twice as much sulfur goes for these purposes as is used in the chemical industry.

The most commercially desirable deposits of sulfur are in the native form. There is some production from volcanic areas where native sulfur may accumulate. Quantitatively much more important are the sulfur deposits related to some salt domes (Section 17.3). Associated with the salt beds are accumulations of gypsum ($CaSO_4 \cdot 2H_2O$). Some of these gypsum beds have been reduced to native sulfur by bacteria that use the oxygen in the sulfate. Hydrogen sulfide is associated with some natural gases and when it is, sulfur can be a byproduct. Sulfur is also a byproduct in the extraction of metallic elements from some of the metallic sulfide minerals.

There is no foreseeable shortage of sulfur, although its cost will undoubtedly rise as the more desirable deposits are used up.

MINERALS FOR THE CHEMICAL INDUSTRY

The chemical industry depends on raw materials from the earth. All of the minerals we have discussed in this chapter (as well as the many more that have gone unmentioned) provide elements for the myriad of chemicals used in the industry. We will not pursue this farther here except to point again to the marine evaporites and particularly to one of its members, halite (NaCl). It provides not only salt for our table but serves

18.9 This salt mine lies beneath Lake Erie at Fairport, Ohio. The halite or rock salt (NACl) was deposited in an arm of a Paleozoic sea. [Morton Salt Company.]

as the raw material from which we manufacture such diverse chemical products as chlorine gas, soda ash (Na_2CO_3), and sodium sulfate (Na_2SO_4). The source is both from geologic deposits, the remnant of ancient seas (Figure 18.9), and from the modern ocean. Its supply is unlimited.

DIAMONDS

Before leaving this section, we shall consider diamonds (C), which, as the hardest known natural substance, serve as an important industrial abrasive. And, of course, diamonds of gem quality have been prized for centuries.

Diamonds are a form of pure carbon formed in nature under very high pressures in the upper mantle. They are brought to the surface in conduits called **pipes** that carry a particular type of peridotite called **kimberlite.** The South African kimberlites are the best known and longest exploited, having been discovered in 1870 (Figure 18.10). Other kimberlites occur in various places around the world, some with diamonds and others barren. What caused the kimberlite pipes in the first place is still obscure, but one theory suggests that they are the traces of hot spots or plumes as continental crust passed over them.

Although diamonds originate in kimberlites, weathering sets them free eventually. Because they are heavier than the normal sand grain or pebble and very tough, they collect in placers. Most of the production today is from placer deposits, both stream and beach.

A few diamonds have been discovered in the United States. Some have come from a kimberlite occurrence at Murfreesboro, Arkansas. Others have been found in glacial deposits from places as widely separated as Wisconsin and New England. They have been carried southward by the ice from some yet-to-be discovered sources in Canada.

Artificial diamonds are now available commercially. They are suitable for industrial purposes.

18.4 BUILDING MATERIALS

The crust of the earth provides us with a tremendous variety of materials for construction. There sources are everywhere about us. We see them as finished products in virtually all of our structures, whether they be towering office buildings or simple bungalows, garden walks or superhighways. Some material is used as it comes from the earth, modified only in shape. Other material goes through extensive change as it is converted into essentially new material. All these earth materials are the result of the common geologic processes that give us our igneous, sedimentary, and metamorphic rocks.

18.4 BUILDING MATERIALS

18.10 The "Big Hole" at Kimberley, South Africa, was opened in the 1870s. The mine followed a "pipe" of kimberlite, an igneous intrusion that carried the diamonds. Before this now-flooded mine was closed in 1914, it had produced over 14.5 million carats of diamonds. [De Beers Consolidated Mines.]

Building stone Until the widespread use of concrete in construction, building stone was largely used as the structural support in buildings. The soaring cathedrals of medieval Europe, the stolid temples of Central America, and the forbidding fortresses of bygone military greatness all testify to the versatility of structural stonework. Today, steel and concrete form the framework for most structures where stone might have been used in the past. Building stone is often seen as an exterior facing material and trim, and as interior paneling or as a decorative device.

Most widely used are the sedimentary rocks, limestone and sandstone; the igneous rock, granite; and the metamorphic rocks, slate and marble. But these by no means exhaust the list of stones used in buildings. Gneiss, anorthosite, basalt, indurated volcanic ash,

conglomerate, peridotite, diabase, and many others have found their uses.

Crushed rock, sand, and gravel The dollar value of crushed rock, sand, and gravel produced in this country annually is exceeded only by the value of the petroleum and natural gas from the country's wells. Most crushed rock is limestone, but other common rocks such as granite and well-cemented sandstone are also quarried for this purpose. The primary use is as an aggregate in concrete and as a base for streets and highways.

Sand and gravel are taken directly from stream beds, from the terraces of old streams, from the glacial outwash of now-vanished glaciers, from beaches, and even from the continental shelves. The greatest use for sand and gravel is the same as that for crushed rock—as an aggregate in concrete work and as a base for roadways.

Cement Cement was discovered by the Romans, a discovery that made possible the creation of many of their monumental buildings. The Roman cement was a mix of quick lime (CaO), produced by heating limestone and a particular volcanic ash called **pozzuolana**, named after the town of Pozzuoli near Naples. When water was added to this mix, the result was a hard, strong product—**concrete**. Pozzuolana is still quarried from the volcanic slopes near Rome and Naples and is used in the manufacture of concrete. The use of cement declined along with the Roman Empire and was eventually forgotten until rediscovered in the mid eighteenth century, this time in England.

Today's cement is made from a mixture of limestone and clay or shale, or from an impure limestone with a high clay content. This material is ground, heated almost to fusion, and then ground again to a fine powder. Mixed with water it hardens and forms the binder or cement that holds together the aggregate of crushed rock, sand, or gravel, which makes concrete.

Clay Clay, molded into the desired form and heated to the proper temperature, produces a variety of products from bricks to fine porcelains, from tiles to ceramic pipes. It is used in the manufacture of paper, rubber, and plastics, for oil-drilling muds, and in the chemical industry.

We have found that **clay** is a term that can apply both to a specific group of minerals (the clay minerals) or to any mineral of a certain size (less than 0.0039 mm). The clay minerals are generally sought for fine ceramics, chemical industries, and manufacturing. Minerals of many types, ground to clay size, are used in coarse pottery ware and in brick manufacture.

Thus the brick industry of New England and New York was based on the clay-sized particles ground up by glacial action and deposited in glacial lakes with the retreat of the last ice sheets from the area.

Clay forms from the weathering of rocks and their minerals and is a large component of the most abundant of the sedimentary rocks, shale (see Sections 4.2 and 5.3).

Plaster Plaster is used mainly as a wall-surfacing material, most commonly now in prefabricated sheets of plaster board.

Gypsum ($CaSO_4 \cdot 2H_2O$) is the raw material for plaster. Gypsum is heated, that is, calcined, thus driving off some of the water and changing the composition to $2CaSO_4 \cdot H_2O$. The treated gypsum is then ground to a powder. When water is readded, the powder reverts back to gypsum, and a fine-grained, smooth-surfaced product results.

We have already discussed gypsum as a marine evaporite. It has been deposited in many places within the geologic column. It is very widespread indeed, and we should have no concern about its future supplies.

Glass We need not detail the many uses of glass, and will note only that recent technology has increased the use of glass tremendously, including its use as a structural material.

We found in Section 3.7 that natural glass, obsidian, comes from a glass melt that cooled so rapidly that the atoms were unable to assemble into mineral crystals. Artificial glass is essentially the same, although quartz is the material used. Sources include both sandstone and unconsolidated sand deposits. It needs to be very pure quartz to be used as a glass sand. There is no problem with supply.

Asbestos is used in electrical and thermal insulation and is in addition very resistant to corrosion. Some forms of asbestos are so fibrous that the threads can be woven into fabrics. Its uses range from brake linings to fire-protective suits.

The major commercial source of asbestos is the mineral chrysotile ($Mg_3Si_2O_5(OH)_4$), which forms from the minerals in the rock peridotite. The largest reserves are in Quebec just across the border from Vermont, in the Union of South Africa, and in the Soviet Union. These reserves should be large enough to meet demand for many years to come.

Much concern has arisen recently because of the associated health hazard, since some forms of asbestos are closely linked to certain lung cancers. This brings us to some concluding thoughts about the exploitation of the earth's resources, whether they be fuels or minerals.

18.5 EARTH MATERIALS AND SOCIETY

In this and the preceding chapter, we have considered several natural resources produced by geologic processes. The use of these resources has increased dramatically during the present century (Figure 18.11). In fact, we now recognize that some of these raw materials are being used up at such a rapid rate that they may not be available to us in the near future, or if they are, only at a very high cost. This raises a set of social, economic, and political problems, which are not difficult to recognize.

Recently other questions have been raised about the exploitation of natural resources. Many of these are environmental questions that range from the detrimental effect of strip mining for coal to the health hazards related to the mining and use of asbestos.

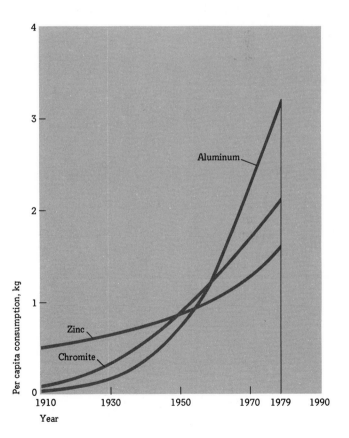

18.11 An example of the rapid increase in the per capita consumption of three different mineral products. [Data from U.S. Bureau of Mines.]

Although the use of natural resources has increased worldwide, there are large discrepancies in the use and availability of these resources. Industrial, developing, and underdeveloped nations do not share equally in the resources on a per capita basis. In fact, many raw materials abundant in the less well-developed countries are largely consumed in the developed countries.

The solutions to the problems raised by these facts are not easily come by. But certainly any adequate solutions will demand a recognition and understanding of the origin and the geologic and geographic distribution of natural resources, which this book has attempted to provide.

OUTLINE

Mineral deposits are generally classed as **metallic** or **nonmetallic deposits.**

Mining involves the removal of economically useful minerals from the earth. **Ore** is an earth material that can be mined at a profit.

Metals can be divided on the basis of their crustal abundance.

Geochemically abundant metals (more than 0.1 percent of the crust) are iron, aluminum, manganese, magnesium, and titanium.

Geochemically scarce metals (less than 0.1 percent of the crust) include copper, lead, zinc, nickel, molybdenum, silver, gold, platinum, tin, and titanium.

Nonmetals include agricultural and chemical minerals and building materials.

Fertilizers use soluble compounds of nitrogen, potassium, and phosphorous. Sulfur is used in superphosphates and ammonium sulfates.

The **chemical industry** relies on minerals and fuels for most of its raw materials.

Diamonds occur in the mantle and come to the surface in kimberlite pipes. Weathering, erosion, and sedimentation produce placers.

Building stone, crushed rock, and **sand** and **gravel** are used in construction more or less as they come from the ground, with the exception of some modification in shape.

Cement, clay, plaster, glass, and **asbestos** are building materials that have been extensively modified from nonmetallic earth materials.

Earth materials involve societal problems, the answers to which must take into account the origin and the geologic and geographic distribution of earth fuels and minerals.

SUPPLEMENTARY READINGS

Brobst, D. A., and W. P. Pratt (eds.) "United States Mineral Resources," *U.S. Geol. Surv. Prof. Paper 820, 1973: An authoritative compendium of what we know we have and what we hope we have in the way of mineral resources.*

Committee on Mineral Resources and the Environment, *Mineral Resources and the Environment,* National Academy of Sciences–National Research Council, Washington, 1975. *A thoughtful, informative, and readable discussion of the subject.*

Douglas, R. J. W. (ed.) *Geology and Economic Minerals of Canada,* Economic Geology Report 1, Geological Survey of Canada, Ottawa, 1970. *A good place to start for the Canadian picture.*

Park, Charles F., Jr., and Roy A. MacDiarmid *Ore Deposits,* 3rd ed., W. H. Freeman & Company Publishers, San Francisco, 1975. *A standard textbook that deals with the principles of deposition of metallic ores, followed by detailed descriptions of various mines representative of the different types of deposits.*

Peters, William C. *Exploration Mining and Geology,* John Wiley & Sons, Inc., New York, 1978. *An overview of the work of the geologist in mineral discovery and production.*

Skinner, Brian J. *Earth Resources,* 2nd ed., Prentice-Hall, Inc., Englewood Cliffs, N. J., 1976. *An excellent compact review of the subject.* (This chapter on "Useful Materials" leans heavily on Chapters 5, 6, and 7 of Skinner's volume.)

U.S. Bureau of Mines, *Minerals Yearbook, Volumes I and II: Metals, Minerals, and Fuels,* Washington, D.C. (issued annually). *Here are the details on the current United States situation.*

1 What is an ore?

2 What are the various ways in which mineral deposits may be formed?

3 What are the geochemically abundant metals?

4 Choose two geochemically abundant metals and discuss their geological origin, their availability, and their uses.

5 Geochemically scarce metals are those whose concentration in the crust is less than 0.1 percent. List at least eight of these with prime economic importance. Choose two and discuss their geology, availability, and use.

6 What are the three most important elements in the fertilizer industry? Where do we get them?

7 What is cement? Where do we get it?

8 What is plaster and what is the source of the raw material from which it is made?

THE PLANETARY SYSTEM

In the 18 chapters of this book, we have examined in some detail the physical nature and processes of the earth, one of nine planets in our solar system. The launching of a Russian earth-orbiting satellite on October 4, 1957, began a new phase in the exploration of our planetary system and of the vast spaces beyond. In the decades that followed, our knowledge has increased dramatically. We have mapped our moon in great detail, put astronauts on its surface, and brought them back with samples of lunar soil and rock. The Russians have landed instrumented space vehicles on Venus. American space probes have brought us vast reams of data on Mercury, Venus, Mars, Jupiter, and Saturn. In 1976, we placed two landing vehicles on Mars and got our first close-up look at the homeland of science fiction. And the vehicle that passed by Saturn in 1981 is headed for encounters with Uranus in January 1986 and Neptune in September 1989. We turn now to some of the things we know about our planetary system.

A.1 THE SUN

We are physically able to exist because of heat and light from our own star, the sun. As we observe it from earth, the sun rises in the east to bring daylight and sets in the west to bring night. This phe-nomenon anciently led to the assumption that the sun moves around a stationary earth. Likewise, the stars appear to rise in the east and set in the west. Therefore, for a long segment of human history, the earth was believed to be the center of the universe, about which everything else revolved.

We now know that the sun is the center of our physical existence. Everything within the sun's gravitational control constitutes what we call our **solar system.** The solar system includes 9 planets, 43 known satellites or "moons," thousands of asteroids, scores of comets, and uncounted millions of meteors. All these revolve around the sun, forming a system 16 billion km in diameter.

The sun is a star. Like all other stars, it is a hot, self-luminous ball of gas. Located 150 million km from the earth, it has a diameter of about 1,391,000 km and a mass 332,000 times that of the earth. It is the source of the solar system's light, heat, and charged particles. Its gravitation holds the system together.

The bulk of the sun is composed of two elements: its volume is 81.76 percent hydrogen and 18.17 percent helium. All other elements total only 0.07 percent. The sun's energy comes from the conversion of hydrogen to helium—the process of fusion that we are currently attempting to develop into a source of energy for our earthly uses with the hydrogen of this planet (see Section 17.5). The sun is "burning" hydrogen at an estimated rate of 4 million t per second, producing an "ash" of helium. At this rate, the sun has an estimated life of about 30 billion years. The continuous output of solar energy is estimated at 38×10^{16} W (watts).

PLANETS AND SATELLITES

The name **planet** was given to certain celestial bodies that appear to wander about the sky, in contrast to the seemingly fixed stars. It came from the Greek *planēs*, meaning "wanderer." Our earth is a planet, one of nine that circle the sun. All travel around it from west to east in nearly the same plane, the **ecliptic.** And they rotate on their respective axes in a "forward" sense, or from west to east, with the exception of Venus and Uranus.

In order of distance from the sun the first four planets are Mercury, Venus, the earth, and Mars. These planets are about the same size and fairly dense, as though they were made of iron and stone. They are called the **terrestrial planets** because of their similarity to the earth. Next in order of distance from the sun are Jupiter, Saturn, Uranus, Neptune, and Pluto. The first four of these are of relatively large size and low density (see Figure A.1). Little is known about Pluto, whose discovery was announced on March 12, 1930, but it is

A.1 The four small terrestrial planets are closest to the sun. Beyond the asteroid belt lie the giant planets and finally the terrestrial-like planet of Pluto. In this diagram, distances from the sun are given by a logarithmic scale and the diameters of the planets by an arithmetic scale.

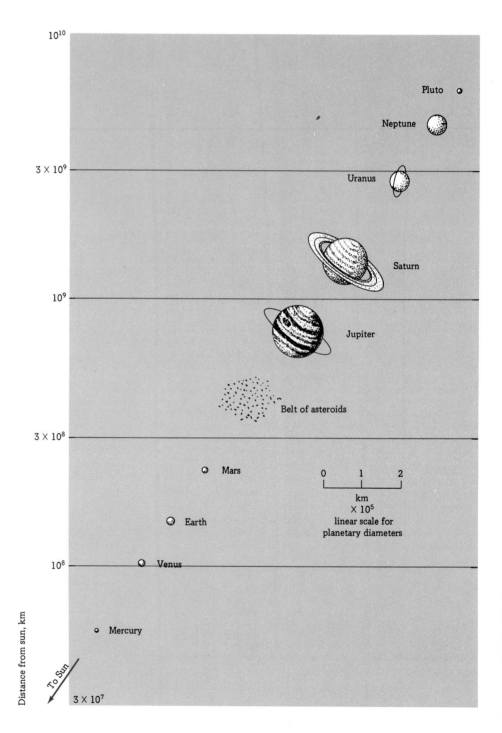

more like the terrestrial planets than the others. A uniform pattern of spacing outward from the sun is broken between Mars and Jupiter. In the "gap" are the thousands of asteroids, ranging in size from 2 km to about 771 km in diameter. These asteroids are believed to be either the remains of a planet that exploded or matter that never completed the planet-forming process.

Tables A.1 and A.2 list some of the characteristics of the planets and our moon.

TABLE A.1

Orbital, rotational, dimensional, and mass characteristics of the planets[a]

Planet	Mean distance to sun km × 10⁶	Diameter at equator, km	Mass, kg	Mean density, g/cm³	Rotation period in earth units	Length of year in earth units	Orbital Velocity, km/sec	Escape Velocity, km/sec	Angle between ecliptic and rotational axis (°)	Orbital eccentricity
Mercury	57.9	5,000	3.30×10^{23}	5.43	59 days	88 days	48	4.2	7	0.206
Venus	108.2	12,180	4.87×10^{24}	5.25	243 days	224.7 days	35	10	6	0.0068
Earth	149.6	12,756	5.98×10^{24}	5.52	23.9 h	365.3 days	29.8	11.2	23.45	0.0167
Mars	227.7	6,760	6.44×10^{24}	3.96	24.6 h	687.0 days	24.1	5.0	25.0	0.0933
Jupiter	778.1	142,700	1.90×10^{27}	1.34	9.8 h	~12 yr[b]	13.1	61	3	0.0484
Saturn	1428.3	120,900	5.69×10^{26}	0.68	10.4 h	29.5 yr	9.65	37	27	0.0557
Uranus	2872.7	46,700	8.76×10^{25}	1.55	10.8 h	84.0 yr	6.85	22	98	0.0472
Neptune	4498.1	49,500	1.03×10^{26}	2.23	15.7 h	164.8 yr	5.47	25	29	0.0086
Pluto	5914.3	~3,000[b]	1.01×10^{22}	<1[c]	6.4 days	249.9 yr	4.84	?	?	0.249

[a] Compiled from various sources.
[b] ~ = approximately.
[c] < = less than.

TABLE A.2

Some characteristics of the planets[a]

Planet	Number of satellites	Diameter of largest satellites, km	Magnetic field	Surface pressure in bars	Surface temp., °C	Major gases	Atmospheric gases, percent	Interior composition	Albedo	Acceleration of gravity, cm/sec²
Mercury	0	—	Present	2×10^{-15}	170	He H	98 2	Silicates Fe core	0.07	3.95×10^2
Venus	0	—	Present	90	457	CO_2 N^2	96 3.5	Silicates Fe-Ni core	0.7	8.8×10^3
Earth	1	3,476	0.3–0.7 gauss	1	25	N_2 O_2 H_2O Ar	77 21 ~1[b] 0.93	Silicates Fe-Ni core	~0.39[b]	9.78×10^2
Mars	2	18	Weak	0.007	−60	CO_2 N_2 Ar	95 2.7 1.6	Silicates Fe-FeS core	0.15	3.73×10^2
Jupiter	16	5,600	Strong	>>100[c]	−140	H_2 He	~89[b] ~11[b]	H_2 He Core?	0.51	2.32×10^3
Saturn	16	4,990	Present	>>100[c]	−175?	Similar to Jupiter		Similar to Jupiter	~0.50[b]	8.77×10^2
Uranus	5	2,600	?	>>100[c]	−215?	Similar to Jupiter		Similar to Jupiter	~0.66[b]	9.46×10^2
Neptune	2	4,200	?	>>100[c]	−220?	Similar to Jupiter		Similar to Jupiter	~0.62[b]	1.37×10^3
Pluto	1	?	?	?	?			H_2O? NH_4?	~0.16[b]	?

[a] Compiled from various sources.
[b] ~ = approximately.
[c] >> = much greater than.

A.2 These reproductions from *The Starry Messenger* record part of what Galileo saw of the moon through his telescope. North is to the top, the last quarter is to the left, and the first quarter is to the right. Compare with Figure A.3. [Willard Starks.]

A.3 The moon as seen through a present-day optical telescope. North is to the top, the last quarter is to the left, and the first quarter is to the right. The lighter zones are the highlands of the moon, and the darker zones are the topographically lower, younger maria. [Lick Observatory.]

EARTH'S MOON

We probably know more about the moon than any of the other bodies in the planetary system with the exception of earth. The first man to see the moon with other than the naked eye was the Italian scientist Galileo Galilei (1564–1642). A Dutch lens grinder had just constructed the first telescope, and Galileo, imitating it, built his own instrument. Early in 1609, Galileo, then a mathematics professor at the University of Padua, turned his telescope on, among other celestial objects, the moon. What he saw is published in a treatise, *Sidereus nuncius* (*The Starry Messenger*), one of the great landmarks of scientific discovery. For the first time, man found that, as Galileo wrote, the "moon is not robed in a smooth and polished surface, but is rough and uneven" (Figure A.2). In the last three and one-half centuries, better telescopes, the camera, spacecraft and their instruments, and direct observation have given us a very good idea of what the moon is like and how it got that way (Figure A.3).

Central peaks

Floor of main crater

Slumped crater wall

Rim-interior zone; ridges concentric to crater

Rim-intermediate zone; branching ridges radial to crater

Rim-exterior zone; faint discontinuous ridges radial to crater

0   50   100
km

N

(b)

(a)

A.4 (a) Map of Copernicus, showing the extent of the major zones of the central crater. (b) Telescopic view of Copernicus. The rays and smaller craters can be seen in the photograph. [Mount Wilson Observatory. Map based on H. H. Schmitt and others, "Geologic Map of the Copernicus Quadrangle of the Moon," *U.S. Geol. Surv. Atlas of the Moon,* map I-515, LAC-58, 1967.]

The moon's diameter is 3,476 km and its mass is 7.35×10^{22} kg, two orders of magnitude less than that of the earth. Its mean density is 3.34.

The major surface divisions of the moon are the dark, topographically lower **maria** or "seas" and the higher, lighter-toned **highlands.** A heavily cratered surface is the trademark of the moonscape. These craters range in size from a few micrometers to the gigantic Mare Imbrium over 1,000 km in diameter. There are two types of craters: (1) the **primary craters,** formed by the original impact of meteorites, and (2) the **secondary,** or **satellite craters,** made by falling fragments ejected from a primary crater.

Copernicus, a well-known, fairly large primary crater, shows many of the features characteristic of moon craters. It measures about 90 km in diameter and over 3 km in depth. The relatively level floor has a raised center portion, a feature that is more pronounced as a central peak in somewhat smaller craters. The walls facing into the crater are terraced or stepped and represent the slumping of large slices of the crater walls into the central depression. The rim of the crater is about 1 km high and slopes away from the crest at the crater's edge, to feather out at the edge of an irregular zone up to 150 km wide and concentric to the crater.

The material that forms the rim is zoned into three rings. The interior zone is very hummocky, with ridges concentric to the crater. Outward beyond this is a less hummocky intermediate zone, and the topography is one of branching ridges more or less radial to the central crater. The outermost zone is still less hummocky, but the radial ridges of the intermediate zone still persist in subdued form. This outermost zone is pocked by many small satellite craters thought to have been formed by fragments thrown out of the primary crater. Superimposed on the rim material and spreading far beyond it to a distance of 500 km from the crater are light-colored, discontinuous streaks, the **rays** of Copernicus (see Figure A.4).

Copernicus is interpreted by most observers as an **impact crater.** The central rise on the floor of the crater is thought to result from upward adjustment following impact. The rim is thought to have been pushed upward and outward as a result of the impact

A.5 (left) A lunar gabbro as seen in a thin section magnified under a microscope. Minerals present are pyroxene (light gray), plagioclase feldspar (nearly white), and ilmenite (black). Width of section, about 1,300 μm. [NASA.]

A.6 (right) A breccia from the moon as seen in a thin section magnified under the microscope. The small fragments of minerals and crystalline rock are set in a fine-grained matrix. Width of section, about 875 μm. [U.S. Geological Survey.]

explosion, and the hummocky ridges of the interior zone of the rim material are thought to stem from the same cause. Outward from this zone the areas of radial ridges are interpreted as material ejected from the crater and arranged in ridges subparallel to the direction of blast. The ray material is believed to be finely crushed matter thrown out from the primary crater and, to a lesser extent, similar material kicked up during formation of the satellite craters. Copernicus is one of a number of **ray craters** on the moon. Most craters, however, lack rays. Their absence is attributed to the greater age—hence longer weathering—of the craters.

Some craters show only a low lip for a rim. In general, these are old craters all but buried by material ejected from younger craters or by lava flows that filled the maria.

Moon Rocks The several hundred kilograms of material returned to the earth from the moon by the *Apollo* missions have shown us that four general rock materials dominate the lunar surface.

Basaltic rocks are common and seem to form the greater part of the filling of the many maria. These basalts differ slightly from those we find on earth. They contain, for instance, more titanium than our earthly basalts and are somewhat lower in silica, alumina, and sodium. Like all lunar samples they contain essentially no water.

The early *Apollo* missions brought samples of basalt and its plutonic equivalent, **gabbro** (Figure A.5). In addition, they brought us a third rock type, a **breccia** (Figure A.6). These breccias contain fragments of basalt, gabbro, and a fourth rock type, **anorthosite**. We theorize that these rocks, composed of the fragments of other rocks, were consolidated by compaction arising from meteoritic impact.

The discovery of anorthosite fragments early in the *Apollo* program led to the speculation that there were large masses of this material on the moon. Anorthosite, a rock composed largely of the feldspar mineral anorthite, $Ca(Al_2Si_2O_8)$, usually contains a small amount of olivine and pyroxene. Anorthosite has a somewhat lower specific gravity than the basalt-gabbro group. Various lines of evidence lead us to believe that the highland areas are composed of anorthosite.

In addition to the rocks already mentioned, there is a lunar "soil," which covers much of the moon's surface. This is not a soil as we think of soil on earth. It is made up of fragments of crystalline rocks, small glassy objects, and some meteoritic material. The assemblage appears to be the result of debris created by meteoritic impact.

Lunar history Even though we cannot yet determine the origin and earliest history of the moon to the satisfaction of most students of that body, we have a fairly good idea of a great deal of the moon's history. Much of this story has been worked out by geologists of the United States Geological Survey working from photographs taken through earth-based telescopes and cameras carried in space probes. Rock units have been mapped on the basis of relative age, surface characteristics, and genetic types. Ages are established by cross-cutting relationships and superposition, stratigraphic principles discussed in Chapter 7 on geologic time.

Age determinations on lunar materials now give us some absolute ages that can be fitted into the relative chronology of events established by these other methods. The first samples returned to earth were of maria material, which turned out to have an age of about 3.6 billion years, at that time older than any known earth rock. Subsequent age dating has produced ages from about 3.2 billion years back to 4.6 billion years for different materials and events.

As we put together what we know about lunar history, we can construct the following thumbnail biography of the moon:

The moon formed some 4.6 billion years ago, along with the earth and the other planets. Once formed, the moon heated up quickly and differentiation began. A crust of anorthosite rose to the surface, and heavier material formed a core that seems to have been molten according to the testimony provided by remanent magnetism. Massive bombardment by meteorites characterized the first 1.5 billion years. The great basins—the maria—were excavated by the impact of enormous meteorites. The undestroyed highland sections of anorthosite crust were deeply pocked by smaller meteorites. The large impact basins were eventually filled by basaltic materials, the bulk of the eruptions being between 3.2 to 3.8 billion years ago. Thereafter major volcanism ceased, and most students of the question believe that this marks the cooling of the moon and the solidification of a small liquid core. This cooling also marks (but did not cause) a decrease in the numbers and sizes of meteorites impacting the moon. True, impressive craters have formed since then (as, for example, Copernicus and Tycho), but the period of mare-producing impacts had ceased.

Today the moon is a very quiet celestial body. There are few moonquakes (these are tidally induced or represent small meteoritic impacts), little if any volcanism, no mountain building, no atmosphere, no water, no life.

A.2 MERCURY

Hermes was the messenger of the gods and the god of dawn and twilight in the Greek Pantheon; Sabuku to the ancient Egyptians; Woden, chief god of the hunt to pagan Anglo-Saxons (hence Woden's day—our Wednesday); and Mercury to the Romans, their messenger of the gods. It is the planet closest to the sun, and the smallest of the inner terrestrial planets, being intermediate in size between our moon and Mars. Appropriate to its name, it has the quickest orbit around the sun of all the planets, just under 88 days.

Its surface is very much like that of the moon, very heavily cratered throughout, as shown in Figure A.7. We see also the division between highlands and maria that we encountered on the moon. Very large impact craters created early in the planet's history are similar to those on the moon, and like those of the moon have been later filled with lavas.

Unlike the moon, however, Mercury has a weak magnetic field and this suggests a liquid core. Various lines of evidence indicate that the planet contains a very high percentage by weight of metal, probably iron. This may be as high as about 70 percent, with lighter weight silicate phases of only about 30 percent. We believe that the planet was chemically differentiated early in its history. Today it is thought to have a metallic, liquid core with a radius of approximately 1,800 km, 75 percent of the radius of the planet. Outside of this is a rind of lower density, very probably similar to that of the outer part of the moon.

A.7 The cratered surface of Mercury is similar to that of our moon and to some of the moons of Jupiter and Saturn. This detail of a portion of the planet also shows a cliff extending for over 300 km from lower left to upper right. [NASA, *Mariner 10.*]

Like the other planets, Mercury first took form 4.6 billion years ago. Its subsequent history is probably very similar to that of our moon, a history we have already sketched.

A.3 VENUS

Named for the Roman goddess of love and beauty, Venus, like Mercury, also appears as a morning and evening "star." It lies between earth and Mercury and is the earth's nearest planetary neighbor. It has been called the earth's "twin" because of its almost identical size to that of the earth. Space probes, however, have shown us how different the two planets are.

The planet is shrouded in a thick blanket of clouds (Figure A.8). This represents a heavy atmosphere composed chiefly of CO_2 (96 percent) and N_2 (about 3.5 percent), with minor amounts of H_2, SO_2, Ar, Ne, CO, HCl, HF, and some aerosols of sulfuric acid.

At the surface of Venus the atmosphere exerts a pressure more than 90 times that at the earth's surface. There the temperatures are extremely high. On the side facing the sun, temperatures exceed

500°C, and on the dark side of the planet they drop to something over 300°C. These elevated surface temperatures, the highest of all the planets, result from the "greenhouse effect" of the CO_2 in the Venusian atmosphere. Short-wave solar radiation passes through the atmosphere to the surface where it is reradiated as long-wave radiation. This is in part absorbed by the CO_2, thus heating the atmosphere. Eventually, a temperature balance is reached as incoming and outgoing radiation equalize. The end results are the very high surface and atmospheric temperatures.'

The permanent cover of clouds has made it impossible to photograph the surface of Venus with conventional earth-based or spacecraft cameras. But radar imagery from orbiting spacecraft has provided a tantalizing view of the planet's surface, and we have a few surface views from the Russian *Venera* landers.

The surface pictures of Venus show a litter of angular blocks separated by deposits of finer sandlike material. This is very similar to the Martian surface discussed in the next section.

Radar imagery from orbiting spacecraft has provided a blurred look at some of the major surface features of Venus, features known only vaguely from earth-based observations. We seem to be seeing, among other things, a high plateau standing 5 to 10 km above the surrounding terrain. Features that include volcanoes, large impact craters, long ridges, and rift valleys are all tentatively identified. Positive resolution of these and other features will come only when and if radar-imaging equipment better-suited to the task is orbited around the planet.

A.4 MARS

The red planet was named for Mars, the Roman god of war, who gave his name also to our month of March. Mars is the outermost of the terrestrial planets.

Telescopic observation of Mars in the nineteenth century led some to believe in the presence there of long, straight "canals" that some unknown civilization had built to bring water from the poles to the arid regions of lower latitudes. Although twentieth-century spacecraft have laid this thought to rest, they have confirmed the presence of polar ice caps and of planet-obscuring dust storms, which earlier telescopic work had reported. The biological experiments conducted by Martian landers have not identified past or present life.

Our first spacecraft pictures of Mars came from *Mariner 4*, which photographed the red planet in a flyby mission in 1965. By November 1971, *Mariner 9* had been inserted into orbit around Mars, and after the subsidence of a violent dust storm, it relayed to us images that revealed a much more varied planet than we had expected. More information arrived from *Viking 1* and *2* orbiters and from their landing vehicles, which were successfully set down in July and September 1976.

A.8 This computer-enhanced photomosaic of Venus' cloud tops was taken from images obtained March 19, 1974, at a range of 720,000 km. The mottled patterns in the center and to the left of center are believed to be convection cells in the Venusian atmosphere. [NASA, *Mariner 10*.]

THE MARTIAN LANDSCAPE

We can divide Mars into two hemispheres: one of old, heavily cratered plains and the other of younger, lower, and lightly cratered plains (Figure A.9). The boundary between the two is inclined at about 50° to the Martian equator. By extrapolating from cratered surfaces of known age on the moon, we believe that the heavily cratered surface of Mars is about 3.5 billion years old. The lightly cratered plains are estimated to be generally younger than 1.5 billion years.

Before spacecraft had circled Mars, the telescope had already

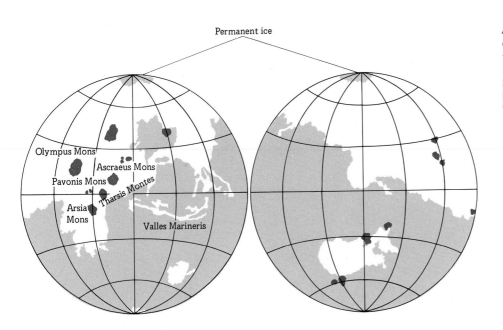

Permanent ice

A.9 Generalized physiographic map of Mars. The clear area is lightly cratered plains. The lighter-colored area includes old, highly cratered terrain. Solid color marks young volcanoes. [Adapted from more detailed map by T. A. Mutch and R. S. Saunders.]

told us of the existence of polar ice caps that waxed and waned with the seasons. At that time they were thought to be made of solid CO_2 (dry ice). So one surprise provided by the *Viking* orbiters was that the northern cap is H_2O ice, and only the southern cap may contain CO_2 ice.

The major activity on the Martian surface today is wind action. The atmosphere, although very thin, does transport dust in suspension and moves sand grains by saltation. In many places today we are able to identify large fields of sand dunes (Figure A.10).

In the past, a number of other processes had major roles in shaping the Martian surface. These included cratering by the impact of meteorites, the record being particularly extensive in the uplands.

Sometime after the major period of meteorite bombardment, extensive volcanism took place. The largest volcano, Olympus Mons, measures more than 500 km across its base and may stand 23 to 29 km in height. As such, it is several times larger than any volcano on earth. And Olympus Mons is but one of nearly two-score volcanic centers known on Mars. Olympus Mons and several other large volcanoes are located on a broad topographic swell or dome known as Tharsis Montes, a feature that seems to indicate upwarping, perhaps similar to (but larger than) those over the hot spots known on earth. Around the margins of this uplift are fractures and faults. But we see no evidence of plate movement or of the associated mountain belts that we have on earth.

The volcanic slopes are comparatively fresh, and the sparseness of craters suggest that the volcanism can be as young as 100 to 200 million years.

Among the most puzzling and startling features on Mars are those that evidence running water on the planet's surface. Associated with the boundary between the lower sparsely cratered plains and the higher heavily cratered plains are great canyons and sinuous channels (Figure A.11).

The canyons, such as Coprates Canyon and Valles Marineris, are huge. Valles Marineris (Figure A.12) is 125 km wide and about 6 km deep and stretches for nearly 5,000 km. Such a feature seems to be a gigantic rift and points to vertical movements of the Martian crust. Dendritic canyons feed into the main valley, as shown in Figure A.12. In addition, sinuous, even meandering, channels are also found (Figure A.13), as well as wide, braided streamways (Figure A.14). These features, taken together, suggest to many that running water has played an important role in fashioning the Martian surface at one or more times in the history of the planet.

Various lines of evidence point to the presence of ground ice beneath the Martian surface. Its melting has produced such features as karstlike topography and debris flows, and may have provided a source of stream water.

A.10 Particles moved by wind can totally obscure Mars, and in some places large areas of dunes occur on its surface. Here the darker, slightly mottled zone is a dune field in the bottom of the Ganges Chasm near the equator. It measures about 20 km by 50 km; sand motion seems to be from lower left to upper right. [NASA, *Viking 1*.]

A.4 MARS

A.11 A view across the Valles Marineris shows a series of slump blocks along the canyon walls. Below the slumps stretch aprons of material that seem to have behaved as gigantic mudflows. At the bottom of the canyon, obstacles have created wind shadows in which has accumulated wind-driven material. In the cliffs, which measure several hundreds of meters in height, differential erosion has etched out nearly horizontal layers of rock. [NASA, *Viking 1*.]

A.12 Part of Valles Marineris. Its width in area is 480 km. Note the dendritic canyons, particularly in the upper center of the area, which feed into the main valley. [NASA, *Mariner 9*.]

A.13 Sinuous valley on Mars, of which some 310 km are shown, 5 to 6 km wide. [NASA, *Mariner 9*.]

A.14 Braided channels of enormous size are seen in this mosaic of photographs taken over the Chryse area near the Martian equator. The impact crater in the lower part of the figure measures about 15 km in diameter and has served as an obstruction around which waters have flowed to produce a streamlined pattern. More streamlined forms and interconnecting channels are seen in the upper part of the figure. [NASA, *Viking 1*.]

A.5 ASTEROIDS

Beyond Mars and between it and Jupiter, the first of the outer planets, lies a belt of tiny planets called the **asteroids**. There are thousands of known asteroids and presumably many times this number not yet identified. The largest, Ceres, has a diameter of about 770 km. Some of the asteroids represent fragments resulting from collisions within the asteroid belt. But the origin of the belt of small cold bodies is still uncertain. Do the asteroids represent the shattered remnants of a small planet, or are they bits of matter that never accreted into a single planetary mass?

A.6 JUPITER

In Roman mythology, Jupiter was the son of Saturn and king of all the gods. In our planetary system, it is one of the four giant outer planets, the others being Saturn, Uranus, and Neptune.

A.15 Jupiter as seen from a distance of 54 million km through the cameras of *Voyager 1* on January 19, 1979; photo shows a swirling, turbulent atmosphere composed largely of hydrogen. The large, eyelike feature in the lower center—the Great Red Eye of Jupiter—has been known from ground-based observations for over 300 years. Jupiter's thin ring is not visible in this view. [NASA, *Voyager 1*.]

Layers of dense clouds surround the planet (Figure A.15). These are made up primarily of hydrogen gas and secondarily of helium. The patterns seen at the top of these clouds reflect a complex system of atmospheric circulation. We have never penetrated to the surface beneath this cloudy shroud. Indeed, there may very probably be no surface similar to that of the terrestrial planets. Most of the planet could be liquid hydrogen with a core of iron-rich rock.

Although we have not seen below Jupiter's cloud cover, spacecraft have provided us with excellent views of Jupiter's moons. The four largest satellites of the planet are known as the **Galilean moons** because of their discovery by Galileo in 1610. In order of decreasing size, they are Ganymede, Callisto, Io, and Europa (Figure A.16).

Ganymede, 5,600 km in diameter, exhibits a cratered surface imposed on a mixture of rock and ice, a puzzling grooved terrain, and evidence that portions of the surface are cut by transverse faults (Figure A.17).

Callisto, 5,200 km in diameter, like Ganymede, seems to have an outer crust of rock and ice. The surface is heavily cratered and very possibly represents one of the oldest surfaces in the planetary system.

Io, the innermost of the Galilean moons, is 3,800 km in diameter and is marked by a bright yellow-orange equatorial band and reddish brown polar zones. This remarkable body is the only one in the planetary system (with the exception of our earth) to have present-

(a) (b) (c) (d)

A.16 Jupiter has 15 moons. The four largest are known as the Galilean satellites, after Galileo who discovered them in 1610. Here they are shown in their correct relative size. Callisto (a) is farthest from Jupiter. Ganymede (b), Europa (c), and Io (d) are progressively closer to the planet. Ganymede and Callisto are larger than the planet Mercury, and Europa and Io are about the size of our moon. Both Ganymede and Callisto appear to have a crust of ice and rock. Europa's crust is an intricately fractured rind of ice. Io, volcanically the most active object yet known in the planetary system, has long-lived volcanoes that are apparently fueled by sulfur. [NASA, *Voyagers 1 and 2.*]

A.17 This photo of Ganymede shows a bright zone of grooved terrain about 150 km wide. It is offset as if by faulting. Craters and their ejecta exhibit variations in tone from dark to light, which may reflect variations in layering of surface material. [NASA, *Voyager 2.*]

day volcanic activity (Figure A.18). But the volcanism on Io is very different from that on earth. Individual eruptions are continuous, lasting for several months or longer. Furthermore, they seem to be fueled by burning sulfur. This continued activity has produced a surface that is comparatively young when compared with Ganymede or Callisto or with our moon or Mercury.

Europa, 3,100 km in diameter, exhibits a very young surface crisscrossed by what appear to be long fractures (Figure A.19). A few

A.18 The two bright spots close together on the barely lighted edge or limb of Io represent volcanic eruptions towering about 100 km above the moon's surface. These two eruptions were first seen by *Voyager 1* and were still active 4 months later when *Voyager 2* passed by. A total of six different eruptions were recorded by the *Voyager* missions. We believe that burning sulfur is the cause of Io's volcanism. [NASA, *Voyager 2.*]

impact craters are present, but again their scarcity suggests a young surface. It is speculated that the outer 100 km or so of the planet is an icy rind over a silicate crust. If this is true, the reticulated pattern may be due to fractures in this rind. Furthermore, old crater scars may have been erased by flow of the ice substrate, thereby accounting for their low incidence.

The *Voyager* missions also discovered that Jupiter has a ring, about 30 km thick, in its equatorial plane.

A.7 SATURN

Saturn was named for an early Roman god of fertility, a mythological figure from whom we also have the word **saturnalia** to designate an event of exuberant revelry. Spacecraft flybys in 1979 (*Pioneer 11*) and in 1980 and 1981 (*Voyagers 1* and 2) have added a wealth of information to our understanding of the ringed planet.

Like Jupiter, Saturn has a thick and turbulent atmosphere composed primarily of hydrogen. This atmosphere, some 1,000 km thick, covers a surface of liquid helium, which at greater depths becomes metallic. This in turn may envelop a rocky core.

The moons and rings of Saturn, long known from earth-based observations, provided spectacular shows for the *Voyager* encounters. The three rings of icy particles, first identified from earth, are now seen to be made of 500 to 1,000 or more rings (Figure A.20). The

A.19 Europa's surface is thought to consist predominantly of water ice. Here we see a view of its strangely marked surface of intersecting, cracklike lineations. Adjacent dark-speckled zones may represent cratered terrain. But large craters are absent, suggesting that the surface of Europa is young when compared with the surfaces of Ganymede and Callisto. [NASA, *Voyager 2.*]

A.20 Images of the rings of Saturn by *Voyager* missions reveal details impossible to resolve from earth. [NASA, *Voyager 1.*]

A.21 In this montage, the artist has shown Saturn, its rings, and its moons, which together resemble a miniature solar system. Ringed Saturn is partially obscured by the moon Dione in the foreground. Tethys and Mimas fade away to the right and Enceladus and Rhea to the left. Titan is above and beyond Saturn to the right. [NASA.]

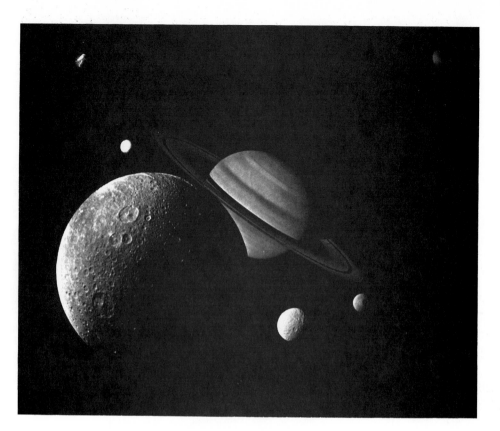

moons of Saturn, from giant Titan to tiny Phoebe, give Saturn the look of a miniature planetary system, even more so if one considers the rings an analogue to the belt of asteroids (Figure A.21). In this regard, Saturn mimics Jupiter, although on a smaller scale.

The moons of Saturn have surfaces of ice mixed with rock. Cratered landscape is common (Figure A.22) and reminds us of our own moon, and of Mercury, portions of Mars, and the outer moons of Jupiter. Only Titan's surface is obscured from us. It is large enough to maintain an atmosphere of nitrogen with about 1 percent methane.

A.8 BEYOND SATURN

Beyond Saturn lie Uranus, Neptune, and Pluto. Hopes are that the *Voyager* probes will report back from Uranus in 1986 and from Neptune in 1989.

Some of the presently known characteristics of these three outer planets are included in Tables A.1 and A.2. These tables show the characteristics of all nine planets.

A.22 Rhea is the most heavily cratered of Saturn's moons. Its surface may be the oldest in the planetary system. [NASA, *Voyager 1*.]

MINERALS

Many of the most common minerals may be identified in hand specimens by physical properties. Among the characteristics useful for this purpose are those discussed in Chapter 2, including crystal form, cleavage, striations, hardness, specific gravity, color, and streak. Still other properties are found to be useful in identifying certain groups of minerals. Some of these properties are summarized in the text and tables that follow.

B.1 MINERAL CHARACTERISTICS

CRYSTAL SYMMETRY AND SYSTEMS

As noted in Chapter 2, when a mineral grows without interference, it is bounded by plane surfaces symmetrically arranged, which give it a characteristic crystal form. This form is the external expression of its definite internal crystalline structure. The faces of crystals are defined by surface layers of atoms.

Every crystal consists of atoms arranged in a three-dimensional pattern that repeats itself regularly. Even in irregular mineral grains, the atoms are arranged according to their typical crystalline structure.

Crystals are classified in six different systems according to the **symmetry** of their faces and the arrangement of their **axes of symmetry.** (An axis is an imaginary straight line that is drawn from the center of a face to the center of the opposite face.) The systems are described and illustrated in Table B.1.

HARDNESS

We can determine the hardness of a mineral by scratching its smooth surface with the edge of another. We must be sure that the mineral tested is actually scratched. Sometimes particles simply rub off the specimen, suggesting that it has been scratched even though it has not.

In Table B.2, ten common minerals have been arranged as examples of the degrees of the Mohs scale of relative hardness. Each of these minerals will scratch all those lower in number on the scale and will be scratched by all those higher. In other words, this is a **relative scale.** In terms of absolute hardness, the steps are approximately uniform up to 9; that is, number 7 is 7 times as hard as 1, and 9 is 9 times as hard as 1. But 10 is about 40 times as hard as 1. A more extensive listing is in Table B.3 (and in Table B.8).

TABLE B.1

Basic crystal systems[a]

Unit-cell shape[b]	Name: description	Typical forms of common-mineral crystals

Isometric: three equal-length axes—all at right angles

Garnet   Magnetite   Halite, pyrite

Tetragonal: two equal axes and a third either longer or shorter—all at right angles

Zircon

Hexagonal: three equal axes in same plane intersecting at 60°, a fourth perpendicular to other three

Quartz   Calcite, dolomite   Hematite

Orthorhombic: three unequal axes—all at right angles

Olivine   Aragonite

Monoclinic: three unequal axes—two at right angles, a third perpendicular to one but oblique to other

Pyroxene (augite)   Mica, clay minerals   Orthoclase   Gypsum

Triclinic: three unequal axes—all at oblique angles

Plagioclase   Calcium aluminum silicate

[a]From Sheldon Judson, Kenneth S. Deffeyes, and Robert B. Hargraves, *Physical Geology*, Prentice-Hall, Inc., Englewood Cliffs, N.J., 1976.
[b]Symmetry axes in color.

TABLE B.2

Mohs scale of hardness

Scale	Mineral	Test
1	Talc	[Softest]
2	Gypsum	
2.5		Fingernail
3	Calcite	Copper coin
4	Fluorite	
5	Apatite	
5.5–6		Knife blade or plate glass
6	Orthoclase	
6.5–7		Steel file
7	Quartz	
8	Topaz	
9	Corundum	
10	Diamond	[Hardest]

SPECIFIC GRAVITY

The ratio of the mass of a mineral to the mass of an equal volume of water at a temperature of 4°C. is known as the **specific gravity** of that mineral. Table B.4 lists the most common minerals in order of increasing specific gravity.

MAGNETISM

Minerals that in their natural state are attracted to a magnet are said to be **magnetic**. Magnetite, Fe_3O_4, and pyrrhotite, $Fe_{1-x}S$, with x between 0 and 0.2, are the only common magnetic minerals, although many others containing iron are drawn to a sufficiently powerful electromagnet.

PYROELECTRICITY

Pyroelectricity is the simultaneous development of positive and negative charges of electricity on different parts of the same crystal under the proper conditions of temperature change. Quartz is a good example. If it is heated to about 100°C, it will, on cooling, develop positive electric charges at three alternate prismatic edges and negative charges at the other three edges.

TABLE B.3

Minerals arranged according to hardness

Hardness	Mineral	Hardness	Mineral	Hardness	Mineral	Hardness	Mineral
1	Talc	3.5-4	Chalcopyrite	5-6	Tremolite	6.5-7	Jadeite
1-2	Graphite	3.5-4	Dolomite	5.5	Chromite	6.5-7	Olivine
1-3	Bauxite	3.5-4	Siderite	5.5	Enstatite	6.5-7	Spodumene
2	Gypsum	3.5-4	Sphalerite	5.5	Uraninite	6.5-7.5	Almandite
2	Stibnite	4	Azurite	5.5-6	Anthophyllite	6.5-7.5	Garnet
2-2.5	Chlorite	4	Fluorite	5.5-6.5	Hematite	7	Kyanite (across crystal)
2-2.5	Kaolinite	5	Apatite	6	Albite	7	Quartz
2-2.5	Muscovite	5	Kyanite (along crystal)	6	Anorthite	7-7.5	Staurolite
2-5	Serpentine	5-5.5	Goethite	6	Arfvedsonite	7-7.5	Tourmaline
2.5	Galena	5-5.5	Limonite	6	Magnetite	7.5	Andalusite
2.5	Halite	5-5.5	Wollastonite	6	Orthoclase	7.5	Zircon
2.5-3	Biotite	5-6	Actinolite	6-6.5	Aegirite	8	Spinel
2.5-3	Chalcocite	5-6	Augite	6-6.5	Pyrite	8	Topaz
3	Bornite	5-6	Diopside	6-7	Cassiterite	9	Corundum
3	Calcite	5-6	Hornblende	6-7	Epidote	10	Diamond
3-3.5	Anhydrite	5-6	Opal	6-7	Sillimanite		

TABLE B.4

Minerals arranged according to specific gravity

Specific gravity	Mineral	Specific gravity	Mineral	Specific gravity	Mineral	Specific gravity	Mineral
1.9-2.2	Opal	2.8-3.2	Biotite	3.3-3.5	Jadeite	4.02	Corundum
2.0-3.0	Bauxite	2.85	Dolomite	3.3-4.37	Goethite	4.1-4.3	Chalcopyrite
2.16	Halite	2.85-3.2	Anthophyllite	3.35-3.45	Epidote	4.25	Almandite
2.2-2.65	Serpentine	2.89-2.98	Anhydrite	3.4-3.55	Aegirite	4.52-4.62	Stibnite
2.3	Graphite	3.0-3.25	Tourmaline	3.4-3.6	Topaz	4.6	Chromite
2.32	Gypsum	3.0-3.3	Actinolite	3.45	Arfvedsonite	4.68	Zircon
2.57	Orthoclase	3.0-3.3	Tremolite	3.5	Diamond	5.02	Pyrite
2.6	Kaolinite	3.15-3.2	Apatite	3.5-4.1	Spinel	5.06-5.08	Bornite
2.6-2.9	Chlorite	3.15-3.2	Spodumene	3.5-4.3	Garnet	5.18	Magnetite
2.62	Albite	3.16	Andalusite	3.56-3.66	Kyanite	5.26	Hematite
2.65	Quartz	3.18	Fluorite	3.6-4.0	Limonite	5.5-5.8	Chalcocite
2.7-2.8	Talc	3.2	Hornblende	3.65-3.75	Staurolite	6.8-7.1	Cassiterite
2.72	Calcite	3.2-3.4	Augite	3.77	Azurite	7.4-7.6	Galena
2.76	Anorthite	3.2-3.5	Enstatite	3.85	Siderite	9.0-9.7	Uraninite
2.76-3.1	Muscovite	3.23	Sillimanite	3.9-4.1	Sphalerite		
2.8-2.9	Wollastonite	3.27-3.37	Olivine	4.0	Carnotite		

PIEZOELECTRICITY

Piezoelectricity is the charge developed in a crystallized body by pressure. Quartz is probably the most important piezoelectric mineral; an extremely slight pressure parallel to its "electric axis" can be detected by the electric charge set up. It is used in specially oriented plates in radio equipment and in sonic sounders.

LUSTER

Luster is the way a mineral looks in reflected light. There are several kinds of luster:

Metallic Of metals
Adamantine Of diamonds
Vitreous Of a broken edge of glass
Resinous Of yellow resin
Pearly Of pearl
Silky Of silk

FLUORESCENCE AND PHOSPHORESCENCE

Minerals that become luminescent during exposure to ultraviolet light, X-rays, or cathode rays are **fluorescent**. If the luminescence continues after the exciting rays are shut off, the mineral is said to be **phosphorescent**.

FUSIBILITY

Minerals can be divided into those fusible and those infusible in a blowpipe flame. Seven minerals showing different degrees of fusibility have been used as a scale to which fusible minerals can be referred. They are listed in Table B.5.

SOLUBILITY

Concentrated hydrochloric acid, HCl, diluted with three parts of water, is commonly used for the solution of minerals being tested. Other wet reagents are used for special tests to help identify minerals.

TABLE B.5

Scale of fusibility

Scale	Mineral	Approx. fusing point, °C	Remarks
1	Stibnite	525	Easily fusible in candle flame
2	Chalcopyrite	800	Small fragment easily fusible in Bunsen-burner flame
3	Garnet (almandite)	1050	Infusible in Bunsen flame but easily fusible in blowpipe flame
4	Actinolite	1200	Sharp-pointed splinter fuses with little difficulty in blowpipe flame
5	Orthoclase	1300	Fragment edges rounded with difficulty in blowpipe flame
6	Bronzite	1400	Only fine splinter ends rounded in blowpipe flame
7	Quartz	1470	Infusible in blowpipe flame

FRACTURE

Many minerals that do not exhibit cleavage (Chapter 2) break, or fracture, in a distinctive manner. Some types of fracture are:

Conchoidal Along smooth curved surfaces like the surface of a shell ("conch"); commonly observed in glass and quartz
Fibrous or splintery Along surfaces roughened by fibers or splinters
Uneven or irregular Along rough, irregular surfaces
Hackly Along a jagged, irregular surface with sharp edges

TABLE B.8

The common minerals[a]

Mineral	Chemical name and/or composition	Form	Cleavage and/or fracture	Hardness	Specific gravity
Actinolite (an *asbestos*; an *amphibole*)	Calcium iron silicate, $Ca_2(Mg, Fe)_5Si_8O_{22}(OH)_2$ (*tremolite* with more than 2% iron)	Slender crystals, usually fibrous	See *amphibole*	5–6	3.0–3.3
Aegirite (a *pyroxene*)	$NaFe^{3+}(Si_2O_6)$	Slender prismatic crystals	Imperfect cleavage at 87° and 93°	6–6.5	3.40–3.55

[a]From Cornelius S. Hurlbut, Jr., *Dana's Manual of Mineralogy,*

Plagioclase feldspars

Species	Albite, %	Anorthite, %
Albite, $Na(AlSi_3O_8)$	100–90	0– 10
Oligoclase	90–70	10– 30
Andesine	70–50	30– 50
Labradorite	50–30	50– 70
Bytownite	30–10	70– 90
Anorthite, $Ca(Al_2Si_2O_8)$	10– 0	90–100

TENACITY

A mineral's cohesiveness, as shown by its resistance to breaking, crushing, bending, or tearing, is known as its **tenacity.** Various kinds of tenacity in minerals include the following:

Brittle Breaks or powders easily

Malleable Can be hammered into thin sheets

Sectile Can be cut by a knife into thin shavings

Ductile Can be drawn into wire

Flexible Bends but does not return to its original shape when pressure is removed

Elastic After being bent, will resume its original position upon release of pressure

TABLE B.7

Ions in common pyroxenes and amphiboles

Ions			
X	Y	Pyroxenes[a]	Amphiboles[b]
Mg	Mg	Enstatite	Anthophyllite
Ca	Mg	Diopside	Tremolite
Li	Al	Spodumene	
Na	Al	Jadeite	Glaucophane
Na	Fe^{3+}	Aegirite	Arfvedsonite
Ca, Na	Mg, Fe, Mn, Al, Fe^{3+}, Ti	Augite	Hornblende

[a]Basic structure: single chain, SiO_3; formula, $XY(Si_2O_6)$.
[b]Basic structure: double chains, Si_4O_{11}; formula, $X_{0-7}Y_{7-14}Z_{16}O_{44}(OH)_4$.

B.2 IMPORTANT MINERALS

SILICATES

More than 90 percent of rock-forming minerals are silicates, with structures based on the $(SiO_4)^{4-}$ tetrahedron. Important classes are listed in Table 2.3 on page 28.

PLAGIOCLASE FELDSPARS

The plagioclase feldspars, also called the **soda-lime feldspars,** form a complete solid-solution series from pure albite to pure anorthite. Sodium substitutes for calcium in all proportions, with accompanying substitution of silicon for aluminum. The series is divided into the six arbitrary species names listed in Table B.6 (also see Table B.8).

PYROXENES AND AMPHIBOLES

The pyroxene family of minerals and the amphibole family of minerals are chain silicates that parallel each other. The amphiboles contain OH^-. The pyroxenes crystallize at higher temperatures than their amphibole analogues. Qualities of the two minerals are shown in Table B.7 (also Table B.8).

Color	Streak	Luster	"Transparency"	Other properties
White to light green	Colorless	Vitreous	Transparent to translucent	A common ferromagnesian metamorphic mineral; a common component of green schists
Brown or green		Vitreous	Translucent	Rare rock former, chiefly in rocks rich in soda and poor in silica

18th ed., John Wiley & Sons, Inc., New York, 1971.

Mineral	Chemical name and/or composition	Form	Cleavage and/or fracture	Hardness	Specific gravity
Albite	Sodic plagioclase *feldspar*, $Na(AlSi_3O_8)$	Tabular crystals; striations caused by twinning	Good in 2 directions at 93°34′	6	2.62
Almandite	$Fe_3Al_2(SiO_4)_3$	12- or 24-sided; massive or granular		6.5–7.5	4.25
Amphibole family	See *anthophyllite, arfvedsonite, hornblende, tremolite*		Perfect prismatic at 56° and 124°, often yielding splintery surface		
Andalusite	Aluminum silicate, Al_2SiO_5	Usually in coarse, nearly square prisms; cross section may show black cross	Not prominent	7.5	3.16
Andesine	A *plagioclase feldspar* 50–70% *albite*				
Anhydrite	Anhydrous calcium sulfate, $CaSO_4$	Commonly in massive fine aggregates not showing cleavage; crystals rare	3 directions at right angles to form rectangular blocks	3–3.5	2.89–2.98
Ankerite	Dolomite in which ferrous iron replaces more than 50% magnesium				
Anorthite	Calcic *plagioclase feldspar*, $Ca(Al_2Si_2O_8)$	Striations caused by twinning; lathlike or platy grains	Good in 2 directions at 94°12′	6	2.76
Anthophyllite (an *amphibole*)	$(Mg, Fe)_7(Si_8O_{22})(OH)_2$	Lamellar or fibrous	See *amphibole*	5.5–6	2.85–3.2
Apatite	Calcium fluophosphate, $Ca_5(F, Cl)(PO_4)_3$	Massive, granular	Poor in 1 direction; conchoidal fracture	5	3.15–3.2
Arfvedsonite (a sodium-rich amphibole)	$Na_3Mg_4Al(Si_8O_{22})(OH, F)_2$	Long prismatic crystals	See *amphibole*	6	3.45
Asbestos	See *actinolite, chrysotile, serpentine*				
Augite (a *pyroxene*)	Ferromagnesian silicate, $Ca(Mg, Fe, Al)(Al, Si_2O_6)$	Short, stubby crystals with 4- or 8-sided cross section; often in granular, crystalline masses	Perfect prismatic along 2 planes at nearly right angles, often yielding splintery surface	5–6	3.2–3.4
Azurite	Blue copper carbonate, $Cu_3(CO_3)_2(OH)_2$	Crystals complex in habit and distorted; sometimes in radiating spherical groups	Fibrous	4	3.77
Bauxite	Hydrous aluminum oxides of indefinite composition; not a mineral	In rounded grains or earthy, claylike masses	Uneven fracture	1–3	2–3
Biotite (black *mica*)	Ferromagnesian silicate, $K(Mg, Fe)_3AlSi_3O_{10}(OH)_2$	Usually in irregular foliated masses; crystals rare	Perfect in 1 direction into thin, elastic, transparent, smoky sheets	2.5–3	2.8–3.2
Bornite (peacock ore; purple copper ore)	Copper iron sulfide, Cu_5FeS_4	Usually massive; rarely in rough cubic crystals	Uneven fracture	3	5.06–5.08

Color	Streak	Luster	"Transparency"	Other properties
Colorless, white, or gray	Colorless	Vitreous to pearly	Transparent to translucent	Opalescent variety, *moonstone*
Deep red	White	Vitreous to resinous	Transparent to translucent	A garnet used to define one of the zones of middle-grade metamorphism; striking in schists
				A group of silicates with tetrahedra in double chains; *hornblende* is the most important; contrast with *pyroxene*
Flesh red, reddish brown, or olive green	Colorless	Vitreous	Transparent to translucent	Found in schists formed by middle-grade metamorphism of aluminous shales and slates; the variety chiastolite has carbonaceous inclusions in the pattern of a cross
				As grains in igneous rock; chief *feldspar* in andesite lavas of the Andes Mountains
White; may have faint gray, blue, or red tinge	Colorless	Vitreous or pearly	Transparent to translucent	Found in limestones and in beds associated with salt deposits; heavier than *calcite*, harder than *gypsum*
				Formed in low-grade regional metamorphism from conversion of *calcite* or *dolomite* or both between 80 and 120°C.
Colorless, white, gray, green, yellow, or red	Colorless	Vitreous to pearly	Transparent to translucent	Occurs in many igneous rocks
Gray to various shades of green and brown		Vitreous		From metamorphism of *olivine*
Green, brown, or red	White	Glassy	Translucent to transparent	Widely disseminated as an accessory mineral in all types of rocks; an important source of fertilizer; a transparent variety is a gem, but too soft for general use
Deep green to black		Vitreous	Translucent	Rock former in rocks poor in silica
				General term for certain fibrous minerals with similar physical characteristics though different composition; *chrysotile* is most common
Dark green to black	Greenish gray	Vitreous	Translucent only on thin edges	An important igneous rock-forming mineral found chiefly in simatic rocks
Intense azure blue	Pale blue	Vitreous to dull or earthy	Opaque	An ore of copper; a gem mineral; effervesces with HCl
Yellow, brown, gray, or white	Colorless	Dull to earthy	Opaque	An ore of aluminum; produced under subtropical to tropical climatic conditions by prolonged weathering of aluminum-bearing rocks; a component of *laterites*; clay odor when wet
Black, brown, or dark green	Colorless	Pearly or glassy	Transparent or translucent	Constructed around tetrahedral sheets; a common and important rock-forming mineral in both igneous and metamorphic rocks
Brownish bronze on fresh fracture; tarnishes to variegated purple and blue, then black	Grayish black	Metallic	Opaque	An important ore of copper

Mineral	Chemical name and/or composition	Form	Cleavage and/or fracture	Hardness	Specific gravity
Bronzite (a *pyroxene*)	*Enstatite* with 5-13% FeO				
Bytownite	A *plagioclase feldspar*, 10-30% albite				
Calcite	Calcium carbonate, $CaCO_3$	Usually in crystals or coarse to fine granular aggregates; also compact, earthy; crystals extremely varied—over 300 different forms	Perfect in 3 directions at 75° to form unique rhombohedral fragments	3	2.72
Carnotite	Potassium uranyl vanadate, $K_2(UO_2)_2(VO_4)_2$	Earthy powder	Uneven fracture	Very soft	4
Cassiterite (tinstone)	Tin oxide, SnO_2	Commonly massive granular	Conchoidal fracture	6-7	6.8-7.1
Chalcocite (copper glance)	Copper sulfide, Cu_2S	Commonly aphanitic and massive; crystals rare but small, tabular with hexagonal outline	Conchoidal fracture	2.5-3	5.5-5.8
Chalcopyrite (copper *pyrites*; yellow copper ore; fool's gold)	Copper iron sulfide, $CuFeS_2$	Usually massive	Uneven fracture	3.5-4	4.1-4.3
Chlorite	Hydrous ferromagnesian aluminum silicate, $(Mg, Fe^{2+})_5(Al, Fe^{3+})_2Si_3O_{10}(OH)_8$	Foliated massive or in aggregates of minute scales	Perfect in 1 direction, like *micas*, but into inelastic flakes	2-2.5	2.6-2.9
Chromite	Iron chromium oxide, $FeCr_2O_4$	Massive, granular to compact	Uneven fracture	5.5	4.6
Chrysotile (*serpentine* asbestos)	See *serpentine*				
Clay	See *illite, kaolinite, montmorillonite*				
Corundum (ruby, sapphire)	Aluminum oxide, Al_2O_3	Barrel-shaped crystals; sometimes deep horizontal striations; coarse or fine granular	Basal or rhombohedral parting	9	4.02
Diamond	Carbon, C	Octahedral crystals, flattened, elongated, with curved faces	Octahedral	10	3.5
Diopside (a *pyroxene*)	$CaMg(Si_2O_6)$	Prismatic crystals; also granular massive	Poor prismatic	5-6	3.2
Dolomite	Calcium magnesium carbonate, $CaMg(CO_3)_2$	Rhombohedral crystals with curved faces; granular cleavable masses or aphanitic compact	Perfect in 3 directions at 73°45′	3.5-4	2.85
Emery	Black granular *corundum* intimately mixed with *magnetite, hematite,* or iron *spinel*				

Color	Streak	Luster	"Transparency"	Other properties
				Primarily as grains in igneous rocks
Usually white or colorless; may be tinted gray, red, green, blue, or yellow	Colorless	Vitreous	Transparent to opaque	A very common rock mineral, occurring in masses as limestone and marble; effervesces freely in cold dilute HCl
Brilliant canary yellow		Earthy	Opaque	An ore of vanadium and uranium
Brown or black; rarely yellow or white	White to light brown	Adamantine to submetallic and dull	Translucent; rarely transparent	Principal ore of tin
Shiny lead gray; tarnishes to dull black	Grayish black	Metallic	Opaque	One of the most important ore minerals of copper; occurs principally as a result of secondary sulfide enrichment
Brass yellow; tarnishes to bronze or iridescence, but more slowly than bornite or chalcocite	Greenish black; also greenish powder in groove when scratched	Metallic	Opaque	An ore of copper; distinguished from *pyrite* by being softer than steel, distinguished from gold by being brittle; like pyrite, known as "fool's gold"
Green of various shades	Colorless	Vitreous to pearly	Transparent to translucent	A common metamorphic mineral characteristic of low-grade metamorphism
Iron black to brownish black	Dark brown	Metallic to submetallic or pitchy	Subtranslucent	The only ore of chromium; a common constituent of peridotites and *serpentines* derived from them; one of the first minerals to crystallize from cooling magma
Brown, pink, or blue; may be white, gray, green, ruby red, or sapphire blue	Colorless	Adamantine to vitreous	Transparent to translucent	Common as an accessory mineral in metamorphic rocks such as marble, mica schist, gneiss; occurs in gem form as *ruby* and *sapphire*; the abrasive *emery* is black granular corundum mixed with *magnetite*, *hematite*, or the magnesian aluminum oxide *spinel*
Colorless or pale yellow; may be red, orange, green, blue, or black	Colorless	Adamantine or greasy	Transparent	Gem and abrasive; 95% of natural diamond production is from South Africa; abrasive diamonds have been made in commercial quantities in the laboratory in the United States
White to light green		Vitreous	Transparent to translucent	Contact metamorphic mineral in crystalline limestones
Pinkish; may be white, gray, green, brown, or black	Colorless	Vitreous or pearly	Transparent to opaque	Occurs chiefly in rock masses of dolomitic limestone and marble or as the principal constituent of the rock named for it; distinguished from limestone by its less vigorous action with cold HCl (the powder dissolves with effervescence, large pieces in hot acid)

Mineral	Chemical name and/or composition	Form	Cleavage and/or fracture	Hardness	Specific gravity
Enstatite (a *pyroxene*)	Magnesium inosilicate, $Mg_2(Si_2O_6)$	Usually massive	Good at 87° and 93°	5.5	3.2–3.5
Epidote	Hydrous calcium aluminum iron silicate, $Ca_2(Al, Fe)Al_2O(SiO_4)$-$(Si_2O_7)(OH)$	Prismatic crystals striated parallel to length; usually coarse to fine granular; also fibrous	Good in 1 direction	6–7	3.35–3.45
Fayalite (an *olivine*)	$Fe_2(SiO_4)$ (see *olivine*)				
Feldspars	Aluminosilicates		Good in 2 directions at or near 90°	6	2.55–2.75
Fluorite	Calcium fluoride, CaF_2	Well-formed interlocking cubes; also massive, coarse or fine grains	Good in 4 directions parallel to faces of an octahedron	4	3.18
Forsterite (an *olivine*)	$Mg_2(SiO_4)$ (see *olivine*)				
Galena	Lead sulfide, PbS	Cube-shaped crystals; also in granular masses	Good in 3 directions parallel to faces of a cube	2.5	7.4–7.6
Garnet	$R''_3 R'''_2 (SiO_4)_3$ (R'' may be Ca, Mg, Fe, or Mn; R''' may be Al, Fe, Ti, or Cr)	Usually in 12- or 24-sided crystals; also massive granular, coarse or fine	Uneven fracture	6.5–7.5	3.5–4.3
Glaucophane (a sodium-rich *amphibole*)	Variety of *arfvedsonite*				
Goethite (bog-iron ore)	$HFeO_2$	Massive, in radiating fibrous aggregates; foliated	Perfect 010	5–5.5	3.3–4.37
Graphite (plumbago; black lead)	Carbon, C	Foliated or scaly masses common; may be radiated or granular	Good in 1 direction, folia flexible but not elastic	1–2	2.3
Gypsum	Hydrous calcium sulfate, $CaSO_4 \cdot 2H_2O$	Crystals prismatic, tabular, diamond-shaped; also in granular, fibrous, or earthy masses	Good in 1 direction, yielding flexible but inelastic flakes; fibrous fracture in another direction; conchoidal fracture in a third direction	2	2.32
Halite (rock salt; common salt)	Sodium chloride, NaCl	Cubic crystals; massive granular	Perfect cubic	2.5	2.16
Hematite	Iron oxide, Fe_2O_3	Crystals tabular; botryoidal; micaceous and foliated; massive	Uneven fracture	5.5–6.5	5.26
Hornblende (an *amphibole*)	Complex ferromagnesian silicate of Ca, Na, Mg, Ti, and Al	Long, prismatic crystals; fibrous; coarse- to fine-grained masses	Perfect prismatic at 56° and 124°	5–6	3.2
Hypersthene	*Enstatite* with more than 13% FeO				
Illite (*clay*)					

Color	Streak	Luster	"Transparency"	Other properties
Grayish, yellowish, or greenish white to olive green and brown		Vitreous	Translucent	Common in pyroxenites, peridotites, gabbros, and basalts; also in both stony and metallic meteorites
Pistachio green or yellowish to blackish green	Colorless	Vitreous	Transparent to translucent	A metamorphic mineral often associated with *chlorite*; derived from metamorphism of impure limestone; characteristic of contact metamorphic zones in limestone
				The most common igneous rock-forming group of minerals; weather to clay minerals
Variable: light green, yellow, bluish green, purple, etc.	Colorless	Vitreous	Transparent to translucent	A common, widely distributed mineral in *dolomites* and limestone; an accessory mineral in igneous rocks; used as a flux in making steel; some varieties fluoresce
Lead gray	Lead gray	Metallic	Opaque	The principal ore of lead; so commonly associated with silver that it is also an ore of silver
Red, brown, yellow, white, green, or black	Colorless	Vitreous to resinous	Transparent to translucent	Common and widely distributed, particularly in metamorphic rocks; brownish red variety, *almandite*
Yellowish brown to dark brown	Yellowish brown	Adamantine to dull	Subtranslucent	An ore of iron; one of the commonest minerals formed under oxidizing conditions as a weathering product of iron-bearing minerals
Black to steel gray	Black	Metallic or earthy	Opaque	Feels greasy; common in metamorphic rocks such as marble, schists, and gneisses
Colorless, white, or gray; with impurities, yellow, red, or brown	Colorless	Vitreous, pearly, or silky	Transparent to translucent	A common mineral widely distributed in sedimentary rocks, often as thick beds; *satin spar* is a fibrous gypsum with silky luster; *selenite* is a variety that yields broad, colorless, transparent folia; *alabaster* is a fine-grained massive variety
Colorless or white; with impurities, yellow, red, blue, or purple	Colorless	Glassy to dull	Transparent to translucent	Salty taste; permits ready passage of heat rays; a very common mineral in sedimentary rocks; interstratified in rocks of all ages to form a true rock mass
Reddish brown to black	Light to dark bloodred; blackens on heating	Metallic	Opaque	The most important ore of iron; red earthy variety known as *red ocher*; botryoidal form known as *kidney ore*; micaceous form, *specular*; widely distributed in rocks of all types and ages
Dark green to black	Colorless	Vitreous; fibrous variety often silky	Translucent on thin edges	Distinguished from *augite* by cleavage; a common rock-forming mineral that occurs in both igneous and metamorphic rocks
				A general term for clay minerals that resemble *micas*; the chief constituent in many shales

Mineral	Chemical name and/or composition	Form	Cleavage and/or fracture	Hardness	Specific gravity
Jadeite (a *pyroxene*)	$NaAl(Si_2O_6)$	Fibrous in compact massive aggregates	87° and 93°	6.5–7	3.3–3.5
Kaolinite (*clay*)	Hydrous aluminum silicate, $Al_2Si_2O_5(OH)_4$	Claylike masses	None	2–2.5	2.6
Kyanite	Aluminum silicate, Al_2SiO_5	In bladed aggregates	Good in 1 direction	5 along, 7 across, crystals	3.56–3.66
Labradorite	A *plagioclase feldspar*, 30–50% *albite*				
Limonite (brown *hematite*; bog-iron ore; rust)	Hydrous iron oxides; not a mineral	Amorphous; mammillary to stalactitic masses; concretionary, nodular, or earthy	None	5–5.5 (finely divided, apparent *H* as low as 1)	3.6–4
Magnetite	Iron oxide, Fe_3O_4	Usually massive granular, granular or aphanitic	Some octahedral parting	6	5.18
Mica	See *biotite, muscovite*				
Montmorillonite (*clay*)	Hydrous aluminum silicate				
Muscovite (white *mica*; potassium *mica*; common *mica*)	Nonferromagnesian silicate, $KAl_3Si_3O_{10}(OH)_2$	Mostly in thin flakes	Good in 1 direction, giving thin, very flexible, and elastic folia	2–2.5	2.76–3.1
Oligoclase	A *plagioclase feldspar*, 70–90% *albite*				
Olivine (peridot)	Ferromagnesian silicate, $(Mg, Fe)_2SiO_4$	Usually in embedded grains or granular masses	Conchoidal fracture	6.5–7	3.27–3.37
Opal	$SiO_2 \cdot nH_2O$; a mineraloid	Amorphous; massive; often botryoidal or stalactitic	Conchoidal fracture	5–6	1.9–2.2
Orthoclase	Potassium *feldspar*, $K(AlSi_3O_8)$	Prismatic crystals; most abundantly in rocks as formless grains	Good in 2 directions at or near 90°	6	2.57
Plagioclase	Soda-lime *feldspar*				
Pyrite (iron pyrites; fool's gold)	Iron sulfide, FeS_2	Cubic crystals with striated faces; also massive	Uneven fracture	6–6.5	5.02

Color	Streak	Luster	"Transparency"	Other properties
Apple green, emerald green, or white		Vitreous		Occurs in large masses in *serpentine* by metamorphism of a nepheline-*albite* rock
White	Colorless	Dull earthy	Opaque	Usually unctuous and plastic; other clay minerals similar in composition and physical properties but different in atomic structure are *illite* and *montmorillonite*; derived from weathering of *feldspars*
Blue; may be white, gray, green, or streaked	Colorless	Vitreous to pearly	Transparent to translucent	Characteristic of middle-grade metamorphism; compare with *andalusite*, which has same composition and is formed under similar conditions but has different crystal habit; contrast with *sillimanite*, which has same composition but different crystal habit and forms at highest metamorphic temperatures
				Widespread as a rock mineral; the only important constituent in large masses of rocks called *anorthosite*
Dark brown to black	Yellow brown	Vitreous	Opaque	Always of secondary origin from alteration or solution of iron minerals; mixed with fine clay, it is a pigment, *yellow ocher*
Iron black	Black	Metallic	Opaque	Strongly magnetic; may act as a natural magnet, known as *lodestone*; an important ore of iron; found in black sands on the seashore; mixed with *corundum*, it is a component of *emery*
				Unique capacity for absorbing water and expanding
Thin, colorless; thick, light yellow, brown, green, or red	Colorless	Vitreous, silky, or pearly	Thin, transparent; thick, translucent	Widespread and very common rock-forming mineral; characteristic of sialic rocks; also very common in metamorphic rocks such as gneiss and schist; the principal component of some mica schists; sometimes used for stove doors, lanterns, etc., as transparent *isinglass*; used chiefly as an insulating material
				Found in various localities in Norway with inclusions of *hematite*, which give it a golden shimmer and sparkle; this is called *aventurine* oligoclase, or *sunstone*
Olive to grayish green or brown	Pale green or white	Vitreous	Transparent to translucent	A common rock-forming mineral found primarily in simatic rocks; the principal component of peridotite; actually a series grading from *forsterite* to *fayalite*; the most common olivines are richer in magnesium than in iron; the clear-green variety *peridot* is sometimes used as a gem
Colorless; white; pale yellow, red, brown, green, gray, or blue; opalescent		Vitreous or resinous	Transparent to translucent	Many varieties; lines and fills cavities in igneous and sedimentary rocks, where it was deposited by hot waters
White, gray, or pink	White	Vitreous	Translucent to opaque	Characteristic of sialic rocks
				A continuous series varying in composition from pure *albite* to pure *anorthite*; important rock-forming minerals; characteristic of simatic rocks
Brass yellow	Greenish or brownish black	Metallic	Opaque	The most common of the sulfides; used as a source of sulfur in the manufacture of sulfuric acid; distinguished from *chalcopyrite* by its paler color and greater hardness; from gold by its brittleness and hardness

Mineral	Chemical name and/or composition	Form	Cleavage and/or fracture	Hardness	Specific gravity
Pyroxene family	Inosilicates: see *aegirite, augite, diopside, enstatite, jadeite, spodumene*				
Quartz (silica)	Silicon oxide, SiO_2, but structurally a silicate with tetrahedra sharing oxygens in 3 dimensions	Prismatic crystals with faces striated at right angles to long dimension; also massive forms of great variety	Conchoidal fracture	7	2.65
Serpentine	Hydrous magnesium silicate, $Mg_3Si_2O_5(OH)_4$	Platy or fibrous	Conchoidal fracture	2–5	2.2–2.65
Siderite (spathic iron; chalybite)	Iron carbonate, $FeCO_3$	Granular, compact, or earthy	Perfect rhombohedral	3.5–4	3.85
Sillimanite (fibrolite)	Aluminum silicate, Al_2SiO_5	Long, slender crystals without distinct terminations; often in parallel groups; frequently fibrous	Good in 1 direction	6–7	3.23
Sphalerite (zinc blende; blackjack)	Zinc sulfide, ZnS	Usually massive; crystals many-sided, distorted	Perfect in 6 directions at 120°	3.5–4	3.9–4.1
Spinel	$MgAl_2O_4$	Octahedral crystals		8	3.5–4.1

Spinel group	(XY_2O_4)			
	X	Y = Al	Y = Fe	Y = Cr
	Mg	Spinel, $MgAl_2O_4$	Magnesioferrite, $MgFe_2O_4$	Magnesiochromite, $MgCr_2O_4$
	Fe	Hercynite, $FeAl_2O_4$	Magnetite, $FeFe_2O_4$	Chromite, $FeCr_2O_4$
	Zn	Gahnite, $ZnAl_2O_4$	Franklinite, $ZnFe_2O_4$	
	Mn	Galaxite, $MnAl_2O_4$	Jacobsite, $MnFe_2O_4$	

Mineral	Chemical name and/or composition	Form	Cleavage and/or fracture	Hardness	Specific gravity
Spodumene (a *pyroxene*)	Lithium aluminum inosilicate, $LiAl(Si_2O_6)$	Prismatic crystals; coarse, some large	Perfect at 87° and 93°	6.5–7	3.15–3.20
Staurolite	Iron aluminum silicate, $Fe^{+2}Al_5Si_2O_{12}(OH)$	Usually in crystals, prismatic, twinned to form a cross; rarely massive	Not prominent	7–7.5	3.65–3.75
Stibnite	Antimony trisulfide, Sb_2S_3	Slender prismatic habit; often in radiating groups	Perfect in 1 direction	2	4.52–4.62
Taconite	Not a mineral				
Talc (soapstone; steatite)	Hydrous magnesium silicate, $Mg_3Si_4O_{10}(OH)_2$	Foliated, massive	Good cleavage in 1 direction, thin, flexible, but inelastic folia	1	2.7–2.8
Topaz	Aluminum fluosilicate, $Al_2SiO_4(F, OH)_2$	Usually in prismatic crystals, often with striations in direction of greatest length	Good in 1 direction	8	3.4–3.6

Color	Streak	Luster	"Transparency"	Other properties
				A group of silicates with tetrahedra in single chains; *augite* is the most important; contrast with *amphibole*
Colorless or white; with impurities, any color	Colorless	Vitreous, greasy, or splendent	Transparent to translucent	An important constituent of sialic rocks; coarsely crystalline varieties: *rock crystal*, *amethyst* (purple), *rose quartz*, *smoky quartz*, *citrine* (yellow), *milky quartz*, *cat's eye*; cryptocrystalline varieties: *chalcedony*, *carnelian* (red chalcedony), *chrysoprase* (apple-green chalcedony), *heliotrope*, or *bloodstone* (green chalcedony with small red spots), *agate* (alternating layers of chalcedony and opal); granular varieties: *flint* (dull to dark brown), *chert* (like flint but lighter in color), *jasper* (red from hematite inclusions), *prase* (like jasper but dull green)
Variegated shades of green	Colorless	Greasy, waxy, or silky	Translucent	Platy variety, *antigorite*; fibrous variety, *chrysotile*, an asbestos; an alteration product of magnesium silicates such as *olivine*, *augite*, and *hornblende*; common and widely distributed
Light to dark brown	Colorless	Vitreous	Transparent to translucent	An ore of iron; an accessory mineral in *taconite*
Brown, pale green, or white	Colorless	Vitreous	Transparent to translucent	Relatively rare but important as a mineral characteristic of high-grade metamorphism; contrast with *andalusite* and *kyanite*, which have the same composition but form under conditions of middle-grade metamorphism
White or green; with iron, yellow to brown and black; red	White to yellow and brown	Resinous	Transparent to translucent	A common mineral; the most important ore of zinc; the red variety is called *ruby zinc*; streak lighter than corresponding mineral color
White, red, lavender, blue, green, brown, or black	White	Vitreous	Usually translucent; may be clear and transparent	A common metamorphic mineral imbedded in crystalline limestone, gneisses, and *serpentine*; when transparent and finely colored, it is a gem; the red is spinel ruby, or *balas ruby*; some are blue
White, gray, pink, yellow, or green		Vitreous	Transparent to translucent	A source of lithium, which improves lubricating properties of greases; some gem varieties
Red brown to brownish black	Colorless	Fresh, resinous or vitreous; altered, dull to earthy	Translucent	A common accessory mineral in schists and slates; characteristic of middle-grade metamorphism; associated with *garnet*, *kyanite*, *sillimanite*, *tourmaline*
Lead gray to black	Lead gray to black	Metallic	Opaque	The chief ore of antimony, which is used in various alloys
				Unleached iron formation in the Lake Superior district, consists of chert (see *quartz*) with *hematite*, *magnetite*, *siderite*, and hydrous iron silicates; an ore of iron
Gray, white, silver white, or apple green	White	Pearly to greasy	Translucent	Of secondary origin, formed by the alteration of magnesium silicates such as *olivine*, *augite*, and *hornblende*; most characteristically found in metamorphic rocks
Straw yellow, wine yellow, pink, bluish, or greenish	Colorless	Vitreous	Transparent to translucent	Represents 8 on Mohs scale of hardness; a gem stone

Mineral	Chemical name and/or composition	Form	Cleavage and/or fracture	Hardness	Specific gravity
Tourmaline	Complex boron-aluminum silicate with Na, Ca, F, Fe, Li, or Mg	Usually in crystals; common with cross section of spherical triangle	Not prominent; black variety fractures like coal	7–7.5	3–3.25
Tremolite (an *amphibole*)	$Ca_2Mg_5(Si_8O_{22})(OH)_2$	Often bladed or in radiating columnar aggregate	Good at 56°	5–6	3.0–3.3
Uraninite (pitchblende)	Complex uranium oxide with small amounts of Pb, Ra, Th, Y, N, He, and A	Usually massive and botryoidal	Not prominent	5.5	9–9.7
Wollastonite	Calcium silicate, $CaSiO_3$	Commonly massive, fibrous, or compact	Good in 2 directions at 84° and 96°	5–5.5	2.8–2.9
Zircon	Zirconium nesosilicate, $Zr(SiO_4)$	Tetragonal prism and dipyramid		7.5	4.68

Color	Streak	Luster	"Transparency"	Other properties
Varied: black or brown; red, pink, green, blue or yellow	Colorless	Vitreous to resinous	Translucent	Gem stone; an accessory mineral in pegmatites, also in metamorphic rocks such as gneisses, schists, marbles
White to light green		Vitreous	Transparent to translucent	Frequently in impure, crystalline *dolomitic* limestones where it formed on recrystallization during metamorphism; also in *talc* schists
Black	Brownish black	Submetallic or pitchy	Opaque	An ore of uranium and radium; the mineral in which helium and radium were first discovered
Colorless, white, or gray	Colorless	Vitreous or pearly on cleavage surfaces	Translucent	A common contact-metamorphic mineral in limestones
Brown; also gray, green, red, or colorless	Colorless	Adamantine	Translucent; sometimes transparent	Transparent variety is a gem stone; a source of zirconium metal, which is used in the construction of nuclear reactors

PERIODIC CHART

Legend

```
16 S          ← Atomic number and symbol
32.0          ← Mass compared to carbon 12 (approximate in parentheses)
-2   1.84     ← Most common electric charge / Ion radius (×10⁻¹⁰ m) with that charge
+6   0.30     ← Next most common charge / Ion radius with that charge
sulfur
```

$16\ S$ — Atomic number and symbol
32.0 — Mass compared to carbon 12 (approximate in parentheses)
-2, 1.84 — Most common electric charge; Ion radius ($\times 10^{-10}$ m) with that charge
$+6$, 0.30 — Next most common charge; Ion radius with that charge

Element with only radioactive species in the earth's crust

Radioactive element not found in the crust

Z	Symbol	Mass	Charge / radius	Charge / radius	Name
1	H	1.01	+1 0.01		hydrogen
2	He	4.00	0 1.08		helium
3	Li	6.94	+1 0.68		lithium
4	Be	9.01	+2 0.35		beryllium
5	B	10.8	+3 0.23		boron
6	C	12.0	+4 0.16		carbon
7	N	14.0	+5 0.13	-3 1.71	nitrogen
8	O	16.0	-2 1.32		oxygen
9	F	19.0	-1 1.33		fluorine
10	Ne	20.2	0		neon
11	Na	23.0	+1 0.97		sodium
12	Mg	24.3	+2 0.66		magnesium
13	Al	27.0	+3 0.51		aluminum
14	Si	28.1	+4 0.42		silicon
15	P	31.0	+5 0.35		phosphorus
16	S	32.0	-2 1.84	+6 0.30	sulfur
17	Cl	35.5	-1 1.81		chlorine
18	Ar	40.0	0		argon
19	K	39.1	+1 1.33		potassium
20	Ca	40.1	+2 0.99		calcium
21	Sc	45.0	+3 0.73		scandium
22	Ti	47.9	+4 0.68		titanium
23	V	50.9	+5 0.59		vanadium
24	Cr	52.0	+3 0.63	+2 0.52	chromium
25	Mn	54.9	+4 0.60	+2 0.80	manganese
26	Fe	55.8	+2 0.74	+3 0.64	iron
27	Co	58.9	+2 0.72	+3 0.63	cobalt
28	Ni	58.7	+2 0.69		nickel
29	Cu	63.5	+2 0.72	+1 0.96	copper
30	Zn	65.4	+2 0.74		zinc
31	Ga	69.7	+3 0.62		gallium
32	Ge	72.6	+4 0.53		germanium
33	As	74.9	+3 0.58		arsenic
34	Se	79.0	-2 1.91		selenium
35	Br	79.9	-1 1.96		bromine
36	Kr	83.8			krypton
37	Rb	85.5	+1 1.47		rubidium
38	Sr	87.6	+2 1.12		strontium
39	Y	88.9	+3 0.89		yttrium
40	Zr	91.2	+4 0.79		zirconium
41	Nb	92.9	+5 0.69		niobium
42	Mo	95.9	+4 0.70	+6 0.62	molybdenum
43	Tc	(99)			technetium
44	Ru	101.	0 1.32		ruthenium
45	Rh	103.	1.34		rhodium
46	Pd	106.	1.37		palladium
47	Ag	108.	+1 1.26		silver
48	Cd	112.	+2 0.97		cadmium
49	In	115.	+3 0.81		indium
50	Sn	119.	+4 0.71		tin
51	Sb	122.	+3 0.76		antimony
52	Te	128.	-2 2.11		tellurium
53	I	127.	-1 2.20		iodine
54	Xe	131.			xenon
55	Cs	133.	+1 1.67		cesium
56	Ba	137.	+2 1.34		barium
57	La	139.	+3 1.02		lanthanum
58	Ce	140.	+3 1.03		cerium
59	Pr	141.	+3 1.01		praseodymium
60	Nd	144.	+3 0.99		neodymium
61	Pm	(147)			promethium
62	Sm	150.	+3 0.96		samarium
63	Eu	152.	+3 0.95	+2 1.09	europium
64	Gd	157.	+3 0.94		gadolinium
65	Tb	159.	+3 0.92		terbium
66	Dy	162.	+3 0.91		dysprosium
67	Ho	165.	+3 0.89		holmium
68	Er	167.	+3 0.88		erbium
69	Tm	169.	+3 0.87		thulium
70	Yb	173.	+3 0.86		ytterbium
71	Lu	175.	+3 0.85		lutetium
72	Hf	178.	+4 0.78		hafnium
73	Ta	181.	+5 0.68		tantalum
74	W	184.	+6 0.62		tungsten
75	Re	186.	0 1.37		rhenium
76	Os	190.	0 1.33		osmium
77	Ir	192.	0 1.36		iridium
78	Pt	195.	0 1.37		platinum
79	Au	197.	0 1.44		gold
80	Hg	201.	+2 1.10		mercury
81	Tl	204.	+3 0.95		thallium
82	Pb	207.	+2 1.20		lead
83	Bi	209.	+3 0.96		bismuth
84	Po	(210)			polonium
85	At	(210)			astatine
86	Rn	(222)			radon
87	Fr	(223)			francium
88	Ra	(226)			radium
89	Ac	(227)			actinium
90	Th	232.	+4 1.02		thorium
91	Pa	(231)			protactinium
92	U	238.	+4 0.97	+6 0.80	uranium
93	Np	(237)			neptunium
94	Pu	(242)			plutonium
95	Am	(243)			americium
96	Cm	(247)			curium
97	Bk	(247)			berkelium
98	Cf	(249)			californium
99	Es	(254)			einsteinium
100	Fm	(253)			fermium
101	Md	(256)			mendelevium
102	No	(254)			nobelium
103	Lw	(257)			lawrencium

[a] Lanthanides, or "rare earths"

[b] Actinides

Appendix D

EARTH DATA

TABLE D.1

Distribution of world's estimated supply of water[a]

	Area, thousands of		Volume, thousands of		
	km^2	mi^2	km^3	mi^3	Total volume, %
World (total area)	510,000	197,000	—	—	—
Land area	149,000	57,500	—	—	—
Water in land areas:					
Freshwater lakes	850	330	125	30	0.009
Saline lakes and inland seas	700	270	104	25	0.008
Rivers (average instantaneous volume)	—	—	1.25	0.3	0.0001
Soil moisture and vadose water	—	—	67	16	0.005
Groundwater to depth of 4,000 m	—	—	8,350	2,000	0.61
Ice caps and glaciers	19,400	7,500	29,200	7,000	2.14
Atmospheric moisture	—	—	13	3.1	0.001
World ocean	361,000	139,500	1,320,000	317,000	97.3
Total water volume (rounded)			1,360,000	326,000	100

[a]In part after R. L. Nace, *U.S. Geol. Surv. Circ.* 536, Table 1, 1967.

TABLE D.2

Composition of seawater at 35 parts per thousand salinity[a]

Element	$\mu g/l$	Element	$\mu g/l$	Element	$\mu g/l$
Hydrogen	1.10×10^8	Nickel	6.6	Praesodymium	0.00064
Helium	0.0072	Copper	23	Neodymium	0.0023
Lithium	170	Zinc	11	Samarium	0.00042
Beryllium	0.0006	Gallium	0.03	Europium	0.000114
Boron	4,450	Germanium	0.06	Gadolinium	0.0006
Carbon (inorganic)	28,000	Arsenic	2.6	Terbium	0.0009
(dissolved organic)	2,000	Selenium	0.090	Dysprosium	0.00073
Nitrogen (dissolved N_2)	15,500	Bromine	6.73×10^4	Holmium	0.00022
(as NO_3^-, NO_2^-, NH_4^+)	670	Krypton	0.21	Erbium	0.00061
Oxygen (dissolved O_2)	6,000	Rubidium	120	Thulium	0.00013
(as H_2O)	8.83×10^8	Strontium	8,100	Ytterbium	0.00052
Fluorine	1,300	Yttrium	0.003	Lutetium	0.00012
Neon	0.120	Zirconium	0.026	Hafnium	<0.008
Sodium	1.08×10^7	Niobium	0.015	Tantalum	<0.0025
Magnesium	1.29×10^6	Molybdenum	10	Tungsten	<0.001
Aluminum	1	Ruthenium	—	Rhenium	—
Silicon	2,900	Rhodium	—	Osmium	—
Phosphorus	88	Palladium	—	Iridium	—
Sulfur	9.04×10^5	Silver	0.28	Platinum	—
Chlorine	1.94×10^7	Cadmium	0.11	Gold	0.011
Argon	450	Indium	—	Mercury	0.15
Potassium	3.92×10^5	Tin	0.81	Thallium	—
Calcium	4.11×10^5	Antimony	0.33	Lead	0.03
Scandium	<0.004	Tellurium	—	Bismuth	0.02
Titanium	1	Iodine	64	Radium	1×10^{-13}
Vanadium	1.9	Xenon	0.047	Thorium	0.0015
Chromium	0.2	Cesium	0.30	Protactinium	2×10^{-10}
Manganese	1.9	Barium	21	Uranium	3.3
Iron	3.4	Lanthanum	0.0029		
Cobalt	0.39	Cerium	0.0013		

[a]Adapted from Karl K. Turekian, *Oceans*, p. 92, Prentice-Hall, Inc., Englewood Cliffs, N.J., 1968.

TABLE D.3

Runoff from the continents[a]

	Area, millions of		Annual runoff Total, $10^{15} \times$		Depth/unit area	
	km²	mi²	l	gal	cm	in
Asia	46.6	18.0	11.1	3.0	23.8	9.4
Africa	29.8	11.5	5.9	1.6	19.8	7.8
North America	21.2	8.2	4.5	1.2	21.1	8.3
South America	19.6	7.6	8.0	2.1	41.4	16.3
Europe	10.9	4.2	2.5	0.6	23.1	9.1
Australia	7.8	3.0	0.4	0.1	2.5	1.0
Total or (mean)	135.9	52.5	32.4	8.6	(24.9)	(9.8)

[a]Calculated from data from D. A. Livingstone, *U.S. Geol. Surv. Prof. Paper* 440-G, 1963.

TABLE D.4

Composition of river waters of the world[a]

Substance	Ppm
HCO_3	58.4
Ca	15
SiO_2	13.1
SO_4	11.2
Cl	7.8
Na	6.3
Mg	4.1
K	2.3
NO_3	1
Fe	0.67

[a]From Daniel A. Livingstone, "Data of Geochemistry," *U.S. Geol. Surv. Prof. Paper* 440-G, p. G-41, 1963.

TABLE D.5

Earth volume, density, and mass

	Av. thickness or radius, km	Volume, millions of km³	Mean density, g/cm³	Mass, × 10²⁴ g
Total earth	6,371	1,083,230	5.52	5,976
Oceans and seas	3.8	1,370	1.03	1.41
Glaciers	1.6	25	0.9	0.023
Continental crust	35	6,210	2.8	17.39
Oceanic crust	8	2,660	2.9	7.71
Mantle	2,883	899,000	4.5	4,068
Core	3,471	175,500	10.71	1,881

TABLE D.6

Average composition of the crust[a]

	Av. igneous rock, %	Av. shale, %	Av. sandstone, %	Av. limestone, %	Weighted-av. crust[b], %
SiO_2	59.12	58.11	78.31	5.19	59.07
TiO_2	1.05	0.65	0.25	0.06	1.03
Al_2O_3	15.34	15.40	4.76	0.81	15.22
Fe_2O_3	3.08	4.02	1.08	0.54	3.10
FeO	3.80	2.45	0.30		3.71
MgO	3.49	2.44	1.16	7.89	3.45
CaO	5.08	3.10	5.50	42.57	5.10
Na_2O	3.84	1.30	0.45	0.05	3.71
K_2O	3.13	3.24	1.32	0.33	3.11
H_2O	1.15	4.99	1.63	0.77	1.30
CO_2	0.10	2.63	5.04	41.54	0.35
ZrO_2	0.04	—	—	—	0.04
P_2O_5	0.30	0.17	0.08	0.04	0.30
Cl	0.05	—	Tr^c	0.02	0.05
F	0.03	—	—	—	0.03
SO_3	—	0.65	0.07	0.05	—
S	0.05	—	—	0.09	0.06
$(Ce, Y)_2O_3$	0.02	—	—	—	0.02
Cr_2O_3	0.06	—	—	—	0.05
V_2O_3	0.03	—	—	—	0.03
MnO	0.12	Tr	Tr	0.05	0.11
NiO	0.03	—	—	—	0.03
BaO	0.05	0.05	0.05	0.00	0.05
SrO	0.02	0.00	0.00	0.00	0.02
Li_2O	0.01	Tr	Tr	Tr	0.01
Cu	0.01	—	—	—	0.01
C	0.00	0.80	—	—	0.04
Total	100.00	100.00	100.00	100.00	100.00

[a]After F. W. Clarke and H. S. Washington, "The Composition of the Earth's Crust," *U.S. Geol. Surv. Prof. Paper* 127, p. 32, 1924.
[b]Weighted average: igneous rock, 95%; shale, 4%; sandstone, 0.75%; limestone, 0.25%.
[c]Trace.

Earth size

	Thousands of km
Equatorial radius	6.378
Polar radius	6.357
Mean radius[a]	6.371
Polar circumference	40.009
Equatorial circumference	40.077
Ellipticity [(equatorial radius − polar radius)/ equatorial radius], 1/297	

[a]Term used by geophysicists to designate radius of a sphere of equal volume.

Earth areas

	Millions of km²
Total area	510
Land (29.22% of total)	149
Oceans and seas (70.78% of total)	361
Glacier ice	15.6
Continental shelves	28.4

TOPOGRAPHIC AND GEOLOGIC MAPS

Topography refers to the shape of the physical features of the land. A **topographic map** is a representation of the shape, size, position, and relation of the physical features of an area. In addition to mountains, hills, valleys, and rivers most topographic maps also show the culture of a region, that is, political boundaries, towns, houses, roads, and similar features.

E.1 TOPOGRAPHIC MAPS

Topographic maps are used in the laboratory for the observation and analysis of the geological processes that are constantly changing the face of the earth.

DEFINITIONS

Elevation, or altitude The vertical distance between a given point and the datum plane.

Datum plane The reference surface from which all altitudes on a map are measured. This is usually mean sea level.

Height The vertical difference in elevation between an object and its immediate surroundings.

Relief The difference in elevation of an area between tops of hills and bottoms of valleys.

Bench mark A point of known elevation and position—usually indicated on a map by the letters B.M., with the altitude given (on American maps) to the nearest foot. (Some maps now use meters.)

Contour line A map line connecting points representing places on the earth's surface that have the same elevation. It thus locates the intersection with the earth's surface of a plane at any arbitrary elevation parallel to the datum plane. Contours represent the vertical, or third, dimension on a map, which has only two dimensions. They show the shape and size of physical features such as hills and valleys. A depression is indicated by an ordinary contour line except that hachures, or short dashes, are used on one side and point toward the center of the depression.

Contour interval The difference in elevation represented by adjacent contour lines.

Scale The ratio on a map of the distance between two points on the ground and the same two points on the map. It may be expressed in three ways:

Fractional scale If two points are 1 km apart in the field, they may be represented on the map as separated by some fraction of that distance, say, 1 cm. In this instance, the scale is 1 cm to the kilometer. There are 100,000 cm in 1 km; so this scale can be expressed as the fraction, or ratio, 1:100,000. Many topographic maps of the United States Geological Survey have a scale of 1:62,500; and many recent maps have a scale of 1:31,250, and others of 1:24,000.

Graphic scale This scale is a line printed on the map and divided into units that are equivalent to some distance, such as 1 km or 1 mi.

Verbal scale This is an expression in common speech, such as "four centimeters to the kilometer," "an inch to a mile," or "two miles to the inch."

CONVENTIONAL SYMBOLS

An explanation of the symbols used on topographic maps is printed on the back of each topographic sheet, along the margin, or for newer maps on a separate legend sheet. In general, culture (artificial works) is shown in black. All water features, such as streams, swamps, and glaciers, are shown in blue. Relief is shown by contours in brown. Red may be used to indicate main highways, and green overprints may be used to designate areas of woods, orchards, vineyards, or scrub.

The United States Geological Survey distributes free of charge a single sheet, entitled "Topographic Maps," that includes an illustrated summary of topographic-map symbols. (Apply to Branch of Distribution, United States Geological Survey, 1200 South Eads Street, Arlington, Virginia 22202.)

LOCATING POINTS

Any particular point or area may be located in several ways on a topographic map. The three most commonly used are:

In relation to prominent features A point may be referred to as being so many kilometers in a given direction from a city, mountain, lake, river mouth, or other easily located feature on the map.

By latitude and longitude Topographic maps of the Geological Survey are bounded on the north and south by parallels of latitude and on the east and west by meridians of longitude. These intersecting lines form the grid into which the earth has been divided. Latitude is measured north and south from the equator, and longitude is measured east and west from the prime meridian that passes through Greenwich, England. Thus maps of the United States are within north latitude and west longitude.

By township and range Most of the area of the United States has been subdivided by a system of land survey in which a square 6 mi on a side is the basic unit, called a **township** when measured north and south of a given base line, and called a **range** when measured east and west of a given principal meridian. Not included in this system are all the states along the eastern seaboard (with the exception of Florida), West Virginia, Kentucky, Tennessee, Texas, and parts of Ohio. Townships are laid off north and south from a base line and east and west from a principal meridian. Each township is divided into 36 sections, usually 1 mi on a side. Each section may be further subdivided into half sections, quarter sections, or sixteenth sections. Thus in Figure E.1 the point *x* is located in the northeast quarter of the northwest quarter of section 3, township 9 north, range 5 west, abbreviated as NE¼NW¼Sec 3, T9N, R5W, or NE NW Sec 3-9N-5W.

CONTOUR SKETCHING

Many contour maps are now made from aerial photographs. Before this can be done, however, the position and location of a number of reference points, or bench marks, must be determined in the field. If the topographic map is prepared from field surveys rather than from

A section

E.1 Subdivision by township and range (see text for discussion).

aerial photographs, the topographer first determines the location and elevation of bench marks and a large number of other points selected for their critical position. They may be on hilltops, on the lowest point in a saddle between hills, along streams, or at places where there is a significant change in slope. On the basis of these points, contours may be sketched through loci of equal elevation. The contours are preferably sketched in the field in order to include minor irregularities that are visible to the topographer.

Because contours are not ordinary lines, certain requirements must be met in drawing them to satisfy their definition. These are listed below:

1 All points on one contour line have the same elevation.

2 Contours separate all points of higher elevation than the contour from all points of lower elevation.

3 The elevation represented by a contour line is always a simple multiple of the contour interval. Every contour line that is a multiple of five times the contour interval is heavier than the others. (An exception is 25-unit contours, in which every multiple of four times the interval is heavier.)

4 Contours never cross or intersect one another.

5 A vertical cliff is represented by coincident contours.

6 Every contour closes on itself either within or beyond the limits of the map. In the latter case, the contours will end at the edge of the map.

7 Contour lines never split.

8 Uniformly spaced contour lines represent a uniform slope.

9 Closely spaced contour lines represent a steep slope.

10 Contour lines spaced far apart represent a gentle slope.

11 A contour line that closes within the limits of the map indicates a hill.

12 A hachured contour line represents a depression. The short dashes, or hachures, point into the depression.

13 Contour lines curve up a valley but cross a stream at right angles to its course.

14 Maximum ridge and minimum valley contours always go in pairs: That is, no single lower contour can lie between two higher ones and vice versa.

TOPOGRAPHIC PROFILES

A topographic profile is a cross section of the earth's surface along a given line. The upper line of this section is irregular and shows the shape of the land along the line of profile, or section.

Profiles are most easily constructed with graph paper. A horizontal scale, usually the map scale, is chosen. Then a vertical scale sufficient to bring out the features of the surface is chosen. The vertical scale is usually several times larger than the horizontal—that is, it is exaggerated. The steps in the construction of a profile are:

1 Select a base (one of the horizontal lines on the graph paper). This may be sea level or any other convenient datum.

2 On the graph paper number each fourth or fifth line above the base, according to the vertical scale chosen.

3 Place the graph paper along the line of profile.

4 With the vertically ruled lines as guides, plot the elevation of each contour line that crosses the line of profile.

5 If great accuracy is not important, plot only every heavy contour and the tops of hills and the bottoms of valleys.

6 Connect the points.

7 Label necessary points along the profile.

8 Give the vertical and horizontal scales.

9 State the vertical exaggeration.

10 Title the profile.

Vertical exaggeration The profile represents both vertical and horizontal dimensions. These dimensions are not usually on the same scale because the vertical needs to be greater than the horizontal to give a clear presentation of changes in level. Thus if the vertical scale is 500 m to the centimeter and the horizontal scale is 5,000 m to the centimeter, the vertical exaggeration is 10 times, written 10×. This is obtained by dividing the horizontal scale by the vertical scale. Note that both horizontal and vertical scales must be expressed in the same unit (in this instance meters to the centimeter) before division.

E.2 GEOLOGIC MAPS

Geologic maps show the distribution of earth materials on the surface. In addition they indicate the relative age of these materials and suggest their arrangement beneath the surface.

Formation The units illustrated upon a geologic map are usually referred to as **formations**. We define a formation as a rock unit with upper and lower boundaries that can be recognized easily in the field and that is large enough to be shown on the map. A formation receives a distinctive designation made up of two parts: The first part is geographic and refers to the place or general area where the formation is first described; the second refers to the nature of the rock. Thus **Trenton Limestone** is a formation dominantly composed of limestone and is named after Trenton Falls in central New York, where it was first formally described. **Wausau Granite** designates a body of granite in the Wausau, Wisconsin, area. If the lithology is so variable that no single lithologic distinction is appropriate, the word *formation* may be used. For instance, the **Raritan Formation** is named for the area of the Raritan River and Raritan Bay in New Jersey, and its lithology includes both sand and clay.

Dip and strike The dip and strike of a rock layer refer to its orientation in relation to a horizontal plane. In Chapter 8 we found that the dip is the acute angle that a tilted rock layer makes with an imaginary plane. We also found that the strike is the compass direction of a line formed by the intersection of the dipping surface with an imaginary horizontal plane. The direction of strike is always at right angles to the direction of dip. The dip-and-strike symbol used on a geologic map is in the form of a top-heavy tee. The crossbar represents the direction of the strike of the bed. The short upright represents the direction of the dip of the bed. This sometimes, but not always, has an arrow pointing in the direction of dip. Very often the angle of dip is indicated alongside the symbol. Here is an example (the top of the page is considered to be north):

⊤ 30 = strike E-W; dip 30°S

⟋ 25 = strike N45°E; dip 25°SE

Contact A contact is the plane separating two rock units. It is shown on the geologic map as a line that is the intersection of the plane between the rock units and the surface of the ground.

Outcrop An outcrop is rock material that is exposed at the surface through the cover of soil and weathered material. In areas of abundant rainfall soil and vegetation obscure the underlying rock and only a small fraction of 1 percent of the surface may be in outcrop. In dry climates, where soils are shallow or absent and the plant cover is discontinuous, bedrock usually crops out much more widely.

Legend and symbols A legend is an explanation of the various symbols used on the map. There is no universally accepted set of standard symbols, but some that are widely used are given in Figure E.2. In addition to the graphic symbols in Figure E.2, letter symbols are sometimes used to designate rock units. Such a symbol contains a letter or letters referring to the geologic column, followed by a letter or letters referring to the specific name of the rock unit. Thus in the symbol Ot the O stands for Ordovician and the t for the Trenton Limestone of central New York. The letters or abbreviations generally used for the geologic column are given in Table E.1.

Sometimes different colors are used to indicate different rock systems. There is no standardized color scheme, but many of the geologic maps of the Geological Survey use the colors given in Table E.1, combined with varying patterns, for systems of sedimentary rocks. No specific colors are designated for igneous rocks, but when colors are used, they are usually purer and more brilliant than those used for sedimentary rocks.

Rock symbols

Sedimentary

Conglomerate

Sandstone

Mudstone

Shale

Limestone

Dolomite

Coal

Igneous

Massive igneous, such as granite

Lava flows

Foliated metamorphic, such as schist and gneiss

Structure symbols

—— Contact

⌐60 Dip and strike of inclined beds

⊕ Horizontal beds

90 Strike of vertical beds

Syncline

D/U Fault, showing up (U) and down (D) blocks

50 Fault, showing dip of fault plane

Fault, showing relative movement

Anticline

E.2 Symbols commonly used on geologic maps.

CONSTRUCTION OF A GEOLOGIC MAP

The basic idea of geologic mapping is simple. We are interested first in showing the distribution of the rocks at the earth's surface. Theoretically all we need to do is plot the occurrence of the different rocks on a base map, and then we have a geologic map. Unfortunately the process is not quite so simple.

In most areas, the bedrock is more or less obscured in one way or another, and only a small amount of outcrop is available for observation, study, and sampling. From the few exposures available the geologist must extrapolate the general distribution of rock types. In this extrapolation the field data are obviously of prime importance. But one will also be guided by changes in soil, vegetation, and landscape as well as by patterns that can be detected on aerial photographs. Furthermore one may be aided by laboratory examination of field samples and by the records of both deep and shallow wells. The geologist may also have available geophysical data that help determine the nature of obscured bedrock. Eventually, when one has marshaled as many data as possible, the geologist draws the boundaries delineating the various rock types.

In addition to the distribution of rock types, the geologist is also concerned with depicting, as accurately as one can, the ages of the various rocks and their arrangement beneath the surface. These goals will also be realized in part through direct observations in the field and in part through other lines of evidence. The preparation of

TABLE E.1

Letter symbols and colors commonly used to designate units in the geologic column

Period	Symbol	Color
Pleistocene	Q	Yellow and gray
Pliocene	Tpl	Yellow ocher
Miocene	Tm	Yellow ocher
Oligocene	To	Yellow ocher
Eocene	Te	Yellow ocher
Paleocene	Tp	Yellow ocher
Cretaceous	K	Olive green
Jurassic	J	Blue green
Triassic	T_R	Light peacock blue or bluish gray green
Permian	P	Blue
Pennsylvanian	Cp	Blue
Mississippian	Cm	Blue
Devonian	D	Gray purple
Silurian	S	Purple
Ordovician	O	Red purple
Cambrian	∈	Brick red
Precambrian	P_e	Terra-cotta and gray brown

an accurate, meaningful, geologic map demands experience, patience, and judgment.

GEOLOGIC CROSS SECTIONS

A geologic map tells us something of how rocks are arranged in the underground. Often, to show these relations more clearly, we find it convenient to draw geologic cross sections. Such a section is really a diagram showing a side view of a block of the earth's crust as it would look if we could lift it up to view. We have used cross sections in many illustrations throughout this book.

A geologic cross section is drawn, insofar as possible, at right angles to the general strike of the rocks. The general manner in which a geologic cross section is projected from a geologic map is shown in Figure E.3. If the projection is made onto a topographic profile in which the vertical scale has been exaggerated, then the angle of dip of the rocks should be exaggerated accordingly.

E.3 Construction of a geologic cross section from a geologic map.

30 Dip and strike

Fault:
D, downthrown block
U, upthrown block
60, dip of fault plane

Contact

A——B Line of section

Direction of relative movement

Mudstone

Sandy mudstone

Conglomerate

Shale

Limestone

Sandstone

GLOSSARY

In the definitions here words in *italics* are often separately defined in their alphabetic place.

Å Abbreviation for *angstrom*, a unit of length (10^{-8} cm).

aa lava *Lava* whose surface is covered with random masses of angular jagged blocks.

ablation As applied to glacier ice, process by which ice below snow line is wasted by evaporation and melting.

abrasion Erosion of rock material by friction of solid particles moved by gravity, water, ice, or wind.

absolute time Geologic time measured in years. Compare with *relative time*.

acid rain Rain higher in acid than normal. Forms from sulfur dioxide (SO_2) from the burning of fossil fuels; combines with water to form sulfuric acid.

acidic lava Containing 70 percent or more of silica.

actinolite A metamorphic ferromagnesian mineral; an *asbestos*.

adiabatic rate In a body of air moving upward or downward, change in temperature without gain or loss of temperature to air through which it moves.

aftershock Earthquake that follows a larger earthquake and originates at or near focus of larger earthquake. Major shallow earthquakes are generally followed by many aftershocks, which decrease in number as time goes on but may continue for days or even months.

agate Variety of *chalcedony* with alternating layers of chalcedony and *opal*.

A horizon Soil zone immediately below surface, from which soluble material and fine-grained particles have been moved downward by water seeping into soil. Varying amounts of organic matter give *A* horizon color ranging from gray to black.

Airy hypothesis Explains *isostasy* by assuming earth's crust has same density everywhere and differences in elevation result from differences in thickness of outer layer.

albite Feldspar in which diagnostic positive ion is Na^+; sodic feldspar, $Na(AlSi_3O_8)$. One of plagioclase feldspars.

alkali rocks Igneous rocks in which abundance of alkalies is unusually high, generally indicated by soda pyroxenes, soda amphiboles, and/or feldspathoids.

alluvial fan Land counterpart of a delta: an assemblage of sediments marking place where a stream moves from a steep gradient to a flatter gradient and suddenly loses transporting power. Typical of arid and semiarid climates but not confined to them.

almandite A deep-red garnet of iron and aluminum formed during regional metamorphism.

alpha decay Radioactive decay taking place by loss of an alpha particle from nucleus. Mass of element decreases by 4, and atomic number decreases by 2.

alpha particle A helium atom lacking electrons and therefore having a double positive charge.

Alpine, or mountain, glacier See *valley glacier.*

amorphous A state of matter in which there is no orderly arrangement of atoms.

amphibole group Ferromagnesian silicates with a double chain of silicon-oxygen tetrahedra. Common example: hornblende. Contrast with *pyroxene group.*

amphibolite A faintly foliated metamorphic rock developed during regional metamorphism of *simatic* rocks. Composed mainly of hornblende and plagioclase feldspars.

amphibolite facies An assemblage of minerals formed during regional metamorphism at moderate to high pressures between 450 and 700°C.

andalusite A silicate of aluminum built around independent tetrahedra, Al_2SiO_5. Characteristic of middle-grade metamorphism. Compare with *kyanite,* which has same composition and forms under similar conditions but which has different crystal habit. Contrast with *sillimanite,* which has same composition but different crystal habit and forms at highest metamorphic temperatures.

andesite A fine-grained igneous rock with no quartz or orthoclase, composed of about 75 percent plagioclase feldspars, balance ferromagnesian silicates. Important as lavas; possibly derived by fractional crystallization from basaltic magma. Widely characteristic of mountain-making processes around borders of Pacific Ocean. Confined to continental sectors.

andesite line A map line designating the petrographic boundary of Pacific Ocean. Extrusive rocks on Pacific side of line are basaltic; on other side, andesitic.

angstrom A unit of length equal to one hundred-millionth of a centimeter (10^{-8} cm). Abbreviation, Å.

angular momentum A vector quantity, product of mass times radius of orbit times velocity. The energy of motion of solar system.

angular unconformity An *unconformity* in which older strata dip at different angle from that of younger strata.

anhydrite Mineral calcium sulfate, $CaSO_4$, which is *gypsum* without water.

anion A negatively charged atom or group of atoms produced by gain of electrons. Compare *cation.*

anorthite Feldspar in which diagnostic positive ion is Ca^{2+}; calcic feldspar, $Ca(Al_2Si_2O_8)$. One of plagioclase feldspars.

anorthosite A plutonic igneous rock composed of 90 percent or more of feldspar mineral anorthite. Pyroxene and some olivine usually make up balance of the rock.

Antarctic bottom water Seawater that sinks to ocean floor off Antarctica and flows equatorward beneath North Atlantic deep water.

Antarctic intermediate water Seawater that sinks at about 50°S and flows northward above North Atlantic deep water.

anthracite Metamorphosed bituminous coal of about 95 to 98 percent carbon.

anticline A configuration of folded, stratified rocks in which rocks dip in two directions away from a crest, as principal rafters of a common gable roof dip away from ridgepole. Reverse of *syncline.* The "ridgepole," or crest, is called *axis.*

aphanitic texture Individual minerals present but in particles so small that they cannot be identified without a microscope.

aquifer A permeable material through which groundwater moves.

arête A narrow, saw-toothed ridge developed between cirques.

arkose A detrital sedimentary rock formed by cementation of individual grains of sand size and predominantly composed of quartz and feldspar. Derived from disintegration of granite.

arroyo Flat-floored, vertically walled channel of an intermittent stream typical of semiarid climates; often applied to such features of southwestern United States. Synonymous with *wadi* and *wash.*

artesian water Water under pressure when tapped by a well and able to rise above level at which first encountered. It may or may not flow out at ground level.

asbestos A general term applied to certain fibrous minerals displaying similar physical characteristics although differing in composition. Some asbestos has fibers long enough to be spun into fabrics with great heat resistance, such as those for automobile brake linings. Types with shorter fibers are compressed into insulating boards, shingles, etc. Most common asbestos mineral (95 percent of United States production) is *chrysotile,* a variety of *serpentine,* a metamorphic mineral.

aseismic ridge An oceanic ridge without extensive seismic activity as, for example, the Hawaiian Island-Emperor Seamount ridge in the Pacific Ocean.

ash Volcanic fragments of sharply angular glass particles, smaller than cinders.

asphalt A brown to black, solid or semisolid bituminous substance. Occurs in nature but is also obtained as residue from refining of certain hydrocarbons ("artificial asphalt").

asteroids Orbiting small bodies believed to be either fragments of a disintegrated planet or of matter that never completed planet-forming process.

asthenosphere A zone within earth's *mantle* where plastic movements occur to permit *isostatic* adjustments. Begins 50 to 100 km below surface and extends perhaps to 500 km.

astrogeology Geology of celestial bodies other than earth. Generally restricted in meaning to geology of extraterrestrial bodies of our planetary system.

asymmetric fold A fold in which one limb dips more steeply than the other.

Atlantic Ridge Belt of mountains under Atlantic Ocean from Iceland to Antarctica and nearly midway between continents; also called *mid-Atlantic Ridge.*

atoll A ring of low coral islands arranged around a central lagoon.

atom A building block of matter; combination of protons, neutrons, and electrons, of which 103 kinds are now known.

atomic energy Energy associated with nucleus of an atom. It is released when nucleus is split.

atomic mass The nucleus of an atom contains 99.95 percent of its mass. Total number of protons and neutrons in nucleus is called *mass number.*

atomic number Number of positive charges on nucleus of an atom; number of protons in nucleus.

atomic reactor A huge apparatus in which a radioactive core heats water under pressure and passes it to a heat exchanger.

atomic size Radius of an atom (average distance from center to outermost electron of neutral atom). Commonly expressed in angstroms.

augite A rock-forming ferromagnesian silicate mineral built around single chains of silicon-oxygen tetrahedra.

aureole A zone in which contact metamorphism has taken place.

axial plane A plane through a rock fold that includes the axis and divides the fold as symmetrically as possible.

axis Ridge or place of sharpest folding of an anticline or syncline.

backset beds Inclined layers of sand developed on gentler dune slope to windward. These beds may constitute a large part of total volume of a dune, especially if there is sufficient vegetation to trap most sand before it can cross over to slip face.

bank-full stage Stage of flow at which a stream fills its channel up to level of its bank. Recurrence interval averages 1.5 to 2 years.

barchan A crescent-shaped dune with wings, or horns, pointing downwind. Has gentle windward slope and steep lee slope inside horns; about 30 m high and 300 m wide from horn to horn. Moves with wind about 15 m/year across flat, hard surface where limited supply of sand is available.

barrier island A low, sandy island near shore and parallel to it, on a gently sloping offshore bottom.

barrier reef A reef separated from a landmass by a lagoon of varying width and depth opening to sea through passes in reef.

basal plane Lower boundary of zone of concentric folding.

basal slip Movement of an entire glacier over underlying ground surface.

basalt A fine-grained igneous rock dominated by dark-colored minerals, consisting of plagioclase feldspars (over 50 percent) and ferromagnesian silicates. Basalts and andesites represent about 98 percent of all extrusive rocks.

base flow That portion of stream flow attributable to groundwater flow.

base level Level below which a stream cannot erode. There may be temporary base levels along a stream's course, such as those established by lakes or resistant layers of rock. Ultimate base level is sea level.

basement complex Undifferentiated rocks underlying oldest identifiable rocks in any region. Usually sialic, crystalline, metamorphosed. Often, but not necessarily, Precambrian.

basic lava Containing less than 50 percent silica.

basin See *dome.*

batholith A *discordant pluton* that increases in size downward, has no determinable floor, and shows an area of surface exposure exceeding 100 km².

bauxite Chief ore of commercial aluminum. A *mineraloid* mixture of hydrous aluminum oxides.

bay barrier A sandy beach, built up across mouth of a bay so that it is no longer connected to main body of water.

bedding (1) A collective term used to signify existence of beds, or layers, in sedimentary rocks. (2) Sometimes synonymous with *bedding plane.*

bedding plane Surface separating layers of sedimentary rocks. Each bedding plane marks termination of one deposit and beginning of another of different character, such as surface separating a sand bed from a shale layer. Rock tends to separate, or break, readily along bedding planes.

bed load Material in movement along stream bottom or, if wind is moving agency, along surface. Contrast with material carried in *suspension* or solution.

belt of soil moisture Subdivision of *zone of aeration.* Belt from which water may be used by plants or withdrawn by soil evaporation. Some water passes down into intermediate belt, where it may be held by molecular attraction against influence of gravity.

bench mark See Appendix E.

Benioff zone A seismic zone dipping beneath a continental margin and having a deep-sea trench as surface expression.

bergschrund Gap, or crevasse, between glacier ice and headwall of a cirque.

berms In coastline terminology berms are storm-built beach features that resemble small terraces; on seaward edges are low ridges built up by storm waves.

beta decay Radioactive decay taking place by loss of a beta particle (electron) from a neutron in nucleus. Mass of element remains same, but atomic number increases by 1.

B **horizon** Soil zone of accumulation below *A* horizon. Here is deposited some material moved down from *A* horizon.

big-bang theory Theory that presently expanding universe originated as primeval cosmic fireball in very short period of time 10 to 12 billion years ago. Compare *steady-state theory.*

binding energy Amount of energy that must be supplied to break an atomic nucleus into its component fundamental particles. It is equivalent to mass that disappears when fundamental particles combine to form a nucleus.

biochemical rock A sedimentary rock made up of deposits resulting directly or indirectly from life processes of organisms.

biotite "Black mica," ranging in color from dark brown to green. Rock-forming ferromagnesian silicate mineral with tetrahedra in sheets.

bituminous coal Soft coal, containing about 80 percent carbon and 10 percent oxygen.

blowout A basin scooped out of soft, unconsolidated deposits by process of deflation. Ranges from a few meters to several kilometers in diameter.

body wave *Push-pull* or *shake* earthquake wave travelling through body of a medium; distinguished from waves travelling along free surface.

bornite A mineral, Cu_5FeS_4; an important ore of copper.

bottomset bed Layer of fine sediment deposited in a body of standing water beyond advancing edge of a growing delta, which eventually builds up on bottomset beds.

boulder size Volume greater than that of a sphere with diameter of 256 mm.

boulder train Series of glacier erratics from the same bedrock source, usually with some property that permits easy identification. Arranged across country in a fan, with apex at source and widening in direction of glacier movement.

Bowen's reaction series Series of minerals for which any early-formed phase tends to react with melt that remains to yield a new mineral further along in the series. Thus early-formed crystals of olivine react with remaining liquids to form augite crystals; these in turn may further react with liquid then remaining to form hornblende. See also *continuous reaction series* and *discontinuous reaction series.*

braided stream Complex tangle of converging and diverging stream channels separated by sand bars or islands. Characteristic of flood plains where amount of debris is large in relation to discharge.

breccia Clastic rock made up of angular fragments of such size that an appreciable percentage of rock volume consists of particles of granule size or larger.

brittle A property of material whereby strength in *tension* is greatly different from strength in *compression.* Substance ruptures easily with little or no flow.

brown clay An extremely fine-grained deposit found on the deep ocean floors, most abundantly in the Pacific Ocean. Derived from the continents and drifted to the open ocean, where the clay-sized particles settle.

burial metamorphism Changes resulting from pressures and temperatures in rocks buried to depth of several kilometers.

calcic feldspar Anorthite, $Ca(Al_2Si_2O_8)$.

calcite A mineral composed of calcium carbonate, $CaCO_3$.

caldera Roughly circular, steep-sided volcanic basin with diameter at least three or four times depth. Commonly at summit of a volcano. Contrast with *crater.*

caliche Whitish accumulation of calcium carbonate in soil profile.

capillary fringe Belt above *zone of saturation*, in which underground water is lifted against gravity by surface tension in passages of capillary size.

capillary size "Hairlike," or very small, such as tubes from 0.0025 to 0.25 cm in diameter.

carbohydrate Compound of carbon, hydrogen, and oxygen. Carbohydrates are chief products of life process in plants.

carbonate mineral Mineral formed by combination of complex ion $(CO_3)^{2-}$ with a positive ion. Common example: calcite, $CaCO_3$.

carbon cycle Process in sun's deep interior by which radiant energy is generated in formation of helium from hydrogen.

carbon 14 Radioactive isotope of carbon, $^{14}_{6}C$, which has half-life of 5,730 years. Used to date events up to about 50,000 years ago.

carbon ratio A number obtained by dividing amount of fixed carbon in coal by sum of fixed carbon and volatile matter and multiplying by 100. This is same as percentage of fixed carbon, assuming no moisture or ash.

cassiterite A mineral, tin dioxide, SnO_2. Ore of tin with specific gravity 7; nearly 75 percent of world's tin production is from placer deposits, mostly from cassiterite.

cataclastic metamorphism Textural changes in rocks in which brittle minerals and rocks are broken and flattened as a result of intense folding or faulting; produces fragmentation of rocks as coarse-grained breccias and fine-grained mylonites.

cation A positively charged atom or group of atoms produced by loss of electrons. Compare *anion*.

cementation Process by which a binding agent is precipitated in spaces among individual particles of an unconsolidated deposit. Most common cementing agents are calcite, dolomite, and quartz; others include iron oxide, opal, chalcedony, anhydrite, and pyrite.

cement rock Clayey limestone used in manufacturing hydraulic cement. Contains lime, silica, and alumina in varying proportions.

chalcedony General name applied to fibrous cryptocrystalline silica with waxy luster. Deposited from aqueous solutions and frequently found lining or filling cavities in rocks. *Agate* is a variety with alternating layers of chalcedony and opal.

chalcocite A mineral, copper sulfide, Cu_2S; sometimes called *copper glance*; one of most important ore minerals of copper.

chalcopyrite A mineral, a sulfide of copper and iron, $CuFeS_2$; sometimes called *copper pyrite* or *yellow copper ore*.

chalk Variety of limestone made up in part of biochemically derived calcite, in form of skeletons or skeletal fragments of microscopic oceanic plants and animals mixed with very fine-grained calcite deposits of biochemical or inorganic-chemical origin.

chemical rock In terminology of sedimentary rocks chemical rock is composed chiefly of material deposited by chemical precipitation, whether organic or inorganic (compare with *detrital sedimentary rock*). Chemical sedimentary rocks may have either clastic or nonclastic (usually crystalline) texture.

chemical weathering Weathering of rock material by chemical processes that transform original material into new chemical combinations. Thus chemical weathering of orthoclase produces clay, some silica, and a soluble salt of potassium.

chert Granular cryptocrystalline silica, similar to *flint* but usually light in color. Occurs as compact massive rock or as nodules.

chlorite Family of tetrahedral sheet silicates of iron, magnesium, and aluminum, characteristic of low-grade metamorphism. Green color, with cleavage like mica except that chlorite small scales are not elastic.

C horizon Soil zone that contains partially disintegrated and decomposed parent material. Lies directly under *B* horizon and grades downward into unweathered material.

chromite Mineral oxide of iron and chromium, $FeCr_2O_4$, only ore of commercial chromium. One of first minerals to crystallize from magma; is concentrated within the magma.

chrysotile Metamorphic mineral; an asbestos, fibrous variety of *serpentine*. Silicate of magnesium, with tetrahedra arranged in sheets.

chute or chute cutoff Applied to stream flow, *chute* refers to new route taken by a stream when main flow is diverted to inside of a bend, along trough between low ridges formed by deposition on inside of bend, where water velocity is reduced. Compare with *neck cutoff*.

cinder cone Structure built exclusively or predominantly of pyroclastic ejecta dominated by cinders. Parasitic to a major volcano, seldom exceeds 500 m in height. Slopes up 30° to 40°. Example: Parícutin.

cinders Volcanic fragments; small, slaglike, solidified pieces of magma 0.5 to 2.5 cm across.

cirque A steep-walled hollow in a mountainside at high elevation, formed by ice-plucking and frost action and shaped like a half bowl or half amphitheater. Serves as principal gathering ground for ice of a valley glacier.

clastic texture Texture shown by sedimentary rocks from deposits of mineral and rock fragments.

clay minerals Finely crystalline, hydrous silicates formed from weathering of such silicate minerals as feldspar, pyroxene, and amphibole. Most common clay minerals belong to *kaolinite*, *montmorillonite*, and *illite* groups.

clay size Volume less than that of a sphere with diameter of $\frac{1}{256}$ mm (0.004 mm).

cleavage (1) *Mineral cleavage:* property possessed by many minerals of breaking in certain preferred directions along smooth plane surfaces. Planes of cleavage are governed by atomic pattern and represent directions in which atomic bonds are relatively weak. (2) *Rock cleavage:* property possessed by certain rocks of breaking with relative ease along parallel planes or nearly parallel surfaces. Rock cleavage is designated as *slaty*, *phyllitic*, *schistose*, and *gneissic*.

coal Sedimentary rock composed of combustible matter derived from partial decomposition and alteration of plant cellulose and lignin.

cobble size Volume greater than that of a sphere with diameter of 64 mm and less than that of a sphere with diameter of 256 mm.

coesite High-pressure form of quartz, with density of 2.92. Associated with impact craters and cryptovolcanic structures.

col Pass through a mountain ridge. Created by enlargement of two cirques on opposite sides of ridge until headwalls meet and are broken down.

cold glacier One in which no surface melting occurs during summer months and whose temperature is always below freezing.

colloidal size Between 0.2 and 1 μm (0.0002 to 0.001 mm).

color Sensation resulting from stimulation of retina by light waves of certain lengths.

column Column, or post, of dripstone joining floor and roof of a cave; result of joining of stalactite and stalagmite.

columnar jointing Pattern of jointing that blocks out columns of rock. Characteristic of tabular basalt flows or sills.

compaction Reduction in pore space between individual grains from pressure of overlying sediments or pressures of earth movement.

composite volcanic cone Composed of interbedded lava flows and pyroclastic material and characterized by slopes of close to 30° at summit, progressively reducing to 5° near base. Example: Mayon.

compression Squeezing stress that tends to decrease volume of a material.

concentric folding Elastic bending of an originally horizontal sheet with all internal movements parallel to a *basal plane* (lower boundary of the fold).

conchoidal fracture A mineral's habit of fracturing to produce curved surfaces like interior of a shell (*conch*). Typical of glass and quartz.

concordant pluton An intrusive igneous body with contacts parallel to layering or foliation surfaces of rocks into which it has intruded.

concretion An accumulation of mineral matter formed around a center, or axis, of deposition after a sedimentary deposit has been laid down. Cementation consolidates the deposit as a whole, but the concretion is a body within host rock that represents local concentration of cementing material: enclosing rock is less firmly cemented than the concretion. Commonly spheroidal or disk-shaped and composed of such cementing agents as calcite, dolomite, iron oxide, or silica.

cone of depression A dimple in the water table, which forms as water is pumped from a well.

cone sheet A dike, part of a concentric set dipping inward, like an inverted cone.

conglomerate Detrital sedimentary rock made up of more or less rounded fragments of such size that an appreciable percentage of volume of rock consists of particles of granule size or larger.

connate water Water trapped in a sedimentary deposit when the deposit was laid down.

contact metamorphism Metamorphism at or very near contact between magma and rock during intrusion.

continental drift Slow, lateral movement of continents; involves rigid plates that may carry both continental and oceanic areas as they move.

continental glacier An ice sheet that obscures mountains and plains of a large section of a continent. Continental glaciers exist on Greenland and Antarctica.

continental margin Continental rise, continental slope, and continental shelf together make up the continental margin, particularly on the trailing edge of a continent.

continental rise In some places base of continental slope is marked by somewhat gentler continental rise, which leads downward to deep ocean floor.

continental shelf Shallow, gradually sloping zone extending from sea margin to a depth at which there is marked or rather steep descent into ocean depths down continental slope. Seaward boundary of shelf averages about 130 m in depth.

continental slope Portion of ocean floor extending downward from seaward edge of continental shelves. In some places, such as south of Aleutian Islands, slopes descend directly to ocean deeps. In other places, such as off eastern North America, they grade into somewhat gentler continental rises, which in turn lead to deep ocean floors.

continuous reaction series Branch of Bowen's reaction series comprising plagioclase feldspars, in which reaction of early-formed crystals with later liquids takes place continuously—that is, without abrupt phase changes.

contour interval See Appendix E.

contour line See Appendix E.

convection Mechanism by which material moves because its density differs from that of surrounding material. Density differences are often brought about by heating.

convection cell Pair of *convection currents* adjacent to each other.

convection current Closed circulation of material sometimes developed during convection. Convection currents normally develop in pairs, each pair called *convection cell*.

convergent plate boundary Boundary between two plates moving toward each other. Compare *divergent plate boundary*.

coquina A coarse-grained, porous, friable variety of clastic limestone made up chiefly of fragments of shells.

core Innermost zone of earth, surrounded by *mantle*.

Coriolis effect Tendency of any moving body, on or starting from surface of earth, to continue in direction in which earth's rotation propels it. Direction in which the body moves because of this tendency combined with direction in which it is aimed determines ultimate course of the body relative to earth's surface. In the northern hemisphere Coriolis effect causes a moving body to veer or try to veer to right of its direction of forward motion; in the southern hemisphere, to left. Magnitude of effect is proportional to velocity of a body's motion. This effect causes cyclonic storm-wind circulation to be counterclockwise in the northern hemisphere and clockwise in the southern hemisphere and determines final course of ocean currents relative to trade winds.

correlation Process of establishing contemporaneity of rocks or events in one area with rocks or events in another area.

covalent bond Bond in which atoms combine by sharing their electrons.

crater Roughly circular, steep-sided volcanic basin with diameter less than three times depth. Commonly at summit of a volcano (contrast with *caldera*). Applied also to depressions caused by meteorites by either direct or explosive impact.

craton Relatively immobile part of the earth, generally large central portion of a continent.

creep Applied to soils and surficial material, slow downward plastic movement. As applied to elastic solids, slow permanent yielding to stresses less than yield point if applied for a short time only.

crevasse (1) Deep crevice, or fissure, in glacier ice. (2) Breach in a natural levee.

cross bedding See *inclined bedding*.

crosscutting relationships, law of A rock is younger than any rock across which it cuts.

crust Outermost shell of the earth. Continental crust averages 35 km thick, density 2.6 t/m³; oceanic crust, about 5 km thick, density 3 t/m³.

cryptocrystalline State of matter in which there is actually orderly arrangement of atoms characteristic of crystals but in units so small (material is so fine grained) that crystalline nature cannot be determined with an ordinary microscope.

cryptovolcanic structure Geologic structure in which rocks appear disrupted by volcanic activity but in which there are no associated volcanic rocks.

crystal Solid with orderly atomic arrangement, which may or may not develop external faces that give it crystal form.

crystal form Geometrical form taken by a mineral, giving external expression to orderly internal atomic arrangement.

crystalline structure Orderly arrangement of atoms in a crystal. Also called *crystal structure*.

crystallization Process through which crystals separate from fluid, viscous, or dispersed state.

Curie temperature Temperature above which ordinarily magnetic material loses magnetism. On cooling below this temperature, it regains magnetism. Example: Iron loses magnetism above 760°C and regains it while cooling below this temperature. Thus 760°C is the Curie temperature of iron.

current bedding See *inclined bedding*.

cutoff See *chute cutoff, neck cutoff*.

cyclosilicate Mineral with crystal structure containing silicon-oxygen, $(SiO_4)^{4-}$, tetrahedra arranged as rings.

datum plane See Appendix E.

debris slide Small, rapid movement of largely unconsolidated material that slides or rolls downward to produce irregular topography.

decomposition Synonymous with *chemical weathering.*

deep focus Earthquake focus deeper than 300 km. Greatest depth of focus known is 700 km.

deep-sea trenches See *island-arc deeps.*

deflation Erosive process in which wind carries off unconsolidated material.

deformation of rocks Any change in original shape or volume of rock masses; produced by mountain-building forces. *Folding, faulting,* and *plastic flow* are common modes of rock deformation.

delta Plain underlain by an assemblage of sediments accumulated where a stream flows into a body of standing water, its velocity and transporting power suddenly reduced. Originally named after Greek letter *delta* (Δ) because many are roughly triangular in plan, with apex pointing upstream.

dendritic pattern An arrangement of stream courses that, on a map or viewed from the air, resemble branching habit of certain trees, such as oaks or maples.

density Measure of concentration of matter, expressed as mass per unit volume.

density current Current due to differences in density of water from place to place caused by changes in temperature and variations in salinity or amount of material held in suspension.

depositional remanent magnetism Magnetism resulting from tendency of magnetic particles such as hematite to orient themselves in earth's magnetic field as they are deposited. Orientation is maintained as soft sediments are lithified and thus records earth's field when particles were laid down. Abbreviation, DRM.

desert varnish Thin, shiny layer of iron and manganese oxides that coats some desert-rock surfaces.

detrital sedimentary rock Rock formed from accumulation of minerals and rocks derived from erosion of previously existing rocks or from weathered products of these rocks.

diabase Rock of basaltic composition, essentially labradorite and pyroxene, characterized by *ophitic* texture.

diamond A mineral composed of element carbon; hardest substance known. Used as a gem and, in industry, for cutting tools.

differential erosion Process by which different rock masses or different parts of same rock erode at different rates.

differential weathering Process by which different rock masses or different parts of same rock weather at different rates.

dike Tabular *discordant pluton.*

dike swarm Group of approximately parallel dikes.

dilatancy Tendency of rocks to expand along minute fractures immediately prior to failure; stress may be from earth movements or from controlled laboratory experiments.

diorite Coarse-grained igneous rock with composition of andesite (no quartz or orthoclase), composed of about 75 percent plagioclase feldspars and balance ferromagnesian silicates.

dip (1) Acute angle that a rock surface makes with a horizontal plane. Direction of dip is always perpendicular to *strike.* (2) See *magnetic declination.*

dipole Any object oppositely charged at two points. Most commonly refers to a molecule that has concentrations of positive or negative charge at two different points.

dipole magnetic field Portion of earth's magnetic field best described by a dipole passing through earth's center and inclined to earth's rotation axis. See also *nondipole magnetic field* and *external magnetic field.*

dip pole See *magnetic pole.*

dip-slip fault Fault in which displacement is in direction of fault's dip.

discharge With reference to stream flow, quantity of water that passes a given point in unit time. Measured in cubic meters per second or, often, cubic feet per second (abbreviation, cfs).

disconformity *Unconformity* in which beds on opposite sides are parallel.

discontinuity Within earth's interior, sudden or rapid changes with depth in one or more of physical properties of materials constituting the earth, as evidenced by seismic data.

discontinuous reaction series Branch of Bowen's reaction series including minerals olivine, augite, hornblende, and biotite, for which each series change represents abrupt phase change.

discordant pluton An intrusive igneous body with boundaries that cut across surfaces of layering or foliation in rocks into which it has intruded.

disintegration Synonymous with *mechanical weathering.*

divergent plate boundary Boundary between two plates moving apart. New oceanic-type lithosphere is created at the opening. Compare *convergent plate boundary.*

divide Line separating two drainage basins.

dolomite Mineral composed of carbonate of calcium and magnesium, $CaMg(CO_3)_2$. Also used as rock name for formations composed largely of mineral dolomite.

dome Anticlinal fold without clearly developed linearity of crest so that beds involved dip in all directions from a central area, like an inverted but usually distorted cup. Reverse of *basin.*

drainage basin Area from which a given stream and its tributaries receive water.

drift Any material laid down directly by ice or deposited in lakes, oceans, or streams as result of glacial activity. Unstratified glacial drift is called *till* and forms *moraines;* stratified forms *outwash plains, eskers, kames,* and *varves.*

dripstone Calcium carbonate deposited from solution by underground water entering a cave in *zone of aeration.* Sometimes called *travertine.*

DRM See *depositional remanent magnetism.*

drumlin Smooth, streamlined hill composed of *till.* Long axis oriented in direction of ice movement: Blunt nose points upstream, and gentler slope tails off downstream. In height drumlins range from 8 to 60 m, with average somewhat less than 30 m. Most drumlins are between 0.5 and 1 km in length, the length commonly several times width. Diagnostic characteristics are shape and composition of unstratified glacial drift, in contrast to *kames,* of stratified glacial drift and random shapes.

dune Mound or ridge of sand piled by wind.

dust-cloud hypothesis Hypothesis that solar system was formed from condensation of interstellar dust clouds.

dust size Volume less than that of sphere with diameter 0.06 mm; used in reference to particles carried in suspension by wind.

earthflow Combination of *slump* and *mudflow.*

earth waves Mechanism for transmitting energy from earthquake focus.

ecliptic Apparent path of the sun in the heavens; plane of planets' orbits.

eclogite facies Metamorphic rocks of *gabbroic* composition, consisting primarily of *pyroxene* and *garnet.*

elastic deformation Nonpermanent deformation, after which body returns to original shape or volume when deforming force removed.

elastic energy Energy stored within a solid during elastic deformation and released during elastic rebound.

elastic limit Maximum *stress* that produces only *elastic deformation.*

elastic rebound Recovery of elastic *strain* when material breaks or deforming force is removed.

elastic solid A solid that yields to applied force by changing shape, volume, or both but returns to original condition when force is removed. Amount of yield is proportional to force.

elasticity A property of materials that defines extent to which they resist small deformations, from which they recover completely when deforming force is removed. Elasticity equals *stress* divided by *strain*.

electric charge Property of matter resulting from imbalance between number of *protons* and number of *electrons* in given piece of matter. The electron has negative charge; the proton, positive charge. Like charges repel each other; unlike charges attract each other.

electric current Flow of electrons.

electric turbine generator Apparatus that uses steam from heat exchanger to drive turbine and generate electricity.

electrical energy Energy of moving *electrons*.

electron Fundamental particle of matter, the most elementary unit of negative electrical charge. Mass, 0.00055 u (*atomic mass unit*).

electron shell Imaginary spherical surface representing all possible paths of electrons with same average distance from nucleus and with approximately same energy.

electron-capture decay Radioactive decay that takes place as an orbital electron is captured by a proton in nucleus. Mass of element remains constant, but atomic number decreases by 1.

element Unique combinations of *protons*, *neutrons*, and *electrons* that cannot be broken down by ordinary chemical methods. Fundamental properties of an element are determined by number of protons, each element assigned a number corresponding to its number of protons. Combinations containing from 1 through 103 protons are now known.

end moraine Ridge or belt of *till* marking farthest advance of a glacier. Sometimes called *terminal moraine*.

energy Capacity for producing motion. Energy holds matter together and can become *mass* or be derived from mass. Takes such forms as kinetic, potential, heat, chemical, electrical, and atomic energy; one form of energy can be changed to another.

energy level Distance from atomic nucleus at which electrons orbit. May be thought of as shell surrounding nucleus.

epicenter Area on surface directly above *focus* of earthquake.

epidote Silicate of aluminum, calcium, and iron characteristic of low-grade metamorphism and associated with *chlorite* and *albite* in *greenschist facies*. Built around independent tetrahedra.

epidote-amphibolite facies Assemblage of minerals formed between 250 and 450°C during *regional metamorphism*.

erg Unit of energy expressing capacity for doing work. Equal to energy expended when force of 1 dyne acts through distance of 1 cm.

erosion Movement of material from one place to another on earth's surface. Agents of movement include gravity, water, ice, and wind.

erosional flood plain Flood plain created by lateral erosion and gradual retreat of valley walls.

erratic In terminology of glaciation a stone or boulder carried by ice to a place where it rests on or near bedrock of different composition.

escape velocity Minimum velocity an object must have to escape from gravitational field. For moon this is about 2.38 km/s and for earth about 11.2 km/s.

esker Winding ridge of stratified glacial *drift*, steep-sided, 3 to 15 m high, and from a fraction of a kilometer to over 160 km long.

eugeosyncline Part of *geosyncline* in which volcanism is associated with clastic sedimentation, generally located away from *craton*.

eustatic change Change in sea level produced entirely by increase or decrease in amount of water in oceans; hence of world-wide proportions.

eutrophic In lake aging pertains to an old-age lake and indicates high supply of nutrients supporting high biologic productivity.

evaporation Process by which liquid becomes vapor at temperature below boiling point.

evaporite Rock composed of minerals precipitated from solutions concentrated by evaporation of solvents. Examples: *rock salt*, *gypsum*, *anhydrite*.

exfoliation Process by which rock plates are stripped from larger rock mass by physical forces.

exfoliation dome Large, rounded domal feature produced in homogeneous coarse-grained igneous rocks and sometimes in conglomerates by process of exfoliation.

external magnetic field Component of earth's field originating from activity above earth's surface. Small when compared with dipole and nondipole components of field, which originate beneath surface.

extrusive rock Rock solidified from mass of magma poured or blown out upon earth's surface.

facies Assemblage of mineral, rock, or fossil features reflecting environment in which rock was formed. See *sedimentary facies*, *metamorphic facies*.

fault Surface of rock rupture along which has been differential movement.

fault-block mountain Mountain bounded by one or more faults.

feldspars Silicate minerals composed of silicon-oxygen and aluminum-oxygen tetrahedra linked together in three-dimensional networks with positive ions fitted into interstices of negatively charged framework of tetrahedra. Classed as aluminosilicates. When positive ion is K^+, mineral is *orthoclase*; when Na^+, mineral is *albite*; when Ca^{2+}, mineral is *anorthite*.

felsite General term for light-colored, fine-grained igneous rocks.

ferromagnesian silicate Silicate in which positive ions are dominated by iron, magnesium, or both.

fibrous fracture Mineral habit of breaking into splinters or fibers.

fiery cloud Avalanche of incandescent *pyroclastic debris* mixed with steam and other gases, heavier than air, and projected down a volcano's side. Also called *nuée ardente*.

fiord Glacially deepened valley now flooded by the sea to form long, narrow, steep-walled inlet.

firn Granular ice formed by recrystallization of snow. Intermediate between snow and glacier ice. Sometimes called *névé*.

fissility Property of splitting along closely spaced planes more or less parallel to bedding. Presence distinguishes shale from mudstone.

fission Process by which atomic nucleus breaks down to form nuclei of lighter atoms.

fissure eruption Extrusion of lava from fissure in earth's crust.

flashy stream Stream with high flood peak of short duration, which may be caused by urbanization.

flint Granular cryptocrystalline silica, usually dull and dark. Often occurs as lumps or nodules in calcareous rocks, such as Cretaceous chalk beds of southern England.

flood basalt Basalt poured out from fissures in floods that tend to form great plateaus. Sometimes called *plateau basalt*.

flood frequency Time within which a flood of a given size can be expected to occur.

flood plain Area bordering a stream, over which water spreads in time of flood.

flood plain of aggradation Flood plain formed by building up of valley floor by sedimentation.

flotation Process that begins concentration of ore minerals from *gangue*.

fluid Material that offers little or no resistance to forces tending to change its shape.

focus Source of given set of earthquake waves.

fold Bend, flexure, or wrinkle in rock produced when rock was in a plastic state.

fold mountains Mountains consisting primarily of elevated, folded sedimentary rocks.

foliation Layering in some rocks caused by parallel alignment of minerals; textural feature of some metamorphic rocks. Produces rock cleavage.

footwall One of blocks of rock involved in fault movement. One that would be under feet of person standing in tunnel along or across fault; opposite *hanging wall*.

fore dune *Dune* immediately behind shoreline of ocean or large lake.

foreset beds Inclined layers of sediment deposited on advancing edge of growing *delta* or along lee slope of advancing sand *dune*.

foreshock Relatively small earthquake that precedes larger earthquake by a few days or weeks and originates at or near *focus* of larger earthquake.

fossil Evidence of past life, such as dinosaur bones, ancient clam shell, footprint of long-extinct animal, or impression of leaf in rock.

fossil fuels Organic remains used to produce heat or power by combustion. Include *coal, petroleum,* and *natural gas.*

fractional distillation Recovery—one or more at a time—of fractions of complex liquid, each of which has different density.

fractionation Process whereby crystals that formed early from magma have time to settle appreciably before temperature drops much further. They are thus effectively removed from environment in which they formed.

fracture As mineral characteristic, way in which mineral breaks when it does not have cleavage. May be *conchoidal* (shell-shaped), *fibrous, hackly,* or *uneven.*

fracture cleavage System of joints spaced fraction of centimeter apart.

freeboard of continents Relative elevation of continents with respect to sea level.

fringing reef Reef attached directly to landmass.

frost action Process of mechanical weathering caused by repeated cycles of freezing and thawing. Expansion of water during freezing cycle provides energy for process.

fumarole Vent for volcanic steam and gases.

fundamental particles *Protons, neutrons,* and *electrons,* which combine to form atoms. Each particle is defined in terms of its *mass* and its *electric charge.*

fusion Process by which nuclei of lighter atoms join to form nuclei of heavier atoms.

gabbro Coarse-grained igneous rock with composition of basalt.

galaxy Family of stars grouped in space. The earth belongs to Milky Way galaxy, which contains about 100 billion stars.

galena A mineral; lead sulfide, PbS. Principal ore of lead.

gangue Commercially valueless material remaining after ore-mineral extraction from rock.

garnet Family of silicates of iron, magnesium, aluminum, calcium, manganese, and chromium, which are built around independent tetrahedra and appear commonly as distinctive 12-sided, fully developed crystals. Characteristic of *metamorphic rocks;* generally cannot be distinguished from one another without chemical analysis.

gas (1) State of matter that has neither independent shape nor volume, can be compressed readily, and tends to expand indefinitely. (2) In geology "gas" is sometimes used to refer to *natural gas,* gaseous hydrocarbons that occur in rocks, dominated by methane. Compare "oil," referring to *petroleum.*

geode Roughly spherical, hollow or partially hollow accumulation of mineral matter from a few centimeters to nearly 0.5 m in diameter. Outer layer of chalcedony lined with crystals that project toward hollow center. Crystals, often perfectly formed, usually quartz although calcite and dolomite also found and—more rarely—other minerals. Geodes most commonly found in limestone and more rarely in shale.

geographic poles Points on earth's surface marked by ends of axis of rotation.

geologic column Chronologic arrangement of rock units in columnar form, with oldest units at bottom and youngest at top.

geologic-time scale Chronologic sequence of units of earth time.

geology Organized body of knowledge about the earth, including *physical geology* and *historical geology,* among others.

geomagnetic poles *Dipole* best approximating earth's observed field is one inclined 11.5° from axis of rotation. Points at which ends of this imaginary magnetic axis intersect earth's surface are known as *geomagnetic poles.* They should not be confused with *magnetic,* or *dip, poles* or *virtual geomagnetic poles.*

geophysical prospecting Mapping rock structures by methods of experimental physics. Includes measuring magnetic fields, force of gravity, electrical properties, seismic-wave paths and velocities, radioactivity, and *heat flow.*

geophysics Physics of the earth.

geosyncline Literally, "earth syncline." Term now refers, however, to a *basin* in which thousands of meters of sediments have accumulated, with accompanying progressive sinking of basin floor explained only in part by load of sediments. Common usage includes both accumulated sediments themselves and geometrical form of basin in which they are deposited. All folded mountain ranges were built from geosynclines, but not all geosynclines have become mountain ranges.

geothermal field Area where wells drilled to obtain elements contained in solution in hot brines and to tap heat energy.

geothermal gradient See *thermal gradient.*

geyser Special type of thermal spring that intermittently ejects its water with considerable force.

glacier A mass of ice, formed by recrystallization of snow, that flows forward or has flowed at some past time under influence of gravity. By convention we exclude icebergs even though they are large fragments broken from seaward end of glaciers.

glacier ice Unique form of ice developed by compression and recrystallization of snow and consisting of interlocking crystals.

glass Form of matter that exhibits properties of a solid but has atomic arrangements, or lack of order, of a liquid.

***Glossopteris* flora** A late-Paleozoic assemblage of fossil plants named for seed fern *Glossopteris,* a plant in the flora. Widespread in South America, South Africa, Australia, India, and Antarctica.

gneiss Metamorphic rock with *gneissic cleavage.* Commonly formed by metamorphism of granite.

gneissic cleavage Rock cleavage where surface may be a few hundredths of a millimeter to a centimeter or more apart.

goethite Hydrous iron oxide, FeO(OH).

Gondwanaland Hypothetical continent thought to have broken up in Mesozoic. Resulting fragments are postulated to form present-day South America, Africa, Australia, India, and Antarctica.

graben Elongated, trenchlike, structural form bounded by parallel normal faults created when block that forms trench floor moves downward relative to blocks that form sides.

grade Term used to designate extent to which *metamorphism* has advanced. Found in such combinations as "high-grade" or "low-grade metamorphism." Compare with *rank*.

graded bedding Type of bedding shown by sedimentary deposit when particles become progressively finer from bottom to top.

gradient Slope of stream bed or land surface; expressed in percent, as feet per mile, or meters per kilometer, or in degrees.

granite Coarse-grained *igneous rock* dominated by light-colored minerals, consisting of about 50 percent *orthoclase*, 25 percent *quartz*, and balance of plagioclase *feldspars* and *ferromagnesian silicates*. Granites and *granodiorites* comprise 95 percent of all intrusive rocks.

granitization Special type of *metasomatism* by which solutions of magmatic origin move through solid rocks, change ions with them, and convert them into rocks that achieve granitic character without having passed through magmatic stage.

granodiorite Coarse-grained igneous rock intermediate in composition between *granite* and *diorite*.

granular texture Composed of mineral grains large enough to be seen by unaided eye; also called *phaneritic texture*.

granulite facies *Gneissic* rocks produced by deep-seated high-grade *regional metamorphism*.

graphic structure Intimate intergrowth of *potassic feldspar* and *quartz* with long axes of quartz crystals lining up parallel to feldspar axis. Quartz part is dark and feldspar is light in color; so pattern suggests Egyptian hieroglyphs. Commonly found in *pegmatites*.

graphite "Black lead." A mineral composed entirely of carbon. Very soft because of crystalline structure; diamond, in contrast, has same composition but is hardest substance known.

gravity anomaly Difference between observed and computed values of gravity.

gravity fault Fault in which *hanging wall* appears to have moved downward relative to *footwall*. Also called *normal fault*.

gravity meter An instrument for measuring force of gravity. Also called *gravimeter*.

gravity prospecting Mapping force of gravity at different places to determine differences in *specific gravity* of rock masses and, through this, distribution of masses of different specific gravity. Done with *gravity meter* (gravimeter).

graywacke A variety of *sandstone* generally characterized by hardness, dark color, and angular grains of *quartz*, *feldspar*, and small rock fragments set in matrix of clay-sized particles. Also called *lithic sandstone*.

greenschist Schist characterized by green color. Product of *regional metamorphism* of *simatic rocks*. (Green color is imparted by mineral *chlorite*.)

greenschist facies Assemblage of minerals formed between 150 and 250°C during *regional metamorphism*.

groundmass Finely crystalline or glassy portion of porphyry.

ground moraine *Till* deposited from a glacier as veneer over landscape and forming gently rolling surface.

groundwater Underground water within *zone of saturation*.

groundwater table Upper surface of *zone of saturation* for underground water. An irregular surface with slope or shape determined by quantity of groundwater and permeability of earth materials. In general, highest beneath hills and lowest beneath valleys. Also referred to as *water table*.

guyot Flat-topped *seamount* rising from ocean floor like a volcano but planed off on top and covered by appreciable water depth. Synonymous with *tablemount*.

gypsum Hydrous calcium sulfate, $CaSO_4 \cdot 2H_2O$. A soft, common mineral in sedimentary rocks, where it sometimes occurs in thick beds interstratified with limestones and shales. Sometimes occurs as layer under bed of *rock salt* since it is one of first minerals to crystallize on evaporation of seawater. Alabaster is a fine-grained massive variety of gypsum.

H Symbol for mineral *hardness*.

hackly fracture Mineral's habit of breaking to produce jagged, irregular surfaces with sharp edges.

half-life Time needed for one-half of nuclei in sample of radioactive element to decay.

halide Compound made from a halogen, such as chlorine, iodine, bromine, or fluorine.

halite A mineral; *rock salt* or common salt, NaCl. Occurs widely disseminated or in extensive beds and irregular masses precipitated from seawater and interstratified with rocks of other types as true sedimentary rock.

hanging valley A valley that has greater elevation than the valley to which it is tributary, at point of junction. Often (but not always) created by deepening of main valley by a glacier. Hanging valley itself may or may not be glaciated.

hanging wall One of blocks involved in fault movement. One that would be hanging overhead for person standing in tunnel along or across fault; opposite *footwall*.

hardness Mineral's resistance to scratching on a smooth surface. Mohs scale of relative hardness consists of 10 minerals, each scratching all those below it in scale and being scratched by all those above it: (1) talc, (2) gypsum, (3) calcite, (4) fluorite, (5) apatite, (6) orthoclase, (7) quartz, (8) topaz, (9) corundum, (10) diamond.

head Difference in elevation between intake and discharge points for a liquid. In geology most commonly of interest in connection with movement of *underground water*.

heat energy Special manifestation of *kinetic energy* in atoms. Temperature of a substance depends on average kinetic energy of component particles. When heat is added to a substance, average kinetic energy increases.

heat exchanger Unit in atomic-power generation that uses water (heated under pressure by atomic reactor) to form steam from water in another system and to drive a turbine for electricity generation.

heat flow Product of thermal gradient and thermal conductivity of earth materials. Average over whole earth, $1.2 \pm 0.15 \ \mu cal/cm^2$-s.

height See Appendix E.

hematite Iron oxide, Fe_2O_3. Principal ore mineral for about 90 percent of commercial iron produced in United States. Characteristic red color when powdered.

hinge fault Fault in which displacement perceptibly dies out along strike and ends at definite point.

historical geology Branch of *geology* that deals with history of the earth, including record of life on earth as well as physical changes in earth itself.

horn Spire of bedrock left where *cirques* have eaten into a mountain from more than two sides around a central area. Example: Matterhorn of the Swiss Alps.

hornblende A rock-forming *ferromagnesian silicate* mineral with double chains of silicon-oxygen tetrahedra. An *amphibole*.

hornfels Dense, granular metamorphic rock. Since this term is commonly applied to metamorphic equivalent of any fine-grained rock, composition is variable.

hornfels facies Assemblage of minerals formed at temperatures greater than 700°C during *contact metamorphism*.

horst Elongated block bounded by parallel *normal faults* in such a way that it stands above blocks on both sides.

hot spot Localized melting region in mantle near base of lithosphere, a few hundred kilometers in diameter and persistent over tens of millions of years. Existence of heat is assumed from volcanic activity at surface.

hot spring *Thermal spring* that brings hot water to surface. Water temperature usually 6.5°C or more above mean air temperature.

hydration Process by which water combines chemically with other molecules.

hydraulic gradient *Head* of underground water divided by distance of travel between two points: If head is 10 m for two points 100 m apart, hydraulic gradient is 0.1, or 10 percent. When head and distance of flow are same, hydraulic gradient is 100 percent.

hydraulic mining Use of strong water jet to move deposits of sand and gravel from original site to separating equipment, where sought-for mineral is extracted.

hydrocarbon Compound of hydrogen and carbon that burns in air to form water and oxides of carbon. There are many hydrocarbons. The simplest, methane, is chief component of natural gas. Petroleum is a complex mixture of hydrocarbons.

hydroelectric power Conversion of energy to electricity by free fall of water. This method supplies about 4 percent of world's electrical energy.

hydrograph Graph of variation of stream flow over time.

hydrologic cycle General pattern of water movement by evaporation from sea to atmosphere, by precipitation onto land, and by return to sea under influence of gravity.

hydrothermal solution Hot, watery solution that usually emanates from *magma* in late stages of cooling. Frequently contains, and deposits in economically workable concentrations, minor elements that, because of incommensurate *ionic radii* or electronic charges, have not been able to fit into atomic structures of common minerals of igneous rocks.

hysteresis Retardation of recovery from *elastic deformation* after *stress* is removed.

icecap Localized *ice sheet*.

ice sheet Broad, moundlike mass of glacier ice of considerable extent that has tendency to spread radially under own weight. Localized ice sheets are sometimes called *icecaps*.

igneous rock Aggregate of interlocking silicate minerals formed by cooling and solidification of magma.

illite Clay mineral family of hydrous aluminous silicates. Structure is similar to that of *montmorillonite*, but aluminum replaces 10 to 15 percent of silicon, which destroys montmorillonite's property of expanding with addition of water because weak bonds replaced by strong potassium-ion links. Structurally illite intermediate between montmorillonite and *muscovite*. Montmorillonite converts to illite in sediments; illite, to muscovite under conditions of low-grade metamorphism. Illite is commonest clay mineral in clayey rocks and recent marine sediments and is present in many soils.

ilmenite Iron titanium oxide. Accounts for much of unique abundance of titanium on moon.

inclined bedding Bedding laid down at angle to horizontal. Also referred to as *cross bedding* or *current bedding*.

index minerals *Chlorite*, low-grade metamorphism; *almandite*, middle-grade metamorphism; *sillimanite*, high-grade metamorphism.

induced magnetism In terminology of rock magnetism one of components of rock's *natural remanent magnetism*. It is parallel to earth's present field and results from it.

inertia member Central element of a seismograph, consisting of weight suspended by wire or spring so that it acts like pendulum free to move in only one plane.

infiltration Soaking into ground of water on surface.

inosilicate Mineral with crystal structure containing silicon-oxygen tetrahedra in single or double chains.

intensity Measure of effects of earthquake waves on human beings, structures, and earth's surface at particular place. Contrast with *magnitude*, which is measure of total energy released by an earthquake.

intermediate belt Subdivision of *zone of aeration*. *Belt that lies between belt of soil moisture and capillary fringe.*

intermediate lava Lava composed of 60 to 65 percent silica.

intrusive rock Rock solidified from mass of magma that invaded earth's crust but did not reach surface.

ion Electrically unbalanced form of an atom or group of atoms, produced by gain or loss of electrons.

ionic bond Bond in which ions are held together by electrical attraction of opposite charges.

ionic radius Average distance from center to outermost electron of an ion. Commonly expressed in *angstroms*.

island-arc deeps Arcuate trenches bordering some continents; some reach depths of 9,000 m or more below sea surface. Also called *deep-sea trenches* or *trenches*.

isoclinal folding Beds on both *limbs* nearly parallel, whether fold upright, *overturned*, or *recumbent*.

isograd Line connecting those rocks comprising the same metamorphic facies; line connecting similar temperature-pressure values; line marking the boundary between two metamorphic facies.

isoseismic line Line connecting all points on surface of earth where intensity of shaking from earthquake waves is same.

isostasy Ideal condition of balance that would be attained by earth materials of differing densities if gravity were the only force governing heights relative to each other.

isotope Alternative form of an element produced by variations in number of neutrons in nucleus.

jasper Granular, cryptocrystalline silica usually colored red by *hematite* inclusions.

joint Break in rock mass with no relative movement of rock on opposite sides of break.

joint system Combination of intersecting joint sets, often at approximately right angles.

juvenile water Water brought to surface or added to underground supplies from magma.

kame Steep-sided hill of stratified glacial *drift*. Distinguished from *drumlin* by lack of unique shape and by stratification.

kame terrace Stratified glacial drift deposited between wasting glacier and adjacent valley wall. When ice melts, this material stands as a terrace along valley wall.

kaolinite Clay mineral, hydrous aluminous silicate, $Al_4Si_4O_{10}(OH)_8$. Structure consists of one sheet of silicon-oxygen tetrahedra each sharing three oxygens to give ratio of $(Si_4O_{10})^{4-}$ linked with one sheet of aluminum and hydroxyl. Composition of pure kaolinite does not vary as for other clay minerals, *montmorillonite* and *illite*, in which ready addition or substitution of ions takes place.

karst topography Irregular topography characterized by *sink holes*, streamless valleys, and streams that disappear underground —all developed by action of surface and underground water in soluble rock such as limestone.

Kerguelen-Gaussberg Ridge Belt of mountains under Indian Ocean between India and Antarctica.

kettle Depression in ground surface formed by melting of a block of ice buried or partially buried by glacial *drift*, either *outwash* or *till*.

kinetic energy Energy of movement. Amount possessed by an object or particle depends on mass and speed.

kyanite A silicate mineral characteristic of temperatures of middle-grade metamorphism. Al_2SiO_5 in bladed blue crystals is softer than a knife along the crystal. Its crystalline structure is based on independent tetrahedra. Compare with *andalusite*, which has same composition and forms under similar conditions but has different crystal habit. Contrast with *sillimanite*, which has same composition but different crystal habit and forms at highest metamorphic temperatures.

L Symbol for earthquake *surface waves*.

laccolith *Concordant pluton* that has domed up strata into which it intruded.

lag time On stream *hydrograph* time interval between center of mass of precipitation and center of mass of resulting flood.

laminar flow Mechanism by which fluid (such as water) moves slowly along a smooth channel or through a tube with smooth walls with fluid particles following straight-line paths parallel to channel or walls. Contrast with *turbulent flow*.

landslide General term for relatively rapid mass movement, such as *slump*, *rock slide*, *debris slide*, *mudflow*, and *earthflow*.

lapilli *Pyroclastic debris* in pieces about walnut size.

large waves Earthquake *surface waves*.

latent heat of fusion Number of calories per unit volume that must be added to a material at melting point to complete process of melting. These calories do not raise temperature.

lateral moraine Ridge of *till* along edge of valley glacier. Composed largely of material fallen to glacier from valley walls.

laterite Tropical soil rich in hydroxides of aluminum and iron and formed under conditions of good drainage.

lava *Magma* poured out on surface of earth or rock solidified from such magma.

leading edge That margin that occurs on the side of a continent which is riding on a plate in motion toward another plate; commonly this margin is very active tectonically.

left-lateral fault *Strike-slip fault* where ground opposite you appears to have moved left when you face it.

levee (natural) Bank of sand and silt built by river during floods, where suspended load deposited in greatest quantity close to river. Process of developing natural levees tends to raise river banks above level of surrounding *flood plains*. Break in natural levee sometimes called *crevasse*.

lignite Low-grade coal, with about 70 percent carbon and 20 percent oxygen. Intermediate between peat and bituminous coal.

limb One of two parts of *anticline* or *syncline*, on either side of axis.

limestone Sedimentary rock composed largely of mineral *calcite*, $CaCO_3$, formed by either organic or inorganic processes. Most limestones have clastic texture, but nonclastic, particularly crystalline, textures are common. Carbonate rocks, limestone and *dolomite*, constitute estimated 12 to 22 percent of sedimentary rocks exposed above sea level.

limonite Iron oxide with no fixed composition or atomic structure; a *mineraloid*. Always of secondary origin, not a true mineral. Is encountered as ordinary rust or coloring material of yellow clays and soils.

liquefaction Process of changing soil and unconsolidated sediments into water mixture immediately following earthquake; often results in foundation failure, with sliding of ground under building structures.

liquid State of matter that flows readily so that the mass assumes form of container but retains independent volume.

lithic sandstone See *graywacke*.

lithification Process by which unconsolidated rock-forming materials are converted into consolidated or coherent state.

lithosphere Rigid outer layer of earth; includes *crust* and upper part of *mantle*. Relatively strong layer in contrast to underlying *asthenosphere*.

loess Unconsolidated, unstratified aggregation of small, angular mineral fragments, usually buff in color. Generally believed to be wind-deposited; characteristically able to stand on very steep to vertical slopes.

longitudinal dune Long ridge of sand oriented in general direction of wind movement. A small one is less than 3 m high and 60 m long. Very large ones are called *seif dunes*.

longitudinal wave *Push-pull wave*.

lopolith Tabular *concordant pluton* shaped like spoon bowl, with both roof and floor sagging downward.

magma Naturally occurring silicate melt, which may contain suspended silicate crystals, dissolved gases, or both. These conditions may be met in general by a mixture containing as much as 65 percent crystals but no more than 11 percent dissolved gases.

magmatic segregation The process by which heavy, early crystallized minerals settle out from a *magma* and become concentrated.

magnetic declination Angle of divergence between geographic meridian and magnetic meridian. Measured in degrees east and west of geographic north.

magnetic inclination Angle that magnetic needle makes with surface of earth. Also called *dip* of magnetic needle.

magnetic pole North magnetic pole is point on earth's surface where north-seeking end of a magnetic needle free to move in space points directly down. At south magnetic pole the same needle points directly up. These poles are also known as *dip poles*.

magnetic reversal Shift of 180° in earth's magnetic field such that north-seeking needle of magnetic compass would point south rather than to north magnetic pole.

magnetite A mineral; iron oxide, Fe_3O_4. Black; strongly magnetic. Important ore of iron.

magnetosphere Region 1,000 to 64,000 km above earth, where magnetic field traps electrically charged particles from sun and space. First believed to consist of two bands, Van Allen belts.

magnetostratigraphy Use of magnetized rocks to determine history of events in record of changes in earth's magnetic field in past geologic ages.

magnitude Measure of total energy released by an earthquake. Contrast with *intensity*, which is measure of effects of earthquake waves at particular place.

manganese nodule A *nodule* found on the deep ocean floors and composed largely of iron and manganese, with smaller amounts of such metals as cobalt, titanium, copper, and nickel.

mantle Intermediate zone of earth. Surrounded by *crust*, it rests on *core* at depth of about 2,900 km.

marble Metamorphic rock of granular texture, with no rock cleavage, and composed of *calcite*, *dolomite*, or both.

maria Dark-toned "seas" of moon. Mark moon's topographically low areas.

marl Calcareous clay or intimate mixture of clay and particles of *calcite* or *dolomite*, usually shell fragments.

marsh gas *Methane*, CH_4; simplest paraffin hydrocarbon. Predominant component of *natural gas*.

mascons Concentrations of mass located beneath surfaces of lunar *maria*.

mass A number that measures quantity of matter. It is obtained on earth's surface by dividing weight of a body by acceleration due to gravity.

mass movement Surface movement of earth materials induced by gravity.

mass number Number of protons and neutrons in atomic nucleus.

mass unit One-twelfth mass of carbon atom. Approximately mass of hydrogen atom.

matter Anything that occupies space. Usually defined by describing its states and properties: solid, liquid, or gas; possesses mass, inertia, color, density, melting point, hardness, crystal form, mechanical strength, or chemical properties. Composed of *atoms*.

meander (1) Turn or sharp bend in stream's course. (2) To turn, or bend, sharply. Applied to stream courses in geological usage.

mechanical weathering Process by which rock is broken down into smaller and smaller fragments as result of energy developed by physical forces. Also known as *disintegration*.

medial moraine Ridge of *till* formed by junction of two *lateral moraines* when two valley glaciers join to form single ice stream.

mélange Heterogeneous mixture of rock materials. Mappable body of deformed rocks that may be several kilometers in length and consists of highly sheared clayey matrix, thoroughly mixed with angular native and exotic blocks of diverse origin and geologic ages.

Mercalli intensity scale Scale to evaluate intensity of earthquake shaking on basis of effects at given place.

mesotrophic In lake aging, stage between *oligotrophic* and *eutrophic*.

metal Substance fusible and opaque, good conductor of electricity, and with characteristic luster. Examples: gold, silver, aluminum. Of the elements 77 are metals.

metallic bonding Special kind of bonding in atoms of metallic elements whereby outermost electrons are not shared or exchanged but are free to move around and connect to any atoms in solid. Relative freedom of movement of electrons accounts for high level of electrical conductivity in metals.

metalloid Element of some metallic and some nonmetallic characteristics. There are nine metalloids. See also Appendix C.

metamorphic facies Assemblage of minerals that reached equilibrium during metamorphism under specific range of temperature and pressure.

metamorphic rock "Changed-form rock." any rock changed in texture or composition by heat, pressure, or chemically active fluids after original formation.

metamorphic zone Area subjected to *metamorphism* and characterized by certain metamorphic facies formed during process.

metamorphism A process whereby rocks undergo physical or chemical changes or both to achieve equilibrium with conditions other than those under which they were originally formed (weathering arbitrarily excluded from meaning). Agents of metamorphism are heat, pressure, and chemically active fluids.

metasomatism Process whereby rocks are altered when *volatiles* exchange ions with them.

meteor Transient celestial body that enters earth's atmosphere with great speed, becoming incandescent from heat generated by air resistance.

meteoric water Groundwater derived primarily from precipitation.

meteorite Stony or metallic body fallen to earth from outer space.

methane Simplest paraffin hydrocarbon, CH_4. Principal constituent of *natural gas*. Sometimes called *marsh gas*.

micas Group of silicate minerals characterized by perfect sheet or scale cleavage resulting from atomic pattern, in which silicon-oxygen tetrahedra linked in sheets. *Biotite* is ferromagnesian black mica. *Muscovite* is potassic white mica.

microseism "Small shaking." Specifically limited in technical usage to earth waves generated by sources other than earthquakes and, most frequently, to waves with periods of from 1 to about 9 s, from sources associated with atmospheric storms.

mid-Atlantic Ridge See *Atlantic Ridge*.

midocean ridge Continuous, seismically active, median mountain range extending through North and South Atlantic, Indian, and South Pacific Oceans.

migmatite Mixed rock produced by intimate interfingering of *magma* and invaded rock.

mineral Naturally occurring solid element or compound, exclusive of biologically formed carbon components. Has definite composition or range of composition and orderly internal atomic arrangement (crystalline structure), which gives unique physical and chemical properties, including tendency to assume certain geometrical forms known as *crystals*.

mineral deposit Occurrence of one or more minerals in such concentration and form as to make possible removal and processing for use at profit.

mineraloid Substance that does not yield definite chemical formula and shows no sign of crystallinity. Examples: *bauxite*, *limonite*, and *opal*.

miogeosyncline That part of a geosyncline in which volcanism is absent, generally located near *craton*.

modulus of elasticity Slope of graph line relating *stress* to *strain* in *elastic deformation*.

Mohorovičić discontinuity (Moho) Base of crust marked by abrupt increases in velocities of earth waves.

molecule Smallest unit of compound that displays properties of that compound.

Monel metal Steel containing 68 percent nickel.

monocline Double flexure connecting strata at one level with same strata at another level.

montmorillonite Clay mineral family, hydrous aluminous silicate with structural sandwich of one ionic sheet of aluminum and hydroxyl between two $(Si_4O_{10})^{4-}$ sheets. Sandwiches piled on each other with water between and with nothing but weak bonds to hold them together. As result, additional water can enter lattice readily, causing mineral to swell appreciably and further weakening attraction between structural sandwiches. Consequently a lump of montmorillonite in a bucket of water slumps rapidly into a loose, incoherent mass. Compare with other clay minerals, *kaolinite* and *illite*.

moon A natural satellite.

moraine General term applied to certain landforms composed of *till*.

mountain Any part of landmass projecting conspicuously above its surroundings.

mountain chain Series or group of connected mountains having well-defined trend or direction.

mountain range Series of more or less parallel ridges, all of which formed within a single *geosyncline* or on its borders.

mountain structure Structure produced by deformation of rocks.

mudcracks Cracks caused by shrinkage of drying deposit of silt or clay under surface conditions.

mudflow Flow of well-mixed mass of rock, earth, and water that behaves like a fluid and moves down slopes with consistency similar to that of newly mixed concrete.

mudstone Fine-grained, detrital sedimentary rock made up of *silt-* and *clay-sized* particles. Distinguished from *shale* by lack of fissility.

muscovite "White mica." Nonferromagnesian rock-forming silicate mineral with tetrahedra arranged in sheets. Sometimes called *potassic mica*.

mylonite Fine-grained rock formed by grinding during intense folding or faulting associated with cataclastic metamorphism.

native state State in which an element occurs uncombined in nature. Usually applied to metals, as in "native copper," "native gold," etc.

natural gas Gaseous hydrocarbons that occur in rocks. Dominated by *methane*.

natural remanent magnetism Magnetism of rock. May or may not coincide with present magnetic field of earth. Abreviation, NRM.

natural resources Energy and materials made available by geological processes.

neck cutoff Breakthrough of a river across narrow neck separating two meanders, where downstream migration of one has been slowed and next meander upstream has overtaken it. Compare with *chute cutoff*.

negative charge Condition resulting from surplus of electrons.

nesosilicate Mineral with crystal structure containing silicon-oxygen tetrahedra arranged as isolated units.

neutron *Proton* and *electron* combined and behaving like fundamental particle of matter. Electrically neutral with mass of 1.00896 u. If isolated, decays to form proton and electron.

névé Granular ice formed by recrystallization of snow. Intermediate between snow and glacier ice. Sometimes called *firn*.

Nichrome Steel alloy with 35 to 85 percent nickel.

nickel steel Steel containing 2.5 to 3.5 percent nickel.

nodule Irregular, knobby-surfaced mineral body that differs in composition from rock in which formed. Silica in form of *chert* or *flint* is common component of nodules. They are commonly found in limestone and dolomite.

nonclastic texture Applied to sedimentary rocks in which rock-forming grains are interlocked. Most sedimentary rocks with nonclastic texture are crystalline.

nonconformity *Unconformity* separating younger rocks from distinctly different older highly metamorphosed or igneous rocks.

nondipole magnetic field Portion of earth's magnetic field remaining after dipole field and external field are removed.

nonferromagnesians Silicate minerals that do not contain iron or magnesium.

nonmetal Element that does not exhibit metallic luster, conductivity, or other features of metal. Of the elements 17 are nonmetals.

normal fault Fault in which *hanging wall* appears to have moved downward relative to *footwall*; opposite of *thrust fault*. Also called *gravity fault*.

North Atlantic deep water Seawater in Arctic that sinks in North Atlantic and drifts southward as far as 60°S.

northeast trades Winds blowing from the northeast toward the equator in the northern hemisphere.

NRM See *natural remanent magnetism*.

nucleus *Protons* and *neutrons* constituting central part of an atom.

nuée ardente "Hot cloud." French term applied to highly heated mass of gas-charged lava ejected from vent or pocket at volcano summit more or less horizontally onto an outer slope, down which it moves swiftly, however slight the incline, because of its extreme mobility.

obduction Process whereby part of the subducted plate and/or associated igneous rocks and deep-sea sediments are broken off and pushed up onto the overriding plate.

oblique slip fault Fault with components of relative displacement along both *strike* and *dip*.

obsidian Glassy equivalent of granite.

oil In geology refers to *petroleum*.

oil shale Shale containing such proportion of hydrocarbons as to be capable of yielding *petroleum* on slow distillation.

oligotrophic In lake aging pertains to a youthful lake and indicates water low in accumulated nutrients and high in dissolved oxygen.

olivine Rock-forming ferromagnesian silicate mineral that crystallizes early from magma and weathers readily at earth's surface. Crystal structure based on isolated $(SiO_4)^{4-}$ ions and positive ions of iron, magnesium, or both. General formula: $(Mg, Fe)_2SiO_4$.

oölites Spheroidal grains of sand size, usually composed of calcium carbonate, $CaCO_3$, and thought to have originated by inorganic precipitation. Some limestones largely made up of oölites.

ooze Deep-sea deposit consisting of 30 percent or more by volume of hard parts of very small, sometimes microscopic, organisms. If particular organism dominant, its name is used as modifier, as in *globigerina* ooze, or *radiolarian* ooze.

opal Amorphous silica, with varying amounts of water; a mineral gel.

open-pit mining Surface mining represented by sand and gravel pits, stone quarries, and copper mines of some western states.

ophitic Rock texture in which lath-shaped plagioclase crystals are enclosed wholly or in part in later-formed augite, as commonly occurs in diabase.

order of crystallization Chronological sequence in which crystallization of various minerals of an assemblage takes place.

ore deposit Metallic minerals in concentrations that can be worked at profit.

orogeny Process by which mountain structures develop.

orthoclase Feldspar in which K^+ is diagnostic positive ion; $K(AlSi_3O_8)$.

orthoquartzite Sandstone composed completely—or almost completely—of *quartz* grains. *Quartzose sandstone* is synonym.

outwash Material carried from a glacier by meltwater and laid down in stratified deposits.

outwash plain Flat or gently sloping surface underlain by outwash.

overbank deposits Sediments (usually clay, silt, and fine sand) deposited on flood plain by river overflowing banks.

overturned fold Fold with at least one *limb* rotated through more than 90°.

oxbow Abandoned *meander* caused by a *neck cutoff*.

oxbow lake Abandoned *meander* isolated from main stream channel by deposition and filled with water.

oxide mineral Mineral formed by direct union of an element with oxygen. Examples: ice, corundum, hematite, magnetite, cassiterite.

P Symbol for earthquake *primary waves*.

pahoehoe lava *Lava* whose surface is smooth and billowy, frequently molded into forms resembling huge rope coils. Characteristic of basic lavas.

paired terraces *Terraces* that face each other across stream at same elevation.

paleomagnetism Study of earth's magnetic field as has existed during geologic time.

paleosol Soil formed in past environment; often buried.

Pangaea Hypothetical continent from which all others are postulated to have originated through process of fragmentation and drifting.

parabolic dune *Dune* with long, scoop-shaped form that, when perfectly developed, exhibits parabolic shape in plan, with horns pointing upwind. Contrast *barchan*, in which horns point downwind. Characteristically covered with sparse vegetation; often found in coastal belts.

paternoster lakes Chain of lakes resembling string of beads along glaciated valley where ice plucking and gouging have scooped out series of basins.

peat Partially reduced plant or wood material, containing approximately 60 percent carbon and 30 percent oxygen. An intermediate material in process of coal formation.

pebble size Volume greater than that of a sphere with diameter of 4 mm and less than that of a sphere of 64 mm.

pedalfer *Soil* characterized by accumulation of iron salts or iron and aluminum salts in *B* horizon. Varieties of pedalfers include red and yellow soils of southeastern United States and *podsols* of northeastern quarter of United States.

pedocal *Soil* characterized by accumulation of calcium carbonate in its profile. Characteristic of low rainfall. Varieties include black and chestnut soils of northern Plains states and red and gray desert soils of drier western states.

pedology Science that treats of *soils*—origin, character, and utilization.

pegmatite Small *pluton* of exceptionally coarse texture, with crystals up to 12 m in length, commonly formed at margin of *batholith*. Nearly 90 percent of all pegmatites are simple pegmatites of *quartz*, *orthoclase*, and unimportant percentages of *micas*; others are extremely rare ferromagnesian pegmatites and complex pegmatites. Complex pegmatites have as major components *sialic* minerals of simple pegmatites but also contain a variety of rare minerals.

pelagic deposit Material formed in deep ocean and deposited there. Example: ooze.

pendulum Inertia member so suspended that, after displacement, restoring force will return it to starting position. If displaced and then released, oscillates, completing one to-and-fro swing in time called *period*.

perched water table Top of *zone of saturation* that bottoms on impermeable horizon above level of general *water table* in area. Is generally near surface and frequently supplies a hillside spring.

peridotite Coarse-grained igneous rock dominated by dark-colored minerals, consisting of about 75 percent *ferromagnesian silicates* and balance *plagioclase feldspars*.

period For oscillating system, length of time required to complete one oscillation.

permeability For rock or earth material, ability to transmit fluids. Permeability equal to velocity of flow divided by hydraulic gradient.

petroleum An oily mixture of hydrocarbons extracted from subsurface earth structures. Thought to result from physical and chemical conversion of remains of animals and plants. A fuel in natural or refined state, yielding on distillation such products as gasoline, kerosene, naphtha.

phaneritic texture Individual grains large enough to be identified without the aid of a microscope; also called *granular texture*.

phase (1) Homogeneous, physically distinct portion of matter in physical-chemical system not homogeneous, as in three phases of ice, water, and aqueous vapor. (2) Group of seismic waves of one type.

phenocryst A crystal significantly larger than crystals of surrounding minerals.

phosphate rock Sedimentary rock containing calcium phosphate.

photosynthesis Process by which carbohydrates are compounded from carbon dioxide and water in presence of sunlight and chlorophyll.

phyllite Clayey metamorphic rock with rock cleavage intermediate between *slate* and *schist*. Commonly formed by the regional metamorphism of *shale* or *tuff*. Micas characteristically impart a pronounced sheen to rock cleavage surfaces. Has phyllitic cleavage.

phyllitic cleavage Rock cleavage in which flakes are produced barely visible to unaided eye. Coarser than *slaty* and finer than *schistose cleavage*.

phyllosilicate Mineral with crystal structure containing silicon-oxygen tetrahedra arranged as sheets.

physical geology Branch of *geology* that deals with nature and properties of material composing the earth, distribution of materials throughout globe, processes by which they are formed, altered, transported, and distorted, and nature and development of landscape.

piedmont glacier Glacier formed by coalescence of *valley glaciers* and spreading over plains at foot of mountains from which valley glaciers came.

placer A concentration of relatively heavy and resistant minerals in stream or beach deposits; two examples are some deposits of gold and of diamonds.

planet Heavenly body that changes position from night to night with respect to background of stars.

planetology Organized body of knowledge about planetary system.

plastic deformation Permanent change in shape or volume not involving failure by rupture and, once started, continuing without increase in deforming force.

plastic solid Solid that undergoes deformation continuously and indefinitely after stress applied to it passes a critical point.

plate Earth's *lithosphere*, varying in thickness from several tens of kilometers to as much as 100 km and including crust and part of upper *mantle* above asthenosphere.

plate tectonics Theory of worldwide dynamics involving movement and interactions of the many rigid plates of earth's *lithosphere*.

plateau basalt Basalt poured out from fissures in floods that tend to form great plateaus. Sometimes called *flood basalt*.

playa Flat-floored center of undrained desert basin.

playa lake Temporary lake formed in a *playa*.

pleochroic halo Minute, concentric-spherical zones of darkening or coloring that form around inclusions of radioactive minerals in *biotite*, *chlorite*, and a few other minerals. About 0.075 mm in diameter.

plume Pipelike convection cells thought to carry heat and *mantle* material from lower mantle up to *crust*, producing hot spots at surface.

plunge Acute angle that axis of folded rock mass makes with horizontal plane.

pluton A body of igneous rock formed beneath earth surface by consolidation from magma. Sometimes extended to include bodies formed beneath surface by metasomatic replacement of older rock.

plutonic igneous rock Rock formed by slow crystallization, which yields coarse texture. Once believed to be typical of crystallization at great depth, but not a necessary condition.

pluvial lake Lake formed during a *pluvial period*.

pluvial period Period of increased rainfall and decreased evaporation; prevailed in nonglaciated areas during time of ice advance elsewhere.

podsol Ashy-gray or gray-brown soil of *pedalfer* group. Highly bleached soil, low in iron and lime, formed under moist and cool conditions.

point bars Accumulations of sand and gravel deposited in slack waters on inside of bends of winding, or meandering, river.

Poisson's ratio Ratio of change of diameter per unit diameter to change of length per unit length in elastic stretching or compression of cylindrical specimen.

polar compound Compound, such as water, with a molecule that behaves like small bar magnet with positive charge on one end and negative charge on other.

polarity epoch Interval of time during which earth's magnetic field has been oriented dominantly in either normal or reverse direction. May be marked by shorter intervals of opposite sign, called *polarity events*.

polarity event See *polarity epoch*.

polar wandering, or migration Movement of position of magnetic pole during past time in relation to present position.

polymorphism Existence of several different morphologic kinds occurring in species or mineral.

porosity Percentage of open space or interstices in rock or other earth material. Compare with *permeability*.

porphyritic Textural term for igneous rocks in which large crystals, called *phenocrysts*, are set in finer groundmass, which may be crystalline or glassy or both.

porphyry Igneous rock containing conspicuous *phenocrysts* in fine-grained or glassy *groundmass*.

Portland cement Hydraulic cement consisting of compounds of silica, lime, and alumina.

positive charge Condition resulting from deficiency of electrons.

potassic feldspar Orthoclase, $K(AlSi_3O_8)$.

potential energy Stored energy waiting to be used. Energy that a piece of matter possesses because of position or because of arrangement of parts.

prairie soils Transitional soils between *pedalfers* and *pedocals*.

Pratt hypothesis Explains isostasy by assuming all portions of the crust have same total mass above certain elevation, called *level of compensation*. Higher sections would have proportionately lower density.

precipitation Discharge of water, in rain, snow, hail, sleet, fog, or dew, on land or water surface. Also, process of separating mineral constituents from solution by evaporation (*halite*, *anhydrite*) or from *magma* to form igneous rocks.

precursor Relating to earthquakes, refers to events immediately preceding actual shaking of ground. Includes changes in seismic velocities, groundwater levels, and tilt of ground surface.

pressure Force per unit area.

prevailing westerlies Winds blowing from the west and characteristic of the middle latitudes both north and south of the equator.

primary wave Earthquake body wave that travels fastest and advances by *push-pull* mechanism. Also known as *longitudinal*, compressional, or *P wave*.

proton Fundamental particle of matter with positive electrical charge of 1 unit (equal in amount, but opposite in effect, to the charge of *electron*) and mass of 1.00758 u.

proton-proton fusion Rapidly moving *protons* in hot interior of stars collide and fuse to form atoms of helium from atoms of hydrogen in continuous buildup of higher elements.

pumice Pieces of magma up to several centimeters across that have trapped bubbles of steam or other gases as they were thrown out in eruption. Sometimes they have sufficient buoyancy to float on water.

push-pull wave Wave that advances by alternate compression and rarefaction of medium, causing particles in path to move forward and backward along direction of wave's advance. In connection with earth waves, also known as *primary wave*, compressional wave, *longitudinal wave*, or *P wave*.

pyrite A sulfide mineral, iron sulfide, FeS_2.

pyroclastic debris Fragments blown out by explosive volcanic eruptions and subsequently deposited on ground. Include *ash*, *cinders*, *lapilli*, blocks, bombs, and *pumice*.

pyroxene group Ferromagnesian silicates with a single chain of silicon-oxygen tetrahedra. Common example: augite. Compare with *amphibole group* (example: hornblende), which has a double chain of tetrahedra.

pyrrhotite A mineral, iron sulfide. So commonly associated with nickel minerals that has been called "world's greatest nickel ore."

quartz A silicate mineral, SiO_2, composed exclusively of silicon-oxygen tetrahedra, with all oxygens joined in a three-dimensional network. Crystal form is six-sided prism tapering at end, with prism faces striated transversely. An important rock-forming mineral.

quartzite Metamorphic rock commonly formed by metamorphism of sandstone and composed of quartz. No rock cleavage. Breaks *through* sand grains in contrast to sandstone, which breaks *around* grains.

radar Ultrahigh-frequency electromagnetic radiation.

radial drainage Arrangement of stream courses in which streams radiate outward in all directions from central zone.

radiant energy Electromagnetic waves travelling as wave motion.

radioactivity Spontaneous breakdown of atomic nucleus, with emission of radiant energy.

rain wash Water from rain after it has fallen on ground and before concentrated in definite stream channels.

range Elongated series of mountain peaks considered to be a part of one connected unit, such as Appalachian Range or Sierra Nevada Range.

rank Term used to designate extent to which *metamorphism* has advanced. Compare with *grade*. Rank is more commonly employed in designating stage of metamorphism of *coal*.

ray craters Lunar craters marked by *rays*. Young on lunar time scale.

rays Light-toned streaks that spread outward from such lunar craters as Tycho, Kepler, and Copernicus.

reaction series See *Bowen's reaction series*.

recessional moraine Ridge or belt of *till* marking period of moraine formation, probably in period of temporary stability or slight readvance, during general wastage of a glacier and recession of its front.

recorder Part of a *seismograph* that makes record of ground motion.

rectangular pattern Arrangement of stream courses in which tributaries flow into larger streams at angles approaching 90°.

recumbent fold Fold with axial plane more or less horizontal.

reflection seismic prospecting Uses reflected waves and places seismographs at distances only a fraction of depths investigated.

refraction seismic prospecting Uses travel times of refracted waves and spreads seismographs over lines roughly four times depth being investigated.

refractory Mineral or compound that resists action of heat and chemical reagents.

regional metamorphism Metamorphism occurring over tens or scores of kilometers.

rejuvenation Change in conditions of erosion that causes a stream to begin more active erosion and a new cycle.

relative time Dating of events by place in chronologic order of occurrence rather than in years. Compare with *absolute time*.

relief See Appendix E.

reverse fault Fault in which *hanging wall* appears to have moved upward relative to *footwall*; contrast with *normal*, or *gravity fault*. Also called *thrust fault*.

rhyolite Fine-grained igneous rock with composition of granite.

rift zone System of fractures in earth's crust. Often associated with lava extrusion.

right-lateral fault *Strike-slip fault* in which ground opposite you appears to have moved right when you face it.

rigidity Resistance to elastic *shear*.

rill Miniature stream channel which forms along axis of broad, shallow trough carrying *sheet wash*, or sheet flow.

rilles Trenchlike depressions on moon's surface. Some are straight walled, others sinuous.

ring dike Arcuate (rarely circular) *dike* with steep *dip*.

ripple marks Small waves produced in unconsolidated material by wind or water. See *ripple marks of oscillation*.

ripple marks of oscillation *Ripple marks* formed by oscillating movement of water such as may be found along sea coast outside surf zone. Symmetrical, with sharp or slightly rounded ridges separated by more gently rounded troughs.

rock Aggregate of minerals of one or more kinds in varying proportions.

rock cycle Concept of sequences through which earth materials may pass when subjected to geological processes.

rock flour Finely divided rock material pulverized by glacier and carried by streams fed by melting ice.

rock-forming silicate minerals Minerals built around framework of silicon-oxygen tetrahedra: olivine, augite, hornblende, biotite, muscovite, orthoclase, albite, anorthite, quartz.

rock glacier Tongue of rock waste found in valleys of certain mountainous regions. Characteristically lobate and marked by series of arcuate, rounded ridges that give it aspect of having flowed as viscous mass.

rock melt Liquid solution of rock-forming mineral ions.

rock salt *Halite*, or common salt, NaCl.

rock slide Sudden and rapid slide of bedrock along planes of weakness.

Rossi-Forel scale Scale for rating earthquake intensities, devised in 1878.

runoff Water that flows off land.

rupture Breaking apart or state of being broken apart.

S Symbol for *secondary wave*.

salt In geology this term usually refers to *halite*, or *rock salt*, NaCl, particularly in such combinations as salt water and *salt dome*.

saltation Mechanism by which a particle moves by jumping from one point to another.

salt dome Mass of NaCl generally of roughly cylindrical shape and with diameter of about 2 km near top. Such mass has been pushed through surrounding sediments into present position. Reservoir rocks above and alongside salt domes trap *oil* and *gas*.

saltwater wedge Body of water, found in some estuaries, which thins toward head of estuary and is overridden by fresh water from land.

sand Clastic particles of *sand size*, commonly but not always composed of mineral *quartz*.

sand size Volume greater than that of a sphere with diameter of 0.0625 mm and less than that of a sphere with diameter of 2 mm.

sandstone Detrital *sedimentary rock* formed by cementation of individual grains of sand size and commonly composed of mineral *quartz*. Sandstones constitute estimated 12 to 28 percent of sedimentary rocks.

sapropel Aquatic ooze or sludge rich in organic matter. Believed to be source material for *petroleum* and *natural gas*.

satellite crater *Crater* formed by impact of a fragment ejected during creation of a primary crater. Also called *secondary crater*.

scale See Appendix E.

schist Metamorphic rock dominated by fibrous or platy minerals. Has *schistose cleavage* and is product of *regional metamorphism*.

schistose cleavage Rock cleavage with grains and flakes clearly visible and cleavage surfaces rougher than in *slaty* or *phyllitic cleavage*.

sea-floor spreading Process by which ocean floors spread laterally from crests of main ocean ridges. As material moves laterally from ridge, new material is thought to replace it along ridge crest by welling upward from mantle.

seamount Isolated, steep-sloped peak rising from deep ocean floor but submerged beneath surface. Most have sharp peaks, but some have flat tops and are called *guyots*, or *tablemounts*. Seamounts are probably volcanic in origin.

secondary crater See *satellite crater*.

secondary wave Earthquake *body wave* slower than *primary wave*. *Shear*, *shake*, or *S wave*.

secular variation of magnetic field Change in inclination, declination, or intensity of earth's magnetic field. Detectable only from long historical records.

sedimentary facies Accumulation of deposits that exhibits specific characteristics and grades laterally into other sedimentary accumulations that were formed at same time but exhibit different characteristics.

sedimentary rock Rock formed from accumulations of sediment, which may consist of rock fragments of various sizes, remains or products of animals or plants, products of chemical action or of evaporation, or mixtures of these. *Stratification* is single most characteristic feature of sedimentary rocks, which cover about 75 percent of land area.

sedimentation Process by which mineral and organic matters are laid down.

seif dune Very large *longitudinal dune*. As high as 100 m and as long as 100 km.

seismic prospecting Method of determining nature and structure of buried rock formations by generating waves in ground (commonly by small explosive charges) and measuring length of time these waves require to travel different paths.

seismic sea wave Large wave in ocean generated at time of earthquake. Popularly but incorrectly known as *tidal wave*. Sometimes called *tsunami*.

seismogram Record obtained on a *seismograph*.

seismograph Instrument for recording vibrations, most commonly employed for recording earth vibrations during earthquakes.

seismology Scientific study of earthquakes and other earth vibrations.

serpentine Silicate of magnesium common among metamorphic minerals. Occurs in two crystal habits: platy, known as *antigorite*; fibrous, known as *chrysotile*, an *asbestos*. "Serpentine" comes from mottled shades of green on massive varieties, suggestive of snake markings.

S.G. Symbol for *specific gravity*.

shake wave Wave that advances by causing particles in path to move from side to side or up and down at right angles to direction of wave's advance, a shake motion. Also called *shear wave*, or *secondary wave*.

shale Fine-grained, detrital sedimentary rock made up of *silt*-

and *clay-sized* particles. Contains clay minerals as well as particles of *quartz, feldspar, calcite, dolomite,* and other minerals. Distinguished from *mudstone* by presence of fissility.

shear Change of shape without change of volume.

shear wave Wave that advances by shearing displacements (which change shape without changing volume) of medium. This causes particles in path to move from side to side or up and down at right angles to direction of wave's advance. Also called *shake wave,* or *secondary wave.*

sheet flow See *sheet wash.*

sheeting Joints essentially parallel to ground surface. More closely spaced near surface and become progressively farther apart with depth. Particularly well developed in granitic rocks, but sometimes in other massive rocks as well.

sheet wash Water accumulating on a slope in thin sheet of water. May begin to concentrate in *rills.* Also called *sheet flow.*

shield Nucleus of Precambrian rocks around which a continent has grown.

shield volcano Volcano built up almost entirely of lava, with slopes seldom as great as 10° at summit and 2° at base. Examples: five volcanoes on island of Hawaii.

sial A term coined from chemical symbols for silicon and aluminum. Designates composite of rocks dominated by *granites, granodiorites,* and their allies and derivatives, which underlie continental areas of globe. Specific gravity considered to be about 2.7.

sialic rock Igneous rock composed predominantly of silicon and aluminum, from whose chemical symbols term is constructed. Average specific gravity, about 2.7.

siderite A mineral; iron carbonate, $FeCO_3$. An ore of iron.

silicate minerals Minerals with crystal structure containing *silicon-oxygen tetrahedra* arranged as isolated units (*nesosilicates*), single or double chains (*inosilicates*), sheets (*phyllosilicates*), or three-dimensional frameworks (*tectosilicates*).

silicon-oxygen tetrahedron Complex ion composed of silicon ion surrounded by four oxygen ions. Negative charge of 4 units, and represented by symbol $(SiO_4)^{4-}$. Diagnostic unit of silicate minerals, and makes up central building unit of nearly 90 percent of materials of earth's crust.

sillimanite A silicate mineral, Al_2SiO_5, characteristic of highest metamorphic temperatures and pressures. Occurs in long slender crystals, brown, green, white. Crystalline structure based on independent *tetrahedra.* Contrast with *kyanite* and *andalusite,* which have same composition but different crystal habits and form at lower temperatures.

silt size Volume greater than that of a sphere with diameter of 0.0039 mm and less than that of a sphere with diameter of 0.0625 mm.

sima Term coined from silicon and magnesium. Designates worldwide shell of dark, heavy rocks. Sima believed to be outermost rock layer under deep, permanent ocean basins, such as Pacific. Originally sima considered basaltic in composition, with specific gravity of about 3.0. It has been suggested also, however, that it may be *peridotitic* in composition, with specific gravity of about 3.3.

simatic rock Igneous rock composed predominantly of ferromagnesian minerals. Average specific gravity, 3.0 to 3.3.

sink See *sinkhole.*

sinkhole Depression in surface of ground caused by collapse of roof over solution cavern.

slate Fine-grained metamorphic rock with well-developed *slaty cleavage.* Formed by low-grade *regional metamorphism* of *shale.*

slaty cleavage Rock cleavage in which ease of breaking occurs along planes separated by microscopic distances.

slip face Steep face on lee side of a dune.

slope failure See *slump.*

slump Downward and outward movement of rock or unconsolidated material as unit or as series of units. Also called *slope failure.*

snowfield Stretch of perennial snow existing in area where winter snowfall exceeds amount of snow that melts away during summer.

snow line Lower limit of perennial snow.

soapstone See *talc.*

sodic feldspar Albite, $Na(AlSi_3O_8)$.

soil Superficial material that forms at earth's surface as result of organic and inorganic processes. Soil varies with climate, plant and animal life, time, slope of land, and parent material.

soil horizon Layer of soil approximately parallel to land surface with observable characteristics produced through operation of soil-building processes.

solar constant Average rate at which radiant energy received by earth from sun. Equal to little less than 2 cal/cm²-min on plane perpendicular to sun's rays at outer edge of atmosphere, when earth is at mean distance from sun.

solar system Sun with group of celestial bodies held by its gravitational attraction and revolving around it.

sole mark Cast of sedimentary structures such as cracks, tracks, or grooves formed on lower surface or underside of sandstone bed, commonly revealed after original underlying sedimentary layer has weathered away.

solid Matter with definite shape and volume and some fundamental strength. May be crystalline, glassy, or amorphous.

solid solution Single crystalline phase that may vary in composition within specific limits.

solifluction Mass movement of soil affected by alternate freezing and thawing. Characteristic of saturated soils in high latitudes.

sorosilicates Mineral with crystal structure containing *silicon-oxygen tetrahedra* arranged as double units.

southeast trades Winds blowing from the southeast toward the equator in the southern hemisphere.

space lattice In crystalline structure of mineral, three-dimensional array of points representing pattern of locations of identical atoms or groups of atoms constituting a mineral's *unit cell.* There are 230 pattern types.

specific gravity Ratio between weight of given volume of material and weight of equal volume of water at 4°C.

specific heat Amount of heat necessary to raise temperature of 1 g of any material 1°C.

sphalerite A mineral; zinc sulfide, ZnS. Nearly always contains iron, $(Zn, Fe)S$. Principal ore of zinc. (Also known as *zinc blende* or *blackjack.*)

spheroidal weathering Spalling off of concentric shells from rock masses of various sizes as result of pressures built up during chemical weathering.

spit Sandy bar built by currents into a bay from a promontory.

spring Place where *water table* crops out at ground surface and water flows out more or less continuously.

stack Small island that stands as isolated, steep-sided rock mass just off end of promontory. Has been isolated from land by erosion and weathering concentrated behind end of a headland.

stalactite Icicle-shaped accumulation of *dripstone* hanging from cave roof.

stalagmite Post of *dripstone* growing upward from cave floor.

star A heavenly body that seems to stay in same position relative to other heavenly bodies.

staurolite Silicate mineral characteristic of middle-grade *metamorphism.* Crystalline structure based on independent tetrahedra with iron and aluminum. Has unique crystal habit that makes it striking and easy to recognize: six-sided prisms intersecting at 90° to form cross or at 60° to form an X shape.

steady-state theory Theory that universe is developing by continuous creation of matter as newly formed galaxies replace those expanding out of sight, thus keeping mass density of universe constant. Compare with *big-bang theory*.

stock Discordant *pluton* that increases in size downward, has no determinable floor, and shows area of surface exposure less than 100 km². Compare with *batholith*.

stoping Mechanism by which *batholiths* have moved into crust by breaking off and foundering of blocks of rock surrounding magma chamber.

strain Change of dimensions of matter in response to *stress:* Commonly, unit strain, such as change in length per unit length (total lengthening divided by original length), change in width per unit width, change in volume per unit volume.

stratification Structure produced by deposition of sediments in layers or beds.

stratigraphic trap Structure that traps *petroleum* or *natural gas* because of variation in permeability of reservoir rock or termination of inclined reservoir formation on up-dip side.

streak Color of fine powder of mineral; may be different from color of hand specimen. Usually determined by rubbing mineral on piece of unglazed porcelain (*hardness* about 7) known as a "streak plate," which is, of course, useless for minerals of greater hardness.

stream order Hierarchy in which segments of a stream system are arranged.

stream terrace Surface representing remnants of stream's channel or flood plain when stream was flowing at higher level. Subsequent downward cutting by stream leaves remnants of old channel or *flood plain* standing as *terrace* above present stream level.

strength *Stress* at which rupture occurs or plastic deformation begins.

stress Force applied to material that tends to change dimensions: commonly, unit stress or total force divided by the area over which applied. Contrast with *strain*. •

striations (1) Scratches, or small channels, gouged by glacial action. Bedrock, pebbles, and boulders may show striations produced when rocks trapped by ice were ground against bedrock or other rocks. Striations along bedrock surface are oriented in direction of ice flow across that surface. (2) In minerals, parallel, threadlike lines, or narrow bands, on face of mineral. Reflect internal atomic arrangement.

strike Direction of line formed by intersection of a rock surface with a horizontal plane. Strike is always perpendicular to direction of *dip*.

strike-slip fault *Fault* in which movement is almost in direction of fault's *strike*.

strip mining Surface mining in which soil and rock covering sought-for commodity are moved to one side. Some coal mining is pursued in this manner.

structural relief Difference in elevation of parts of deformed stratigraphic horizon.

structure Attitudes of deformed masses of rock

subduction Act of one tectonic unit's descending under another (commonly slab of *lithosphere*).

subduction zone Elongate region along which lithospheric block descends relative to another lithospheric block.

sublimation Process by which solid material passes into gaseous state without first becoming liquid.

subsequent stream Tributary stream flowing along beds of less erosional resistance and parallel to beds of greater resistance. Course determined subsequent to uplift that brought more resistant beds within sphere of erosion.

subsurface water Water below ground surface. Also referred to as *underground water* and *subterranean water*.

subterranean water Water below ground surface. Also referred to as *underground water* and *subsurface water*.

subtropical high pressure zones Zones of atmospheric high pressure located about 30° north and 30° south of the equator.

sulfate mineral Mineral formed by combination of complex ion $(SO_4)^{2-}$ with positive ion. Common example: gypsum, $CaSO_4 \cdot 2H_2O$.

sulfide mineral Mineral formed by direct union of element with sulfur. Examples: argentite, chalcocite, galena, sphalerite, pyrite, and cinnabar.

superposition Law by which, if series of sedimentary rocks has not been overturned, topmost layer is always youngest and lowermost always oldest.

surface wave Wave that travels along free surface of medium. Earthquake surface waves sometimes represented by symbol *L*.

surge Applied to glaciers; rapid and sometimes catastrophic advance of ice.

suspended water Underground water held in *zone of aeration* by molecular attraction exerted on water by rock and earth materials and by attraction exerted by water particles on one another.

suspension Process by which material is buoyed up in air or water and moved about without making contact with surface while in transit. Contrasts with *traction*.

symmetrical fold *Fold* in which axial plane is essentially vertical. Limbs dip at similar angles.

syncline A configuration of folded, stratified rocks in which rocks dip downward from opposite directions to come together in a trough. Reverse of *anticline*.

tablemount See *guyot*.

tabular Shape with large area relative to thickness.

taconite Unleached iron formation of Lake Superior District. Consists of chert with hematite, magnetite, siderite, and hydrous iron silicates. Ore of iron, averaging 25 percent iron, but natural leaching turns it into ore with 50 to 60 percent iron.

talc Silicate of magnesium common among metamorphic minerals. Crystalline structure based on tetrahedra arranged in sheets; greasy and extremely soft. Sometimes known as *soapstone*.

talus Slope established by accumulation of rock fragments at foot of cliff or ridge. Rock fragments that form talus may be rock waste, slide rock, or pieces broken by frost action. Actually, term "talus" widely used to mean rock debris itself.

tarn Lake formed in bottom of *cirque* after glacier ice has disappeared.

tectonic change of sea level Change in sea level produced by land movement.

tectosilicate Mineral with crystal structure containing *silicon-oxygen tetrahedra* arranged in three-dimensional frameworks.

temporary base level Nonpermanent *base level*, such as that formed by lake.

tension Stretching stress that tends to increase volume of a material.

terminal moraine Ridge or belt of *till* marking farthest advance of a glacier. Sometimes called *end moraine*.

terminal velocity Constant rate of fall eventually attained by grain or body when acceleration caused by influence of gravity is balanced by resistance of fluid through which grain falls or air through which body falls.

terrace Nearly level surface, relatively narrow, bordering a stream or body of water and terminating in a steep bank. Commonly term is modified to indicate origin, as in *stream terrace* and *wave-cut terrace*.

terrae Light-toned highlands of moon.

terrigenous deposit Material derived from above sea level and deposited in deep ocean. Example: volcanic ash.

tetrahedron A four-sided solid. Used commonly in describing silicate minerals as shortened reference to *silicon-oxygen tetrahedron*.

texture General physical appearance of rock, as shown by size, shape, and arrangement of particles that make it up.

thermal gradient In earth, rate at which temperature increases with depth below surface.

thermal pollution Increase in normal temperatures of natural waters caused by intervention of human activities.

thermal spring Spring that brings warm or hot water to surface. Temperature usually 6.5°C or more above mean air temperature. Sometimes called *warm spring*, or *hot spring*.

thermoremanent magnetism Magnetism acquired by igneous rock as it cools below Curie temperatures of magnetic minerals in it. Abbreviation, TRM.

thin section Slice of rock ground so thin as to be translucent.

tholeiite Group of basalts primarily composed of *plagioclase* (approximately An$_{50}$), *pyroxene*, and iron oxides as *phenocrysts* in glassy groundmass of *quartz* and alkali *feldspar*; little or no *olivine* present.

thrust fault Fault in which *hanging wall* appears to have moved upward relative to *footwall*; opposite of *gravity*, or *normal*, *fault*. Also called *reverse fault*.

tidal current Water current generated by tide-producing forces of sun and moon.

tidal inlet Waterway from open water into a lagoon.

tidal wave Popular but incorrect designation for *tsunami*.

tide Alternate rising and falling of surface of ocean, other bodies of water, or earth itself in response to forces resulting from motion of earth, moon, and sun relative to each other.

till Unstratified, unsorted glacial *drift* deposited directly by glacier ice.

tillite Rock formed by lithification of *till*.

time-distance graph Graph of travel time against distance.

tombolo Sand bar connecting an island to mainland or joining two islands.

topographic deserts Deserts deficient in rainfall because they are either located far from oceans toward center of continents or cut off from rain-bearing winds by high mountains.

topography Shape and physical features of land.

topset bed Layer of sediment constituting surface of *delta*. Usually nearly horizontal and covers edges of inclined *foreset beds*.

tourmaline Silicate mineral of boron and aluminum with sodium, calcium, fluorine, iron, lithium, or magnesium. Formed at high temperatures and pressures through agency of fluids carrying boron and fluorine. Particularly associated with *pegmatites*.

township and range See Appendix E.

traction Process of carrying material along bottom of a stream. Traction includes movement by saltation, rolling, or sliding.

trailing edge The margin of a continent that occurs on the side opposite the margin moving toward another continental or oceanic plate (leading edge); commonly is less active tectonically, behaving in a passive manner. The margin of a continent moving away from the locus of spreading along a midoceanic ridge as, for example, the eastern coasts of North and South America.

transcurrent fault See *strike-slip fault*.

transducer Device that picks up relative motion between mass of seismograph and ground and converts this into form that can be recorded.

transform fault Point at which strike-slip displacements stop and another structural feature, such as a ridge, develops.

transition element Element in series in which inner shell is being filled with electrons after outer shell has been started. All transition elements metallic in free state.

transpiration Process by which water vapor escapes from a living plant and enters atmosphere.

transverse dune Dune formed in areas of scanty vegetation and in which sand has moved in ridge at right angles to wind. Exhibits gentle windward slope and steep leeward slope characteristic of other dunes.

transverse wave See *shear*, or *shake*, *wave*.

trap rock Popular synonym for *basalt*.

travel time Total elapsed time for wave to travel from source to designated point.

travertine Form of calcium carbonate, $CaCO_3$, which forms stalactites, stalagmites, and other deposits in limestone caves or incrustations around mouths of hot and cold calcareous springs. Sometimes known as *tufa*, or *dripstone*.

trellis pattern Roughly rectilinear arrangement of stream courses in pattern reminiscent of garden trellis, developed in region where rocks of differing resistance to erosion have been folded, beveled, and uplifted.

trenches See *island-arc deeps*.

TRM See *thermoremanent magnetism*.

tropical deserts Deserts lying between 5° to 30° north and south of equator.

truncated spur Beveled end of divide between two tributary valleys where they join a main valley that has been glaciated. Glacier of main valley has worn off end of divide.

tsunami Large wave in ocean generated at time of earthquake. Popularly but incorrectly known as *tidal wave*. Sometimes called *seismic sea wave*.

tufa Calcium carbonate, $CaCO_3$, formed in stalactites, stalagmites, and other deposits in limestone caves, as incrustations around mouths of hot and cold calcareous springs, or along streams carrying large amounts of calcium carbonate in solution. Sometimes known as *travertine* or *dripstone*.

tuff Rock consolidated from volcanic ash.

tundra Stretch of Arctic swampland developed on top of permanently frozen ground. Extensive tundra regions have developed in parts of North America, Europe, and Asia.

turbidites Sedimentary deposits settled out of turbid water carrying particles of widely varying grade size. Characteristically display *graded bedding*.

turbidity current Current in which limited volume of turbid or muddy water moves relative to surrounding water because of current's greater density.

turbulent flow Mechanism by which fluid (such as water) moves over or past a rough surface. Fluid not in contact with irregular boundary outruns that slowed by friction or deflected by uneven surface. Fluid particles move in series of eddies or whirls. Most stream flow is turbulent; turbulent flow is important in both erosion and transportation. Contrast with *laminar flow*.

ultimate base level Sea level, lowest possible *base level* for a stream.

unconformity Buried erosion surface separating two rock masses, older exposed to erosion for long interval of time before deposition of younger. If older rocks were deformed and not horizontal at time of subsequent deposition, surface of separation is *angular unconformity*. If older rocks remained essentially horizontal during erosion, surface separating them from younger rocks is called *disconformity*. Unconformity that develops between massive igneous or metamorphic rocks exposed to erosion and then covered by sedimentary rocks is called *nonconformity*.

underground water Water below ground surface. Also referred to as *subsurface water* and *subterranean water*.

uneven fracture Mineral habit of breaking along rough, irregular surfaces.

uniformitarianism Concept that present is key to past. This means that processes now operating to modify earth's surface have also operated in geologic past, that there is uniformity of processes past and present.

unit cell In crystalline structure of mineral, three-dimensional grouping of atoms arbitrarily selected so that mineral's structure represented by periodic repetition of this unit in *space lattice*.

unpaired terrace A *terrace* formed when an eroding stream, swinging back and forth across a valley, encounters resistant rock beneath unconsolidated alluvium and is deflected, leaving behind single terrace with no corresponding terrace on other side of stream.

valley glacier Glacier confined to stream valley. Usually fed from a *cirque*. Sometimes called *Alpine glacier* or *mountain glacier*.

valley train Gently sloping plain underlain by glacial outwash and confined by valley walls.

varve Pair of thin sedimentary beds, one coarse, one fine. This couplet has been interpreted as representing a cycle of 1 year or interval of thaw followed by interval of freezing in lakes fringing a glacier.

ventifact Pebble, cobble, or boulder that has had its shape or surface modified by wind-driven sand.

vertical exaggeration See Appendix E.

vesicle Small cavity in aphanitic or glassy igneous rock, formed by expansion of bubble of gas or steam during solidification of rock.

virtual geomagnetic pole Pole consistent with magnetic field as measured at any one locality. Refers to magnetic-field direction of single point, in contrast to *geomagnetic pole*, which refers to best fit of geocentric dipole for entire earth's field. Most *paleomagnetic* readings expressed as virtual geomagnetic poles.

viscosity An internal property of rock that offers resistance to flow. Ratio of deforming force to rate at which changes in shape are produced.

volatile components Materials in magma, such as water, carbon dioxide, and certain acids, whose vapor pressures are high enough to cause them to become concentrated in any gaseous phase that forms.

volcanic ash *Dust-sized pyroclastic* particle: volume equal to, or less than, that of sphere with diameter of 0.06 mm.

volcanic block Angular mass of newly congealed *magma* blown out in eruption. Contrast with *volcanic bomb*.

volcanic bomb Rounded mass of newly congealed *magma* blown out in eruption. Contrast with *volcanic block*.

volcanic breccia Rock formed from relatively large blocks of congealed lava embedded in mass of *ash*.

volcanic dust *Pyroclastic* detritus consisting of particles of *dust* size.

volcanic earthquakes Earthquakes caused by movements of *magma* or explosions of gases during volcanic activity.

volcanic eruption Explosive or quiet emission of *lava, pyroclastics*, or volcanic gases at earth's surface, usually from volcano but rarely from fissures.

volcanic mountains Mountains built up from extrusion of *lava* and *pyroclastic debris*.

volcanic neck Solidified material filling vent, or pipe, of dead volcano.

volcanic tremor Continuous shaking of ground associated with certain phase of volcanic eruption.

volcano Landform developed by accumulation of magmatic products near central vent.

vug Small unfilled cavity in rock, usually lined with crystalline layer of different composition from surrounding rock.

wadi See *arroyo*.

warm glacier Reaches melting temperature throughout thickness during summer season.

warm spring *Thermal spring* that brings warm water to surface. Temperature usually 6.5°C or more above mean air temperature.

warp Large section of continent composed of horizontal strata gently bent upward or downward.

wash See *arroyo*.

water gap Gap cut through resistant ridge by superimposed or antecedent stream.

water table Upper surface of *zone of saturation* for underground water. An irregular surface with slope or shape determined by quantity of groundwater and permeability of earth materials. In general, highest beneath hills and lowest beneath valleys.

wave Configuration of matter that transmits energy from one point to another.

weathering Response of materials once in equilibrium within earth's crust to new conditions at or near contact with water, air, or living matter. See also *chemical weathering* and *mechanical weathering*.

wrinkle ridges Ridges found on surfaces of lunar *maria* and flooded craters. May be caused by uplift due to volcanism or to compression.

xenolith Rock fragment foreign to igneous rock in which it occurs. Commonly inclusion of country rock intruded by igneous rock.

yazoo-type river Tributary unable to enter main stream because of *natural levees* along main stream. Flows along backswamp zone parallel to main stream.

yield point Maximum stress that solid can withstand without undergoing permanent deformation, either by plastic flow or by rupture.

zone of aeration Zone immediately below ground surface, in which openings partially filled with air and partially with water trapped by molecular attraction. Subdivided into (1) *belt of soil moisture*. (2) *intermediate belt*, and (3) *capillary fringe*.

zone of saturation Underground region within which all openings filled with water. Top of zone of saturation is called *water table*. Water contained within zone of saturation is called *groundwater*.

zones of regional metamorphism High-grade, above 700°C; middle-grade, 400 to 700°C; low-grade, 150 to 400°C.

INDEX

Å, 17, 441
aa lava, 40–41, 441
Abajo Mountains, 54
ablation, 307, 441
abrasion, 441
  glaciers and, 310–311, 315
  in streams, 271
  by wind, 357
absolute time, 2–3, 133–137, 139, 441
abyssal plains, 339
acceleration due to gravity, 154
acid lava, 40, 441
acid rain, 370, 441
actinolite, 418, 441
adiabatic rate, 441
aegirite, 418
aeration, zone of, 289–290
Africa, 225, 227
African rift zone, 167

aftershocks of earthquakes, 194–195, 441
Agassiz, Louis, 322–323
agate, 95, 441
aggradation, flood plain of, 279
agriculture, sediment production and, 268
A horizon, 86–88, 441
Airy, G. B., 152
Airy hypothesis, 152, 441
Alaska, 252, 375
  earthquakes in, 179–182, 188–189, 199, 245–246
Alaska Range, 246
albite, 31–32, 77, 78, 420, 441
alcohol, 368, 382
Aleutian Islands, 195–196
alfisols, 85
Algonquin, Lake, 348

A

alkalic basalt, 39
alkali rocks, 441
alluvial fans, 283, 441
almandite, 115, 420, 441
alpha decay, 134, 441
alpha particles, 17, 442
Alpine glaciers, 305, 442
aluminosilicates, 115, 116
aluminum, 388–389
aluminum silicates, 75, 76
Amazon River, 81
American plate, 10–11, 48, 199, 231, 238
ammonium sulfate, 394
amorphous matter (mineraloids), 33, 442
amphibole family, 420
amphiboles, 30, 442
amphibolite facies, 116, 442
amphibolites, 122, 442
amphitheaters (see cirques)
andalusite, 115, 116, 420, 442
Anderson, D. L., 208
andesine, 420
andesite, 39, 61, 442
andesite line, 39, 210, 442
andesitic lava flows, 39
Andes Mountains, 238, 285, 352
angstrom (Å), 17, 441, 442
angular momentum, 442
angular unconformity, 170–171, 442
anhydrite, 32, 100, 420, 442
animals (see also fossils; reptiles):
    chemical weathering and, 74
    mechanical weathering and, 72
anion, 17, 442
ankerite, 420
anorthite, 31–32, 77, 78, 406, 420, 442
anorthosite, 63, 406
Antarctica, 306, 307, 325, 327
Antarctic bottom water, 335, 442
Antarctic intermediate water, 335, 442
Antarctic Ocean, 342
anthophyllite, 420
anthracite, 102, 127, 370, 442
anticlines, 159, 442
Antrim Plateau, 51
apatite, 394, 420
aphanitic texture, 58, 442

Apollo missions, 406
Appalachian Mountains, 230, 274
aquifer, 291, 442
Arctic Ocean, 337, 342
arête, 312, 442
arfvedsonite, 420
aridisols, 85
Aristotle, 196
Arkansas River, 258
arkose, 99, 442
arroyo, 442
artesian water, 294, 442
Asama, Mount, 42
asbestos, 397, 420, 442
aseismic ridges, 196, 339, 442
ash, volcanic, 41, 42, 442
asphalt, 378, 442
astatine, 18
asteroids, 411, 442
asthenosphere, 211, 442
astrogeology, 442
asymmetric folds, 159, 442
asymmetric rock knobs and hills, 315
Athabascan tar sands, 378
Atlantic Ocean, 337
    surface currents of, 335
Atlantic Ridge (see Mid-Atlantic Ridge)
atoll, 442
atomic energy, 442 (see also nuclear power)
atomic mass, 17, 442
atomic number, 17, 442
atomic reactor, 442
atomic size, 17, 442
atomic structure, 17, 18
atoms, 16–19, 442
    covalent bond of, 26–27
    metallic bond of, 27
    radii of, 25–26
augite, 28–29, 420, 442
aureoles, 114, 442
Austin Post, 246
authigenic minerals, 98
avalanches, 310
axes of symmetry, 415
axial plane, 159, 442
axis, of an anticline or syncline, 159, 442
azurite, 420

backset beds, 363, 442–443
backshore, 345
back swamps, 280
Baltic Sea, 348
banded iron formation, 387–388
bank-full stage, 261, 264, 443
barchans, 362, 443
Barnegat Bay, 347
barrier islands, 348, 443
barrier reef, 443
basal plane, 443
basal slip, 309, 443
basaltic lava flows, 39
basalts, 8, 39, 61, 443
    alkalic, 39
    moon, 406
    plateau, 51
    tholeiitic, 39
base flow, 259, 443
base level of a stream, 265, 443
basement complex, 443
basic lava, 40, 443
Basin and Range Physiographic
        Province, 167
basins, drainage, 80–81, 273,
        446
batholiths, 55, 443
bauxite, 33, 88, 388–389, 420, 443
bay barriers, 347, 443
bedding, 102–103, 443
    graded, 335
bedding planes, 102–103, 443
bed load, 80–81, 267, 443
bedrock, 68
belt of soil moisture, 289, 443
bench mark, 437, 443
Benioff zone, 238, 443
bergschrund, 310, 443
berms, 345, 443
beryl, 62
beta decay, 134, 443
B horizon, 85–88, 443
big-bang theory, 19, 20, 443
Big Horn River, 276
binding energy, 443

biochemical rocks (organic rocks),
        93, 101–102, 443
biotite, 30, 78, 420, 443
bishop's ring, 46
bituminous coal, 102, 127, 370, 443
Blackhawk slide, 246
Black Hills, 294
blocks, volcanic, 41
blowouts, 357, 443
blueschist facies, 116
body waves, 202–203, 443
Bogosloff Island, 49
bombs, volcanic, 41
bornite, 390, 420, 443
Bothnia, Gulf of, 348
bottomset beds, 282, 443
Bouger, Pierre, 152
boulders, spheroidally weathered,
        71–72
boulder size, 97, 443
boulder tills, 315
boulder trains, 318, 443
Bowen's reaction principle, 57–58,
        443
Box Canyon, 171
braided streams, 280, 443
Bråsellbreen Glacier, 309
Brazil current, 336
breccia, 99, 406, 443
    fault, 114
breeder reactors, 380
Bridgman, Percy, 198
Britoarctic Plateau, 51
brittleness, 151, 443
Brongniart, Alexandre, 132
bronzite, 422
brown clay, 342, 443
Bryce Canyon, 106
Buenguela current, 336
building materials, 395–397
building stone, 396
burial metamorphism, 113, 443
Burke, Kevin, 235
Bushveld complex, 392
bytownite, 422

B

Calabria earthquakes, 201
Calaveras fault, 138
calcic feldspar (see anorthite)
calcite, 32, 77, 79, 95, 114, 422, 433
    caves and, 297–299
calcium bicarbonate, 95, 100

calcium carbonate, 393
    in pedocals, 87–88
calderas, 37–38, 443
    formation of, 42–44
caliche, 88, 443
Callisto, 411

C

calving, 307
Cambrian rocks, 138
Canada, 375
Canadian shield, 173
canyons:
    of Mars, 409
    ocean, 338–339
capacity of streams, 266
Cape Mountains, 230
capillary fringe, 289, 444
capping bed, 374
carbonates, 32, 444
carbon cycle, 19, 444
carbon 14 dating, 135–136, 444
carbon ratio, 444
Caribbean plate, 199
carnalite, 393
carnotite, 422
cassiterite, 24, 32, 392, 422, 444
Castile Formation, 100
cataclastic metamorphism, 114, 444
cation, 17, 444
caves, 297–300
    sea, 346
Cedar Range fault, 167
cement, 396
cementation, 98, 444
cement rock, 444
Ceres, 411
chalcedony, 95, 444
chalcocite, 32, 390, 444
chalcopyrite, 390, 422, 444
chalk, 101–102, 444
Charpentier, Jean de, 322
chemically active fluids, metamor-
    phism and, 113–114
chemical sedimentary rocks, 93, 96,
    100–101, 444
chemical weathering, 73–78, 444
    of igneous rocks, 74–78
    particle size and, 73
chert, 95, 444
Chilean earthquakes, 195
Chilean seismic sea wave of 1960,
    182
China, earthquakes in, 199–200
chlorine, 18
chlorite, 115–116, 422, 444
C horizon, 85–88, 444
chromite, 392, 422, 444
chromium, 392
chrysotile, 397, 422, 444
chute cutoff, 280, 444
cinder cones, 41, 44–45, 444
cinders, volcanic, 41, 444
Cirillo, Nicholas, 200–201

cirques, 310–312, 444
clastic texture, 96–97, 444
clay, 95, 96, 396–397 (see also illite;
    kaolinite; montmorillonite):
    brown, 342, 443
clay minerals, 396–397, 444
    chemical weathering of feldspars
        and, 74–76
clay size, 96, 97, 444
clay tills, 315
cleavage:
    of minerals, 22, 444
    of rocks, 120–121, 444
cliff glaciers, 305
cliffs, wave-cut, 346
climates:
    ancient, continental drift and,
        227–229
    chemical weathering and, 73–74
    glaciation and changes in, 326–
        328
coal, 96, 102, 370–371, 444
    temperature of formation of, 126–
        127
coast, 345
cobble size, 97, 444
Cocos plate, 199
coesite, 444
col, 312, 444
Cold Spring, 295
colloidal size, 76, 444
Colombian earthquake, 188
color, 444
    of minerals, 23
Colorado Plateau, 51, 54, 157
Colorado River, 365
columnar jointing, 160–164, 444
columns, 299, 444
compaction, 98, 444
compensation, level of, 152
competence of streams, 266
composite volcano, 44, 444
compounds:
    covalent bond of, 26–27
    defined, 25
    dipolar, 27
    inorganic, 25
    ionic, 25–26
    organic, 25
compression, 444
    convergent plate boundaries and,
        233
compression strength, 151
compression stress, 151
Comrie earthquakes, 201
concentric folding, 158, 445

conchoidal fracture, 418, 445

concordant plutons, 53–56, 445

concrete, 396

concretions, 105, 445

cone of depression, 290, 445

cone sheets, 53, 164, 445

conglomerate, 99, 445

Congo River, 81

connate water, 445

conodont color changes, 128

contact, 439

contact metamorphism, 114–115, 445

continental crust, 8–11, 209–210 (see
    also crust of the earth):
  oldest, 218

continental drift, 8–11, 215, 237, 445
  evidence for, 224–231
    ancient climates, 227–229
    paleomagnetism, 225, 230
    plants and reptiles, 229–230
    sea-floor spreading, 225–226
    shape of continents, 225
    Wegener's theory, 224–225
  glaciation and, 327

continental glaciers, 306, 445

continental margins, 337–339, 445
  deposits on, 340–341

continental rises, 337, 445

continental shelves, 337, 338, 445
  deposits on, 341

continental slopes, 337, 445

continents:
  board of, 218
  crust of the (see continental crust)
  glaciation and, 325–327
  origin and permanence of,
    215–218
  shape of, continental drift and,
    225
  streams and, 285
  trailing edges of, 337

continuous degassing of the earth,
    332

continuous reaction series, 57, 445

contour interval, 437

contour line, 437

contour sketching, 438–439

contraction and expansion, me-
    chanical weathering by, 70

convection, 445

convection currents (or cells), 235–
    237, 445

convergent-boundary metamor-
    phism, 123–124

convergent plate boundaries, 109,
    232, 233, 445

Copernicus crater, 405

copper, 390
  crystal of, 21

coquina, 102, 445

coral reefs, continental drift and, 229

core of the earth, 211, 445
  magnetism of the earth and, 221

Coriolis effect, 336, 445

correlation, 445
  of sedimentary rocks, 131–133

corundum, 32, 422

cosmic rays, 135

covalent bond, 26–27, 445

Cox, Allan, 222

Crater Lake, 43–44

craters, 37, 445
  moon, 405–406

Craters of the Moon National Monu-
    ment, 51

craton, 445

creep, 249–250, 445

crevasse fillings, 320

crevasses, 308, 445
  bergschrund, 310

cross bedding, 103 (see *also* in-
    clined bedding)

crosscutting relationships, law of,
    139, 445

cross sections, geologic, 440

cross-valley profiles, 273, 276

crushed rock, 396

crustal plates, 211 (see *also* plates)

crust of the earth, 8–11, 208–210,
    445 (see *also* continental
    crust; ocean crust; plates;
    plate tectonics):
  average composition of, 435
  deformation of, 144, 149
  earliest, 218
  horizontal movements of, 9–11
  isostasy of, 9
  plates of, 9–11
  vertical movements of, 8–9

crust of the oceans (see ocean crust)

cryptocrystalline form, 95, 445

cryptovolcanic structure, 445

crystal form, 20–22, 445

crystalline structure, 24–27, 445

crystallization, 445
  of magma, 56–58
  of sedimentary rocks, 98

crystals, 20–22, 445

crystal symmetry, 415

crystal systems, 415, 416

Cumberland overthrust, 167

Curie temperature, 221, 445

current bedding (see inclined bedding)

current ripple marks, 104

currents, 333–337
  coastal, wave refraction and, 344–345
  density, 334–335, 446
  major surface, 335–337

currents (cont)
  tidal, 333
  turbidity, 335

cut-and-fill terrace, 283–285

cutoff (see chute cutoff; neck cutoff)

Cuvier, Georges, 132

cyclosilicates, 445

Cyprus, 124

D

dams, 379
  flood-control, 262

Danan, 46

Darcy's law, 292–294

dating:
  fission-track, 136
  plate movements, 235–236
  radioactive, 3, 133–136

Davis, William Morris, 249

debris, pyroclastic, 41–42, 44, 455

debris slides, 246–247, 446

Deccan Plateau, 51

decomposition (see chemical weathering)

deep-sea deposits, 341, 342

deep-sea trenches, 339, 340

deflation, 358, 446

deformation:
  of the earth's crust, 144–149
  of rocks, 149–174, 446
    broad surface features, 173–174
    faults, 165–169
    folds, 157–161
    gravity, 153–155
    isostasy and equilibrium, 152–153
    joints, 162–165
    strength of rocks, 151–152
    stress and strain, 149–151
    unconformities, 169–173

deglaciation, shoreline changes and, 348–349

Delaware Water Gap, 82

deltas, 282–283, 446

dendritic stream pattern, 274, 446

density, 446
  of the earth, 435

density currents, 334–335, 446

deposition (see sedimentation)

depositional remanent magnetism (DRM), 222, 446

deposits, on flood plains, 281–282

desert pavement, 358

deserts, 352–367

deserts (cont)
  definition of, 352
  deposition by the wind in, 358–364
  distribution of, 352–353
  erosion by the wind in, 358
  soils of, 365
  subtropical, 352–353
  topographic, 352
  transportation by wind in, 354–358
  water in, 365–366
  weathering in, 365

desert varnish, 365, 446

dessication, 98

detrital deposits, 93, 94

detrital sedimentary rocks, 93, 99–100, 446

deuterium, 18, 380

Devil's Post Pile National Monument, 164

Devil's Tower, 164

Devonian sedimentary rocks, 171

diabase, 446

diagenesis, 98, 116

diamond, 24, 395, 422, 446

diatomaceous mud, 342

diatomite, 101

differential erosion, 81–82, 446

differential weathering, 446

differentiation, into continental and oceanic crusts, 21

dikes, 53, 446
  formed during movements of magma, 164–165
  ring, 164

dike swarms, 54, 164–165, 446

dilatancy model, 207–208, 446

diopside, 115, 422

diorite, 61, 446

dip, 156, 439, 446

dipolar compound, 27

dipole, 27, 446

dipole magnetic field, 220, 446

dip poles, 219, 220
dip-slip faults, 165, 166, 446
direct lifting, 270
disappearing islands, 49–50
discharge, 446
  meanders and, 280
  of streams, 259–265
disconformity, 171, 446
discontinuity, 208, 446
  Mohorovičić (Moho or M), 209
  surface of, 360
discontinuous reaction series, 57, 446
discordant plutons, 53, 55–56, 446
disintegration (mechanical weather-
    ing), 70–72, 365, 452
dissolution, 98
dissolved load, 268
distributary channels, 283
disturbed bedding, 103
divergent-boundary metamorphism,
    124–125
divergent plate boundaries, 109,
    232–233, 446
divides, 272, 446
Dokuchaev, V. V., 82
dolomite, 95, 100, 114–115, 389, 422,
    446
dolostone, 95, 100

domes, 446
  exfoliation, 71
  structural, 159
drainage basins, 80–81, 272, 446
drift, 315, 446
dripstone, 100, 297, 446
DRM (depositional remanent mag-
    netism), 222, 446
drowned valleys, 338
drumlin fields, 317–318
drumlins, 317–318, 446
ductility, 151
Duluth lopolith, 54
dunes, 360–364, 446
  longitudinal, 362
  on Mars, 409
  parabolic, 362
  shoreline, 362
  slip face of, 361
  stratification in, 363
  wind shadow of, 361, 362
dust:
  in space, 327, 446
  volcanic, 41, 42
dust size, 446
dust storms, 354–358
Dutton, C. E., 152
dynamo theory of earth magnetism,
    221, 224

earth, the:
  core of, 211, 221, 445
  crust of (see continental crust;
      crust of the earth)
  magnetism of, 218–221
  mantle of, 8, 210, 451
  statistics, 433–436
  structure of, 208–212
  temperature of, 8
earthflows, 248, 446
Earthquake Lake, 182
earthquakes, 177–208 (see also seis-
    mology)
  cause of, 196–198
  control of, 208
  deformation of the earth's crust
      and, 145–148
  distribution of, 188–196
    Alaska and Aleutian Islands,
        189–195
    volcanic activity, 195–196
  effects of, 177–184
    cracks in the ground, 182–184

earthquakes (cont)
  effects of (cont)
    damage to structures, 179
    fire, 179
    land movements, 184
    landslides, 182–184
    liquefaction, 184
    seismic sea waves, 179–182
    sounds, 184
  focus of, 185
  intensity of, 185–187
  locating, 206–207
  magnitude and energy of, 187–188
  measurement and recording of,
      184–188
    early instrumental observations,
        200–201
    modern seismology, 201
  plates and, 231
  plate movements as cause of,
      198–200
  plate tectonics and, 9
  prediction of, 207–208

E

earthquakes (cont)
   volcanic activity and, 195–196
   volcanic eruptions associated
      with, 51–52
earthquake sounds, 184
earthquake waves:
   records of, 103–107
   seismic sea waves, 179–182
earth waves, 202, 446
eccentricity of the earth's orbit, 328
ecliptic plane, 402, 446
eclogite facies, 116–117, 446
elastic deformation, 150, 446
elastic energy, 197, 446
elasticity, 447
elastic limit, 446
elastic-rebound hypothesis, 197–198,
   447
elastic solid, 447
electric charges, 16, 447
electric current, 27, 447
electricity (electrical energy), 16, 27,
   447
electron-capture decay, 134, 447
electrons, 16–19, 447
elements, 17, 447
   abundance of, 20
   distribution in nature of, 19–20
   formation of new minerals by re-
      shuffling of, 34
   in minerals, 20
   origins of, 19–20
emery, 422
Emperor seamount chain, 48
emphiboles, 419
end moraine, 316, 447
energy, 368–384, 447
   alcohol, 382
   coal, 370–371
   defined, 68
   geothermal, 380–382
   geothermal gradient, 382
   kinetic, 69
   methane, 382
   nuclear, 379–380
   oil and natural gas, 371–379
   potential, 69
   solar, 382
   sources of, 68–70

energy (cont)
   tidal, 382
   water power, 379
   weathering and, 69–70
   wind, 382
enstatite, 424
entisols, 85
eolian rocks, 93
epicenter, of earthquakes, 185, 447
epidote, 115, 424, 447
epoch, 139
Epstein, A. G., 128
Epstein, J. B., 128
equator, magnetic, 219
equatorial currents, 335
equilibrium, deformation of rocks
   and, 152–153
equinoxes, precession of the, 328
era, 139
erg, 447
erosion, 68, 447
   differential, 81–82
   glaciers and, 311–315
   meanders and, 279–280
   rainwater and, 253
   rates of, 80–82
   shoreline features caused by, 346
   streams and, 270–271
   by wind, 357, 358
erosional flood plain, 278–279, 447
erratics, 318, 447
escape velocity, 447
eskers, 319, 447
Etna, Mount, 41, 44
eugeosyncline (eugeocline), 109, 447
Europa, 412–413
eustatic change, 447
eutrophic lake, 447
evaporation, 257, 447
evaporites, 100–101, 447
   continental drift and, 228–229
   marine, 393
exfoliation, 71–72, 447
exfoliation domes, 71, 447
expansion and contraction, mechanical weathering by, 70
external magnetic field, 220, 221,
   447
extrusive rock, 447

facies, 447
  defined, 116
  metamorphic, 116–117
  sedimentary, 107–108
Falcon Island, 49
fault breccias, 114
fault creep, 184
faults, 159, 165–169, 447
  dip-slip, 165, 166, 446
  normal, 166, 167
  strike-slip, 458
  thrust, 166, 167, 459
  transform, 169, 234–235
  transform boundaries and, 234
fault scarp, 182
fayalite, 28, 424
feldspars, 31–32, 73, 424, 447
  chemical weathering of, 75–77
  plagioclase, 31–32, 77, 78, 419
  potassium, 62
  in sedimentary rocks, 96
felsic rocks (felsite), 39
Fernow Experimental Forest, 265, 268
ferromagnesians, 28
  chemical weathering of, 77–78
ferromagnesium silicate, 447
fertilizers, 393–394
fibrous fracture, 447
fiery clouds, 42, 447
Finger Lakes, 315
fiords, 314–315, 337
fire, after earthquakes, 179
firn, 305, 447
fissility, 447
fission, 368, 379–380, 447
fission-track dating, 136
fissure eruption, 50, 447
flash floods, 366
flashy stream, 263, 447
flint, 95, 447
flood basalts, 51, 447
flood-control plans, 262
flood frequency, 447
flood plain, 273, 448
  of aggradation, 279, 448
  deposits on, 281–282
  erosional, 278, 447
floods, 261
  desert, 366
  flash, 366

flotation, 448
flow, 158
fluid, 448
fluorescence, 418
fluorite, 424
fluvial rocks, 93
focus of earthquake waves, 185, 448
fold mountains, 448
folds, 157–161, 448
foliated textures, 120–121
foliation, 120, 448
footwall, 166, 448
foraminiferal muds, 342
Forbes, James, 201
fore dunes, 361, 448
Forel, F. A., 185
foreset beds, 103, 282, 363, 448
foreshock, 448 (see also precursors of earthquakes)
foreshore, 345
forest cutting, stream flow and changes in, 265, 268
formations, 439
forsterite, 28, 424
fossil fuels, 106, 368–369, 379, 448 (see also coal; natural gas; petroleum)
fossils, 106, 171, 448
  continental drift and, 230
  correlation by, 132
fractional distillation, 448
fractionation, 57, 448
fracture, 418, 448
fracture cleavage, 448
Franciscan formation, 124
francium, 18
Franklin, Benjamin, 42
Frank rock slide, 243–245
free-air correction, 154
freeboard of continents, 218, 448
fringing reef, 448
frost action, 70–71, 448
  glaciers and, 310
  solifluction and, 251–252
frost heaving, 70
frost wedging, 70
fumaroles, 42, 448
fundamental particles, 448
fusibility, 418
fusion, 368, 379, 380, 448
  proton-proton, 19

G

gabbro, 39, 62, 63, 406, 448
Gal, 154
galaxy, 448
galena, 32, 390, 424, 448
Galilean moons, 411
Galileo Galilei, 404
Gamow, George, 19
gangue, 448
Ganymede, 411
Garlock fault, 169
Garmany, J. D., 208
garnets, 115, 424, 448
gases, 448
  volcanic, 41
Gauss, K. F., 222
Gay Head, 243
Geer, Baron de, 136
geodes, 105, 448
geographic poles, 219, 223–224, 448
geologic column, 3, 137–139, 448
geologic cross sections, 440
geologic maps, 439–440
geologic time (see time)
geologic-time scale, 138–141, 448
geology, 448
  defined, 1–2
  practical applications of, 11–12
geomagnetic poles, 219–220, 448
  paleomagnetic studies of, 222–224
  virtual, 222
geophysical prospecting, 448
geophysics, 448
Georgia Warm Springs, 295
geosynclines, 109, 448
geothermal energy, 368
geothermal field, 119, 448
geothermal gradient, 52, 382, 448
geothermal power, 380–382
geothermometers, 126–128
geysers, 295–296, 380–381, 448
Giant's Causeway, 164
Gibraltar, Strait of, 334
Gilbert, William, 16, 221
glacial deposits, 359
glacial marine deposits, 342
glacial theory, 322–323
glaciated valleys, 312–313
glaciation:
  causes of, 325–328
  deserts and, 366
  implications for the future, 328–329
  indirect effects of, 321–322
  late Paleozoic, continental drift
      and, 227–229
  loess deposits and, 359
  multiple, theory of, 323

glaciation (cont):
  Nebraskan, 323
  Pleistocene, 323–328, 349
  pre-Pleistocene, 325
  results of, 310–322
  shoreline changes and, 348–349
  valley versus continental, 320–321
  Wisconsin, 323
glacierets, 305
glacier ice, 448
  formation of, 304–305
glaciers, 8, 227–228, 304–333, 448
  classification of, 305–306
  distribution of, 307
  erosional effects of, 311–315
  movement of, 307–310
  movement of earth materials by,
      210–211
  nourishment and wastage of, 307
  rock, 252–253
  temperatures of, 310
  valley, 305–306, 310
glass, 59, 397, 448
glassy texture, 58–59
glauconite, 136
glaucophane, 424
glaucophane schist facies, 116
Glossopteris flora, 229–230, 448
gneiss, 122, 448
gneissic cleavage, 121, 448
Gobi Desert, 360
goethite, 77, 96, 424, 449
gold, 391–392
Goldau rock slide, 247
Gondwanaland, 227, 236, 325, 449
Gondwana System, 227
Gorda plate, 48
graben, 166, 449
grade, 449
graded bedding, 103, 335, 449
gradient of stream bed, 258–259,
      449
Grand Canyon, 106, 277
  unconformities in the, 171–173
Grand Canyon Series, 171, 173
granite, 8, 55, 60, 171, 449
  origin of, 125–126
granite pegmatite, 62–63
granitic front, 126
granitization, 55, 125, 449
granodiorite, 55, 449
granular texture, 58, 449
granule size, 97
granulite facies, 116, 449
graphic structure, 62, 449
graphite, 24, 115, 424, 449

gravel, 396
  outwash, 318–319
gravity, 153–154
  mass movement and, 241, 242
gravity anomalies, 154, 449
gravity fault, 449
gravity meter (gravimeter), 154, 449
gravity prospecting, 449
graywacke, 99, 449
Great Basin, 174
Great Glen fault, 164, 167
Great Lakes area, 348
greenhouse effect, on Venus, 408
Greenland, 306, 307
Greenland current, 337
Green River shales, 137, 376–378
greenschist, 121–122, 449
greenschist facies, 116, 119, 449
grooves, glacial, 311
groundmass, 59, 449

ground moraine, 316, 449
groundwater, 289, 292–297, 449 (see
    also underground water)
  artesian, 294
  caves and, 297–300
  geysers, 295–296
  problems caused by human use
    of, 301
  recharge of, 296–297
  springs, 292–293
  thermal springs, 294–295
  wells, 293–294
groundwater table, 449 (see also
    water table)
Guatemalan earthquakes, 199
Gulf Stream, 335
Gutenberg, B., 187, 189
guyots, 340, 449
gypsum, 32, 96, 100–101, 393, 394,
    397, 424, 449

H, 449
hackly fracture, 449
Halema'uma'u, 41
half-life, 133, 449
halides, 32, 449
halite, 20, 25, 96, 100, 393–395, 424,
    449 (see also rock salt)
  crystals of, 21
Halley, Edmund, 256–257
Hall, James, 109
hanging glaciers, 305
hanging valleys, 314–315, 449
hanging wall, 165, 449
hardness of minerals, 23, 415–417,
    449
harmonic tremor, 52
Harris, L. D., 128
Harvard seismograph station, 209
Haute Feuille, J. de la, 200
Hawaii, 48, 180–182, 339
  crust under, 210
Hawaiian Island-Emperor Seamount
    Ridge, 196, 339
Hawaiian Islands, 48–49
  seismic sea waves, 179–182
Hays, James D., 328
Hayward fault, 148
head, 291, 449
headstones, weathering of, 79–80
heat energy, 449
heat exchanger, 449
heat flow, earth's, 52, 449

heat, metamorphism and, 113 (see
    also temperature)
Hebgen Reservoir, 182
helium, 17–19
  abundance of, 20
hematite, 24, 32, 77, 96, 106, 424, 449
Hêng, Chang, 200
Henry Mountains, 54
Herculaneum, 45–46
Highland Boundary fault, 164
High Plains, 358
hills, asymmetric, 315
hinge fault, 449
historical geology, 449
histosols, 85
Hollis Foundation, 295
horizons, 85–86
horizontal displacements, 147–148
horizontal movements, of the earth's
    crust, 9–11
hornblende, 29–30, 424, 449
hornfels, 449
hornfels facies, 450
horns, 312, 449
horse latitudes, 304–305
horst, 166, 450
Horton, Robert, 272
hot spots, 48, 450
hot springs, 294–295, 380–381, 450
Hoyle, Fred, 19
Hualalai, 48
Hubbard Glacier, 305

H

Hudson River, 338, 339
humus, 68
Hurley, P. M., 230
Hutton, James, 3, 7, 260
hydration, 75, 450
hydraulic gradient, 291, 450
hydraulic mining, 450
hydrocarbon, 450
hydroelectric power, 450 (see also
    water power)
hydrogen, 17–20
  abundance of, 20

hydrogen (cont)
  origin of the elements and, 19, 20
hydrograph, 259–260, 262–265, 450
hydrologic cycle, 255–257, 450
hydrothermal activity, mineral de-
    posits and, 386
hydrothermal solutions, 62, 450
  metamorphism and, 113–114
hydrous aluminum silicates, 75, 76
hypersthene, 424
hysteresis, 450

ice, 32
Ice Age, 227, 278, 323–328, 359, 366
    (see also Pleistocene glacia-
    tion)
icebergs, 307
icecap, 306, 450
Iceland, 42, 50, 51
ice sheets, 227, 306, 450
igneous activity:
  causes of, 52
  mineral deposits and, 386
igneous rocks, 5, 53–63, 450
  chemical weathering of, 74–78
  formation of, 56–58, 386
  geologic-time scale and, 139
  geothermal power from, 381–382
  magnetism of, 221–222
  masses of, 53–55
  on the moon, 63
  in rock cycle, 6–7
  texture of, 60–63
  types of, 60–63
Illinoian glaciation, 323
illite, 76, 424, 450
ilmenite, 389, 450
Imbrie, John, 328
Imperial Valley, 117
inceptisols, 85
inclined bedding, 103, 450
index minerals, 450
Indian Ocean, 337

induced magnetism, 221–222, 450
industry, sediment production and,
    268
inequilibrium, 154
inertia member, 202, 450
infiltration, 257, 450
inosilicate, 450
intensity, 450
interglacial period, 327
intermediate belt, 289, 450
intermediate lava, 450
intrusive rock, 450
Io, 411–412
ionic bond, 450
ionic compounds, 25–26
ions, 17, 450
  radii of, 25–26, 450
iron, 386–388
iron oxides, 87, 106
island-arc deeps, 450 (see also
    deep-sea trenches)
islands:
  barrier, 348, 443
  disappearing, 49–50
isoclinical folding, 450
isograds, 117, 450
isomorphism, of minerals, 28, 32
isoseismic lines, 187, 450
isostasy, 9, 153, 450
  shoreline changes and, 348–349
isotopes, 18, 450

jadeite, 426
Japan, 42, 183
  earthquakes, 197
  seismic sea wave, 182
jasper, 450

Java, 46, 154
jet flow, 258
joints, 71, 78, 162–165, 450
  columnar, 160, 164

I

J

joints (cont)
  formed during movements of
    magma, 164–165
  sheeting, 164
joint systems, 162, 450

Juan de Fuca plate, 48
Juan de Fuca Ridge, 225
Jupiter, 402, 403, 411–413
Jura Mountains faults, 169
juvenile water, 450

kames, 319, 450
kame terraces, 319, 450
Kansan glaciation, 323
Kansu, 207
Kansu landslide, 182
kaolinite, 75–76, 87, 426, 450
karst topography, 300, 450
Katmai, Mount, 41, 42
Kay, Marshall, 109
Kelleys Island, 311
Kenya, Mount, 307
Kerguelen-Gaussberg ridge, 450
kerogen, 372

kettles, 319, 451
Kilauea, 41, 43, 48–49, 52, 210
kimberlites, 395
kinetic energy, 69, 451
Kiruna iron mining district, 387
Knopf, Adolf, 4
Kodiak Island, 194, 195
Kohala, 48
Kosaka, 149
Krakatoa, 46, 79, 179
Kupferschiefer deposits, 390
Kutiah Glacier, 309
kyanite, 115, 116, 451

L, 451
Labrador current, 337
labradorite, 426
laccoliths, 54, 451
lacustrine rocks, 93
lag time, 259, 451
Lake, Mount, 42
lakes:
  base level of stream formed by,
    265
  oxbow, 279
  paternoster, 313
  playa, 366
  pluvial, 366
Lake Superior district, 387, 388
Laki Fissure, 50
laminar flow, 258, 451
laminated bedding, 103
landslides, 243–248, 310, 451
  earthquake-caused, 182–184
  warnings of, 247–248
land subsidence, 301
Langbein, W. B., 257
lapilli, volcanic, 41, 451
Larderello geothermal field, 381
large waves, 451
La Sal Mountains, 54
latent heat of fusion, 451
lateral moraine, 316, 451
laterites (tropical soil), 88, 451

latitude, 438
Laurasia, 236
lava floods, 50–51
lavas, 38–41, 451
  aa, 40–41, 441
  composition of, 39–41
  pahoehoe, 40
  temperature of, 41
leaching, 86
lead, 390
lead-206, 133, 134
leading edge, 451
left-lateral fault, 451
legend, 439
Leopold, Luna B., 257, 263
lepidolite, 63
levees, natural, 280, 451
Libby, Willard F., 135
lightning, 70
lignite, 102, 127, 370, 451
limbs, fold, 159, 451
limestone, 100, 396, 451
  abundance of, 102
  biochemically formed, 101–102
  in caves, 297–300
  marine, 93
limonite, 24, 33, 77–78, 96, 426, 451
liquefaction, 451
liquid, 451
Lisbon earthquake, 182–183, 188

K

L

lithic sandstone, 99 (see also graywacke)
lithification, 97–98, 451
lithosphere, 8, 211, 451
lithostatic pressure, 151
load, of a stream, 266
loess, 358–360, 451
longitude, 438
longitudinal dunes, 362, 451

longitudinal wave, 451
lopoliths, 54, 451
low-grade metamorphism, 115, 117–120
luster, 418
    metallic, 18
L waves, 204, 206, 210
Lystrosaurus, 230

M

Mackenzie Valley, 311
Madison River Canyon rock slide, 245
mafic rocks, 39
magma, 5, 38–41, 386, 451
    crystallization of, 56–58
    depth of source of, 52
    igneous activity and, 52, 56
    joints and dikes formed during movements of, 164–165
    textures of, 58–60
magmatic secretion, 387, 451
magnesium, 389
magnetic declination, 219, 451
magnetic equator, 219
magnetic field:
    secular variation of, 219
    variations in intensity and reversals of, 222
magnetic inclination (dip), 219, 451
magnetic poles, 219–220, 451
magnetic reversal, 451
magnetism, 218 (see also paleomagnetism)
    earth's, 218–221
    depositional remanent (DRM), 222
    induced, 221–222
    of minerals, 416
    natural remanent (NRM), 221–222
    of rocks, 221–222
    thermoremanent (TRM), 221–222
magnetite, 32, 416, 426, 451
magnetosphere, 220, 451
magnetostratigraphy, 139, 141, 451
magnitude of earthquakes, 451
manganese, 389
manganese nodules, 343, 451
Mann, Thomas, 2
mantle of the earth, 8, 210, 451
maps:
    geologic, 439–440
    topographic, 437–439

Maquoketa Formation, 108
marble, 79, 122–123, 451
Marbut, C. F., 82, 84
marcasite, 24
Mare Imbrium, 405
maria, 405, 451
marine evaporites, 393
marine limestone, 93
Mariner spacecraft, 408
Mariotte, Edmé, 256
marl, 451
Mars, 67, 402, 403, 408–410
marsh gas, 451 (see also methane)
Martinique Island, 42
mascons, 451
mass, 452
    atomic, 17, 442
massive plutons, 54
mass movements, 241–254, 452
    behavior of material in, 243
    creep, 249–250
    debris slides, 246–247
    earthflows, 248
    factors of, 241–243
    landslides, 243–248
    mudflows, 248
    rapid, 243–248
    rock slides, 243–246
    slow, 249–253
    slump, 243
    solifluction, 250–252
    talus, 248
mass number, 17, 18, 452
mass unit, 452
matter, 19, 452
    defined, 15
Matterhorn, 312
Matthews, D. H., 225
Mauna Kea, 48
Mauna Loa, 38, 48
Mayon, Mount, 42, 44
Mazama, Mount, 43–44
meanders, 279–282, 452

mechanical weathering, 70–72, 365, 452
Medford dike, 53
medial moraine, 316–317, 452
Mediterranean and trans-Atlantic earthquake zone, 189
Mediterranean Sea, 334
mélange, 124, 452
melt, 56–57
meltwater hypothesis, 312
Menderes River, 279
Mercalli, L., 186
Mercalli intensity scale, 186–187, 452
Mercury, 402, 403, 407–408
Mesosaurus, 230
mesotrophic, 452
metallic bond, 27, 452
metallic luster, 18
metalloids, 18, 452
metals, 18, 452
    abundant, 386–389
    deposits of, 385–393
    scarce, 389–393
metamorphic facies, 452
    regional, 116–117
metamorphic rocks, 5, 112, 120–123, 452
    in rock cycle, 6–7
    textures of, 120–121
    types of, 121–123
metamorphic zones, 452
    regional, 115–116
metamorphism, 112–120, 452
    agents of, 113
    burial, 443
    cataclastic, 114, 444
    contact, 114–115, 445
    convergent-boundary, 123–124
    divergent-boundary, 124–125
    plate boundaries and, 123–125
    regional, 115–120
    transform- (parallel), 125
metasomatism, 113–114, 452
meteoric water, 452
meteorites, 407, 409, 452
meteors, 452
methane, 368, 371, 382, 452
Metis Shoal, 49–50
micas, 452 (see also biotite; muscovite)
    in sedimentary rocks, 96
microscopes, scanning electron, 22
microseism, 452
Mid-Atlantic Ridge, 10, 124, 169, 235, 339

midocean ridges, 236, 237, 339, 452
    (see also ocean ridges)
migmatites, 126, 452
Milankovitch hypothesis, 327
milligal, 154
Milne, John, 202
Mindanao Deep, 155
mineral deposits, 385–395, 452
    geologic processes responsible for, 386
    metallic, 385–393
mineraloids, 33, 452
minerals, 4, 15–36, 452
    associations of, 34
    carbonate, 32, 444
    characteristics of, 20–24, 415–419
    cleavage of, 22, 444
    color of, 23
    common, 27–33, 418–431
    composition of, 20
    contact metamorphic, 114–115
    crystalline structure of, 24–27
    crystal symmetry of, 415
    crystal systems of, 415
    defined, 15, 33–34
    ferromagnesians, 28–30
    fluorescence and phosphorescence of, 418
    fracture of, 418, 448
    fusibility of, 418
    halides, 32, 449
    hardness of, 23, 415–417
    luster of, 418
    magnetism of, 416
    nonferromagnesians, 30–32
    organization of, 33
    oxide, 32
    piezoelectricity of, 417
    pyroelectricity of, 416
    regional metamorphic, 115
    reshuffling of elements and formation of, 34
    rock-forming, 27–28
    in sedimentary rocks, 95–96
    silicates, 28–32
    solubility of, 418
    specific gravity of, 23, 416, 417
    streak of, 24
    striations of, 22
    sulfate, 32
    sulfide, 32
    tenacity of, 419
mining, 385–386
miogeosyncline (miogeocline), 109, 452
Mississippi delta, 280

Mississippi River, 258, 272, 358
modulus of elasticity, 452
Mohorovičić (Moho, or M) discontinuity, 209, 232, 452
Mohs scale of relative hardness, 415, 416
molecules, 25, 452
mollisols, 85
molybdenum, 391
Monel metal, 452
monoclines, 158, 452
montmorillonite, 76, 426, 452
moon(s), 452
    earth's, 404–407
    igneous rocks on, 63
    of Jupiter, 411
    of Saturn, 414
moraines, 315–317, 452
Morgan, W. Jason, 48
Moru Rock, 72

Motagua fault zone, 199
mottled bedding, 103
mountain chain, 452
mountain glaciers, 305 (see also valley glaciers)
mountain range, 452
mountains, 174, 452
    reasons for continued elevation of, 155–156
    transform faults and, 169
mountain structure, 452
mudcracks, 103, 104, 452
mudflows, 248, 452
mud, on ocean floors, 341, 342
mudstones, 99–100, 452
    abundance of, 102
Muir glacier, 79
muscovite, 30–31, 80, 426, 453
mylonites, 114, 453

N

Naples, Bay of, 45, 46
Nashville floods, 263
natural gas, 371–376, 453
natural remanent magnetism (NRM), 221–222, 453
natural resources, 369–370, 397–398, 453
Nazca plate, 10–11
neck cutoff, 279, 453
negative pole, 27
Neptune, 402, 403
Nero Deep, 154
networks of streams, 273
Neumann, Frank, 186
neutrons, 17–19, 453
névé, 305, 453
New Jersey, coastline of, 346–347
Newton, Isaac, 153
New Zealand, 182
New Zealand earthquake, 187
Niagara Falls, 278
nickel, 390–391

Nile River, 282, 365
nitrates, 393
nitrogen, 268, 393
nivation, 311
nodules, 104–105, 453
nonclastic texture, 97, 453
nonconformity, 171, 453
nondipole magnetic field, 220, 453
nonferromagnesians, 30–32, 453
nonmetals, 18, 453
    deposits of, 385, 393–395
norite, 63
North Atlantic current, 336, 337
North Atlantic deep water, 335, 453
North Atlantic plateau, 51
northeast trade winds, 353, 453
northern hemisphere, 331
north magnetic (dip pole), 219
NRM (natural remanent magnetism), 221–222, 453
nuclear power, 379, 380
nucleus, atomic, 17, 453

O

Oahu, 180–182
obduction, 235, 453
oblique slip fault, 453
obliquity of the ecliptic, 328
obsidian, 61, 397, 453
ocean basins, 235, 236, 337

ocean basins (cont)
    age of, 218
    floors of, 339
    origin and permanence of, 215–218
ocean crust, 8–11, 210

ocean crust (cont)
  plate tectonics and, 235–238
ocean floors:
  sediments on, 340–342
  topography of, 337–340
ocean ridges, 234–235, 339 (see also
    midocean ridges)
  aseismic, 442
  transform faults across, 169
oceans, 331–351
  currents of, 333–337
  depth of, 331
  shorelines of, 343–347
ocean water:
  composition of, 434
  constituents dissolved in, 332–333
  origin of, 331–332
octahedral face, 21
offshore, 345
oil (petroleum), 371–379, 453
  accumulation of, 372–375
  origin of, 371–372
  search for, 375–376
oil shale, 376–378, 453
Old Faithful Geyser, 296
oligoclase, 426
oligotrophic lake, 453
olivine, 28, 78, 426, 453
Olympus Mons, 38, 409

oölites, 100, 453
oozes, 342, 453
opal, 33, 95, 426, 453
opalescence, 31
open-pit mining, 386, 453
ophitic, 453
Oquirrh Range fault, 167
orbital hypothesis, 327
ore deposits, 386, 453
organic matter, 96
organic rocks (biochemical rocks),
    93, 101–102, 443
orogeny, 453
orthoclase, 31, 75–77, 87, 426, 453
orthoquartzite, 99, 453
oscillation, ripple marks of, 103–104
osmosis, 106
outcrop, 439
outlet glaciers, 306
outwash, 318–319, 453
outwash plains, 319
overbank deposits, 281, 453
overthrusts, 167
overturned fold, 453
oxbow, 279, 453
oxbow lake, 279, 453
oxide minerals, 32, 453
oxisols, 85

P, 453
Pacific basin, 210
Pacific Ocean, 331, 337, 432
Pacific plate, 199, 231, 238, 339
pahoehoe lava, 40, 453
Painted Desert, 106
paired terraces, 285, 453
paleomagnetic poles, 223–225
paleomagnetism, 218, 221–224, 453
  continental drift and, 225, 230
paleosols, 88–89, 453
Paleozoic Series, 173
Palisades, The, 53
Pangaea, 225, 454
parabolic dunes, 362, 454
parallel bedding, 103
parallel-boundary metamorphism,
    125
parallel plate boundaries, 109
Parícutin, Mount, 45
particles, 15–17
particle size, chemical weathering
    and, 73

paternoster lakes, 313, 454
patterns of stream systems, 273, 274
peat, 102, 127, 454
pebble size, 97, 454
pedalfers, 86–87, 454
pedocals, 87–88, 454
pedology, 82, 454
pegmatites, 62–63, 454
  complex, 63
  granite (simple), 62–63
pelagic deposits, 342, 454
Pelée, Mount, 42
pelites, 99–100
pendulum, 454
penlandite, 391
Perboewatan, 46
perched water table, 293, 454
peridotite, 61, 454
period, 139, 454
periodic chart, 18, 432
permeability, 454
  of earth materials, 291–292

Perrault, Pierre, 256
Perret, Frank A., 41
Peruvian earthquakes, 188
petroleum (see oil)
phaneritic texture, 58, 454
phase, 454
phenocrysts, 59, 454
phosphate rock, 454
phosphorescence, 418
Phosphoria Formation, 394
phosphorous, 268, 394
photosynthesis, 454
phyllite, 121, 454
phyllitic cleavage, 121, 454
phyllosilicate, 454
Piedmont area of Maryland, 268
Piedmont glaciers, 306, 454
piezoelectricity, 417
pitted outwash plains, 319
placer deposits, 386, 454
placer mining, 386
plagioclase, 87, 426
plagioclase feldspars, 31–32, 77, 78, 419
plains, 174
planetary winds, 230
planetology, 454
planets, 402–403, 454
plants (see also fossils):
    chemical weathering and, 73–74, 75
    continental drift and, 229–230
    mechanical weathering and, 72
plaster, 397
plastic deformation, 454
plastic solid, 454
plateau basalts, 51, 454
plateaus, 174
plate boundaries:
    classification of, 232–234
    convergent, 109, 123–124, 232, 233, 445
    divergent, 109, 124–125, 232–233, 446
    metamorphism and, 123–125
plates, 8–11, 34, 454
    crustal, 211
    dating the movements of, 235–236
    makeup of, 232
    movement of, 234–236
    world system of, 231
plate tectonics, 8–11, 215, 231–238, 454
    erosion and, 285
    rock cycle and, 11
    sedimentation and, 108–109

platinum, 392
playa lakes, 366, 454
playas, 366, 454
Pleistocene glaciation, 323–328, 349
    (see also Ice Age)
pleochroic halos, 134, 454
plucking, 310
Pluhowski, E. J., 270
plume, 48, 454
plume theory, 235
plunge of a fold, 159, 454
Pluto, 402–403
plutons, 53–56, 454
pluvial lakes, 366
pluvial periods, 366, 454
podsol, 87, 454
point bars, 281, 455
Poisson's ratio, 455
polar compound, 455
polarity epochs, 139, 141, 222–223, 455
polarity events, 141, 223
polarity, reversed, 141
polar wandering (or migration), 455
Poldervaart, Arie, 92–93
poles:
    geographic, 219, 223–224, 448
    geomagnetic, 219, 448
        paleomagnetic studies of, 222–224
        virtual, 222
    magnetic (dip), 218–220
    movement (migration) of, 223–225
    paleomagnetic, 223–225
polish, 310
pollution of underground water, 301
polymorphism, 22, 455
Pompeii, 45–46
porosity of rocks, 291, 455
porphyritic texture, 59, 455
porphyry, 59–62, 455
porphyry copper ores, 390
Port Royal landslide, 182
positive charge, 455
positive pole, 27
potassic feldspar, 62, 455
potassium, 393, 394
potassium 40, 134–135
potential energy, 69, 455
pozzuolana, 396
Pratt, J. H., 152
Pratt hypothesis, 455
Precambrian rocks, 138, 171, 173, 230
precession of the equinoxes, 328

precipitation, 94, 455
  stream flow and, 256–260
precursors of earthquakes, 207, 455
pressure, 455
  lithostatic, 151
  metamorphism and, 113
pressure-temperature fields for metamorphic facies, 117
prevailing westerly winds, 353, 455
primary craters, moon, 405
primary wave, 455
Prince William Sound earthquake of
  1964, 146, 191–195
profiles, topographic, 439
promethium, 18

proton-proton fusion, 19, 455
protons, 16–19, 455
psilomene, 389
pumice, 42, 455
push-pull waves, 203, 455
Puyehue, 195
P waves, 204, 206, 210, 211
pyrite, 20, 24, 32, 426, 455
pyroclastic debris, 41–42, 44, 455
  igneous rock formation and, 56
pyroclastic rocks, 96
pyroelectricity, 416
pyrolusite, 389
pyroxenes, 29, 419, 455
pyrrhotite, 416, 455

Q

quarrying, 310, 386
quartz, 32, 62, 78, 87, 95, 114–115,
  416, 428, 455
  chemical weathering of, 75
  crystals of, 21–22

quartz arenite, 99
quartzite, 123, 455
quartzrose sandstone, 99
quick lime, 396

R

radar, 455
radial drainage, 455
radial stream pattern, 274
radioactive dating, 3, 133–136
radioactive decay, 133–136
radioactivity, 3, 133–136, 455
radiocarbon years, 136
radiolarian muds, 342
rain (see also precipitation):
  acid, 370
raindrops, in erosion, 253
rain wash, 253, 455
Rakata, 46
Rance River, 382
range, 438, 455
rank, 455
Raoul Island, 51–52
rapids, 278
Raritan Formation, 439
Rat Islands earthquake, 184
ray craters, 406
rays, 455
rays of Copernicus, 405
reaction principle, Bowen's, 57–58,
  443
reaction series, 57, 78
Rebeur-Paschwitz, Ernst von, 201
recessional moraines, 316, 455

recorder, 455
  seismograph, 202
rectangular stream pattern, 274, 455
recumbent fold, 455
Red Sea, 124–125
reefs, 101
  barrier, 443
  coral, 229
Reelfoot Lake, 146
reflection seismic prospecting, 455
refraction of waves, 344–345
refraction seismic prospecting, 455
refractory minerals or compounds,
  455
regional metamorphic facies,
  116–117
regional metamorphic zones,
  115–116
regional metamorphism, 115–120
  low-grade, 117–120
regolith, 68
Reiche, Parry, 78
Reid, H. F., 197
rejuvenation, 455
relative time, 2–3, 130–133
relief, 156
replacement, 98
reptiles, continental drift and, 230

reservoir rock, 374
reversed polarity, 141
reverse fault, 166
Reykjanes Ridge, 225
rhyolite, 39, 61, 456
Richter, C. F., 187, 189
Richter scale, 188
Rift valley, 10
rift zone, 456
right lateral fault, 456
rigidity, 150, 456
rilles, 456
rills, 253, 456
ring dikes, 53, 164, 456
ripple marks, 103–104, 456
    of oscillation, 456
rivers (see streams)
rock basins, 313
rock cycle, 5–7, 456
    plate tectonics and, 11
rock flour, 310, 456
rock formation, mineral deposits
    and, 386
rock-forming silicate minerals,
    27–28, 456
rock glaciers, 252–253
rock knobs, asymmetric, 315
rock melt, 456
rock(s), 4–7, 456
    crushed, 396

rock(s) (cont)
    igneous (see igneous rocks)
    metamorphic (see metamorphic
        rocks)
    moon, 406
    sedimentary (see sedimentary
        rocks)
rock salt, 100–101 (see also halite)
rock slides, 243–246, 456
rock units, 139
rock waste, 248
rolling of particles in a stream's bed
    load, 267
Rossi-Forel intensity scale, 185–186,
    456
Rossi, M. S. de, 185
rubidium 87, 135
running water, 255–286 (see also
    streams)
    composition of, 435
    on Mars, 409
    work of, 266–273
        deposition, 273
        erosion, 270–273
        transportation, 266–269
runoff, 257–258, 456
    from the continents (table), 434
rupture, 456
rutile, 389

S, 456
Sahara Desert, 325, 360, 362
Saidmarreh landslide, 246
St. Helens, Mount, 6, 47–48
Saint Pierre, 42
Salsano, D. Domenico, 201
salt, 456 (see also halite)
saltation, 356, 456
    of particles in a stream's bed
        load, 267
Salton Trough, 117–120
saltwater invasion, 301
saltwater wedge, 456
San Andreas fault, 109, 147–148,
    169, 184, 197, 231
sand, 93, 341, 396, 456
    outwash, 318–319
sand deposits, 360–364
sand grains, 74
    abrasion by, 357
sand size, 97, 456
sandstone, 99, 171, 456
    abundance of, 102

sandstorms, 354–356
Sandy Hook, 347
San Francisco earthquake (1906),
    179
San Joaquin Valley, 301
Santa Monica Mountains, 248
sapropel, 456
satellite craters, 405, 456
saturation, zone of, 289–290
Saturn, 402, 403, 413–414
scale of a map, 437–438
scanning electron microscopes, 22
scheelite, 393
schist, 121–122, 456
schistose cleavage, 121, 456
Scotland, 164
scratchability of minerals, 23
sea, 344
sea arch, 346
sea caves, 346
sea cliff, 345
sea-floor spreading, 8, 10, 215,
    225–226, 236, 456

S

sea level, 348–349

seamounts, 339–340, 456

seawater, 331–333

  composition of, 434

secondary crater, 456

secondary wave, 456

secular changes, 219

Sederholm, J. J., 126

sedimentary facies, 107–108, 456

sedimentary rocks, 5, 91–109, 456

  abundance of, 102

  amounts of, 92–93

  chemical, 93, 96, 100–101, 444

  classification of, 98–99

  color of, 106

  concretions of, 105

  correlation of, 131–133

  dating of, 136–138

  defined, 91

  detrital, 93, 99–100, 446

  formation of, 93–98

  fossils in, 106

  geodes of, 105

  geologic-time scale and, 139

  lithification of, 97–98

  magnetism of, 222

  mineral composition of, 95–96

  mud cracks of, 103, 104

  nodules of, 104–105

  origin of material that forms, 93

  radioactive dating of, 136

  relative age of, 130–133

  ripple marks of, 103–104

  in rock cycle, 6–7

  size of the individual particles of, 96, 106

  sole marks of, 103, 104

  stratification (bedding) of, 91, 102–103

  texture of, 96–102

  types of, 98–102

sedimentation (deposition), 94–95, 456

  meanders and, 279–280

  of ocean floors, 340–342

  plate tectonics and, 108–109

  rate of, 136–137

  shoreline features caused by, 346–347

  in streams, 272

  by winds, 358–364

sediment production, 268

seif dunes, 362, 456

seismic gap, 199

seismic prospecting, 456

seismic sea waves, 179–182, 456

seismic-sea-wave warning system (SSWWS), 180–181

seismogram, 456

seismograph, 202, 456

seismograph stations, worldwide network of, 184

seismology, 201–202, 456

Seneca, 256

serpentine, 428, 456

serpentinized peridotites, 40

Seward, 194

Seymour Narrows, 333

Shackleton, N. J., 328

shake waves, 203, 456

shales, 99–100, 376–378, 456–457

  abundance of, 102

  Green River, 137

shear, 149, 151, 457

shear strength, 151

shear stress, 151

shear wave, 457

sheeting, 164, 457

sheet wash (sheet flow), 253, 457

Shepard, Francis P., 180

Sherman glacier slide, 246

shields, 173–174, 457

Shield volcano, 44

shooting flow, 258

shoreline dunes, 361–362

shorelines, 343–347

  deposition-caused features of, 346–347

  erosional features along, 346

  features of, 345–347

  waves and, 343–345

sial, 209, 457

sialic rocks, 60, 457

siderite, 428, 457

Sierra Nevada batholith, 55

Sierra Nevada fault, 167

Sierra Nevada range, 352

silica, 95 (see also quartz)

silicates, 28–32, 419, 457

  regional metamorphism and, 115

silicic, 38

silicified geodes, 106

silicon-oxygen tetrahedron, 28, 457

sillimanite, 115, 116, 428, 457

sills, 53

silt size, 97, 457

silver, 391.

sima, 209–210, 457

simatic front, 126

simatic rock, 457

simple pegmatite, 62–63

sinkholes (sinks), 299–300, 457

Skapta, Mount, 50
Skinner, Brian J., 386
slate, 79–80, 121, 457
slaty cleavage, 121, 457
slide rock, 248
sliding of particles in a stream's bed
    load, 267
slip face of dunes, 361, 457
slope failure, 243
slump, 243, 457
slurries, 291
Smith, William, 131–132
Snake River, 51, 164
snowfields, 304, 457
snow, glacier ice formation and,
    304–305
snow line, 304, 457
soapstone (see talc)
soda-lime feldspars (see plagioclase
    feldspars)
sodic feldspar, 457
soil, 68, 82–88, 457
    classification of, 82–86
    creep of, 249–250
    of deserts, 365
soil horizons, 85–86, 457
solar constant, 69, 327, 457
solar energy, 368, 382
solar system, 401
sole marks, 103, 104, 457
solid, 457
solid solution, 28, 32, 457
solifluction, 250–252, 457
solubility of minerals, 418
solution:
    hill slopes reduced by, 253
    in streams, 266, 272
Somma, Mount, 45–46
sorosilicates, 457
sounds, earthquake, 184
sound waves, 203
source rock, 372
South America, 225, 227, 237–238
South Atlantic currents, 335, 336
southeast trade winds, 353, 457
southern hemisphere, 331
Southern Upland fault, 164
south magnetic (dip) pole, 219
space lattice, 457
specific gravity, 457
    of minerals, 23, 416, 417
specific heat, 457
sphalerite, 32, 390, 428, 457
spheroidally weathered boulders,
    71–72
spilites, 39

spinel, 428
spinel group, 428
spit, 347, 457
spodosols, 85
spodumene, 62, 428
springs, 292–293, 457
    thermal, 294–295
stack, 346, 457
stalactites, 299, 457
stalagmites, 299, 457
Stansbury Range fault, 167
stars, 457
    origin of the elements and, 19, 20
staurolite, 115, 428, 457–458
steady-state theories, 19, 20, 458
Steno, Nicolaus, 131
stibnite, 428
stock, 55, 458
stony tills, 315
stoping, 55, 458
storms, sand and dust, 454, 457
strain of rocks, 149–151, 458
strain unit, 149
stratification, 458
    in dunes, 363
    of sedimentary rocks, 91, 102–103
stratigraphic trap, 458
streak, 458
    of minerals, 24
streak plate, 24
stream bed, gradient of, 258–259
stream channels, 261–265
stream flow:
    laminar and turbulent, 258
    precipitation and, 256–260
    velocity of, 258–259
streams (see also valleys):
    adjustments of discharge, veloc-
      ity, and channel of, 260–264
    alluvial fans, 283
    base level of, 265
    bed load of, 267
    braided, 280, 443
    capacity of, 266
    competence of, 266
    deltas of, 282–283
    deposition in, 272
    in desert regions, 365–366
    discharge of, 259–265
    economy of, 260–265
    erosion and, 80–81, 270–271, 285
    flashy, 264
    floods and, 261–262
    hydrograph of, 259–260
    hydrologic cycle and, 255–257
    load of, 266

streams (*see also* valleys) (cont):
  loess deposits and, 359
  meanders, 279–282
  networks of, 272
  patterns of, 272, 274
  plate tectonics and, 285
  sedimentation and, 94
  superimposed, 274–276
  suspension of particles in, 266–267
  transportation of material by,
    266–269
  urbanization and suburbanization
    and, 262–264
  velocity of, 258–259, 261
  yazoo-type, 280
stream terraces, 283–285, 458
strength of rocks, 151–152, 458
stress of rocks, 149–151, 458
stress-to-strain ratio, 150
striations, 22, 458
  glacial, 311
strike, 157, 439, 458
strike-slip faults, 164, 166–169, 458
strip mining, 386, 458
structural domes, 161
structural features, 165–174
structural relief, 156, 458
subduction, 458
subduction zones, 235, 458
sublimation, 305, 458
subsequent stream, 458
subsidence, 301
subsurface water, 458 (*see also*
  underground water)
subterranean water, 458 (*see also*
  underground water)

subtropical deserts, 352–353
subtropical high-pressure zones,
  353, 458
suburbanization, streams and,
  262–264
Sudbury district, 391, 392
Suess, Hans, 135–136
sulfate minerals, 32, 458
sulfide minerals, 32, 458
sulfur, 394
Sumatra, 46, 154
Sumbawa Island, 42
sun, 401–402
  as ultimate source of renewable
    energy, 370
Sunda Strait, 46
superimposed streams, 274–276
superinterglacial period, 329
superphosphate, 394
superposition, law of, 130–131, 458
surf, 344
surface of discontinuity, 360
surface waves, 203, 458
surges of glaciers, 309, 458
Suribachi, Mount, 148–149
suspended water, 289, 458
suspension, 80, 458
  in streams, 266–267
swamps, 279
  back, 280
S waves, 204, 206, 210, 211
swells, 344
sylvite, 32, 393
symmetrical fold, 458
symmetry, crystal, 415
synclines, 159, 458

tablemount (*see* guyots)
tabular, 458
tabular plutons, 53–54
taconite, 387, 428, 458
Takeyo, 149
Takla Makan Desert, 352
talc, 428, 458
talus, 248, 458
Tambora, Mount, 42
tantalite, 63
tarns, 311, 458
tar sands, 378
technetrium, 18
tectonism, 145
tectosilicate, 458
temperature (*see also* heat):
  of coal formation, 126–127

temperature (*see also* heat) (cont)
  conodont color changes and, 128
  Curie, 221, 445
  of the earth, 8
    geothermal gradient, 52
    geothermometers used to deter-
      mine, 126
    of glaciers, 310
  mechanical weathering and, 70
temporary base level, 458
tenacity, 419
tension, 458
  divergent plate boundaries and,
    233
tension strength, 151
tension stress, 151
terminal moraine, 316, 458

terminal velocity, 267, 355, 458
terraces, 458
  kame, 319
  stream, 283–285
  wave-built, 346
  wave-cut, 345
terrae, 459
terrigenous deposits, 342, 459
tetrahedron, 459
texture, 459
  of igneous rocks, 58–61
  of metamorphic rocks, 120–121
  of sedimentary rocks, 96–97
Tharsis Montes, 409
Thellier, Emile, 222
thermal gradient, 459
thermal pollution, 269, 301, 459
thermal springs, 293–295, 459
thermoremanent magnetism (TRM),
  221–222, 459
thin section, 459
tholeiite, 459
tholeiitic basalts, 39
Trailkill, John, 298
thrust faults, 166, 167, 459
Tiber River, 81
tidal currents, 333–335, 459
tidal inlet, 347, 459
tidal power, 368, 382
tidal wave, 459
tide, 459
tillite, 99, 459
tills, 315, 459
time (geologic time), 2–4, 130–141
  absolute, 2–3, 133–137, 139, 441
  deformation of rocks and, 151
  divisions of, 3
  relative, 203, 130–133
time-distance graphs, 204, 459
time scale, 138–141
time units, 139
tin, 392–393
titanium, 389
tombolo, 459
Toncítaro, Mount, 47
Tonga Archipelago, 49–50
topaz, 63, 428
topographic deserts, 352, 459

topographic maps, 437–439
topographic profiles, 439
topography, 459
topset beds, 282–283, 363, 459
tourmaline, 63, 430, 459
townships, 438
traction, 459
traction load, 80–81
trade winds, 353
trailing edges of continents, 337, 459
transcurrent faults (see strike-slip
  faults)
transducer, 202, 459
transform boundary metamorphism,
  125
transform faults, 169, 234–235, 459
transform plate boundaries, 109,
  232–234
transition element, 459
transpiration, 459
transverse dunes, 362, 459
trap rock, 61, 459
traps, 374
travel time, 459
travertine, 100, 459
tree cover, stream flow and
  changes in, 265, 268
trellis stream pattern, 274, 459
tremolite, 430
trenches, deep-sea, 339, 340
Trenton limestone, 439
tritium, 380
tropical deserts, 459
tropical soil (laterites), 88
truncated spurs, 313, 459
tsunami (seismic sea waves),
  179–182, 459
tufa, 100, 459
tuff, 56, 459
Tulare Wasco area, 301
tundra, 459
tungsten, 393
turbidites, 103, 335, 459
turbidity currents, 335, 459
turbulent flow, 270, 459
Turkey, earthquakes in, 200
Turner, Francis J., 113
Turtle Mountain, 243–245

ultimate base level, 459
ultisols, 85
ultramafic rocks, 38
Uncompahgre River, 259
unconformities, 169–173, 459–460
underground mining, 386
underground water, 288–303, 460
    (see also groundwater):
    basic distribution of, 288–290
    caves and, 297–299
    movement of, 290–292
    pollution of, 301
    saltwater invasion of, 301
    water table, 289–290
    zones of saturation and aeration
      of, 288–289
uneven fracture, 460
unfoliated textures, 120
uniformitarianism, doctrine of, 3–4,
    323, 460
unit cell, 21, 460

United States, oil exploration in,
    375–376
United States Coast and Geodetic
    Survey, 181, 184, 192
United States Forest Service, 265
United States Geological Survey,
    184, 186–187, 192, 208, 438,
    439
unit stress, 149
unpaired terraces, 285, 460
Upper Rhine valley, 167
uraninite, 63, 430
uranium 235, 135, 380
uranium 238, 133, 134, 136, 379–380
uranium deposits, 380
Uranus, 402, 403
urbanization:
    sediment production and, 268
    streams and, 262–264
    thermal pollution and, 269–270
Ussher, James, 2

Valdez, 194
Valley and Ridge Physiographic
    Province, 167
valley glaciers, 305–306, 310, 460
Valley of Ten Thousand Smokes, 42
valleys, 273–285
    broad, 278–285
    cross-valley profiles, 273, 276
    drainage basins, 273
    drowned, 338
    enlargement of, 276–277
    glaciated, 312–313
    hanging, 314–315
    narrow, 278
    networks of streams, 273
    patterns of stream systems, 273,
      274
    water gaps, 274–276
valley train, 319, 460
valley walls, 273
Vanderford, 315
varves, 136, 321, 460
Vatna glacier, 306
vegetative cover, stream flow and
    changes in, 265, 268
velocity:
    of a stream, 258–259, 261
    terminal, 267, 355
Venera landers, 408
Venetz, J., 322
ventifacts, 357, 460

Venus, 402, 403, 408
vertical displacements, 146, 184
    of the earth's crust, 8–9
vertisols, 85
vesicle, 460
Vesuvius, Mount, 41, 45–46
Viking spacecraft, 408, 409
Vine, Fred, 225
virtual geomagnetic poles, 222, 460
viscosity, 150–151, 460
    earthquakes and, 198
Vishnu Schist, 171, 173
volatile components, 460
volcanic activity, on Io, 412
volcanic ash, 41, 42, 442, 460
volcanic block, 460
volcanic bomb, 460
volcanic breccia, 56, 460
volcanic conglomerate, 56
volcanic dust, 460
volcanic earthquakes, 460
volcanic eruptions, 39–42, 460 (see
    also lavas; magma):
    disappearing islands produced
      by, 49–50
    earthquakes associated with,
      51–52
    fiery clouds during, 42
    gases in, 41
    pyroclastic debris from, 41–42, 44
    worldwide effects of, 42

U

V

volcanic gases, 41
volcanic mountains, 460
volcanic neck, 460
volcanic tremors, 51–52, 460
volcanism, types of, 39
volcanoes, 37–50, 460
  active, 45
  aseismic ridges and, 339
  calderas of, 37–38
  classification of, 44–45
  craters of, 37

volcanoes (cont)
  distribution of active, 50
  dormant, 45
  earthquakes and, 195–196
  evolution of, 39
  extinct, 45
  history of some, 45–50
  on Mars, 409
Voyager missions, 413, 414
vug, 460

W

warm glacier, 460
warm springs, 294–295, 460
Warm Springs (Georgia), 295
warps, 460
  broad, 174
Wasatch fault, 167
water:
  action on hill slopes, 253
  artesian, 442
  covalent bond of molecule of, 26–27
  in desert areas, 365
  dipolar character of, 27
  distribution of world's estimated supply of, 433
  mass movements and, 241–242
  ocean, 331–333
  running (*see* running water)
  suspended, 289
  underground (*see* underground water)
  world distribution of, 235
waterfalls, 278
water gaps, 274–276, 460
waterpocket fold, 157
water power, 368, 379
water table, 289–290, 460
  cave formation and, 298–299
  perched, 293
Wausau Granite, 439
wave-built terrace, 345
wave crests, 343
wave-cut cliffs, 346
wave-cut terrace, 346
wave height, 344
wavelength, 344
waves, 460
  earth, 202
  refraction of, 344–345
  wind-formed, 343–344

weathering, 6, 67–80, 460
  chemical (*see* chemical weathering)
  defined, 67, 78
  in deserts, 365
  energy and, 69–70
  mechanical, 70–73, 365, 452
  mineral deposits and, 386
  rates of, 78–80
Wegener, Alfred, 11, 224–225
weighing, 153
weight, 153–154
wells, 293–294
Wentworth scale, 96
West Yellowstone earthquake, 245
Whitcomb, J. H., 208
White Mountains batholith, 55, 120
Williams, H., 44
Wilson, J. Tuzo, 48, 169, 235
wind:
  deposition of material by, 358–364
  erosion by, 357
  on Mars, 409
  movement of material by, 354–358
  planetary, 230
  trade, 353
  waves formed by, 343–344
wind power, 368, 382
wind shadows, 360–361
Wisconsin glaciation, 349
Wise, D. U., 218
Witwatersrand, 392
Wizard Island, 43
wolframite, 393
wollastonite, 114, 430
Wolman, M. Gordon, 268
wood, 368
wrinkle ridges, 460

xenoliths, 55, 460

Yazoo River, 280
yazoo-type rivers, 280, 460
yield point, 460
Yosemite Falls, 278
Yuba River, 258–259

zeolite facies, 116
zinc, 390
zircon, 430

zone of accumulation:
    of glaciers, 307, 309
    soil, 85
zone of aeration, 289–290, 460
zone of flow of glaciers, 308, 309
zone of fracture of glaciers, 308
zone of leaching, 86
zone of no movement, 354
zones of regional metamorphism,
    115–116, 460
zone of saturation, 289–290, 460
zone of wastage of glaciers, 307, 309

The following numbers and exponents (and prefixes) are essentially an outline of the metric system:

$$1,000,000,000,000 = 10^{12} \text{ (tera-)}$$
$$1,000,000,000 = 10^{9} \text{ (giga-)}$$
$$1,000,000 = 10^{6} \text{ (mega-)}$$
$$1,000 = 10^{3} \text{ (kilo-)}$$
$$100 = 10^{2} \text{ (hecto-)}$$
$$10 = 10^{1} \text{ (deka-)}$$
$$1 = 10^{0} \text{ [unit]}$$
$$0.1 = 10^{-1} \text{ (deci-)}$$
$$0.01 = 10^{-2} \text{ (centi-)}$$
$$0.001 = 10^{-3} \text{ (milli-)}$$
$$0.000001 = 10^{-6} \text{ (micro-)}$$
$$0.000000001 = 10^{-9} \text{ (nano-)}$$

The fundamental unit of length is the meter, originally defined as one ten millionth of the distance from the equator to the North Pole, later defined as the distance between two marks inscribed on the standard meter bar in Paris, and most recently specified in terms of the wavelength of krypton. So:

$$1 \text{ km (kilometer)} = 10^{3} \text{ m (meter)}$$
$$1 \text{ cm (centimeter)} = 10^{-2} \text{ m}$$
$$1 \text{ mm (millimeter)} = 10^{-3} \text{ m}$$
$$1 \text{ } \mu\text{m (micrometer)} = 10^{-6} \text{ m}$$

The official definition of an inch is based on the length of a meter:

$$1 \text{ in (inch)} = 2.54 \times 10^{-2} \text{ m}$$

Temperature is measured in degrees Celsius, a scale in which the interval between the freezing and boiling points of water is divided into 100 degrees, with $0°$ representing the freezing point and $100°$ the boiling point. So:

$$\frac{5}{9}°\text{C (Celsius)} = 1°\text{F (Fahrenheit)} \quad \text{or}$$

$$°\text{C} = \frac{(°\text{F} - 32°)}{1.8} \quad \text{or} \quad °\text{F} = (°\text{C} \times 1.8) + 32°$$

Conversion of mass

	Grams, g	Kilograms, kg	Pounds, lb	Ounces, oz
g	1	1,000	453.6	28.35
kg	0.001	1	0.4536	2.835×10^{-2}
lb	2.205×10^{-3}	2.205	1	6.25×10^{-2}
oz	3.527×10^{-2}	35.27	16	1

Multiply units in the column heads by the figures in the table to convert to units at left. (For example, to convert pounds to kilograms, multiply the number of pounds by 0.4536).

Conversion of volume

	Cubic meters, m³	Cubic yards, yd³	Cubic centimeters, cm³	Cubic inches, in³	Cubic feet, ft³
m³	1	0.7646	1×10^{-6}	1.639×10^{-5}	2.832×10^{-2}
yd³	1.308	1	1.308×10^{-6}	2.143×10^{-5}	3.704×10^{-2}
cm³	1×10^{6}	7.646×10^{5}	1	16.39	2.832×10^{4}
in³	6.102×10^{4}	46,656	6.102×10^{-2}	1	1,728
ft³	35.31	27	3.531×10^{-5}	5.787×10^{-4}	1

Multiply units in the column heads by the figures in the table to convert to units at left. (For example, to convert cubic inches to cubic centimeters, multiply the number of cubic inches by 16.39).